Comprehensive Asymmetric Catalysis I–III

Springer

Berlin
Heidelberg
New York
Barcelona
Hong Kong
London
Milan
Paris
Singapore
Tokyo

Eric N. Jacobsen · Andreas Pfaltz · Hisashi Yamamoto (Eds.)

Comprehensive Asymmetric Catalysis I–III

With contributions by numerous experts

 Springer

ERIC N. JACOBSEN
Department of Chemistry and Chemical Biology
Harvard University
12 Oxford Street
MA 02138 Cambridge, USA
e-mail: jacobsen@chemistry.harvard.edu

ANDREAS PFALTZ
Department of Chemistry
University of Basel
St. Johanns-Ring 19
CH-4056 Basel, Switzerland
e-mail: pfaltz@ubaclu.unibas.ch

HISASHI YAMAMOTO
School of Engineering
Nagoya University
Chikusa, 464-01 Nagoya, Japan
e-mail: j45988a@nucc.cc.nagoya-u.ac.jp

ISBN 3-540-64336-2 Springer-Verlag Berlin Heidelberg New York

Cip Data applied for   1001703856

Die Deutsche Bibliothek - CIP-Einheitsaufnahme
Comprehensive asymmetric catalysis / Eric N. Jacobsen ... (eds.). With contributions by numerous experts. - Berlin ; Heidelberg ; New York ; Barcelona ; London ; Milan ; Paris ; Tokyo : Springer
     ISBN 3-540-64336-2

3.-(1999)

© Springer-Verlag Berlin Heidelberg 1999
Printed in Germany

The use of general descriptive names, registered names, trademarks, etc. in this publication does not imply, even in the absence of a specific statement, that such names are exempt from the relevant protective laws and regulations and therefore free for general use.

Typesetting: Data conversion by MEDIO GmbH, Berlin
Cover: E. Kirchner, Heidelberg

SPIN: 10538291   02/3020 - 5 4 3 2 1 0 - Printed on acid-free paper

Authors

Varinder K. Aggarwal
Department of Chemistry
University of Sheffield
Sheffield S3 7HF, UK
e-mail: V.Aggarwal@Sheffield.ac.uk

Susumu Akutagawa
Takasago International Corporation
Nissay Aroma Square 5–37–1
Kamata Ohta-ku
Tokyo 144–8721, Japan
e-mail: akutag@bni.co.jp

Tadatoshi Aratani
Organic Synthesis Research Laboratory
Sumitomo Chemical Co., Ltd.
Takatsuki
Osaka 569-1093, Japan
e-mail: aratani@sc.sumitomo-chem.co.jp

Oliver Beckmann
Institut für Organische Chemie
Rheinisch-Westfälische Technische
Hochschule
Professor-Pirlet-Straße 1
D-52074 Aachen, Germany
e-mail: carsten.bolm@oc.rwth-aachen.de

Hans-Ulrich Blaser
Novartis Services AG
Catalysis & Synthesis Services
R 1055.6.28
CH-4002 Basel, Switzerland
e-mail: hans-ulrich.blaser@sn.novartis.com

Carsten Bolm
Institut für Organische Chemie
Rheinisch-Westfälische Technische
Hochschule
Professor-Pirlet-Straße 1
D-52074 Aachen, Germany
e-mail: carsten.bolm@oc.rwth-aachen.de

Klaus Breuer
Institut für Organische Chemie
Rheinisch-Westfälische Technische
Hochschule
Professor-Pirlet-Straße 1
D-52074 Aachen, Germany
e-mail: enders@rwth-aachen.de

John M. Brown
Dyson Perrins Laboratory
South Parks Road
Oxford OX1 3QY, UK
e-mail: bjm@ermine.ox.ac.uk

Stephen L. Buchwald
Department of Chemistry
Massachusetts Institute of Technology
77 Massachusetts Ave.
Cambridge
MA 02139-4307, USA
e-mail: sbuchwal@mit.edu

Erick M. Carreira
Laboratorium für Organische Chemie
ETH Zürich
Universitätsstraße 16
CH-8092 Zürich, Switzerland
e-mail: carreira@org.chem.ethz.ch

Albert L. Casalnuovo
DuPont Agricultural Products
Stine-Haskell Research Center
Newark
Delaware 19714, USA
e-mail: albert l.casalnuovo@usa.dupont.com

André B. Charette
Département de Chimie
Université de Montréal
P.O. Box 6128, Station Downtown,
Montréal (Québec), Canada H3C 3J7
e-mail: charetta@chimie.umontreal.ca

Geoffrey W. Coates
Department of Chemistry, Baker Laboratory
Cornell University, Ithaca
New York 14853-1301, USA
e-mail: gc39@cornell.edu

Scott E. Denmark
Roger Adams Laboratory
Department of Chemistry
University of Illinois
Urbana, Illinois, 61801, USA
e-mail: sdenmark@uiuc.edu

Dieter Enders
Institut für Organische Chemie
Rheinisch-Westfälische Technische
Hochschule
Professor-Pirlet-Straße 1
D-52074 Aachen, Germany
e-mail: enders@rwth-aachen.de

David A. Evans
Department of Chemistry and Chemical
Biology
Harvard University
Cambridge
Massachusetts 02138, USA
e-mail: evans@chemistry.harvard.edu

Harald Gröger
Graduate School of Pharmaceutical Sciences
The University of Tokyo
Hongo 7-3-1, Bunkyo-ku
Tokyo 113, Japan
e-mail: mshibasa@mol.f.u-tokyo.ac.jp

Ronald L. Halterman
Department of Chemistry and Biochemistry
University of Oklahoma
620 Parrington Oval
Norman, OK 73019, USA
e-mail: rhalterman@ou.edu

Tamio Hayashi
Department of Chemistry
Faculty of Science
Kyoto University
Sakyo
Kyoto 606–8502, Japan
e-mail: thayashi@th1.orgchem.ku-chem.
kyoto-u.ac.jp

Yujiro Hayashi
Department of Chemistry
School of Science
The University of Tokyo
Hongo, Bunkyo-ku
Tokyo 113-0033, Japan
e-mail: narasaka@chem.s.u-tokyo.ac.jp

Nicola M. Heron
Department of Chemistry
Merkert Chemistry Center
Boston College
Chestnut Hill
MA 02467, USA
e-mail: nicola.heron@bc.edu

Frederick A. Hicks
Department of Chemistry
Massachusetts Institute of Technology
Cambridge
77 Massachusetts Ave
MA 02139-4307, USA
e-mail: fhicks@email.unc.edu

Jens P. Hildebrand
Institut für Organische Chemie
Rheinisch-Westfälische Technische
Hochschule
Professor-Pirlet-Straße 1
D-52074 Aachen, Germany
e-mail: carsten.bolm@oc.rwth-aachen.de

Amir H. Hoveyda
Department of Chemistry
Merkert Chemistry Center
Boston College
Chestnut Hill
MA 02467, USA
e-mail: amir.hoveyda@bc.edu

David L. Hughes
Merck and Co., Inc.
Mail Drop R80Y-250
Rahway, NJ 07065, USA
e-mail: Dave_Hughes@Merck.com

Shohei Inoue
Department of Industrial Chemistry
Faculty of Engineering
Science University of Tokyo
Kagurazaka, Shinjuku
Tokyo 162–8601, Japan
e-mail: amori@res.titech.ac.jp

Yoshihiko Ito
Department of Synthetic Chemistry and
Biological Chemistry
Graduate School of Engineering
Kyoto University
Sakyo-ku
Kyoto 606–8501, Japan
e-mail: yoshi@sbchem.kyoto-u.ac.jp

Shinichi Itsuno
Department of Materials Science
Toyohashi University of Technology
Tempaku-cho
Toyohashi 441-8580, Japan
e-mail: itsuno@tutms.tut.ac.jp

Eric N. Jacobsen
Department of Chemistry and Chemical
Biology
Harvard University
12 Oxford Street
Cambridge, MA 02138, USA
e-mail: jacobsen@chemistry.harvard.edu

Kim D. Janda
Department of Chemistry
The Scripps Research Institute and
The Skaggs Institute for Chemical Biology
10550 North Torrey Pines Road
La Jolla, CA 92037, USA
e-mail: kdjanda@scripps.edu

Jeffrey S. Johnson
Department of Chemistry and
Chemical Biology
Harvard University
Cambridge
Massachusetts 02138, USA
e-mail: evans@chemistry.harvard.edu

Henri B. Kagan
Laboratoire de Synthèse Asymétrique
Institut de Chimie Moléculaire d' Orsay
Université Paris-Sud
F-91405 Orsay, France
e-mail: kagan@icmo.u-psud.fr

Tsutomu Katsuki
Department of Chemistry
Faculty of Science
Kyushu University 33
Hakozaki, Higashi-ku
Fukuoka 812–8581, Japan
e-mail: katsuscc@mbox.nc.kyushu-u.ac.jp

Ryoichi Kuwano
Department of Synthetic Chemistry and
Biological Chemistry
Graduate School of Engineering
Kyoto University
Sakyo-ku
Kyoto 606–8501, Japan
e-mail: kuwano@sbchem.kyoto-u.ac.jp

Mark Lautens
Department of Chemistry
University of Toronto
Toronto, Ontario, Canada, M5 S 3H6
e-mail: mlautens@alchemy.chem.utoronto.ca

Hélène Lebel
Département de Chimie
Université de Montréal
P.O. Box 6128, Station Downtown
Montréal (Québec), Canada H3C 3J7
e-mail: charetta@chimie.umontreal.ca

T. O. Luukas
Laboratoire de Synthèse Asymétrique
Institut de Chimie Moléculaire d' Orsay
Université Paris-Sud
F-91405 Orsay, France
e-mail: tiluukas@icmo.u-psud.fr

Kevin M. Lydon
School of Chemistry
The Queen's University
David Keir Building
Stranmillis Road
Belfast BT9 5AG, Northern Ireland
e-mail: k.lydon@qub.ac.uk

Istvan E. Markó
Department of Chemistry
University of Louvain
Place Louis Pasteur 1
B-1348 Louvain-la-Neuve, Belgium
e-mail: marko@chor.ucl.ac.be

Keiji Maruoka
Department of Chemistry
Graduate School of Science
Hokkaido University
Sapporo, 060–0810, Japan
e-mail: maruoka@sci.hokudai.ac.jp

M. Anthony McKervey
School of Chemistry
The Queen's University
David Keir Building
Stranmillis Road
Belfast BT9 5AG, Northern Ireland
e-mail: t.mckervey@qub.ac.uk

Koichi Mikami
Department of Chemical Technology
Tokyo Institute of Technology
Meguro-ku
Tokyo 152, Japan
e-mail: kmikami@o.cc.titech.ac.jp

Atsunori Mori
Research Laboratory of Resources
Utilization
Tokyo Institute of Technology
Nagatsuta
Yokohama 226–8503, Japan
e-mail: amori@res.titech.ac.jp

Johann Mulzer
Institut für Organische Chemie
Universität Wien
Währingerstrasse 38
A-1090 Wien, Austria
e-mail: mulzer@felix.orc.univie.ac.at

Kilian Muñiz
Institut für Organische Chemie
Rheinisch-Westfälische Technische
Hochschule
Professor-Pirlet-Straße 1
D-52074 Aachen, Germany
e-mail: carsten.bolm@oc.rwth-aachen.de

Yasuo Nagaoka
Graduate School of Pharmaceutical Sciences
Kyoto University
Yoshida, Sakyo-ku
Kyoto 606-8501, Japan
e-mail: tomioka@pharm.kyoto-u.ac.jp

Koichi Narasaka
Department of Chemistry
School of Science
The University of Tokyo,
Hongo, Bunkyo-ku
Tokyo 113-0033, Japan
e-mail: narasaka@chem.s.u-tokyo.ac.jp

Hisao Nishiyama
School of Materials Science
Toyohashi University of Technology
Tempaku-cho,
Toyohashi 441, Japan
e-mail: hnishi@tutms.tut.ac.jp

Olivier J.-C. Nicaise
Monsanto Hall, Department of Chemistry
Saint Louis University
St. Louis, Missouri, 63103, USA
e-mail: sdenmark@uiuc.edu

R. Noyori
Department of Chemistry and Molecular
Chirality Research Unit
Nagoya University
Chikusa
Nagoya 464–8602, Japan
e-mail: noyori@chem3.chem.nagoya-u.ac.jp

Kyoko Nozaki
Department of Material Chemistry
Graduate School of Engineering
Kyoto University
Yoshida
Sakyo-ku, 606–8501, Japan
e-mail: nozaki@npc05.kuic.kyoto-u.ac.jp

Günther Oehme
Institut für Organische Katalyseforschung
Universität Rostock e.V.
Buchbinderstr. 5–6
D-18055 Rostock, Germany
e-mail: goehme@chemie1.uni-rostock.de

T. Ohkuma
Department of Chemistry and Molecular
Chirality Research Unit
Nagoya University
Chikusa
Nagoya 464–8602, Japan
e-mail: noyori@chem3.chem.nagoya-u.ac.jp

Takashi Ooi
Department of Chemistry
Graduate School of Science
Hokkaido University
Sapporo, 060–0810, Japan
e-mail: maruoka@sci.hokudai.ac.jp

Andreas Pfaltz
Department of Chemistry
University of Basel
St. Johanns-Ring 19,
CH-4056 Basel, Switzerland
e-mail: pfaltz@ubaclu.unibas.ch

Benoît Pugin
Novartis Services AG
Catalysis and Synthesis Services
R-1055.6.29
CH-4002 Basel, Switzerland
e-mail: benoit.pugin@sn.novartis.com

T.V. RajanBabu
Department of Chemistry
The Ohio State University
100 W. 18th Avenue
Columbus, Ohio 43210, USA
e-mail: rajanbabu.1@osu.edu

Tomislav Rovis
Department of Chemistry
University of Toronto
Toronto, Ontario, Canada, M5 S 3H6
e-mail: trovis@alchemy.chem.utoronto.ca

Michelangelo Scalone
Process Research and Catalysis Department
Pharmaceuticals Division
F. Hoffmann-La Roche AG
CH-4070 Basel, Switzerland
e-mail: michelangelo.scalone@roche.com

Rudolf Schmid
Process Research and Catalysis Department
Pharmaceuticals Division
F. Hoffmann-La Roche AG
CH-4070 Basel, Switzerland
e-mail: rudolf.schmid@roche.com

Masakatsu Shibasaki
Graduate School of Pharmaceutical Sciences
The University of Tokyo
Hongo 7-3-1
Bunkyo-ku, Tokyo 113, Japan
e-mail: mshibasa@mol.f.u-tokyo.ac.jp

Takanori Shibata
Department of Applied Chemistry
Faculty of Science
Science University of Tokyo
Kagurazaka, Shinjuku-ku

Tokyo 162–8601 Japan
e-mail: ksoai@ch.kagu.sut.ac.jp

Ken D. Shimizu
Department of Chemistry and Biochemistry
University of South Carolina
Columbia
South Carolina 29208, USA
e-mail: shimizu@psc.sc.edu

Marc L. Snapper
Department of Chemistry
Merkert Chemistry Center
Boston College
Chestnut Hill
Massachusetts 02467, USA
e-mail: marc.snapper@bc.edu

Kenso Soai
Department of Applied Chemistry
Faculty of Science
Science University of Tokyo
Kagurazaka, Shinjuku-ku
Tokyo 162–8601, Japan
e-mail: ksoai@ch.kagu.sut.ac.jp

Felix Spindler
Novartis Services AG
Catalysis & Synthesis Services
R 1055.6.28
CH-4002 Basel, Switzerland
e-mail: felix.spindler@sn.novartis.com

Martin Studer
Novartis Services AG
Catalysis & Synthesis Services
R 1055.6.28
CH-4002 Basel, Switzerland
e-mail: martin.studer@sn.novartis.com

John S. Svendsen
Department of Chemistry
University of Tromsø
N-9037 Tromsø, Norway
e-mail: johns@chem.uit.no

Masahiro Terada
Department of Chemical Technology
Tokyo Institute of Technology
Meguro-ku
Tokyo 152, Japan
e-mail: kmikami@o.cc.titech.ac.jp

Kiyoshi Tomioka
Graduate School of Pharmaceutical Sciences
Kyoto University
Yoshida, Sakyo-ku
Kyoto 606-8501, Japan
e-mail: tomioka@pharm.kyoto-u.ac.jp

Erasmus M. Vogl
Graduate School of Pharmaceutical Sciences
The University of Tokyo
Hongo 7-3-1, Bunkyo-ku
Tokyo 113, Japan
e-mail: mshibasa@mol.f.u-tokyo.ac.jp

Paul Wentworth Jr.
Department of Chemistry
The Scripps Research Institute and
The Skaggs Institute for Chemical Biology
10550 North Torrey Pines Road
La Jolla, CA 92037, USA
e-mail: paulw@scripps.edu

Michael H. Wu
Department of Chemistry and
Chemical Biology
Harvard University
Cambridge
MA 02138, USA
e-mail: jacobsen@chemistry.harvard.edu

Masahiko Yamaguchi
Graduate School of Pharmaceutical Sciences
Tohoku University
Aoba
Sendai 980-8578, Japan
e-mail: yama@mail.pharm.tohoku.ac.jp

Hisashi Yamamoto
Graduate School of Engineering
Nagoya University
CREST, Japan Science and Technology
Corporation (JST)
Chikusa
Nagoya 464–8603, Japan
e-mail: j45988a@nucc.cc.nagoya-u.ac.jp

Akira Yanagisawa
Graduate School of Engineering
Nagoya University
CREST, Japan Science and Technology
Corporation (JST)
Chikusa
Nagoya 464–8603, Japan
e-mail: j45989a@nucc.cc.nagoya-u.ac.jp

Preface

The title of this collection is an accurate reflection of the goals we defined at the outset of the project. Our intention was to bring together all important aspects of the field of asymmetric catalysis and to present them in a format that would be most useful to a wide range of scientists including students of chemistry, expert practitioners, and chemists contemplating the possibility of using an asymmetric catalytic reaction in their own research.

This project was initiated by Joe Richmond, who was one of many to recognize the need for an exhaustive and current treatment of the field of asymmetric catalysis, but was unique in being willing and able to get such an ambitious effort started. Considering that it is a field that is evolving in parallel in laboratories throughout the world, he sought to select editors who were not only authoritative, but also as geographically distributed as the field itself. He approached each of us separately, and in the end we were compelled equally by the significance of the project, and by the exciting prospect of working together.

Given the dramatic growth of activity in the field of asymmetric catalysis over the past few years in particular, it was apparent from the start that a comprehensive treatment would be a ambitious task, especially if we were to succeed in capturing the excitement and challenges in field, as well its basic principles. The field is interdisciplinary by its nature, incorporating organic synthesis, coordination chemistry, homogeneous catalysis, kinetics and mechanism, and advanced stereochemical concepts all at its very heart. We realized that the project would require authors who would be willing not only to commit the effort of writing definitive and compelling chapters, but who would also be capable of analyzing their topic with absolute authority. At a hotel near the Frankfurt airport in the Fall of 1996, we got together and constructed an exhaustive list of topics in asymmetric catalysis, and then we devised a "dream list" of contributors. These were individuals who contributed in defining ways to the topics in question. That this dream list came true hopefully should be evident by surveying the names of the contributing authors. If we have succeeded to any extent in our effort to put forth a comprehensive and useful analysis of the field of asymmetric catalysis, it is thanks to them.

Eric N. Jacobsen, Cambridge   July 1999
Andreas Pfaltz, Basel
Hisashi Yamamoto, Nagoya

Contents

Volume III

Chapter 29   Aldol Reactions

Chapter 29.1
Mukaiyama Aldol Reaction

Erick M. Carreira

Laboratorium für Organische Chemie, ETH Zürich, Universitätsstraße 16, CH-8092 Zürich, Switzerland
e-mail: carreira@org.chem.ethz.ch

Keywords: Aldol, Enol silane, Transition metal

1
Introduction

The asymmetric aldol addition reaction has emerged as one of the most power-ful stereoselective transformations available to the synthetic chemist for com-plex molecule synthesis [1a, 1b, 1c, 1d, 1e, 1f, 1g, 1h, 1i, 1j, 1k, 1l, 1m, 1n]. A new carbon-carbon bond is formed with concomitant generation of up to two new stereogenic centers. Thus, the aldol addition reaction has proved to be of great synthetic utility since it allows for fragment coupling in the assembly of complex structures. Numerous aldol addition methods have been developed for the ster-eocontrolled construction of molecules containing the characteristic β-hydrox-ycarbonyl retron [2]. The reaction methodology spans the range of processes utilizing chiral substrates (aldehydes or enolates), stoichiometric quantities of optically active additives, or, more recently, chiral catalysts. Additionally, im-pressive advances in biocatalysis have afforded enzyme and antibody-catalyzed processes that furnish products in high enantioselectivity and yield [3a3b, 4a, 4b, 4c, 4d]. A large, continually-expanding body of work on asymmetric aldol methodology renders comprehensive coverage of the area well beyond the scope of a single chapter. Consequently, this review is focused on the presentation and discussion of the recent advances specifically related to transition metal-cata-lyzed, enantioselective aldol addition reactions. The success of this methodolo-gy is already evident in the increasing number of applications of catalytic meth-ods to the asymmetric synthesis of stereochemically complex natural products [5a, 5b, 5c, 5d, 5e, 5f].

The rapid evolution of catalytic reaction methods for enantioselective aldol additions affords newer processes that are increasingly practical in their execu-tion for a broad range of substrates prescribing minuscule amounts of catalyst. However, when compared to other catalytic asymmetric processes such as hy-drogenation, dihydroxylation, and epoxidation it is evident that there is much room for further optimization. Without doubt, discovery and innovation in this area of C-C bond-forming reactions will lead to the development of catalysts and processes indispensable to the synthesis of optically active, stereochemically complex structures with applications in materials science and medicine.

2
Background

The discovery of the Lewis acid-mediated addition of enol silanes to aldehydes and acetals by Mukaiyama and coworkers pioneered a novel approach to the construction of molecules via the crossed aldol reaction (Eq. 1) [6a6b]. Impor-tantly, this development proved to be a key lead for the subsequent evolution of this C-C bond forming reaction into a catalytic Si atom-transfer process. Typical enol silanes derived from esters, thioesters, and ketones are unreactive towards aldehydes at ambient temperatures. However, stoichiometric quantities of Lewis acids such as $TiCl_4$, $SnCl_4$, $AlCl_3$, BCl_3, $BF_3 \cdot OEt_2$, and $ZnCl_2$ were found to promote aldehyde addition to give β-hydroxycarbonyl adducts. Innumerable electrophilic promoters and catalysts have been investigated for this reaction including Sn(IV) [7], Sn(II) [8a, 8b, 8c, 8d, 8e, 8f, 8g, 8h, 8i, 8j, 8k, 8l, 8m, 8n, 8o, 8p, 8q, 8r, 8s, 8t, 8u, 8v, 8w, 8x, 8y, 8z], Mg(II) [9], Zn(II) [10a,10b,10c,10d,], Li(I) [11a, 11b, 11c], Bi(III) [12a, 12b, 12c, 12c], In(III) [13], Ln(III) [14a, 14b, 14c, 14d, 14e, 14f, 14g, 14h, 14i, 14j, 14k, 14l, 14m, 14n, 14o], Pd(II) [15], Ti(IV) [16a, 16b, 16c, 16d, 16e, 16f, 16g], Zr(IV) [17a, 17b], Ru(II) [18a, 18b], Rh(II) [19a, 19b], Fe(II) [20a, 20b, 20c, 21a, 21b, 21c, 21d], Al(III) [22a, 22b, 22c], Cu(II)[23a, 23b, 24], Au(I) [25], R$_3$SiX [26a, 26b, 26c, 26d, 26e, 26f], Ar$_3$C$^{(+)}$ [27a, 27b, 27c, 27d, 27e, 27f, 27g], acridinium salts [28], and clay [29]. Additionally, the aldol reaction of silyl enol ethers has also been conducted utilizing Lewis bases as catalysts or promoters. These include fluoride [30a, 30b, 30c, 30d], for which naked enolates are proposed as the reactive species, and, more recently, phosphoramide bases [31a, 31b]. Since the trail-blazing report by Mukaiyama, many examples of stereoselective additions between chiral aldehydes and enol silanes have been documented and applied to the total syntheses of stereochemically complex natural products [32]. The stereochemical features of these processes have been analyzed in detail utilizing transition-state models incorporating steric, dipolar, and stereoelectronic effects [33].

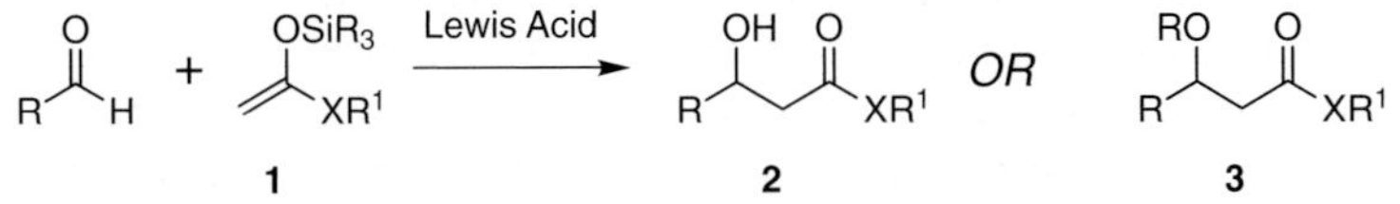

$$\tag{1}$$

The methods that have been reported for the crossed aldol addition reaction may be classified on the basis of the enolate employed and, correspondingly, on the structure of the substituted products generated (Scheme 1). Following this rubric, two general aldol addition processes may be identified: the first includes the reaction of unsubstituted acetate-derived enolates 5 and aldehydes, which generates a single new stereogenic center in the form of a secondary alcohol 6; the second comprises the reactions of α-mono- or disubstituted enolates 7, 8, 9, which furnish adducts containing up to two new stereogenic centers. In this lat-

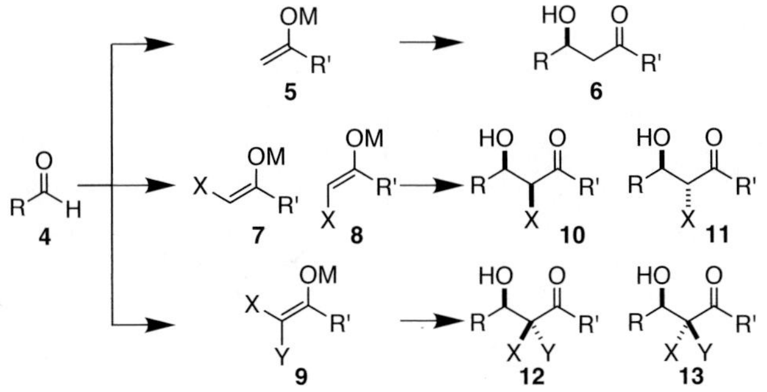

Scheme 1

ter category, the use of a substituted enolate leads to additional complexity since 1,2-*syn* **10** or 1,2-*anti* **11** diastereomeric adducts may be formed, a stereochemical feature referred to as simple diastereoselection [1a]. The extent to which the simple diastereoselection is correlated to the geometric isomer of the starting enolate (*E* versus *Z*) depends on the mechanism of the process and structure of the transition states which, in turn, are a function of reaction parameters including the substrates, metals, counterions, and solvents. Aldol addition methods abound in which the *syn/anti* stereochemistry of products is stereospecifically determined by the choice of enolate used [1g], and, for selected examples, [34a, 34b]; however, examples have been reported in which either *Z*- or *E*-enolates afford the same product in a stereoconvergent manner, for selected examples in which simple diastereoselectivity in the products is not correlated to the enolate geometry, see Refs. [35a, 35b].

The structural details of the transition states in the Mukaiyama aldol addition reaction have been the subject of intense experimental and theoretical investigations [36a, 36b, 36c, 36d, 36e, 36f, 37a, 37b, 37c, 37d, 37e]. The proposed models of the transition-state structures for the addition of enol silanes to an aldehyde-Lewis acid complex may be categorized into two general classes: (1) those for which open, extended transition states **14, 15, 16** have been proposed; and (2) those for which closed, cyclic transition state structures **17** to **20** have been invoked (Fig. 1). It is worthwhile to scrutinize the diverse models; however, it is important to recognize that a classification based on putative transition-structure types has its limitations. An understanding of the catalytic Mukaiyama aldol at such resolution is tenuous as a result of the fact that the rate at which preparative synthetic methodology is being discovered has outpaced the rate of the accompanying structural and mechanistic studies. By contrast, there exists a wealth of empirical and theoretical data for the corresponding non-catalytic, asymmetric aldol addition reactions [1]. While the study of such processes is daunting, it can provide useful insight in the parallel analysis of the catalytic methodology.

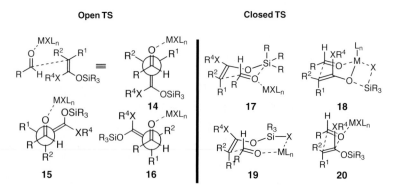

Fig. 1. Proposed transition state structures for the addition of silyl enol ethers and aldehydes

2.1
Analysis of Transition States

2.1.1
Extended, Open Structures

In a landmark study of Mukaiyama aldol addition reactions, Heathcock proposed that the observed stereochemical outcome of the products in the Lewis
acid-mediated addition of silyl ketene acetals to aldehydes was consistent with
extended, open transition-state structures [38a, 38b]. This analysis has gained
wide acceptance as a consequence of its predictive power. Alternative models involving cyclic, closed structures have also been postulated, in particular, the latter have been invoked with increasing regularity in the analyses of catalytic,
enantioselective aldol addition reactions [7, 30b, 39a, 39b].

The conformational degrees of freedom associated with the aldehyde and
enolate in extended, open transition state structures complicates any detailed
analysis of the reaction (Fig. 2). Accordingly, the free energy difference between
synclinal arrangements (**15, 16, 22,** or **24**) and *anticlinal* arrangements (**14 and
23**) are not always readily discerned on the basis of steric interactions [40a, 40b].
Studies aimed at factoring any inherent stereoelectronic preference that may favor *synclinal* or *anticlinal* dispositions have underscored the delicate, intricate
balance of dipolar, inductive, and steric effects which exist in these addition reactions (for an investigation of Lewis acid aldehyde complexes, see Refs. [41a,
41b, 41c]. In this regard, given the diversity of reaction conditions reported for
the Mukaiyama aldol addition, it may not be feasible to reduce this multifarious
reaction into a single mechanistic construct [42a, 42b, 42c].

A number of structural and mechanistic studies of related nucleophilic addition processes deserve close scrutiny since they provide relevant parallels that
are useful in the analysis of the Mukaiyama aldol addition reaction [43, 44].

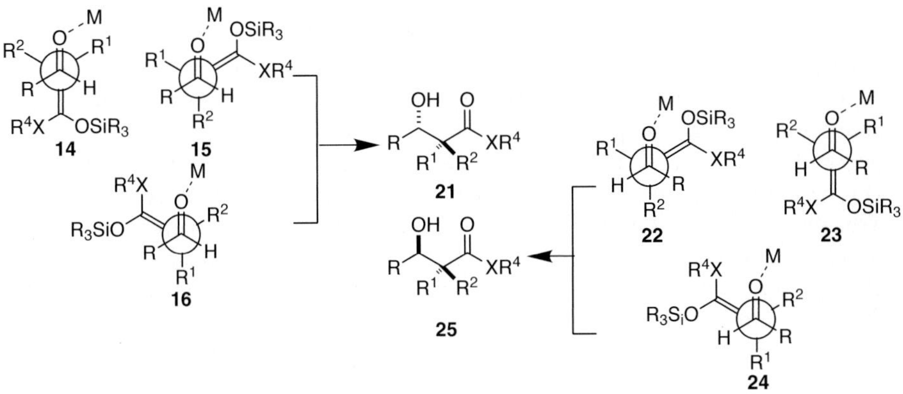

Fig. 2. Analysis of open transition states

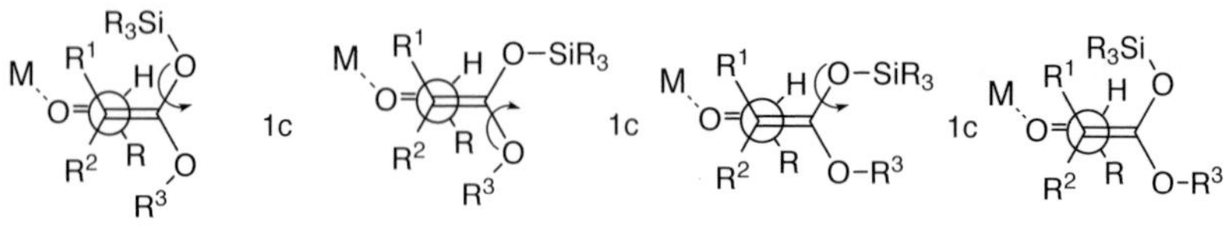

Fig. 3. The diversity of rotamers that may be populated by silyl ketene acetals

Their integration into the study of mechanistic models for the addition of enol silanes to aldehydes may allow a more detailed understanding of the stereochemical features of this reaction. For example, the subtle interplay of steric interactions ($A_{1,3}$) and stereoelectronic effects ($Lp_O \rightarrow \sigma^*_{C-O}$) can bias the conformational profile of enol silanes in the ground state (Fig. 3). When operating in synergy, such effects can result in well-defined minima having pin wheel-type conformations [45]. Wilcox has obtained an X-ray crystal structure of a propionate-derived enol silane that displays structural features similar to those reported for alkyl vinyl ethers and reveals the inherent stereoelectronic preference for the *s-cis* isomer in the ground state in the absence of overriding steric effects [46, 47a, 47b]. The consequences of the various available conformations of the enol silane on the relative energies of the extended transition-state structures that may be populated, **26** to **29**, have received little attention. Structural features such as these will need to be incorporated into models since the energetic consequences of such effects impact the reaction stereoselectivity and are likely to be augmented in the transition state.

Other topological features that are important to consider in the context of enol silane additions to aldehydes were proposed by Seebach to account for the diastereoselective Michael additions of enamines to nitroolefins [48]. These have been generalized in the form of rules for C-C bond forming processes between prochiral centers and can be utilized in an analysis of the Mukaiyama aldol addition reaction. The preferred approach of the two reacting components is such that: (1) all bonds are staggered, (2) a gauche arrangement exists between the enol silane C=C and the aldehyde C=O and C-H bonds (cf. **32** Fig. 4), (3) the smaller of the substituents of the enolate (R^1 vs R^2) is *anti* with respect to the acceptor bond C=O, and (4) the enolate and aldehyde oxygen atoms which develop charges as the reaction progresses are positioned proximal to one another. Importantly, the approach of the two reactants is necessarily governed by a Bürgi-Dunitz alignment between the aldehyde LUMO and enol silane HOMO [49a, 49b, 49c]. Optimization of these orbital interactions conspire to enforce a nonparallel arrangement of the planes defined by each of the trigonal carbons in the electrophilic and nucleophilic partners **30, 31, 32** (Fig. 4). Such an arrangement further augments any steric interactions between the aldehyde substituent and the substituent on the enolate C_α as the reacting partners proceed to product.

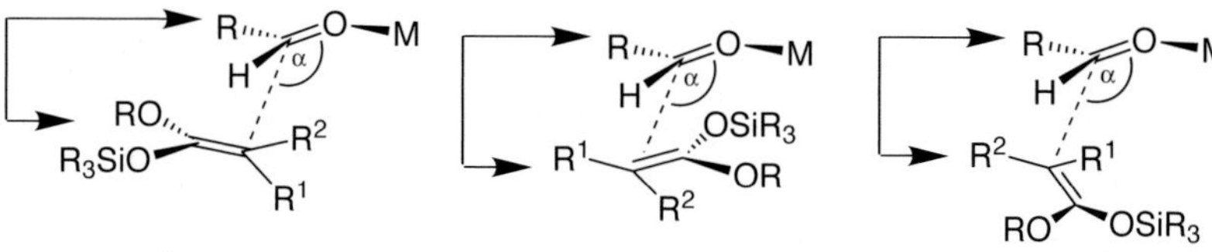

Fig. 4. The Seebach model as applied to enol silane additions to aldehydes emphasizes the incipient steric interactions that result from Bürgi-Dunitz constraints

2.1.2
Closed Structures

The most intensely studied aldol addition mechanisms are those believed to proceed through closed transition structures, which are best understood within the Zimmerman-Traxler paradigm (Fig. 5) [1d]. Superposition of this construct on the Felkin-Ahn model for carbonyl addition reactions allows for the construction of transition-state models impressive in their ability to account for many of the stereochemical features of aldol additions [50a, 50b, 50c, 51]. Moreover, consideration of dipole effects along with remote non-bonding interactions in the transition-state have imparted additional sophistication to the analysis of this reaction and provide a bedrock of information that may be integrated into the further development and refinement of the corresponding catalytic processes [52a, 52b]. One of the most powerful features of the Zimmerman-Traxler model in its application to diastereoselective additions of chiral enolates to aldehydes is the correlation of enolate geometry (Z- versus E-) with simple diastereoselectivity in the products (*syn* versus *anti*). Consequently, the analyses of catalytic, enantioselective variants that display such stereospecificity often invoke closed, cyclic structures. Further studies of these systems are warranted, since it is not clear to what extent such models, which have evolved in the context of diastereoselective aldol additions via chiral auxiliary control, are applicable in the Lewis acid-catalyzed addition of enol silanes and aldehydes.

The closed transition-state structures that have been proposed for the Lewis acid-mediated addition of enol silanes to aldehydes are of two general specifications (Fig. 6). The first includes models in which the metal complex plays an integral role in the closed structure through its incorporation into the cyclic array (Fig. 6, **18** and **19**). The second includes models in which the metal is exocyclic to the ring (Fig. 6, **17** and **20**). These include six-membered rings **17** [36a], fused bicyclic 4- to 6-membered rings **18** [6b, 14b, 53a, 53b, 53c, 54], eight-membered rings **19** [6, 55], and four-membered rings **20** [56a, 56b, 56c, 57]. All of these models share a common mechanistic feature: the intact enol silane is the reactive nucleophilic species in the addition to the aldehyde.

Recently, catalytic, enantioselective aldol addition reactions have been reported which are proposed to proceed through mechanistic pathways involving

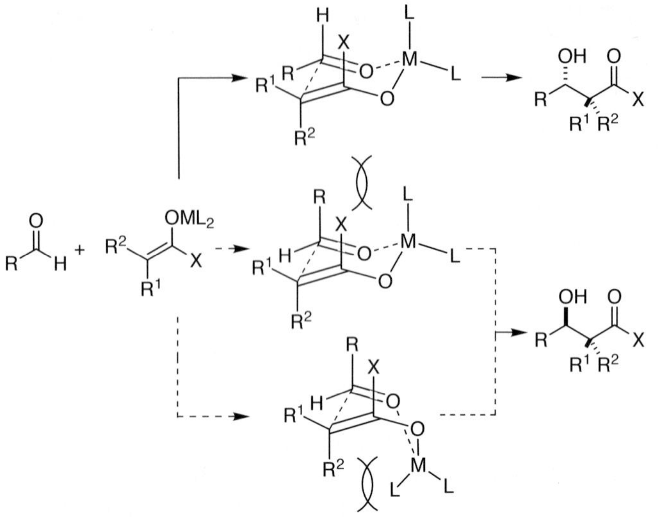

Fig. 5. The Zimmerman-Traxler transition states for aldol addition reactions

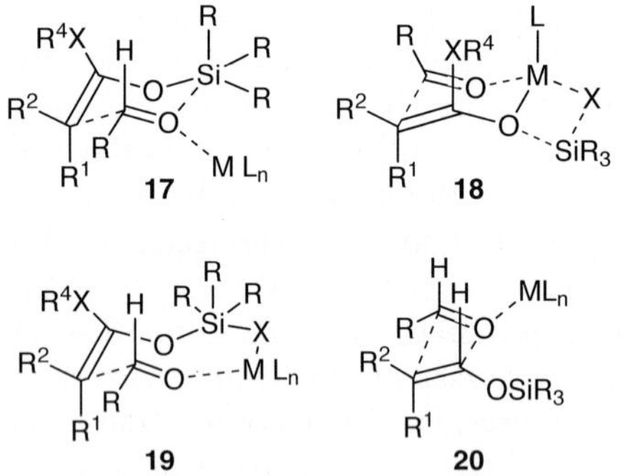

Fig. 6. Proposed closed transition-state structures for the Mukaiyama aldol addition of enol silanes

a metalloenolate intermediate **34** (Eq. 2). The metalloenolate intermediate is formed upon reaction of the enol silane with a metal complex or, alternatively, upon deprotonation of an acidic C-H compound. Regarding the latter, Hayashi and Ito have reported the addition reaction of isocyanoacetates to aldehydes that is mediated by a chiral Au(I) complex [25]. Mechanistic investigations by these researchers implicate an Au-enolate as a catalytically important species. More recently, Shibasaki and Carreira have reported enantioselective processes in which optically active Pd(II) and Cu(II) enolates, respectively, are generated in a catalytic manner and undergo addition to aldehydes [15, 24].

$$\text{(2)}$$

A second distinct process disclosed by Denmark involves the Lewis base-catalyzed addition of enol trichlorosilanes **36** to aldehydes (Eq. 3) [30b]. Remarkably, despite the fact that the uncatalyzed addition of such enol silanes to aldehydes is rapid at –78 °C, the use of optically active phosphoramides substantially accelerates the addition reaction and leads to the formation of optically active products. As a consequence of stereochemical studies involving substituted enol trichlorosilanes, Denmark has proposed a hexacoordinated silicon atom as the organizational locus about which enolate and aldehyde are arranged in a cyclic array **37**.

$$\text{(3)}$$

2.1.3
Aldehyde-Metal Complexes

A number of excellent studies have been conducted that examine the structural features and energetics of bonding in complexes formed between aldehydes and Lewis acid [41a, 58, 59a, 59b, 59c, 59d, 59e]. The work has been meticulously reviewed and represents an important resource in understanding the important complexation phenomena between aldehyde and Lewis acid and its relationship to asymmetric catalysis. The coordination of an aldehyde or ketone to a Lewis acid leads to the enhancement of the electrophilicity of the carbonyl towards nucleophilic addition by the otherwise unreactive enol silane. Of particular importance in the analysis of such processes is the conformation about the M-O dative bond in the active complex formed between aldehyde and Lewis acid (Fig. 7). Models that account for the energetic preference for aldehyde binding in an orientation that exposes or blocks an aldehyde diastereoface have factored in non-bonded steric interactions or attractive dipolar interactions (cf. **40**) and stereoelectronic effects (**39**) [60, 61a, 61b]. It has been suggested that the non-bonded aldehyde lone pair opposite the formyl C-H can interact in an energetically favorable manner with the M-X antibonding orbital σ^*_{M-X} as in **39**. Although there is only scant evidence that corroborates such a model, a stereoelectronic effect would lock in place a conformation about the M-O bond such that the dihedral angle C-O-M-X is 0°. Additional stabilizing features that serve to accentuate the

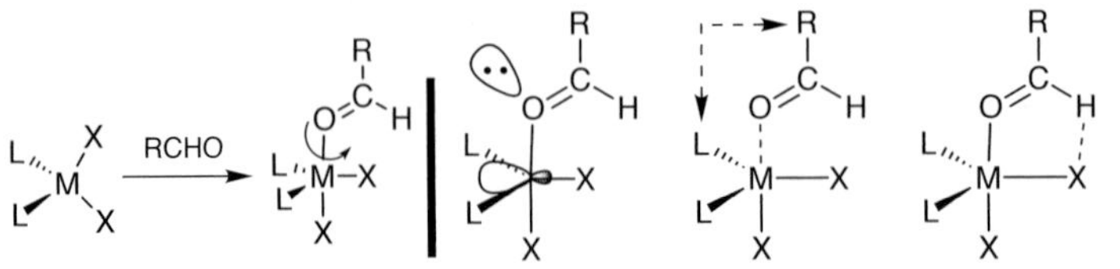

Fig. 7. Postulated bonding models for aldehyde-Lewis acid complex

difference between the aldehyde diastereofaces include dipole-induced, charge-transfer effects that have been generically referred to as π-π or π-stacking interactions. In the prototypical model such energetically stabilizing features are optimal when the polarized metal-bound aldehyde is proximal to a polarizable aromatic moiety [62].

Recently, a fundamentally new approach to understanding the bonding in aldehyde-metal complexes has been proposed and discussed by Corey (Fig. 7) [63a, 63b, 63c, 63d, 63e]. The Corey model incorporates a hydrogen bond between the formyl C-H on the bound, polarized aldehyde moiety and one of the metal-bound, ligating heteroatoms (**41**). This bonding arrangement is suggested to lead to increased organization in the aldehyde-Lewis acid complex which enforces and further augments any inherent steric and dipolar effects of the ligand. In its most generalized form, the construct provides an elegant explanation that accounts for the observation of highly stereoselective processes in a broad range of aldehyde addition reactions.

3
Catalysis of Mukaiyama Aldol Additions

3.1
General Mechanistic Aspects

The nucleophilic addition of enol silanes with aldehydes to produce β-silyloxy carbonyl adducts **47** is an example of a group-transfer process (Scheme 2), for applications in polymer synthesis, see: [64a, 64b, 64c]. In its simplest mechanistic rendition the reaction proceeds upon coordination of the aldehyde to Lewis acid MX_4 to afford an activated electrophilic species **42**. Addition of the nucleophilic enol silane **43** to **42** leads to C-C bond formation and generation of the aldol adduct. Various intermediate structures **44, 45, 46** have been postulated to be formed concomitant with or following C-C bond formation. The generation of intermediates **45** and **46** necessitates subsequent silylation of the β-alkoxide furnishing aldol adduct **47** and regenerating catalyst MX_4.

The initial studies on the reaction of enol silanes and aldehydes implicated the stoichiometric metal promoter, such as $TiCl_4$, as a Lewis acid. Subsequent investigations confirmed this hypothesis, ruling out a reaction between ketone- or ester-derived enol silane and $TiCl_4$ under the typical conditions employed for

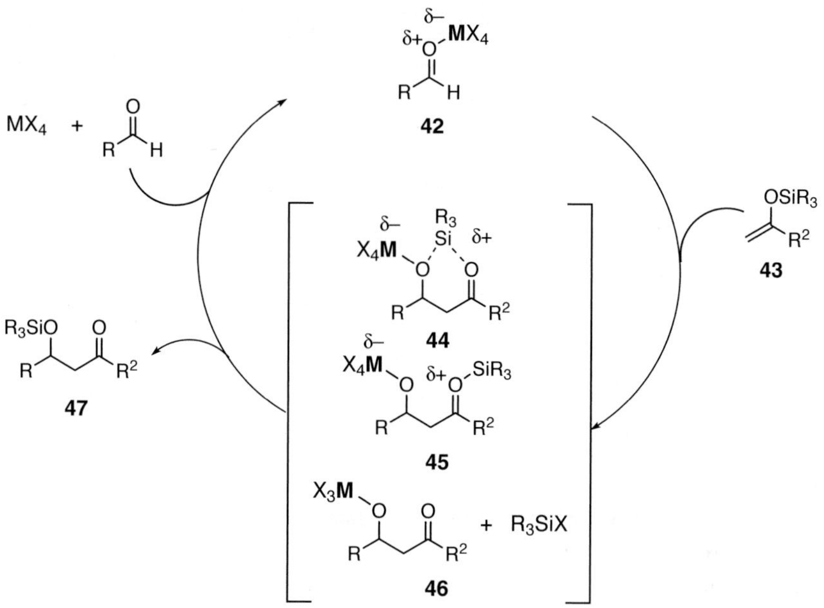

Scheme 2

the process [65]. For example, on the basis of ^{29}Si-NMR experiments Chan concluded that silylketene acetals did not undergo metallation by $TiCl_4$. These results were validated in subsequent investigations by Kuwajima and Nakamura [66a, 66b]. These investigators demonstrated that the stereochemistry of aldol adducts derived from propionate-derived titanium enolates differed from that obtained from $TiCl_4$-mediated addition of silyl enol ethers to aldehydes.

The results of these key studies, however, do not preclude catalytically competent metalloenolate intermediates in a process that complements the traditional Lewis acid-mediated Mukaiyama aldol addition. Evidence for such a process has been rigorously documented, leading to the successful development of catalytic, enantioselective versions (Scheme 3). In this regard, Bergman and Heathcock documented the reaction chemistry of Rh(I), W(I), and Re(I) enolates and their ability to sustain catalytic C-C bond forming addition reactions [67a, 67b, 67c, 67d, 67e]. Importantly, this study documented the ability of metal aldolates such as **46** to undergo O-silylation by the starting enol silane **43**. Additional mechanistic investigations provided insight into the kinetic profile of the individual steps of the process. The formation of the metal aldolate adduct was shown to be reversible with subsequent alkoxide silylation as the rate-determining step. It was suggested that the development of an enantioselective variant of these processes would necessitate that the O-silylation step effect a kinetic resolution of the diastereomeric metal aldolates **46**. Recent disclosures by Shibasaki and Carreira have documented the feasibility of developing enantioselective

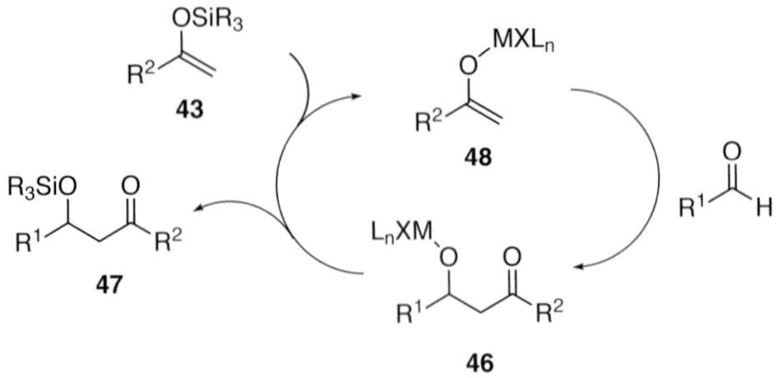

Scheme 3

processes based on this general type of mechanistic construct involving a catalytically competent metalloenolate intermediate.

3.2
The Atom Transfer Step: Silylation of Metal Aldolate

In its original formulation, the procedure for the Mukaiyama aldol prescribed
stoichiometric quantities of a strong Lewis acid such as $BF_3 \cdot OEt_2$, $TiCl_4$, or $ZnCl_2$.
Upon completion, the reaction mixture was typically quenched under condi-
tions that would not allow the exact identity of the aldol adduct to be established:
metal aldolate **46**, O-silyl ether **47**, or a mixture of both. Subsequent mechanistic
studies have suggested that, for some of these processes, a metal aldolate **45** or
46 is the product directly formed from the metal-mediated C-C bond-forming
reaction [36b, 68, 69a, 69b]. This mechanistic feature is of great importance in
understanding the overall catalytic atom-transfer process. In this regard, Reetz's
elegant NMR studies on the $TiCl_4$-mediated addition of pinacolone-derived enol
silane **50** to α-benzyloxypropionaldehyde **49** are informative (Scheme 4) [35]. In
the spectroscopic experiment, consumption of **49** and **50** was observed to be rapid at −78 °C with concomitant formation of a metal aldolate adduct **53** and
Me_3SiCl. In preceding studies, Reetz had documented that additions to substrates such as α-benzyloxypropionaldehyde proceeded from the chelated reactive intermediate **51** formed upon binding $TiCl_4$ by the aldehyde. It is important
to note that **53** was observed as the only product of the addition reaction. The
investigators suggested that although chelate **52** may be the first-formed adduct,
it rapidly rearranged giving **53**. Additionally, an important observation in these
experiments germane to understanding catalysis of this reaction is the fact that
the corresponding β-O-silylated product **54** was not observed within the time-
scale of the experiment. In related studies, Denmark has studied spectroscopi-
cally the $TiCl_4$-promoted reactions of allylsilanes and aldehydes wherein the
only observed product was the Ti(IV)-alkoxide and not the corresponding O-silyl ether [70a, 70b].

The generation of a discreet metal aldolate such as **53** may produce a reactive
silylating agent as co-product that is itself a competent catalyst for the Mukaiya-

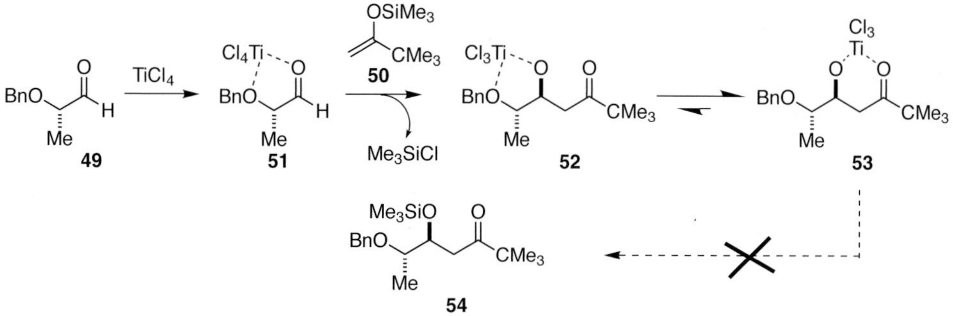

Scheme 4

ma aldol addition reaction. The relative rates of metal aldolate silylation versus the silyl-catalyzed aldol addition reaction becomes critical to the metal mediated enantioselective process and to the extent that metal complex functions as the true catalytic entity.

Recent mechanistic studies of the Mukaiyama aldol addition reaction by Bosnich and Carreira suggest that some of the processes that have been proposed to proceed by metal catalysis may only be metal initiated with the observed rapid reaction catalyzed by a silylating species generated in situ (Scheme 5) [69a, [69b, 70a, 70b]. Thus, in the presence of the metal promoters, the generation of adventitious Brønsted acid or Lewis acidic silyl species can be problematic. This is likely to be particularly important with strong Lewis acid promoters, wherein a strong M-O bond is generated in the aldolate adduct. The production of reactive silylating agents in the reaction mixture can lead to rapid silyl-catalyzed aldol addition reaction 55→56→47 that outpaces the metal catalyzed process. In this regard, caution is warranted in the analyses of these systems since the formation of silylated adducts is not a sufficient condition to validate the claim that metal catalysis is operative. The use of optically active complexes along with the observation of optically active products represents the most direct test of metal complex participation. For such cases, the extent of participation by the metal catalyst can be correlated to the optical purity of the product. The proportion of racemic product generated can result from either the inherent limitation of the chiral metal catalyst or the extent to which a competing, stereorandom Brønsted acid or silane catalyzed process is occurring.

In a metal-catalyzed process, the catalytically active Lewis acid complex is only regenerated upon silylation of the metal aldolate intermediate 45. Silylation of 45 can occur through a variety of mechanisms which are represented in their simplest forms in Scheme 6: (1) direct intramolecular Si-transfer (45→44→47); (2) intramolecular silyl transfer mediated by a transient intermediate which is produced upon silylation of the ligand (45→58→47); or (3) intermolecular silylation (45→46→47).

Experimental evidence that corroborates the existence of a putative zwitterionic intermediate analogous to 45 has been provided by Bosnich and coworkers

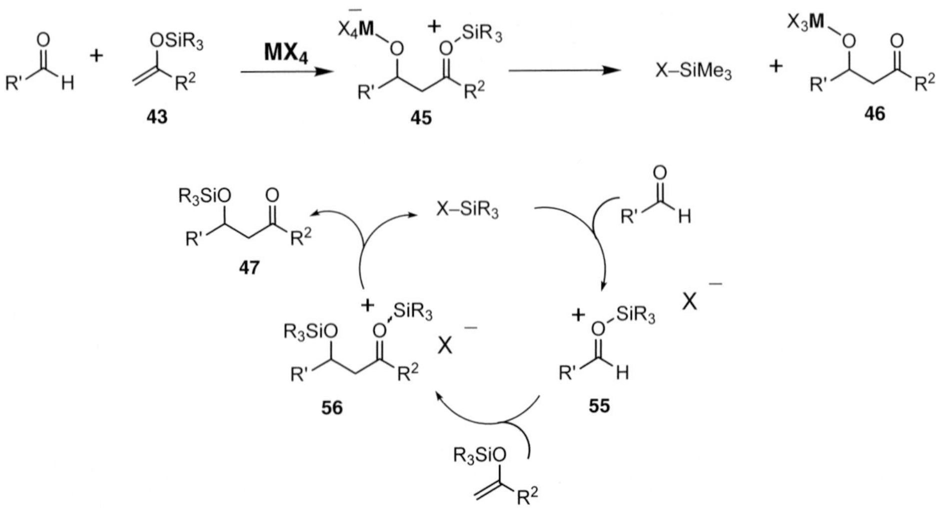

Scheme 5

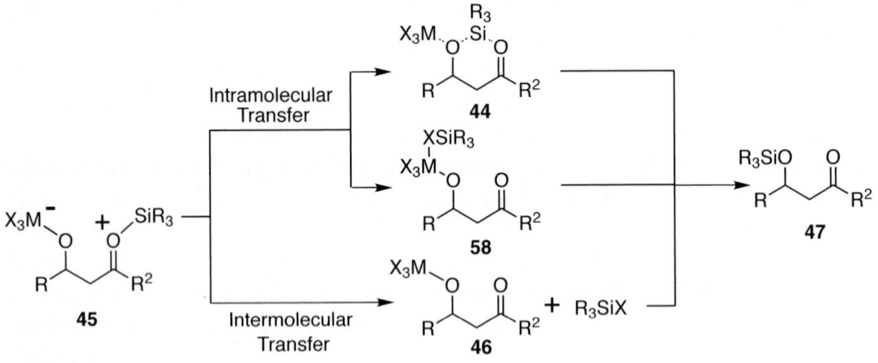

Scheme 6

in a study of the Eu(III)-catalyzed addition of **59** to benzaldehyde (Scheme 7) [71]. Spectroscopic data has been obtained that is consistent with the formation of four-membered ring adducts **61/62** as the kinetic products of the reaction. The step leading to these cyclic products has been shown to be reversible: prolonged exposure of **61/62** to the reaction conditions led to the conversion of these metastable oxetanes to **63**, the thermodynamic product of the reaction. The investigators have speculated that the formation of oxetane adducts in this study is a consequence of a slow silyl transfer step **60**→**63**. Thus, these observations highlight the fine balance that can exist between the various reaction pathways available to the adduct of the C-C bond-forming step (cf **60**).

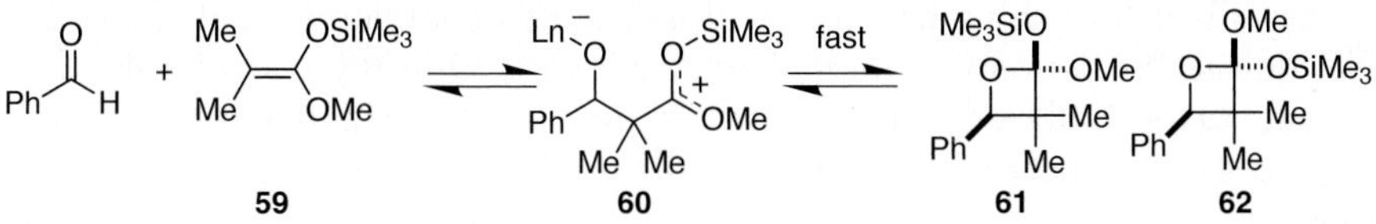

Scheme 7

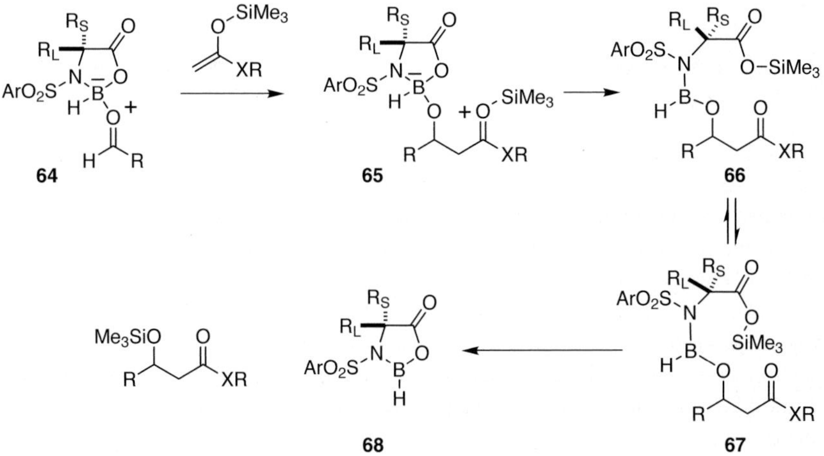

Scheme 8

Inspection of the reported enantioselective catalytic aldol addition reactions reveals some general mechanistic trends that provide useful considerations in the design of catalysts for C=O additions. A number of examples of asymmetric aldol additions have been proposed to proceed through an intermediate in which ligand has undergone silylation. One of the earliest examples is the ox-azaborolidene-catalyzed aldol addition reactions reported by Masamune [72a, 72b, 72c]. A novel mechanistic model was postulated which featured an interme-diate boron aldolate wherein the carboxylate ligand has undergone silylation (**66/67**, Scheme 8). In the course of the study the investigators observed that ex-tent of asymmetric induction (% ee) of the product varied with the addition rate of reactants, with optimal induction observed when enol silane is added over 48 h. Two critical assumptions were made in the interpretation of the data: (1) boronate **66/67** may be a competent Lewis acid catalyst for the aldol addition reaction, albeit furnishing products with attenuated enantioselectivity since the chiral ligand is bound in a monodentate manner; and (2) intramolecular silyl transfer to the β-alkoxyboronate is the rate determining step in the overall process (Eq. 4). The observed sensitivity of the product enantiomeric excess to the rate of addition is elegantly accounted for by Masamune's mechanistic scheme. Thus, under conditions that involve slow addition of substrates, intramolecular or intermolecular aldolate silylation can occur competitively at the expense of the undesired process catalyzed by **66/67**. This mechanistic paradigm may be quite general; in this regard, it is interesting to note that a preponderance of catalytic enantioselective group-transfer process have been reported which utilize optically-active ligands that feature carboxylate donor ligands prominently.

$$ (4) $$

Carreira has utilized a related mechanistic construct in the design of a Ti(IV) complex **73** that is catalytically active (Scheme 9) [68]. The use of a metal-bound salicylate ligand proved critical to the development of a workable catalytic, enantioselective process. Thus, the Ti(IV) complex **73** effects the addition of unsubstituted acetate-derived silyl ketene acetals to a broad range of aldehydes utilizing as little as 0.2 mol % catalyst. Importantly, the operational aspects of the process are greatly simplified since the reaction may be conducted at −10–23 °C without the requirement of slow addition. The efficiency of this process has been attributed to the operation of a silyl-shuttle mediated by a carboxyl ligand. In the context of an octahedral metal complex **72**, the silyl-transfer step is suggested to benefit from the proximity of the silylcarboxylate to the alcoholate with concomitant activation of this silylating agent by coordination to the Lewis acidic metal center.

In contrast to the mechanism discussed in the previous section, catalytic, enantioselective aldol addition processes have been described which proceed through an intermediate aldolate that undergoes subsequent intermolecular silylation. Denmark has discussed this possibility in a study of the triarylmethylcation-catalyzed Mukaiyama aldol reaction (Scheme 10) [73]. The results of exploratory experiments suggested that it would be possible to develop a competent catalytic, enantioselective Lewis-acid mediated process even when strongly Lewis acidic silyl species are generated transiently in the reaction mixture. A system of this type is viable only if the rate of silylation of the metal aldolate is faster than the rate of the competing silyl-catalyzed aldol addition reaction ($k_{Si} \gg k_{Si\text{-}aldol}$ Scheme 10). A report by Chen on the enantioselective aldol addition reaction catalyzed by optically active triaryl cations provides support for the mechanistic conclusions of the Denmark study [74].

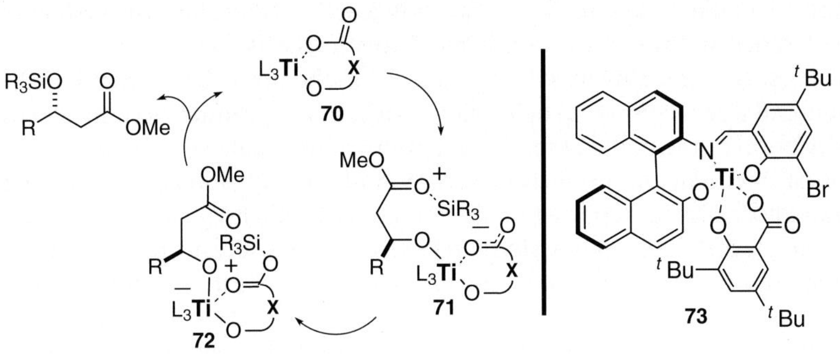

Scheme 9

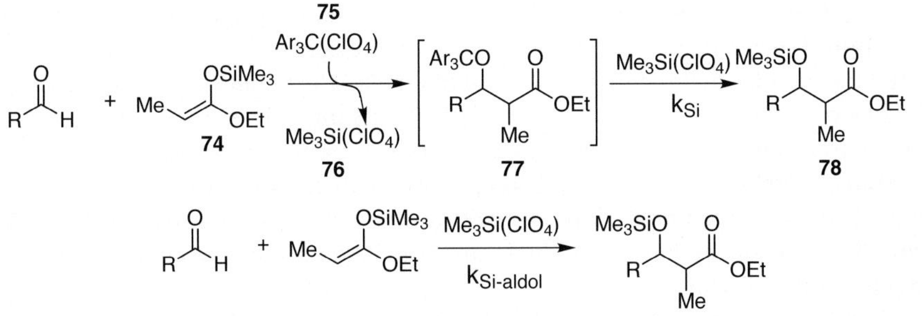

Scheme 10

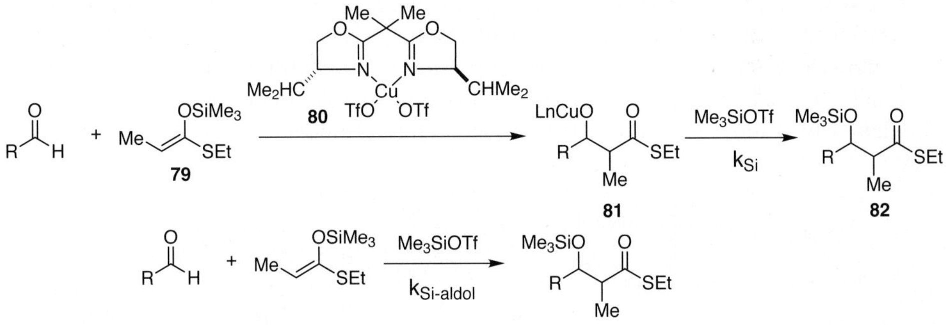

Scheme 11

An enantioselective process which provides a powerful illustration of this phenomena has been documented by Evans (Scheme 11) [24]. In this work, the addition of enol silanes **79** to α-benzyloxyacetaldehyde is catalyzed by the optically active Cu(II)·bis(oxazoline) complex **80**, furnishing adducts **82** in excellent enantio- and diastereoselectivity. Two noteworthy features of this system merit consideration in the context of this mechanistic discussion:

(1) the reaction was shown to be accelerated in the presence of added Me_3SiOTf without a corresponding deleterious effect in the product enantioselectivity;

(2) silylation of the metal aldolate occurs in an intermolecular fashion.

The first phenomenon implicates silylation of a metal aldolate intermediate as the rate-determining step wherein the presence of added silylating agent at –78 °C selectively accelerates metal aldolate silylation. The second phenomenon was demonstrated using double-labeling experiments and attests to the efficiency of the process. Importantly, the viability of this system probably results from the fact that the metal alkoxide generated undergoes rapid silylation as a consequence of a weak Cu-O metal bond. This is to be contrasted to the metal alkoxides generated as intermediates when strong oxophillic Lewis acids are used when silyl transfer has been suggested to be considerably slower.

Advances in the development of metal-catalyzed Mukaiyama aldol addition reactions have primarily relied on a mechanistic construct in which the role of the Lewis acidic metal complex is to activate the electrophilic partner towards addition by the enol silane. Alternate mechanisms that rely on metallation of enol silane to generate reactive enolates also serve as an important construct for the design of new catalytic aldol addition processes. In pioneering studies, Bergman and Heathcock documented that transition-metal enolates add to aldehydes and that the resulting metallated adducts undergo silylation by the enol silane leading to catalyst turnover.

Two systems have been reported that may be operating through the intermediacy of a transition-metal enolate intermediate (Eqs. 5 and 6). Although extensive mechanistic information is lacking, experimental and spectroscopic evidence is consistent with the turn-over step in the catalytic cycle occurring from silylation of a metalloaldolate by the starting enol silane. In this regard, Shibasaki has described a Pd(II)-catalyzed addition of ketone-derived silyl enol ethers to aldehydes [15]. Carreira has also described a process which utilizes a Cu(II) complex that is proposed to initiate the catalytic cycle by metallation of the starting enol silane which subsequently participates in a catalytic aldol addition reaction [24].

(5)

(6)

4
Lewis Acid-Catalyzed Aldol Addition Reactions

4.1
Tin(II)

Pioneering studies of stoichiometric Sn(II)-promoted additions of enol silanes to aldehydes by Mukaiyama and Kobayashi are valuable resources in understanding the catalytic versions of the reaction. Stoichiometric quantities of optically active Sn(II) complexes prepared from diamines mediate a collection of aldol addition reactions (Eqs. 7 and 8) [7, 75a, 75b, 75c]. Thus, the addition of the S-ethyl thioacetate-derived enol silane **90** with benzaldehyde promoted by a complex generated in situ by mixing 1 equivalent each of $Sn(OTf)_2$ and Bu_3SnF along with 1.2 equiv. of diamine **91** produced adduct **92** (2:1 1,3,5-trimethylbenzene/CH_2Cl_2, −78 °C) in 82% ee. Importantly, in the absence of added Bu_3SnF, the product was isolated in racemic form [76]. Subsequent intensive investigation of this process identified the optimal ligands for a broad range of carbonyl additions such as those derived from the aminonaphthalene substituted ligand **93**. Additionally, the aldol addition reactions performed with Sn(II) complex **93** display considerably less sensitivity to the reaction conditions in providing optically active aldol adducts. However, even with this ligand, the high levels of stereoinduction in the products were observed when the reactions were conducted in the presence of additives, such as 1.1 eq of $Bu_2Sn(OAc)_2$ [7 h].

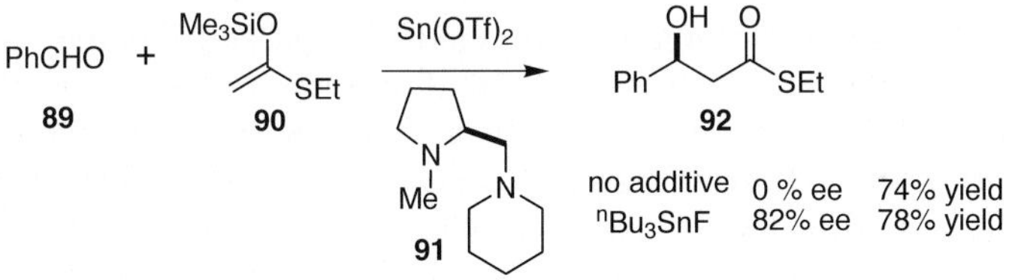

$$(7)$$

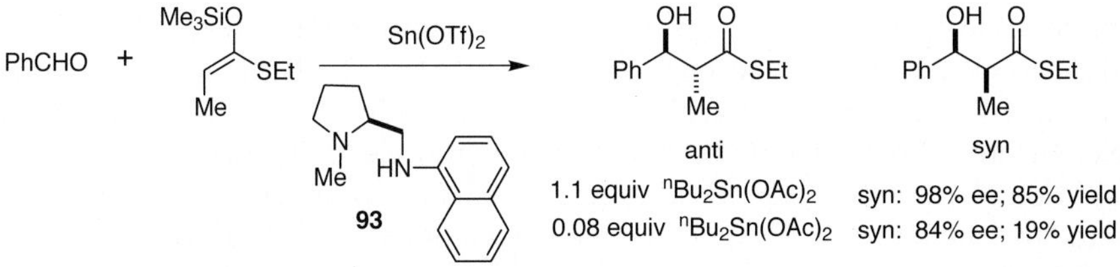

$$(8)$$

A diverse family of optically active diamine ligands have been prepared by Mukaiyama and Kobayashi commencing with proline. The diamine ligands are readily available in either enantiomeric form from (R)-(+)- or (S)-(−)- proline through a short synthetic sequence (Scheme 12) [77a, 77b]. The versatility of this proline-derived ligand class has been elegantly documented by Mukaiyama and Kobayashi. Subtle structural modifications of ligands derived from a single

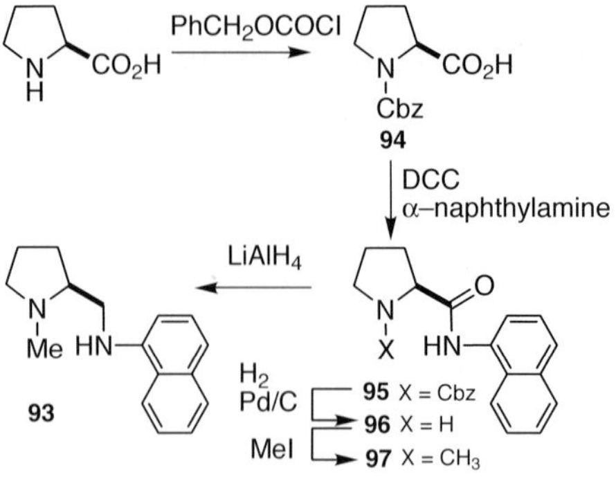

Scheme 12

proline enantiomer allow the preparation of the corresponding Sn(II) complexes which behave as pseudo-enantiomers [7, 78a, 78b, 78c, 78d, 78e, 78f]; for example, in the aldol additions of silyl thioketene acetal **100** with aldehydes in the presence of stoichiometric quantities of the Sn(II) complexes derived from **98** or **99** adducts **101** and their enantiomers **102**, respectively, were isolated in up to 99% ee and >99:1 *syn* diastereoselectivity (Fig. 8).

The working mechanistic model crafted by these investigators invokes an intermediate metal aldolate **104** whose silylation was proposed to be slow. Formation of Me_3SiOTf **105** as a by-product in the reaction was expected to be problematic since **105** is known to function as a catalyst in a competitive, stereorandom aldol addition reaction (Scheme 13). This analysis suggested that the ameliorative effects of Bu_3SnF and $Bu_2Sn(OAc)_2$ as additives in the stoichiometric Sn(II)-catalyzed aldol reactions were due to the ability of such additives to suppress the competing Me_3SiOTf-catalyzed additions. This analysis suggested two critical modifications of the Sn(II)-promoted reaction that would facilitate the subsequent evolution of this stoichiometric process into the corresponding enantioselective Sn(II)-catalyzed version. In this regard, when a dichloromethane solution of cyclohexanecarboxaldehyde and thioketene silyl acetal was added slowly over 3.5 h to a solution of the catalyst **103** at –78 °C, the adduct was isolated in 76% yield and 73% ee. By contrast, when the addition was performed over 6 h, an improvement (80% ee) in the enantioselectivity was observed [8j]. Additionally, the use of a polar, aprotic, Lewis-basic propionitrile as solvent was found to lead to further improvement in the enantioselectivity of the process affording the adduct of cyclohexanecarboxaldehyde in 92% ee. Slow addition of the reactants maintains the concentration of the reactants low, thereby allowing the rate of Sn-aldolate silylation and catalytic turnover to compete with the rate of the competing Si-catalyzed process. Mukaiyama has suggested that the polarity of the solvent propionitrile (dielectric constant=28.86) as well as its ability to function as a coordinating ligand for the active Sn(II) complex **103** is key to the success of the catalytic process. For the Sn(II)-mediated aldol process, it can be speculated that coordination of propionitrile to the metal aldolate complex may

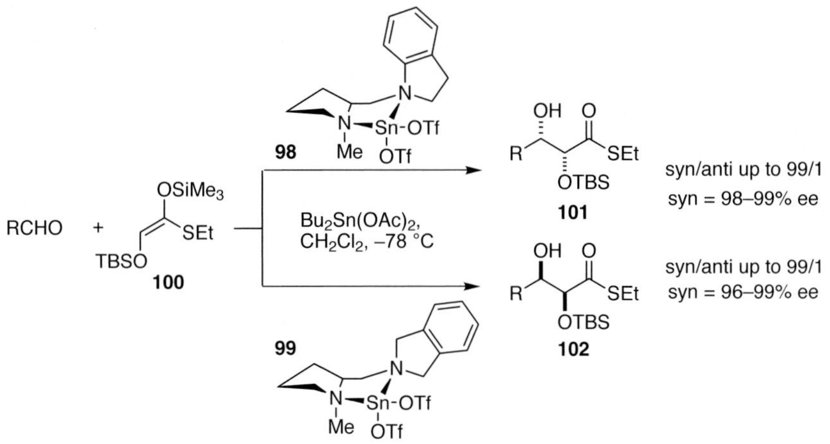

Fig. 8. The use of diamines derived from (*R*)-proline as pseudo-enantiomers

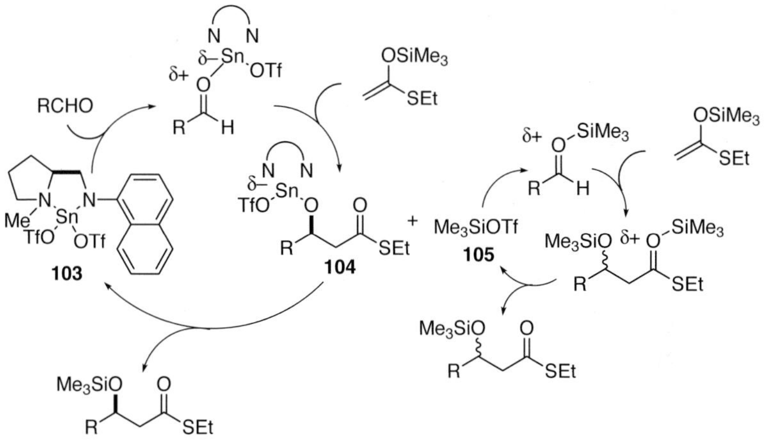

Scheme 13

increase the electron density at the metal and thereby increase the reactivity of the metal alkoxide towards silylation. Working in synergy, these effects lead to selective acceleration of the metal aldolate silylation and catalyst turnover.

The additions of acetate, propionate, and other substituted enolates following the optimized protocol have been reported. The typical set of conditions prescribe the use of 10 to 30 mol % catalyst in propionitrile at −78 °C and slow addition of reactants. For the acetate-derived silyl thioketene acetals **106** adducts are obtained in up to 93% ee and 90% yield (Eq. 9) [8j]. The addition of thiopropionate-derived *Z*-silyl ketene acetal **108** to a range of aldehydes delivered aldol adducts with high levels of simple diastereoselectivity (89/11→99/1 **109/110**) and up to 98% ee (Eq. 10) [79]. Further studies of these processes has led to the observation that the addition of Sn(II) oxides (40 mol %) can lead to improvements in *syn/anti* ratio along with the enantioselectivity of the aldol product [80].

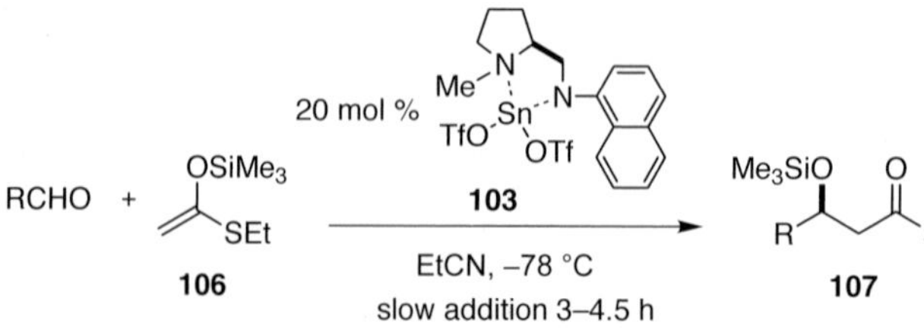

$$\text{RCHO} + \underset{\textbf{106}}{\overset{\text{OSiMe}_3}{\overset{|}{\diagdown}}}\!\!\!\diagdown_{\text{SEt}} \xrightarrow[\substack{\text{EtCN, }-78\,°\text{C} \\ \text{slow addition 3–4.5 h}}]{\substack{20 \text{ mol \%} \\ \textbf{103}}} \underset{\textbf{107}}{\overset{\text{Me}_3\text{SiO} \quad \text{O}}{R\!\diagdown\!\diagup\!\diagdown_{\text{SEt}}}} \qquad (9)$$

Entry	Aldehyde	Yield	% ee
1	H₃C–(CH₂)₅–CHO	79%	93 %
2	c-C₆H₁₁CHO	81%	92 %
3	(CH₃)₂CHCHO	48%	90 %
4	H₃C~~~CHO	65%	72 %
5	Bu—≡—CHO	68%	88 %
6	Me₃Si—≡—CHO	75%	77 %
7	Ph—≡—CHO	71%	79 %
8	C₆F₅CHO	90%	68 %

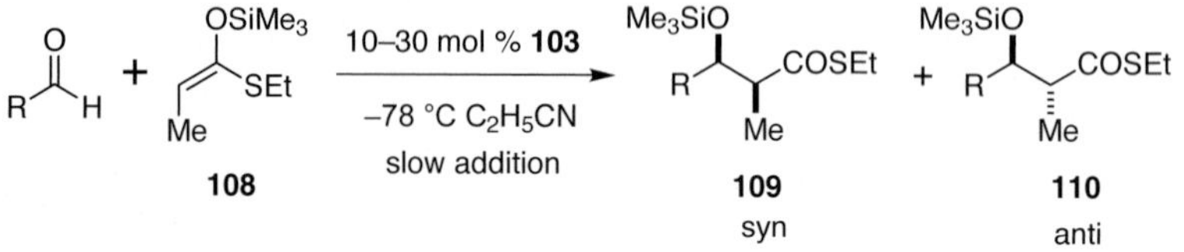

$$\underset{R}{\overset{\text{O}}{\diagdown}}\!\!\diagup^{\text{H}} + \underset{\textbf{108}}{\overset{\text{OSiMe}_3}{\overset{|}{\diagdown}}\!\!\!\diagdown_{\text{SEt}}^{\text{Me}}} \xrightarrow[\substack{-78\,°\text{C C}_2\text{H}_5\text{CN} \\ \text{slow addition}}]{10\text{–}30 \text{ mol \% } \textbf{103}} \underset{\substack{\textbf{109} \\ syn}}{\overset{\text{Me}_3\text{SiO}}{R\!\diagdown\!\diagup\!\diagdown_{\text{COSEt}}^{\text{Me}}}} + \underset{\substack{\textbf{110} \\ anti}}{\overset{\text{Me}_3\text{SiO}}{R\!\diagdown\!\diagup\!\diagdown_{\text{COSEt}}^{\text{Me}}}} \qquad (10)$$

Entry	Aldehyde	Yield	syn/anti	% ee
1	PhCHO	77%	93/7	90%
2	p–ClC₆H₄CHO	83%	87/13	90%
3	p–MeC₆H₄CHO	75%	89/11	91%
4	H₃C~~CHO	76%	96/4	93%
5	H₃C–(CH₂)₅–CHO	80%	>99/1	>98%
6	c-C₆H₁₁CHO	71%	>99/1	>98%

Experiments using optically active aldehyde substrates in combination with optically active catalysts can provide insight into the structure of the active complex. Moreover, the data generated may corroborate and help to further refine the putative catalyst model. By pitting the inherent stereochemical bias of the substrate against that of the catalyst for control of the stereochemical outcome of the reaction such a study provides a measure of the extent to which each dictate the product stereochemistry. For the Sn(II)·diamine-mediated addition reactions of enol silane **111** and enantiomeric aldehydes **112** and **115**, the resident chirality of the substrate had no measurable effect on the magnitude of asymmetric induction (Eq. 11) [81]. Thus, the product stereochemistry is exclusively controlled by the absolute stereochemistry of the chiral Sn(II) complex (**112**→**113**, 96:4 diastereoselectivity; **115**→**116**, 96:4 diastereoselectivity). This feature is of practical significance in the fragment coupling of chiral subunits for complex molecule assembly [8x, 78a, 78b, 78c, 78d, 78e, 78f, 82].

(11)

The structures of the functional catalyst or relevant intermediates in the proposed catalytic cycle in solution are presently unknown; however, the available spectroscopic experiments are informative [78f]. In CD_2Cl_2 at –78 °C, ^{1}H-NMR spectra of a 1:1 mixture of the silyl enol ether and Sn(OTf)$_2$·diamine **102** did not show evidence of metallation of the enol silane to give the corresponding Sn(II) enolate. Although the catalytic reactions were conducted in propionitrile and not dichloromethane, these experiments support the contention that a metalloenolate is not involved as an intermediate in the catalytic cycle.

Mukaiyama and Kobayashi have postulated the Sn(II)·diamine complex **118** to possess square pyramidal geometry. The optically active diamine ligand forms a chelate to tin with the stereochemically relevant lone-pair residing in the sterically demanding position proximal to the pyrrolidine ring. Aldehydes are proposed to bind at the site *trans* to the lone pair and *syn* to the aminonaphthalene moiety. It is possible that a stabilizing dipolar interaction between the bound polarized aldehyde and the electron-rich aminonaphthalene ring may be locking the aldehyde in place and leading to effective blocking of one of the aldehyde diastereofaces.

Evans has recently reported the use of structurally well-defined Sn(II) Lewis acids **119** and **120** (Fig. 9)for the enantioselective aldol addition reactions of α-heterosubstituted substrates [83]. These complexes are easily assembled from Sn(OTf)$_2$ and C$_2$-symmetric bisoxazoline ligands **124** and **126** (Fig. 10). The facile synthesis of these ligands commences with optically active 1,2-amino alcohols **122**, which are themselves readily available from the corresponding α-amino acids **121** [84, 85]. The Sn(II)·bis(oxazoline) complexes were shown to function optimally as catalysts for enantioselective aldol addition reactions with aldehydes and ketone substrates that are suited to putatively chelate the Lewis acid. For example, using 10 mol % of **119**, thioacetate and thiopropionate derived silyl ketene acetals add at –78 °C in CH$_2$Cl$_2$ to glyoxaldehyde to give hydroxy diesters **130** in superb yields and enantioselectivities as well as diastereoselectivities (Eq. 12). The process represents an unusual example wherein 2,3-*anti*-aldol adducts are obtained in a stereoselective manner.

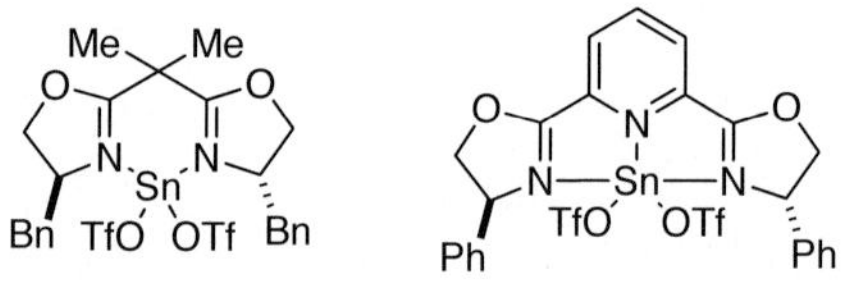

Fig. 9. Chiral bisoxazoline complexes utilized as catalysts for enantioselective additions to chelating aldehydes

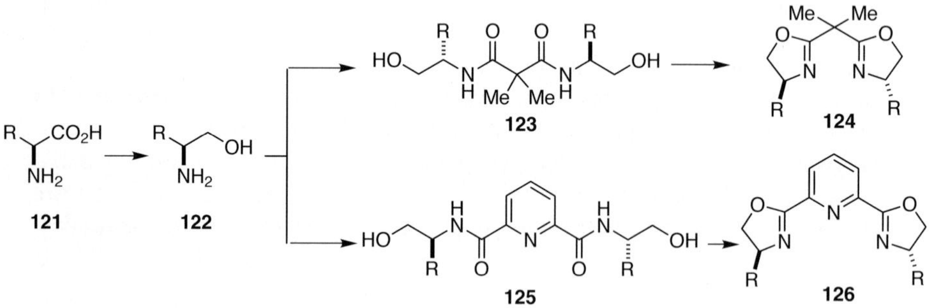

Fig. 10. Synthesis of bisoxazoline ligands

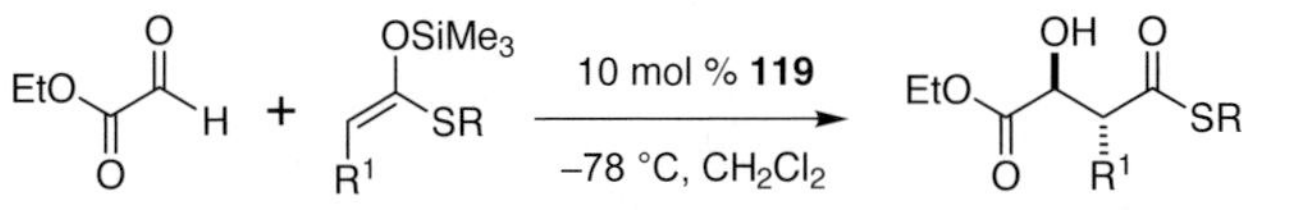

127–129 130

(12)

Entry	SR	R^1	anti:syn	Yield	% ee
1	SPh	H	–	90%	98%
2	SPh	Me	90:10	87%	95%
3	SPh	Et	92:8	90%	95%
4	SPh	^{i}Pr	93:7	72%	95%

Aldol additions to ethyl pyruvate **131** by silyl ketene thioacetals **132, 133, 134** have been shown to proceed in high yields and superb levels of induction (Eq. 13). This process represents an uncommon example of catalytic, asymmetric aldol additions to ketones, providing access to synthetically useful compounds such as **137**. The remarkable ability of the catalyst to differentiate between subtle steric differences of substituents flanking a 1,2-diketone was elegantly demonstrated in the highly enantioselective additions to 2,3-pentanedione **136** (Eq. 14). The aldol adduct of *S-tert*-butyl thiopropionate derived silyl ketene acetal afforded 2,3-*anti*-aldol adduct (>99:1 *anti/syn*) in 98% ee and 97:3 chemoselectivity for the methyl ketone [86].

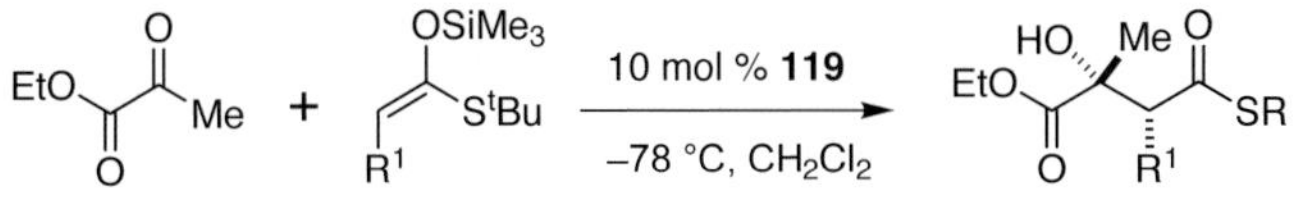

131 132–134 135

(13)

Entry	R^1	anti:syn	Yield	%ee
1	Me	99:1	84%	96%
2	Et	98:2	84%	97%
3	^{i}Bu	99:1	81%	99%

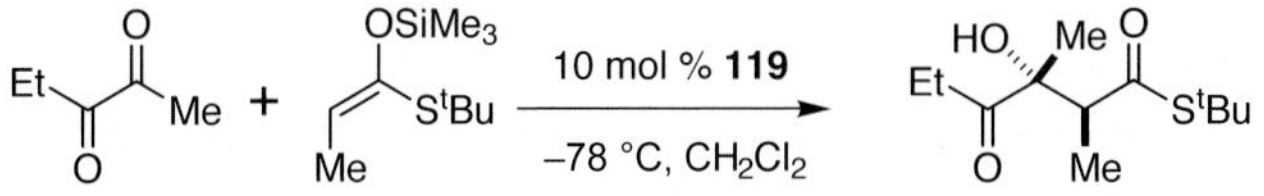

136 137

98 % ee
anti:syn 99:1
regioselectivity 97:3

(14)

An important feature of the Evans system is the ability to provide insight into the catalyst structure. The X-ray crystal structure of the (bisoxazoline)·$Sn(OTf)_2$

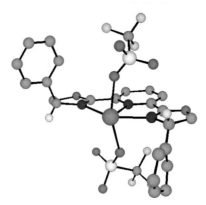

Fig. 11. X-ray crystal structure of the Sn(II)·bisoxazoline adducts

complex displays the central Sn(II) atom in square pyramidal geometry (Fig. 11). The three amino donors occupy a meridional position with the triflate counterions *trans*-diaxially bound. These exhibit some distortion away from the stereochemically relevant, Sn-centered lone pair that resides in the meridional plane. Electrospray ionization studies reveal that the cationic complex is readily generated by dissociation of the triflate counterions, underscoring the kinetic lability of **119** and **120** towards ligand exchange, the sine qua non of a catalytic processes. This structural information has already proven valuable in the design of additional processes, and promises to lead to further advances in the development of other catalytic systems.

4.2
Titanium(IV)

Mukaiyama and coworkers have utilized complexes prepared from (*R*)- or (*S*)-1,1'-2,2'-binaphthol (BINOL) and Ti(IV) precursors generated upon treating Ti(O^{i}Pr)$_4$ with an equivalent of H$_2$O in benzene [87, 88]. The catalytically active species is suggested to be a BINOL·Ti=O complex **138**. However, given the proclivity of related Ti(IV)=O complexes to exist as dimers centered about a (Ti(µ-O))$_2$ core coupled with the inherent instability of group IV oxo complexes, it is likely that the structure of the Ti(IV) species is not monomeric. Using the catalyst generated upon combining 20 mol % each of BINOL, Ti(O^{i}Pr)$_4$, and H$_2$O in toluene, the aldol addition reaction of unsubstituted silyl ketene thioacetals with a variety of aromatic aldehydes afforded products in 91–98% yield and 36–85% enantiomeric excess (Eq. 15). Subtle effects on the enantioselectivity were noted as a function of the aromatic solvent employed with enantioselectivity decreasing along the series: toluene (60% ee), ethylbenzene (54% ee), *m*-xylene (50% ee), 1,3,5-trimethylbenzene (41% ee), and chlorobenzene (16% ee) [89].

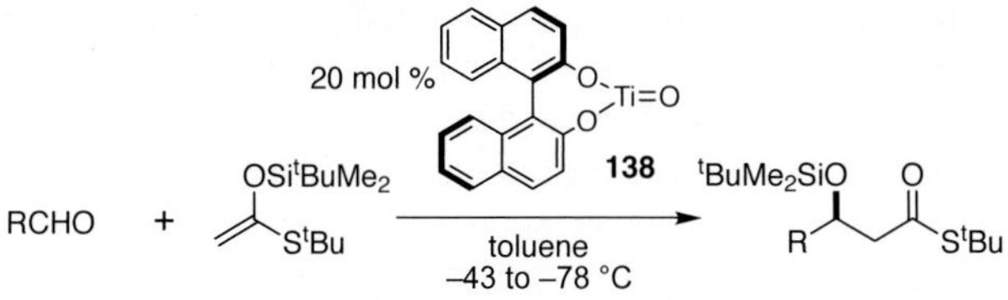

(15)

Entry	Aldehyde	Yield	% ee
1	PhCHO	91%	60%
2	(E)—PhCH=CHCHO	98%	85%
3	β—NapthylCHO	98%	80%
4	α—NapthylCHO	97%	36%

Mikami has examined a BINOL·TiCl$_2$ complex **139** that effectively catalyzes the addition of methyl ketone-, thioacetate-, and thiopropionate-derived enol silanes and aldehydes, giving adducts in impressive yields and enantioselectivity [38a, 90] (Eq. 16). The protocol prescribes the use of 5 mol % of a catalyst generated by treating TiCl$_2$(O^{i}Pr)$_2$ and (R)- or (S)-BINOL in toluene. The synthesis of TiCl$_2$(O^{i}Pr)$_2$ is effected by disproportionation reaction of a 1:1 mixture of Ti(O^{i}Pr)$_4$ and TiCl$_4$ followed by distillation. In general, an important advantage of processes such as these in catalytic asymmetric processes that utilize BINOL as ligand is the fact that both enantiomeric forms are commercially available [91].

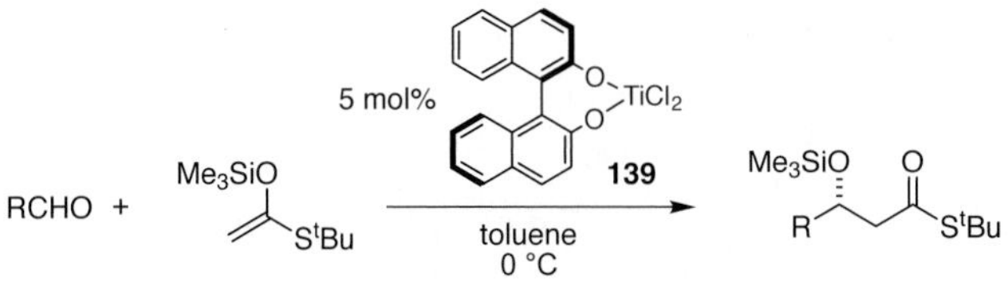

(16)

Entry	Aldehyde	Yield	% ee
1	BnO⌒CHO	81%	96%
2	Cl⌒CHO	61%	91%
3	BocNH⌒CHO	64%	88%
4	C$_6$H$_{17}$⌒CHO	60%	91%
5	Me⌒CHO	61%	81%
6	^{n}BuO$_2$C⌒CHO	84%	95%
7	Me⌒CHO (Me)	61%	85%

The addition reaction of *tert*-butyl thioacetate-derived silyl ketene acetal produces the corresponding aldol adducts in 84% yield and up to 96% enantiomeric excess (Eq. 16). The enantioselectivity of the products was observed to be optimal with toluene as solvent; the use of the more polar dichloromethane consistently produced adducts with 10–15% lower enantiomeric excess. The bulkier *tert*-butylthioacetate-derived enol silane was found to lead to uniformly higher levels of enantioselectivity than the smaller S-ethyl thioketene acetal. This process is impressive in that it tolerates a wide range of aldehyde substrates; for instance, the aldol addition reaction has been successfully conducted with aldehydes substituted with polar functionality such as N-Boc amides, chlorides, esters, and O-benzyl ethers. A key feature of this system when compared to previously reported processes was the ability to achieve high levels of stereoselectivity at 0 °C, in contrast to other processes that commonly prescribe operating temperatures of –78 °C.

The addition of propionate-derived enol silanes **140** delivered 1,2-disubstituted aldol adducts **141** and **142** in useful yields and selectivities (Eq. 17) [90]. As in the acetate-derived additions, the selectivity of the process was dependent on the thioalkyl substituent of the silyl ketene acetal **140**. The 1,2-*syn* adduct was obtained from the addition of E-enolsilane and *n*-butyl glyoxylate (Eq. 17, entry 3). Correspondingly, the formation of 1,2-*anti* adduct was observed in the addition of α-benzyloxy acetaldehyde and the Z-enol silane derived from the *tert*-butyl thioester.

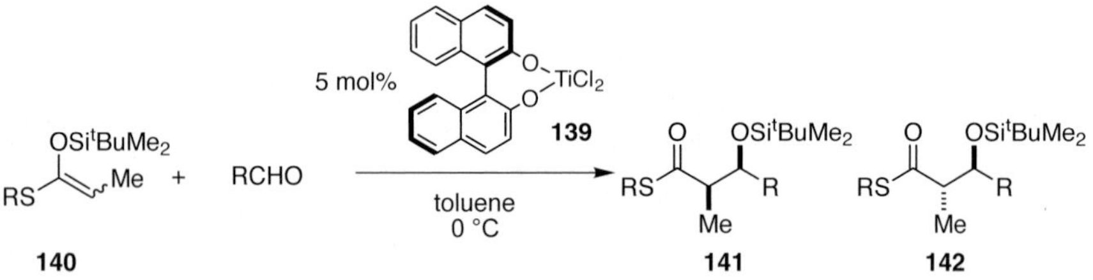

$$(17)$$

Entry	Enol silane	Aldehyde	E/Z	Yield	syn/anti	% ee
1	R = Et	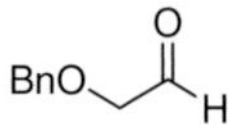 BnO⌿CHO	77% E	85%	72:28	90%
2	R = Et		95% Z	80%	48:52	86%
3	R = Et	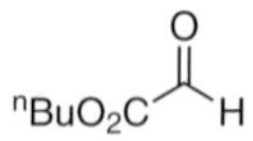 $^nBuO_2C$⌿CHO	77% E	64%	92:8	98%
4	R = ^tBu		93% Z	81%	8:92	90%

The Mikami catalyst (**139**) has been utilized in numerous interesting applications. For example, the addition reaction has been extended to include trifluoroacetaldehyde as substrate, giving optically active fluorinated adducts which are of increasing importance in medicinal chemistry as well as materials science [92]. The addition of thioacetate-derived silyl ketene acetals to trifluoroacetaldehyde affords adduct in 96% ee (Eq. 18). The corresponding aldol addition of substituted enolates produces a mixture of *syn/anti* adducts 55% – 89% enantiomeric excess (Eq. 19).

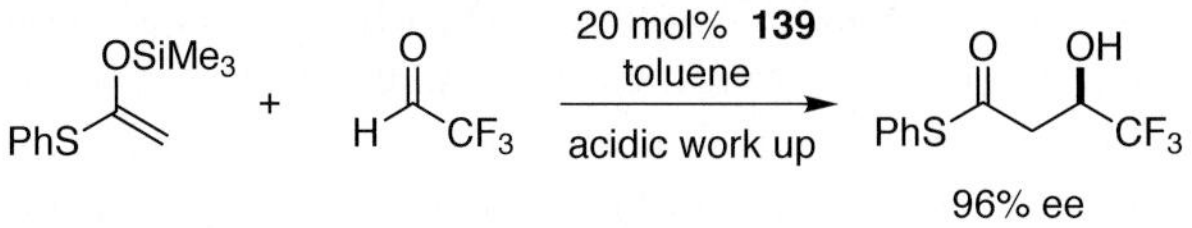

$$(18)$$

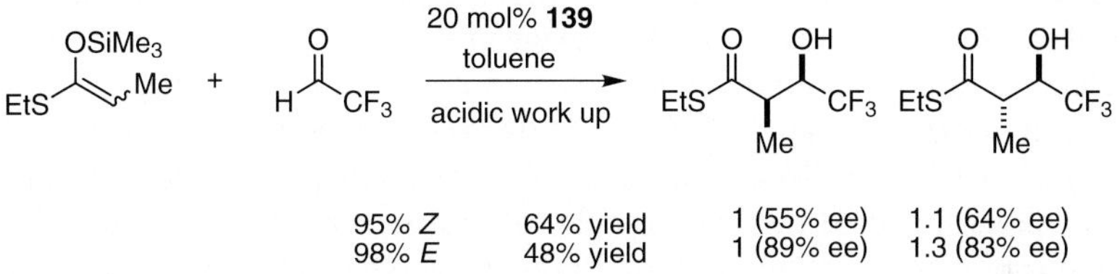

96% ee

95% Z	64% yield	1 (55% ee)	1.1 (64% ee)
98% E	48% yield	1 (89% ee)	1.3 (83% ee)

$$(19)$$

Mikami has conducted a series of mechanistic studies that provide insight into the structural details of the transition-state structure and the Si-atom-transfer step (Fig. 12) [93]. Experiments involving two enol silanes, such as 143 and 144, incorporating minor differences in the *O*-silyl and *O*-alkyl substituents allow the nature of the key atom-transfer step to be probed. Such double-label experiments can effectively determine whether the Si atom-transfer step proceeds via an intra- or intermolecular process. When α-benzyloxyacetaldehyde was allowed to react with 0.5 equiv each of 143 and 144, only aldol adducts 145 and 147 were isolated from the reaction mixture. The absence of adducts 146 and 148 was consistent with a mechanism involving intramolecular metal aldolate silylation. Mikami has suggested a transition-state structure 149 in which C-C bond formation and aldolate silylation occur concomitantly in a silatropic ene-like process. For such an arrangement, the migrating trialkylsilyl group serves to anchor the six-membered ring, with the metal complex residing exocyclic to the array. The results of additional experiments with optically active aldehydes by Mikami is suggested to provide additional validation for the working model 149.

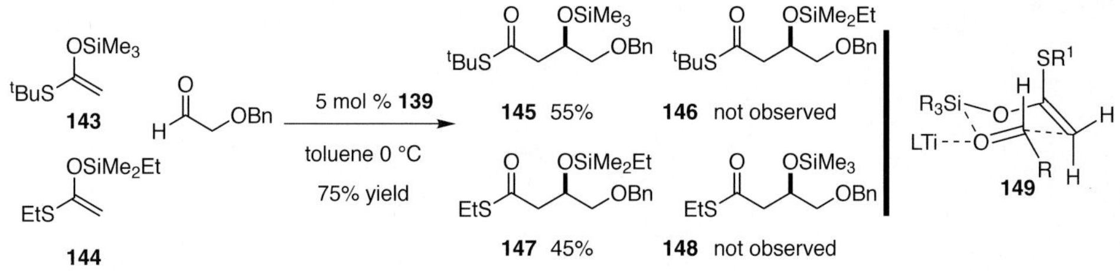

Fig. 12. Double labeling experiments by Mikami supporting a silatropic ene transition-state

   Although this analysis of the transition state structure is consistent with the
experimental observations, other mechanistic pathways may be postulated
(Fig. 13) [94]. The mechanistic analysis of a related carbonyl addition process
involving the BINOL·Ti(O^{i}Pr)$_2$-catalyzed addition of allylstannanes to aldehydes
offers insight to an alternate mechanistic pathway that may account for the results of the double labeling experiments by Mikami. In an in-depth analysis of
the Keck catalytic, enantioselective aldehyde allylation reaction by allyltributyl-
tin, Corey has proposed an intermediate 152 wherein the BINOL ligand is transiently O-stannylated. In a subsequent step, intramolecular trans-stannylation
occurs releasing product 154 with regeneration of catalyst. The operation of an
analogous mechanistic step in the Mikami aldol process wherein the ligand participates as a temporary repository of the silyl group (155) would account for the
observed results in the double-labeling experiments.
   A related Mukaiyama aldol catalyst system reported by Keck prescribes the
use of a complex that is prepared in toluene from (R)- or (S)-BINOL and Ti(O^{i}Pr)$_4$
in the presence of 4 Å molecular sieves. In work preceding the aldol addition reaction, Keck had studied this remarkable catalyst system and subsequently developed it into a practical method for enantioselective aldehyde allylation [95a,
95b, 95c, 96]. Because the performance of the Ti(IV) complex as an aldol catalyst
was quite distinct from its performance as a catalyst for aldehyde allylation, a
careful examination of the reaction conditions was conducted. This meticulous
study describing the use of (BINOL)Ti(O^{i}Pr)$_2$ as a catalyst for aldol additions is
noteworthy since an extensive investigation of reaction parameters, such as temperature, solvent, and catalyst loading and their effect on the enantiomeric excess of the product was documented. For example, when the reaction of benzaldehyde and tert-butyl thioacetate-derived enol silane was conducted in dichloromethane (10 mol % catalyst, –10 °C) the product was isolated in 45% yield and
62% ee; by contrast, the use of toluene as solvent under otherwise identical conditions furnished product of higher optical purity (89% ee), albeit in 54% yield.
For the reaction in toluene, increasing the amount of catalyst from 10 to 20 mol %

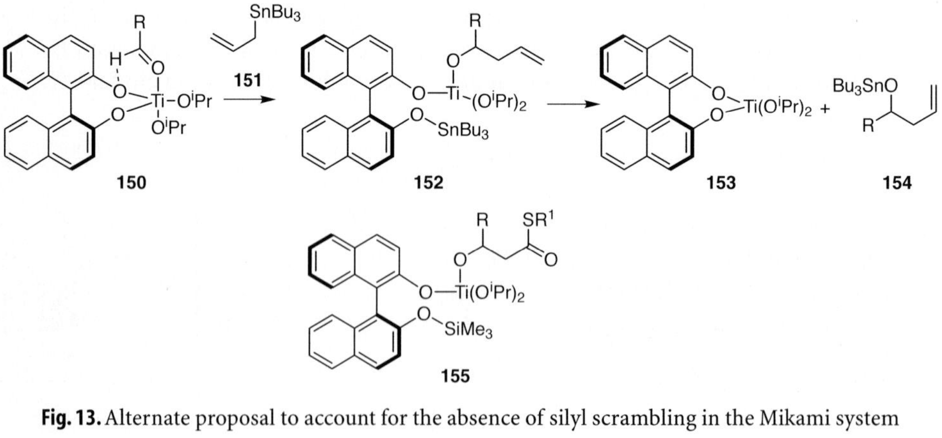

Fig. 13. Alternate proposal to account for the absence of silyl scrambling in the Mikami system

resulted in further amplification in the enantioselectivity of the products (95% ee); however, further increasing in the catalyst load to 50 mol % produced a diminution in asymmetric induction to 87% ee. Additional studies of solvent effects identified Et_2O as the optimal medium for the catalytic process.

Treatment of the thioacetate-derived enol silane with aldehyde and 20 mol % catalyst **153** at –20 °C in Et_2O gave the product in high optical activity (Eq. 20). Hydrolytic work-up of the crude reaction products afforded the β-hydroxy aldol adducts in excellent enantioselectivity (97–99% ee) [97]. The addition reaction in Et_2O also displayed some sensitivity to concentration, temperature, and catalyst loading. Thus, the aldol addition with benzaldehyde at two different substrate concentrations, 0.25 and 0.5 M, gave product in 97 and 92% ee, respectively. Decreasing the catalyst load to 10 mol % afforded the product in identical % ee's, albeit in diminished yields (71% after 16 h) when compared to the corresponding yield (90% after 10 min) with 20 mol % catalyst. An important conclusion from this study is that the variation in product yield and selectivity as a function of catalyst concentration and catalyst loading are correlated to the known non-linear effects in these systems.

(20)

Entry	Aldehyde	Yield	%ee
1	PhCHO	90%	97%
2	$PhCH_2CH_2CHO$	80%	97%
3	2-FurylCHO	88%	>98%
4	c-$C_6H_{11}CHO$	79%	89%
5	PhCH=CHCHO	76%	89%
6	$PhCH_2OCH_2CHO$	82%	>98%

Carreira has reported a series of studies that have led to the development of a family of optically active tridentate ligands **156** and their corresponding derived Ti(IV) complexes **157** and **158** (Fig. 14). Tridentate ligands were selected for study on speculation that greater control over catalyst design and structure would be available with a ligand able to coordinate through more than two heteroatom donors. The ligand family is derived from the unsymmetrical, chiral 2,2'-amino-hydroxy-1,1'-binaphthalene, a ligand fundament not previously investigated in

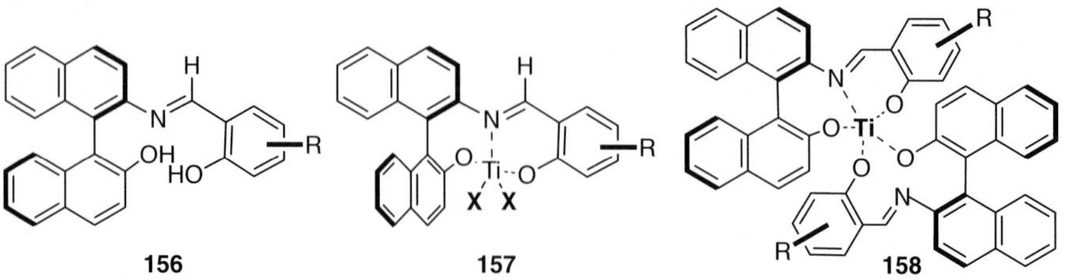

Fig. 14. Tridentate ligand and the derived complexes employed by Carreira

inorganic coordination chemistry. Racemic 2,2'-aminohydroxy-1,1'-binaphthalene is readily prepared by oxidative heterocoupling of 2-aminonaphthalene and 2-hydroxynaphthalene following a procedure developed by Kocovsky and Smcrina [98a, 98b]. Resolution of the racemate is effected with the commercially available (R)- or (S)-camphorsulfonic acids, allowing for either enantiomer of the unsymmetrically substituted binaphthyl to be readily accessed.

The ability to derivatize the amine functionality in **159** confers flexibility in the synthesis of a number of tridentate ligands (**156**) of varying electronic or steric properties (Fig. 15). Specifically, the preparation of Schiff bases derived from **159** and salicylaldehydes **160** would provide a library of tridentate ligands for Ti(IV). The tridentate ligands incorporate two aryloxy donors and a neutral imine nitrogen; consequently, coordination to Ti(IV) would necessarily generate a complex with two additional charged donor atoms **157**. Further substitution of these two remaining ligands allows for additional structural variables in the form of chelating groups whose electronic and steric properties are amenable to alteration. Thus, the overall modular design of the complex **162** allows for independent variation of several ligand variables in the optimization of the process [99]. The results of preliminary experiments that varied the nature of substituents on the salicylimine component led to the identification of the 4-*tert*-butyl-6-bromosalicylaldehyde (**163**) derived ligands as optimal in the asymmetric addition reactions for the broadest scope of aldehyde substrates.

The initial investigations of catalytic carbonyl addition processes were guided by the hypothesis that formation of the silylated aldolate product and release of the metal complex would be facilitated by incorporating into the complex an effective means for transfer of the trialkylsilyl group. Importantly, salicylate ligands were investigated since Masamune had articulated the critical role that carboxylates could play in Si-atom transfer reactions. In this regard, a number of electronically and sterically diverse salicylic acids are commercially available at a nominal price. Moreover, the replacement of the two coordinated isopropoxides in **164** could be easily achieved upon adding chelating ligands to the catalyst solution followed by azeotropic removal of the released isopropanol (Eq. 21).

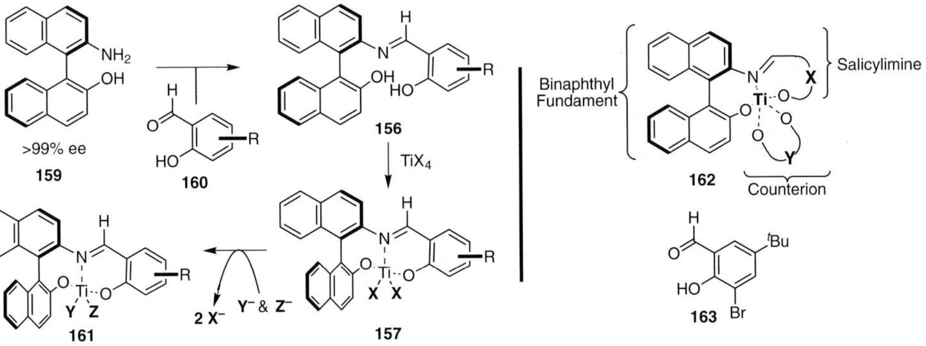

Fig. 15. Synthesis and structural analysis of the Carreira complex

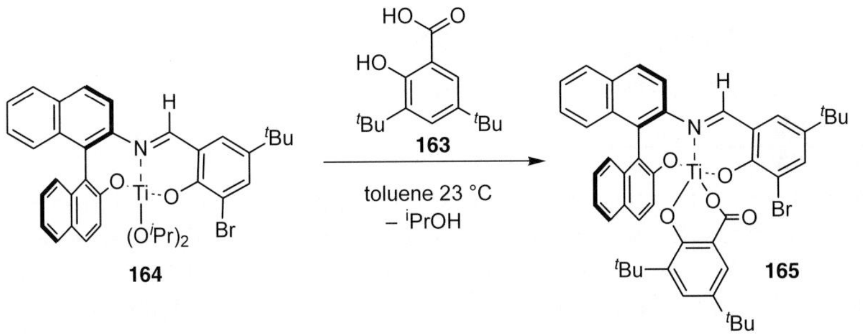

(21)

Treatment of tridentate ligand with $Ti(O^iPr)_4$ and di-*tert*-butylsalicyclic acid (**163**) in toluene followed by evaporation of the solvent afforded an orange complex postulated to be **165**, which was shown to be an effective catalyst for the Mukaiyama aldol reaction. Under optimized conditions, the simple methyl acetate-derived enol silane **166** adds to aldehydes in the presence of as little as 0.5 mol % of **165** at 0 °C to give optically active adducts **167** in high yields and up to 99% ee (Eq. 22) [100]. Importantly, the reaction can be conducted with a wide range of substrates such as aliphatic, aromatic and unsaturated aldehydes as well as functionalized aldehydes.

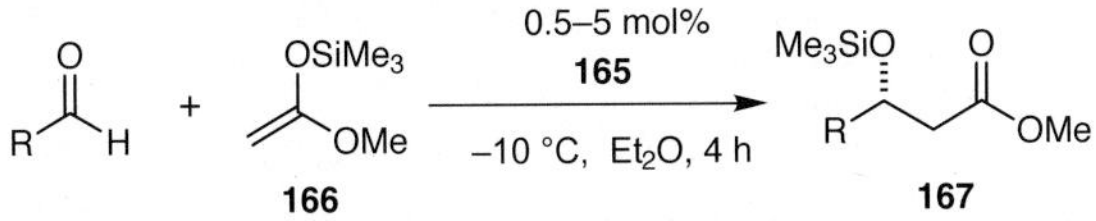

(22)

Entry	Aldehyde	Yield	% ee
1	$Ph(CH_2)_3$—≡—CHO	84%	96%
2	$TBSOCH_2$—≡—CHO	91%	96%
3	Ph—≡—CHO	96%	94%

4	$^{i}Pr_3Si$—≡—CHO	97%	97%
5	Me⌒⌒CHO	82%	98%
6	Me⌒⌒CHO	76%	95%
7	Ph⌒⌒CHO	99%	98%
8	Ph⌒⌒CHO	98%	94%
9	Me PhCHO	84%	96%
10	$^{t}BuMe_2SiO$⌒⌒⌒CHO	87%	95%
11	$^{t}BuPh_2SiO$⌒⌒CHO	89%	91%
12	Ph_3CS(⌒)$_2$⌒CHO	99%	98%
13	$C_6H_{11}CHO$	81%	95%
14	Me⌒⌒⌒CHO	83%	98%
15	Ph⌒⌒CHO Me	92%	95%

The catalytic process has found successful application in several natural product total syntheses. In 1996, Simon reported a synthesis of the antitumor depsipeptide FR-9001,228 in which the aldol addition reaction of **168** and the ethyl acetate-derived enol silane furnished a key synthetic intermediate (Eq. 23). The enantioselective aldol addition reaction of **168** was conducted with **165** and its enantiomer *ent*-**165** to separately provide both enantiomers of the aldol adducts **169** and **170** (Scheme 14). These were then utilized in the preparation of diastereomeric seco acids **171** and **172** [101]. Macrocylization of **172** through a Mitsonobu reaction yielded the desired natural product **173**.

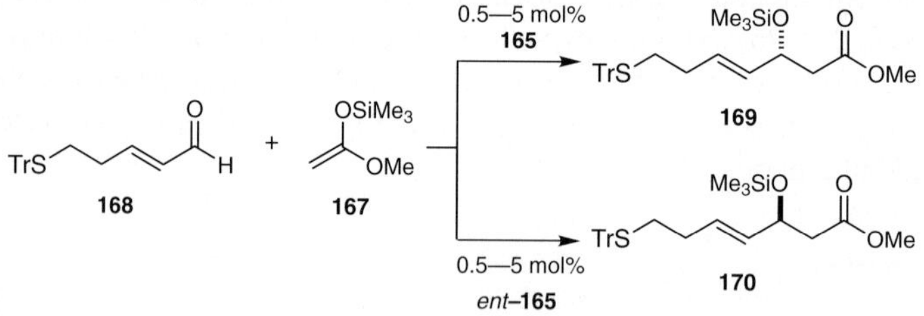

(23)

In a separate, elegant use of **165**, Rychnovsky and coworkers have carried out a diastereoselective addition of methyl acetate-derived silyl ketene acetal to aldehyde **174** to afford adduct **175** in high diastereomeric purity (Scheme 15) [102]. Hydroxy ester **175** was subsequently employed as an intermediate in the total synthesis of the polyene macrolide antibiotic Roflamycoin. This work highlights a novel application of the chiral catalyst system in reagent-controlled coupling of chiral functionalized substrates which by themselves display only mod-

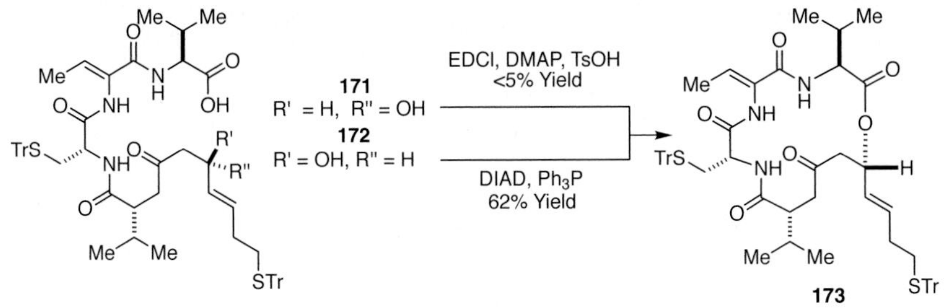

Scheme 14

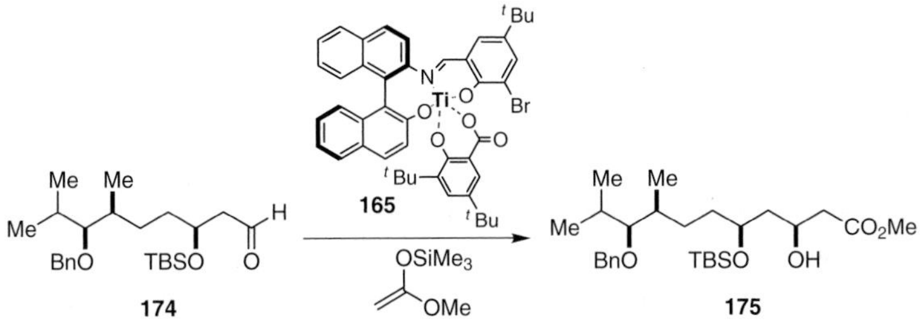

Scheme 15

est levels of asymmetric induction, thus underscoring the ability of the catalyst to function in a distereoselective synthesis.

In addition to the efficiency exhibited by catalyst **165** with a broad spectrum of aldehydes in acetate aldol addition reactions, this catalyst has been shown to function competently in enantioselective additions of dienol silane **87**. The requisite dienolate is readily synthesized from 2,2,6–trimethyl-4H-1,3–dioxin-4-one **84** (diketene+acetone adduct) by deprotonation with LDA and quenching with Me$_3$SiCl (Eq. 24). Dioxinone **84** is commercially available at a nominal price; in addition, the silyl dienolate **87** is easily purified by distillation and stable to prolonged storage. The addition reactions of **87** with aldehydes were conducted with 1–3 mol % of **165** at 0 °C (Eq. 25). A variety of aldehydes serve as substrates and give aldol adducts in 79–97% yields and up to 99% ee after a single recrystallization.

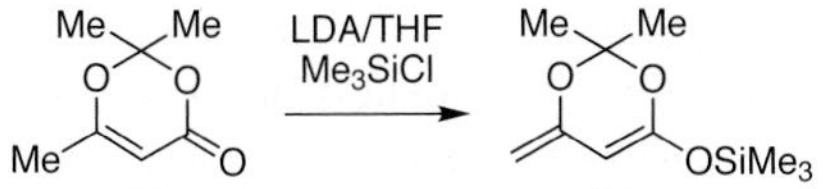

(24)

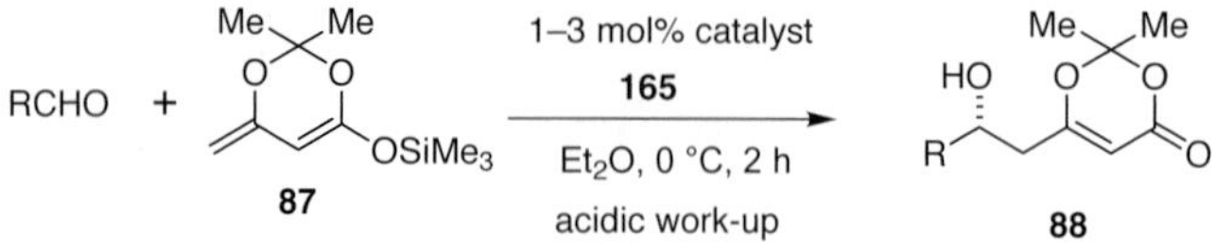

$$\text{RCHO} + \mathbf{87} \xrightarrow[\substack{\text{Et}_2\text{O, 0 °C, 2 h} \\ \text{acidic work-up}}]{\substack{\text{1–3 mol\% catalyst} \\ \mathbf{165}}} \mathbf{88} \tag{25}$$

Entry	Aldehyde	Yield	% ee
1	$^i\text{Pr}_3\text{Si}$—≡—CHO	86%	91%
2	$^t\text{BuMe}_2\text{SiO}$—__CHO	97%	94%
3	Ph\~_CHO	88%	92% (99%)
4	Me\~\~_CHO	95%	92%
5	PhCHO	83%	84% (96%)
6	Ph\~_CHO	97%	80%
7	Bu_3Sn\~_CHO	79%	92%

The catalytic, enantioselective aldol addition reaction generates products that can serve as versatile precursors to useful building blocks for asymmetric synthesis (Eq. 26). For example, treatment of cinnamaldehyde adduct **177** with LiAl(HNBn)$_4$ **178** afforded the crystalline amide **179** (73%). Heating in *n*-BuOH converted **177** to ester **180** (81%). Heating in alkaline methanol yielded (79%) the crystalline lactone **181**. The synthetic utility of adducts **179** and **180** is enhanced by the stereoselective reaction methods that have been developed for their reduction to the corresponding *syn* and *anti* 3,5-diols [103, 104].

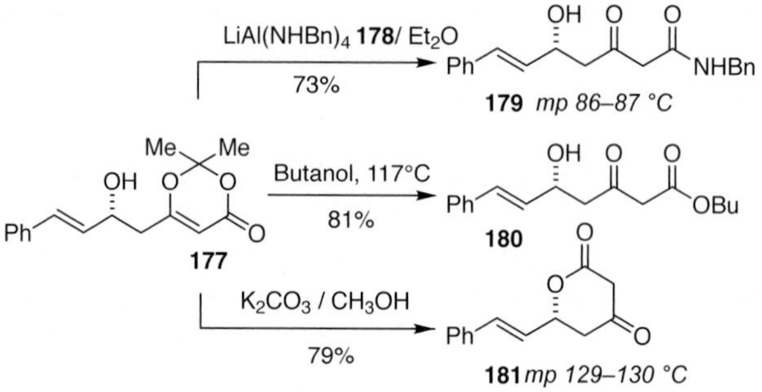

In work concurrent with that of Carreira, Sato and coworkers reported that dienol silane **87** participated in catalytic Mukaiyama aldol addition reactions in the presence of 20–100 mol % of Mikami's Ti(IV) complex **139** or Yamamoto's CAB complex **184** (Eqs. 27 and 28). The addition of **87** to benzaldehyde and pentanal at –78 °C utilizing 20 mol % of **139** afforded the corresponding adducts **182** and **183** in 38% yield/88% ee and 55% yield/92% ee, respectively. The use of **184** at 50 mol % delivered **182** and **183** in 69% yield/67% ee and 52% yield/70% ee, respectively.

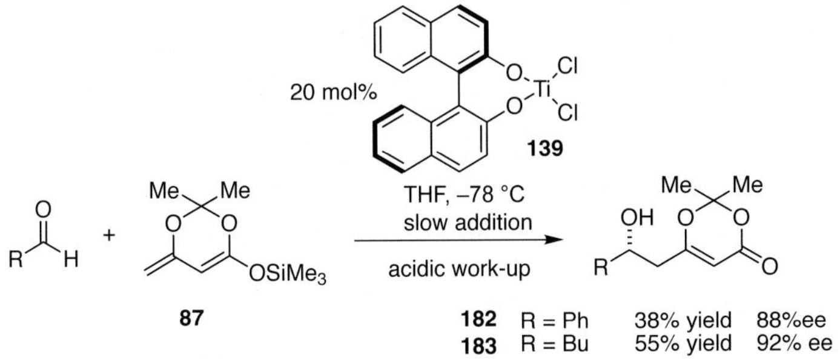

$$(27)$$

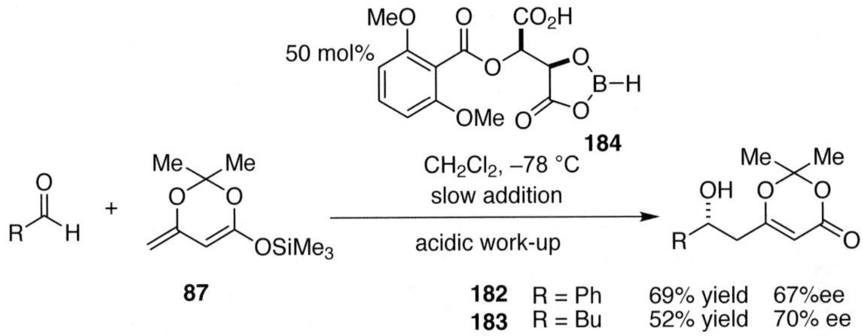

182 R = Ph   38% yield 88%ee
183 R = Bu   55% yield 92% ee

$$(28)$$

182 R = Ph   69% yield 67%ee
183 R = Bu   52% yield 70% ee

4.3
Boron

Numerous boron complexes have been prepared and studied as catalysts for the Mukaiyama aldol addition reaction and related processes [105]. The types of ligands that have been generally utilized in the preparation of these complexes are derived from either N-sulfonyl-α-amino acids or α-hydroxy acids. Both classes share common structural features and form a five-membered chelate involving the carboxylate moiety and the corresponding α-heteroatom (Fig. 16). As discussed above, Masamune has postulated an important mechanistic role for the acyloxy group in facilitating silyl-group transfer [72]. In general, catalytic aldol addition reactions utilizing borane-derived complexes require 20–30 mol % catalyst in propionitrile as solvent at low temperature with slow addition of the reactants.

Yamamoto has documented the use of boryl complexes prepared with a tartaric acid-derived ligand class [106]. The modified tartrate ligands are conven-

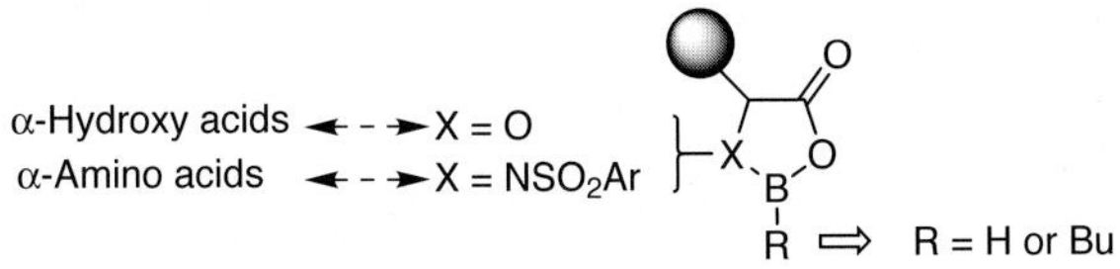

Fig. 16. The general structural characteristics of amino acid and tartrate derived ligands for boron catalysts

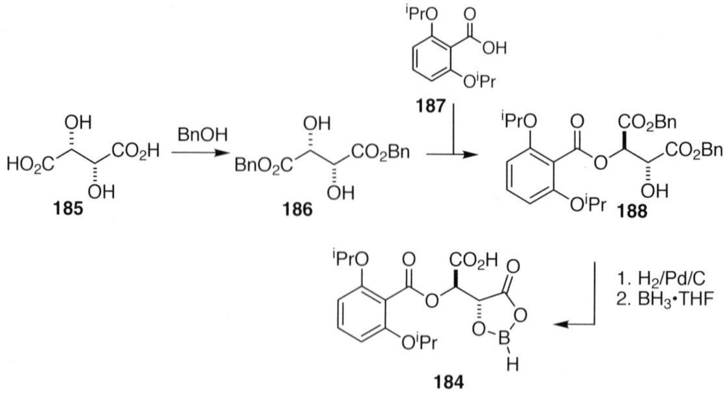

Scheme 16

iently prepared from (R)- or (S)-tartaric acid following a three-step sequence of reactions: (1) formation of the tartrate bisbenzyl ester **186**; (2) monoacylation **186**+**187**→**188**; followed by (3) hydrogenolysis of the O-benzyl esters to afford **184** (Scheme 16). In the initial studies these complexes were employed as catalysts for the Diels-Alder cycloaddition reactions [107, 108]. Subsequent investigations have documented the use of the Lewis-acidic boryl complex **184** in aldehyde addition reactions [109].

The addition of ketone-derived enol silanes and aldehydes in the presence of **184** at –78 °C in propionitrile afforded the aldol adducts in excellent yields as well as diastereo- and enantioselectivity (Eq. 29) [106]. The versatility of this catalyst is evidenced by the fact that enol silanes derived from aliphatic methyl and ethyl ketones as well as acetophenone are substrates for the aldol addition reaction.

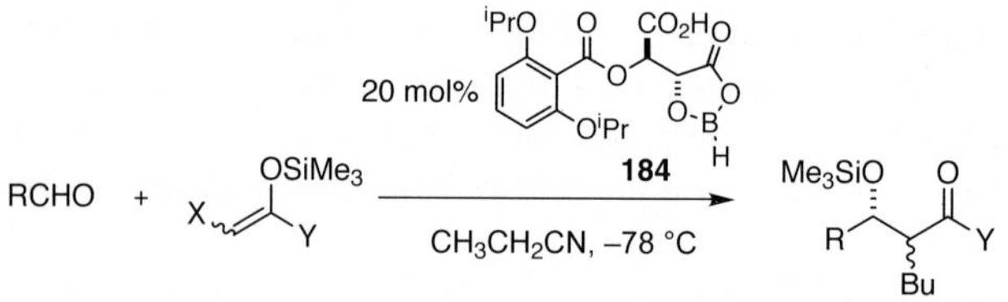

$$(29)$$

Enrty	Enol silane	Aldehyde	Yield	syn/anti	% ee
1	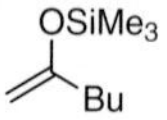 OSiMe₃ / Bu	PhCHO	81%	-	85%
2	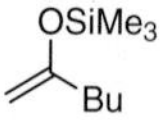 OSiMe₃ / Bu	n—BuCHO	70%	-	80%
3	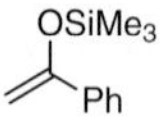 OSiMe₃ / Ph	PhCH=CHCHO	88%	-	83%

4	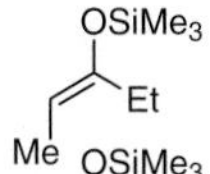 OSiMe$_3$ structure with Et, Me	PhCHO	96%	94:6	96%
5	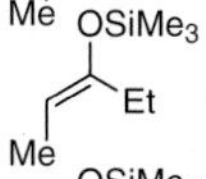 OSiMe$_3$ structure with Et, Me	n—PrCHO	79%	>94:6	93%
6	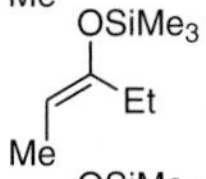 OSiMe$_3$ structure with Et, Me	MeCH=CHCHO	79%	>94:6	93%
7	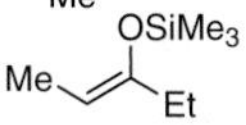 Me, OSiMe$_3$, Et structure	PhCHO	97%	93:7	94%

The use of these boryl complexes in catalytic, enantioselective additions to aldehydes by silyl ketene acetals has also been the subject of intense investigation by Yamamoto (Eq. 30) [108]. Although ethyl and benzyl acetate-derived enol silanes furnished racemic products, the phenyl acetate-derived trimethylsilyl ketene acetals proved optimal, giving adducts in up to 84% ee. Additionally, Yamamoto has documented the use of **184** in aldol addition reactions of propionate- and isobutyrate-derived enol silanes (Eqs. 31 and 32). Thus, the addition of the phenyl acetate derived (E)-enol silane afforded adducts as diastereomeric mixtures with the *syn* stereoisomer displaying up to 97% ee (Eq. 32).

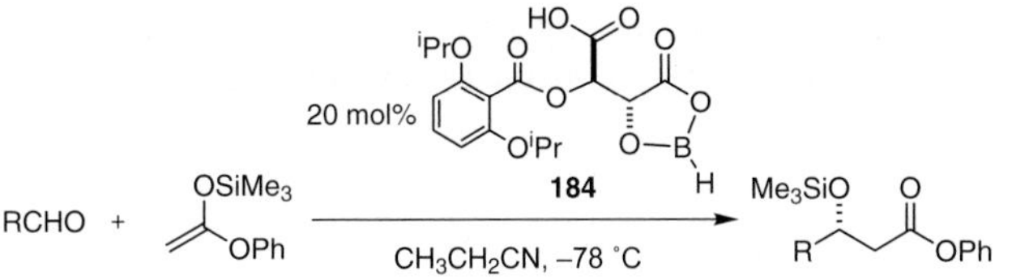

$$\text{(30)}$$

	Yield	% ee
R = Ph	63%	84%
R = n—Bu	49%	76%

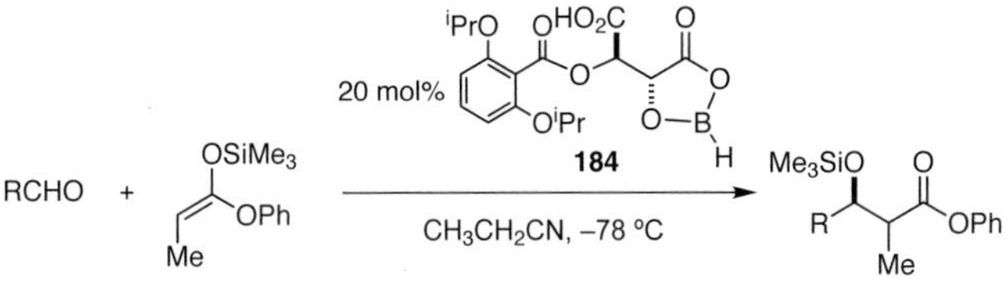

$$\text{(31)}$$

Entry	Aldehyde	Yield	syn/anti	%ee (syn)
1	PhCHO	83%	79/21	92%
2	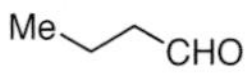 Me, CHO	57%	65/35	88%
3	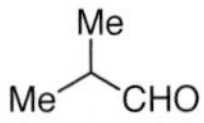 Me, Me, CHO	45%	64/36	79%
4	 Ph, Me, CHO	97%	96/4	97%
5	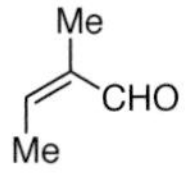 Me, Me, CHO	86%	>95/5	94%

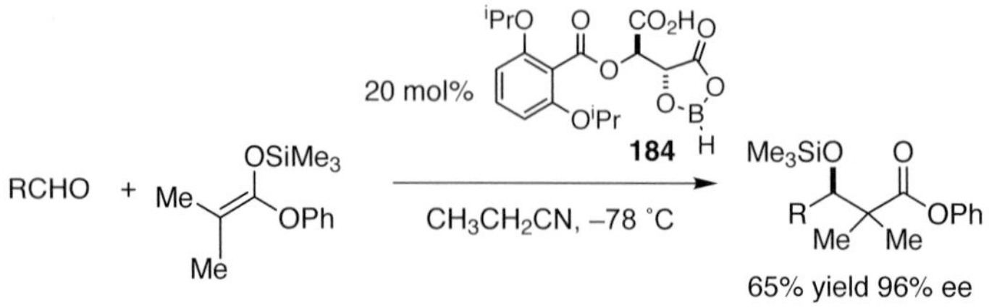

$$(32)$$

Although detailed structural data on the active catalysts in the Mukaiyama aldol addition processes is lacking, related studies of the complex as a catalyst for the Diels Alder cycloaddition reaction provided important insight [110]. In this regard, Yamamoto has conducted ^{1}H-NMR spectroscopic experiments that display a strong NOE between the aromatic ring protons of the aroyl moiety and the enal β-protons. This observation has led Yamamoto to postulate a structure which positions the bound electrophilic component proximally to the 2,6-diisopropoxybenzene ring . The NOE data suggest a structure wherein the bound aldehyde is positioned proximally to the aroyl moiety. Yamamoto has also carried out molecular weight measurements of a solution of the complex generated from phenyl boronic acid; in benzene the formula weight corresponds closely with that for a well defined monomeric species. Additional infrared spectroscopic data are consistent with a complex possessing a five-membered chelate formed between the alkoxy and acyloxy functional groups (Fig. 17).

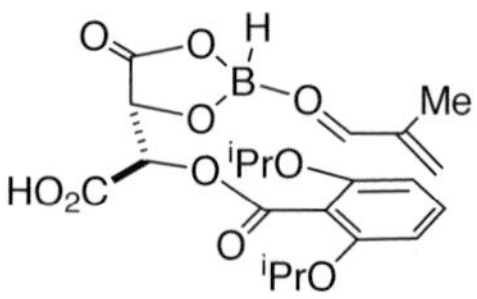

Kiyooka and coworkers have reported a boron catalyst prepared from $BH_3 \cdot THF$ and N-p-toluenesulfonyl-L-valine [111]. These boron complexes were first reported as stoichiometric reagents which promote the Mukaiyama aldol addition reaction. The outcome of the addition reaction exhibits a dramatic dependence

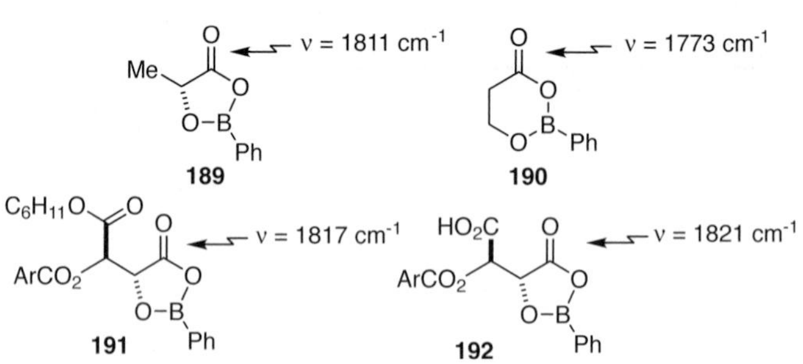

Fig. 17. IR data of a various boryl complexes

on the nature of the trialkylsilyl moiety (Me_3Si versus $^tBuMe_2Si$) of the enol silane. The enol silane derived from the trimethylsilylketene acetal afforded the
expected aldol addition adducts in 77–87% yields with 83–92% ee (Eq. 33). By
contrast, the bulkier *tert*-butyldimethylsilylketene acetal furnished the reduced
adduct **194** (Eq. 34). This latter product is suggested to be formed by addition of
the silyl ketene acetal to the activated aldehyde·promoter complex followed by
hydride transfer by the borane to the putative *O*-silylated ester intermediate.
This result highlights the fine balance that exists in group transfer reactions such
as these between silyl transfer(catalyst turn over) and alternate reaction pathways [112, 113, 114, 115]. Two important modifications made it possible to convert the process into a corresponding catalytic aldol addition: the use of nitromethane as solvent and the *N-p*-nitrobenzenesulfonyl derived ligand (Eq. 35).

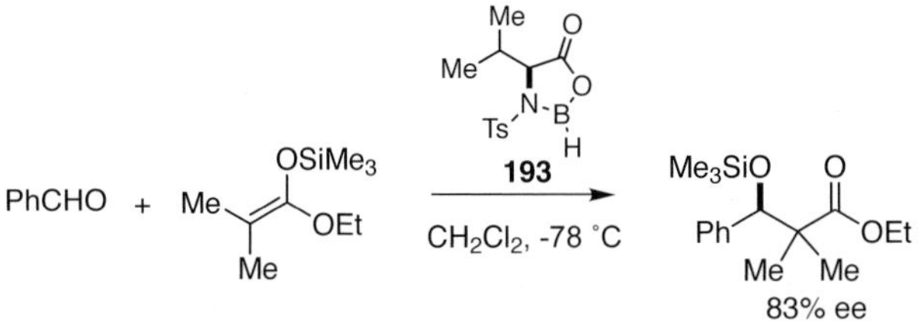

$$(33)$$

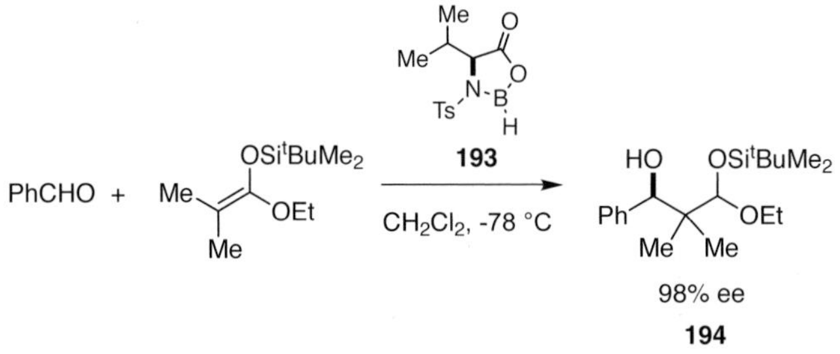

$$(34)$$

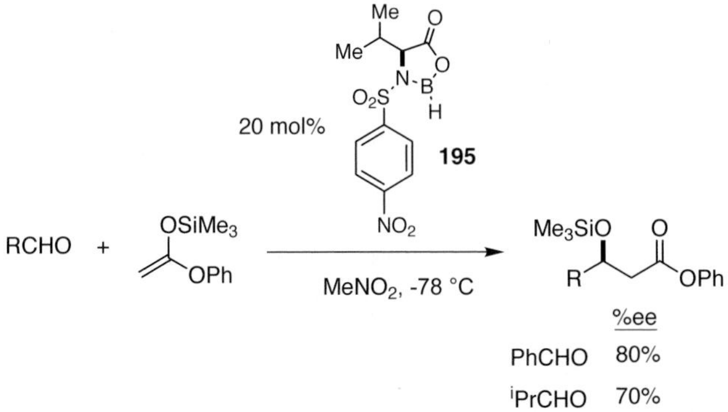

$$(35)$$

	%ee
PhCHO	80%
^{i}PrCHO	70%

The addition of phenyl acetate-derived enol silane to benzaldehyde and *iso*–
butyraldehyde in nitromethane utilizing 20 mol % catalyst **195** provided the silylated aldol adducts in 80 and 70% ee, respectively. Kiyooka has also document-

ed the ability of **195** to mediate the addition of substituted enolates (Eq. 36). The *E*-propionate derived enol silane afforded the adducts in 60–91% yield, albeit in modest levels of simple diastereoselectivity. The enantiomeric excess for the *anti* adduct was observed to be of uniformly higher optical purity. Addition reactions of hetero-substituted enol silanes **196** (Eq. 37) have also been studied. The reductive removal of the dithiane following aldol addition reaction provides an alternative to the acetate aldol addition reaction.

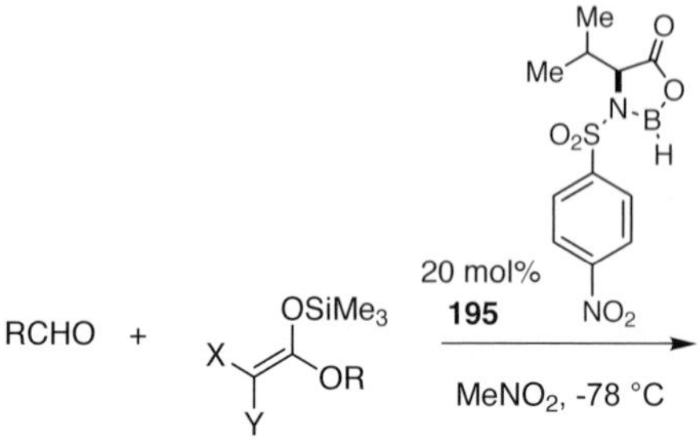

(36)

Entry	Enol silane	Aldehyde	Yield	syn/anti	%ee
1	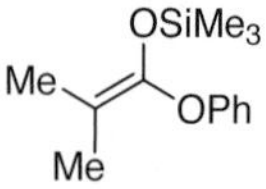 OSiMe₃, Me, OPh, Me	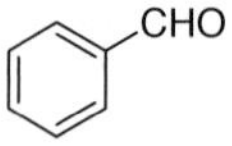 CHO	65%	-	96%
2	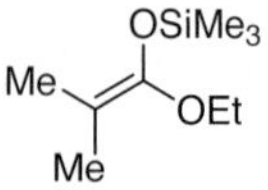 OSiMe₃, Me, OEt, Me	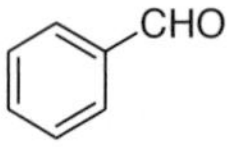 CHO	92%	-	90%
3	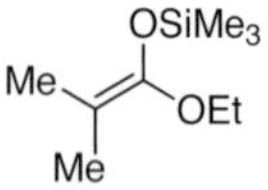 OSiMe₃, Me, OEt, Me	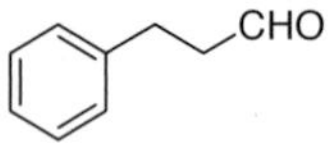 CHO	97%	-	95%
4	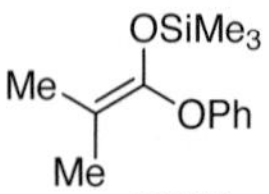 OSiMe₃, Me, OPh, Me	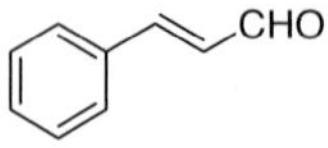 CHO	63%	-	81%
5	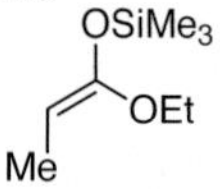 OSiMe₃, OEt, Me	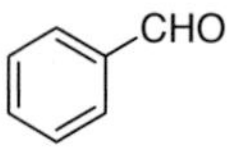 CHO	91%	76:24	syn: 66% anti: 90%
6	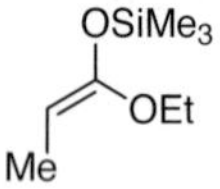 OSiMe₃, OEt, Me	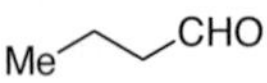 Me CHO	60%	60:40	syn: 60% anti: 91%
7	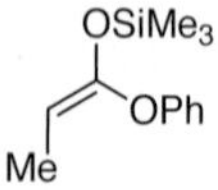 OSiMe₃, OPh, Me	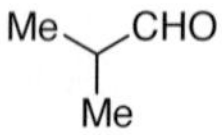 Me CHO Me	64%	65:35	syn: 90% anti: 97%

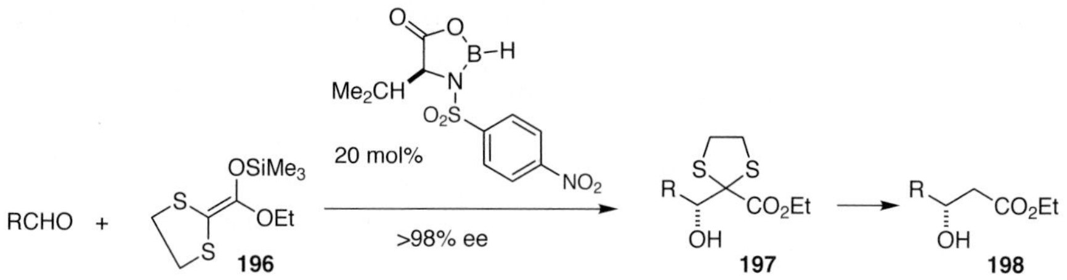

(37)

An AM1 optimized structure of the chiral borane complex has been utilized as the centerpiece of the model that is proposed to account for the stereochemical outcome of the reaction (Fig. 18). The aldehyde is suggested to coordinate to the boron on the face opposite the isopropyl substituent thereby minimizing steric interactions. The Kiyooka model **199** places the formyl-H over the five-membered ring chelate subtending an obtuse H-B-O-C dihedral angle. Analogous modes of binding have been proposed in other chiral Lewis acid boron compounds that have been ingeniously utilized for Diels-Alder cycloaddition reactions [116a, 116b]. The preference for such orientation may result from the presence of a stabilizing anomeric interaction. Alternatively, the bound aldehyde may be locked in the conformation invoked by Kiyooka as a result of a formyl C-H hydrogen bond to the acyloxy donor following the bonding model proposed by Corey (Fig. 18) [63a, 63b, 63c, 63d, 63e].

Masamune has examined a number of oxazaborolidines derived from a series of simple α-amino acid ligands derivatized as the corresponding N-p-toluenesulfonamides [72a, 72b, 72c]. A dramatic improvement in the reaction enantioselectivity was observed when the complex prepared from α,α-disubstituted glycine arylsulfonamides were employed (Eq. 38). Thus, using 20 mol % of **200** the aldol adduct of benzaldehyde and O-phenyl isobutyrate-derived enol silane was isolated in 98% ee. In subsequent studies the product enantioselectivity was optimized as a function of substitution of the arylsulfonamide (Eq. 39). Thus, for complexes possessing the general structure **201** the enantiomeric excess of the benzaldehyde adduct varies along the series: Ar=3,5-bis(trifluoromethyl)phenyl (52% ee); mesityl (53% ee); 1-naphthyl (67% ee); 2-naphthyl (78% ee); 4-tert-butylphenyl (81% ee); phenyl (83% ee); 4-methoxyphenyl (86% ee); 4-acetamidophenyl (86% ee).

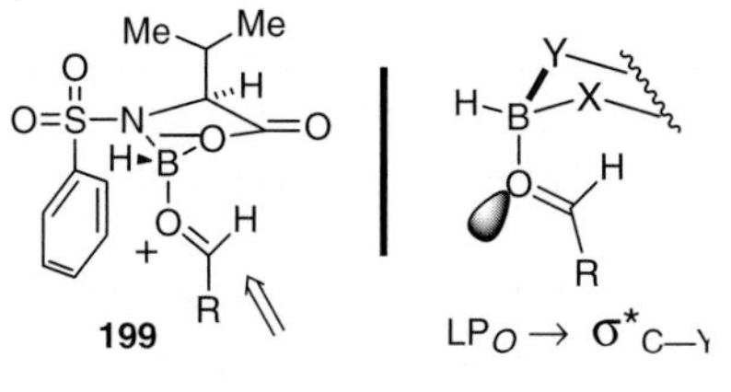

Fig. 18. Kiyooka's model of the aldehyde·Lewis acid complex

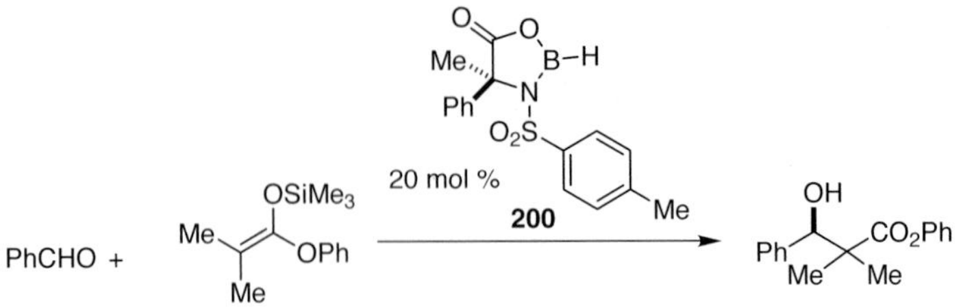

$$(38)$$

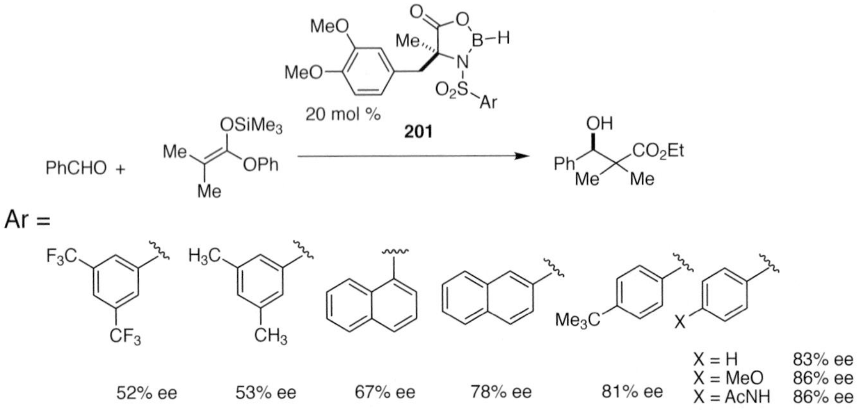

Ar =

52% ee	53% ee	67% ee	78% ee	81% ee	X = H 83% ee
					X = MeO 86% ee
					X = AcNH 86% ee

$$(39)$$

The preparation of the novel ligand **202b** illustrates the general synthetic approach to this class of quaternary α,α-disubstituted glycine sulfonamides (Scheme 17). Menthone (**203**) is converted to **204** using the Strecker amino acid synthesis procedure. A short sequence of synthetic manipulations subsequently yields the desired amino acid ligand **206**.

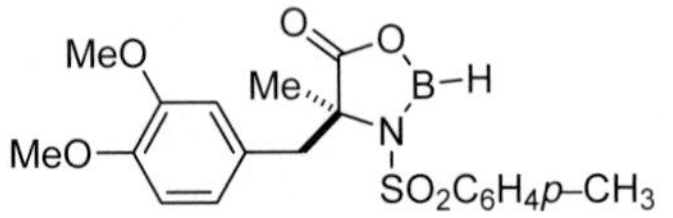

202a 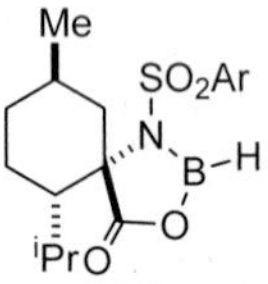   **202b**

In analogy to the Yamamoto and Kiyooka catalysts, Mukaiyama aldol addition reactions catalyzed by **202a** and **202b** are optimal for O-phenyl acetate-derived enol silanes under conditions wherein the aldehyde substrates are added slowly to the reaction mixture in propionitrile at –78 °C. The aldol adducts are isolated for a broad range of aldehydes in excellent yields and up to 92% ee (Eq. 40). The propionate aldol adducts are isolated in good yields, with preference for the *syn* diastereomer in up to 98% ee (Eq. 41) [117].

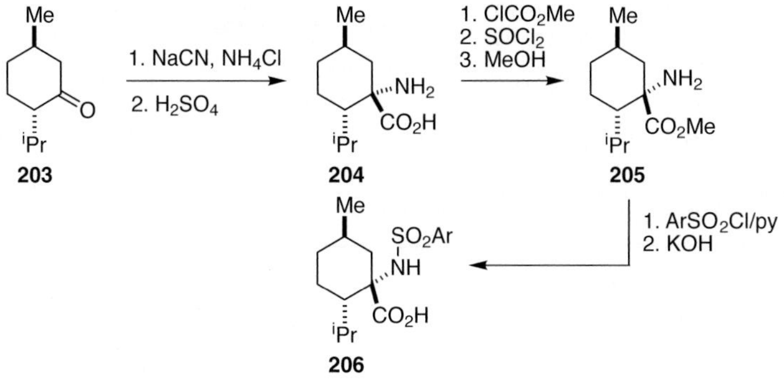

Scheme 17

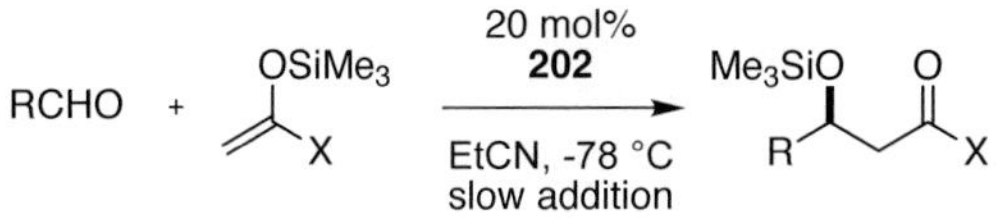

RCHO + (OSiMe$_3$ enol ether, X) $\xrightarrow[\text{EtCN, -78 °C}]{\begin{array}{c}\text{20 mol\%}\\\textbf{202}\\\text{slow addition}\end{array}}$ Me$_3$SiO–R–CH$_2$–C(=O)–X   (40)

Entry	Aldehyde	X	Yield	%ee
1	PhCHO	SEt	86%	87%
2		S^{t}Bu	89%	89%
3		OPh	77%	93%
4	PrCH=CHCHO	S^{t}Bu	91%	82%
5	PhCH$_2$CH$_2$CHO	SEt	82%	89%
6		OPh	78%	85%
7	PrCHO	S^{t}Bu	91%	92%
8	C$_6$H$_{11}$CHO	S^{t}Bu	75%	81%
9		OPh	87%	84%
10	2-furylCHO	S^{t}Bu	98%	85%

$$\text{RCHO} + \underset{\underset{\textbf{207}}{\text{Me}}}{\overset{\text{OSiMe}_3}{\diagdown}}\!\!\!X \xrightarrow[\substack{\text{CH}_3\text{CH}_2\text{CN} \\ -78\,°\text{C} \\ \text{then Bu}_4\text{NF}}]{\substack{20\text{ mol\%} \\ \textbf{202a} \text{ or } \textbf{202b}}} \underset{\text{anti}}{\overset{\text{Me}_3\text{SiO} \quad O}{R\diagup\underset{\text{Me}}{\diagup}\!\!X}} + \underset{\text{syn}}{\overset{\text{Me}_3\text{SiO} \quad O}{R\diagup\underset{\text{Me}}{\diagup}\!\!X}} \tag{41}$$

Entry	Aldehyde	Catalyst	X	Yield	syn/anti	% ee syn	% ee anti
1	PhCHO	202b	SEt	89%	87:13	80%	94%
2		202b	S^{t}Bu	78%	94:6	82%	66%
3		202b	OPh	77%	77:23	87%	>98%
4	PrCH=CHCHO	202b	SEt	80%	80:20	60%	73%
5	PhCH$_2$CH$_2$CHO	202a	SEt	85%	91:9	82%	81%
6		202a	OPh	72%	90:10	75%	>98%
7	PrCHO	202a	SEt	81%	88:12	70%	81%
8	furyl-CHO	202b	SEt	94%	66:33	89%	90%
9	MeO-C$_6$H$_4$-CHO	202b	SEt	78%	89:11	75%	>98%

The Masamune analysis of the acyloxyborane-catalyzed Mukaiyama aldol addition reactions underscores an important structural feature of acyloxy ligands in metal complexes that mediate Si-atom transfer reactions. At short reaction times when the reaction is run with 1 equiv of complex, the β-hydroxy ester is isolated as the major product from the reaction mixture, while with prolonged reaction times the product mixture becomes enriched in the silylated product. The analysis suggests a critical role for the acyloxy ligand wherein it is silylated during the nucleophilic addition of enol silane to the coordinated aldehyde. In a subsequent step, transilylation from the carboxyl group to the metal aldolate occurs, leading to product release and catalyst regeneration. The use of α-alkyl-substituted α-amino acid ligands was postulated to lead to more rapid transilylation reaction as a consequence of the Thorpe-Ingold effect. Moreover, the slow addition of substrates would allow this intermediate to undergo transilylation at a rate that is more rapid than the boryl-catalyzed aldol addition reaction (Scheme 8).

Corey has developed an interesting class of Lewis acidic boron complexes prepared from N-aroylsulfonyl-L-tryptophan 208. The active complex is conveniently prepared from 208 and BuB(OH)$_2$ (Eq. 42). This same complex catalyzes the enantioselective addition of methyl ketone derived trimethylsilylenol ethers and 1-methoxy-3-trimethylsilyloxybutadiene (Eq. 43) [118, 119]. Using 20 mol %

of **209** in propionitrile at –78 °C the acetophenone and 2-hexanone-derived enol silanes add to aldehydes to give aldol adducts in up to 90% ee and useful yields. Moreover, the cyclopentanone-derived enol silane adds to benzaldehyde to give **211** in an impressive 92% ee and 88% diastereoselection (Eq. 44). Upon completion of the reaction, the *N*-tosyltryptophan ligand can be conveniently extracted in alkaline aqueous wash, facilitating isolation of the desired aldol adduct; subsequently, the ligand may be recovered by extraction of the acidified aqueous solution. This same complex has been shown to effectively catalyze the addition of 2-trimethylsilyloxy-4-methoxybutadiene **212** and aldehydes giving the corresponding aldol adducts **213** in good yields and up to 82% ee (Eq. 45). The dienolate adducts can be converted in high yields to the corresponding dihydro-4*H*-pyran-4-ones **214** upon treatment with trifluoroacetic acid in Et$_2$O [116b].

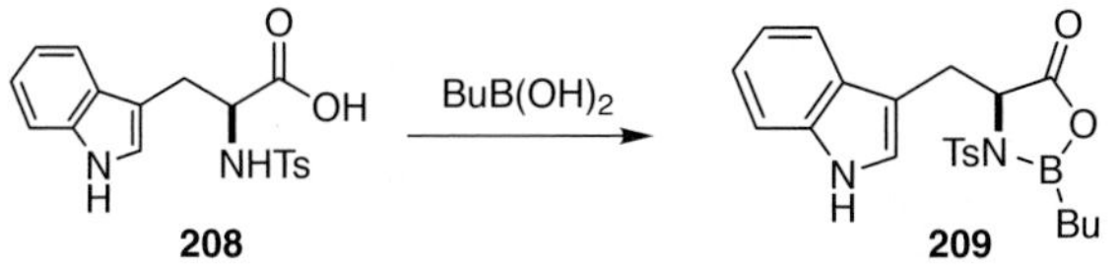

(42)

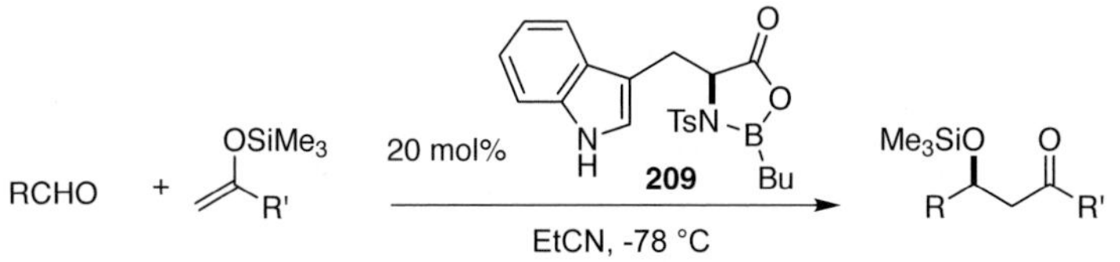

(43)

Entry	R'	Aldehyde	Yield	%ee
1	Ph	PhCHO	82%	89%
2	Ph	c-C$_6$H$_{11}$CHO	67%	93%
3	Ph	n-C$_3$H$_7$CHO	94%	89%
4	Ph	2-furylCHO	100%	92%
5	n-Bu	PhCHO	100%	90%
6	n-Bu	c-C$_6$H$_{11}$CHO	56%	86%

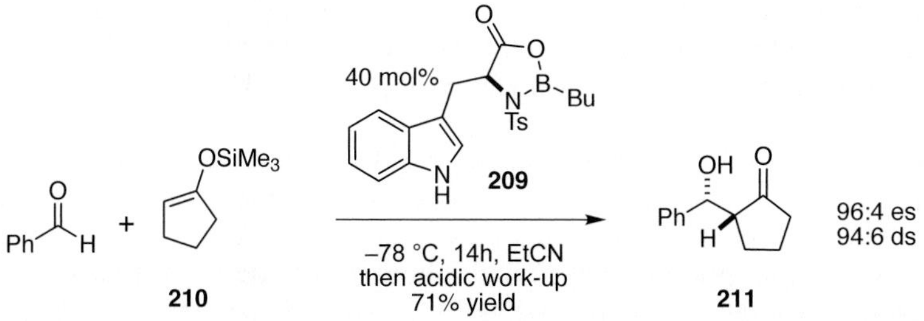

(44)

96:4 es
94:6 ds

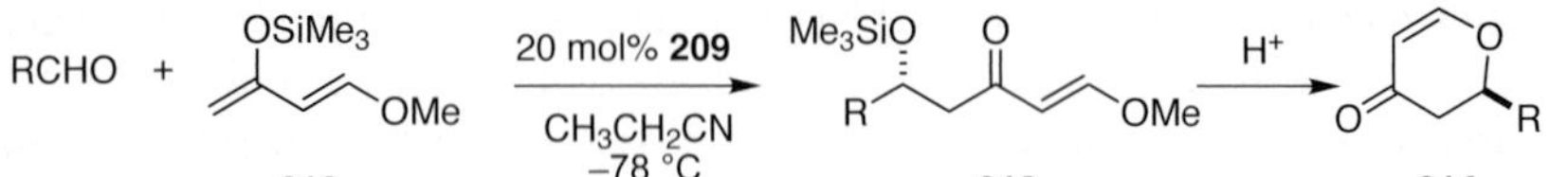

$$RCHO + \text{212} \xrightarrow[\substack{CH_3CH_2CN \\ -78\,°C}]{20\ \text{mol}\%\ \textbf{209}} \text{213} \xrightarrow{H^+} \text{214}$$

(45)

Entry	Aldehyde	Yield (214)	%ee
1	PhCHO	100%	82%
2	PhCH$_2$CH$_2$CHO	57%	69%
3	2-furylCHO	83%	67%
4	c–C$_3$H$_7$CHO	87%	73%
5	c–C$_6$H$_{11}$CHO	80%	76%

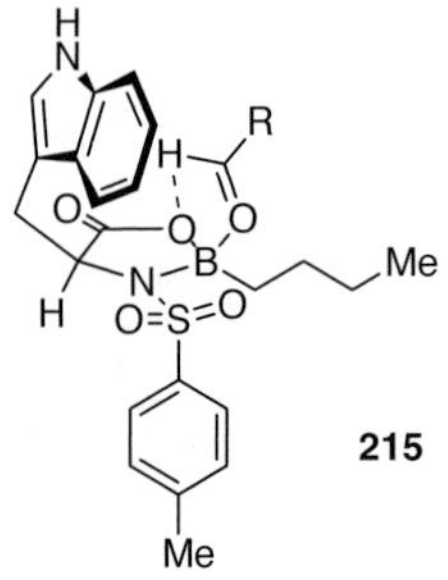

Fig. 19. Corey's model for the aldehyde·Lewis acid complex

In related studies of the tryptophan-derived oxazaborolidene complex as a catalyst for Diels-Alder cycloaddition reactions, Corey has provided insight into the structure of the methacrolein complex **215** (Fig. 19). The results of ^{1}H-NMR NOE experiments indicate the presence of a well-defined aldehyde complex in which the enal is in close contact with the indole ring. The presence of this interaction is supported by the observation of a broad UV absorption band (400–600 nm) that is reversibly formed by successive warming (250 K) and cooling (210 K) of a methylene chloride solution of the aldehyde and complex. Moreover, replacement of the indolyl subunit with a β-naphthyl group leads to a diminution on the product enantioselectivity. When the N-tosyl group is replaced with an N-mesitylsulfonyl, little enantioselectivity is observed, suggesting that the interaction between the indole and aldehyde has been disrupted due to steric demands of the methyl-substituted aromatic group.

4.4
Copper(II)

The use of Cu(II) complexes as Lewis acid catalysts for the Mukaiyama aldol addition reaction has been documented and studied by Evans [120a, 120b, 121a,

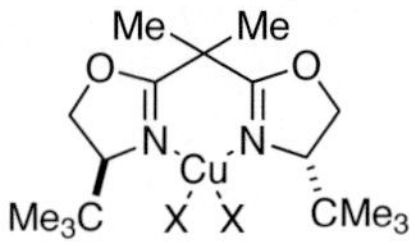

216a X = OTf
216b X = SbF$_6$

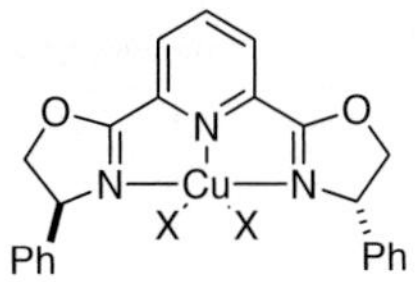

217a X = OTf
217b X = SbF$_6$

121b]. The catalysts **216** and **217** are generated upon treatment of Cu(OTf)$_2$ with bisoxazoline ligands. These have been shown to function effectively in the addition of enol silanes with α-heteroatom-substituted aldehydes and ketones such as benzyloxyacetaldehyde and pyruvates in superb yields and selectivities.

The addition of substituted and unsubstituted enolsilanes at –78 °C utilizing 5 mol % catalyst was shown to be very general for various nucleophiles including silyl dienolates along with enol silanes prepared from butyrolactone as well as acetate and propionate esters (Eqs. 46 and 47). It is noteworthy that the addition of both propionate-derived *Z*- and *E*-silylketene acetals stereoselectively forms the *syn* adduct in 97% and 85% ee, respectively.

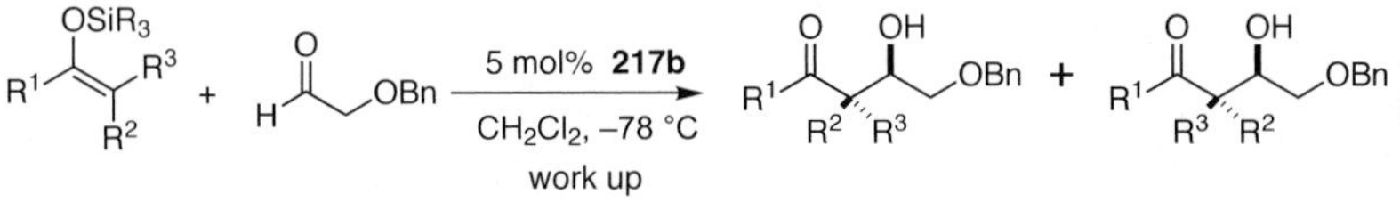

$$(46)$$

Entry	Nucleophile	mol% **217b**	Yield	syn/anti	%ee
1	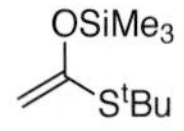 OSiMe$_3$ / S^{t}Bu	0.5	96%	-	99%
2	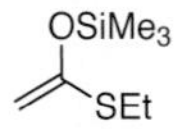 OSiMe$_3$ / SEt	0.5	95%	-	98%
3	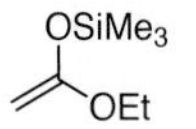 OSiMe$_3$ / OEt	0.5	99%	-	98%
4	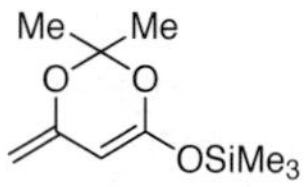 Me,Me O,O / OSiMe$_3$	5	94%	-	92%
5	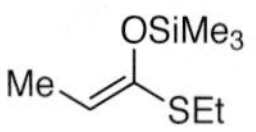 OSiMe$_3$ / Me, SEt	10	90%	97:3	97%
6	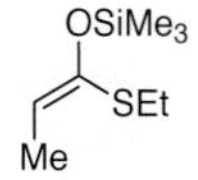 OSiMe$_3$ / SEt, Me	10	48%	86:14	85%
7	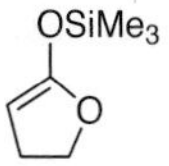 OSiMe$_3$ / O	10	95%	96:4	95%

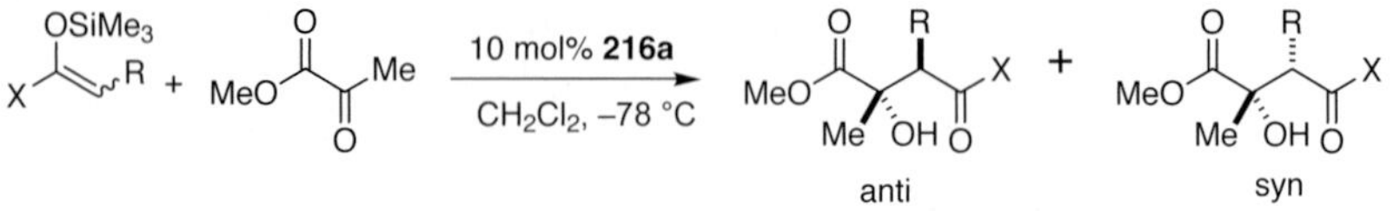

(47)

Entry	Nucleophile	Yield	syn/anti	%ee(syn)
1	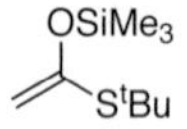 OSiMe₃ / S'Bu	99%	-	99%
2	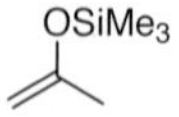 OSiMe₃	76%	-	93%
3	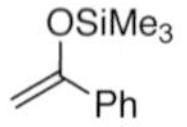 OSiMe₃ / Ph	99%	-	77%
4	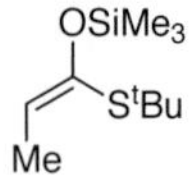 OSiMe₃ / S'Bu, Me	96%	94:6	96%
5	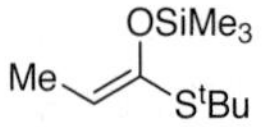 Me OSiMe₃ / S'Bu	98%	95:5	88%

The results of addition reactions with related substrate types provide important insight into the structural and mechanistic aspects of the Cu(II)-catalyzed process. Thus the reaction of α-*tert*-butyldimethylsilyloxyacetaldehyde furnished the corresponding adduct in only 56% ee; moreover, additions to β-(benzyloxy)propionaldehyde yielded only racemic adduct. These two critical observations suggest the important role of the substrate binding to the Lewis-acid center in producing a complex possessing a five-membered ring chelate that leads to aldehyde-face differentiation. This is further underscored in a series of experiments in which aldol addition to enantiomeric α-benzyloxy-propionaldehydes (R)-218 and (S)-218 was investigated (Scheme 18). The addition of *tert*-butyl thioacetate-derived enol silane to (R)-218 furnished adduct in 98.5:1.5 diastereoselectivity; by contrast, addition to the enantiomeric substrate (S)-218

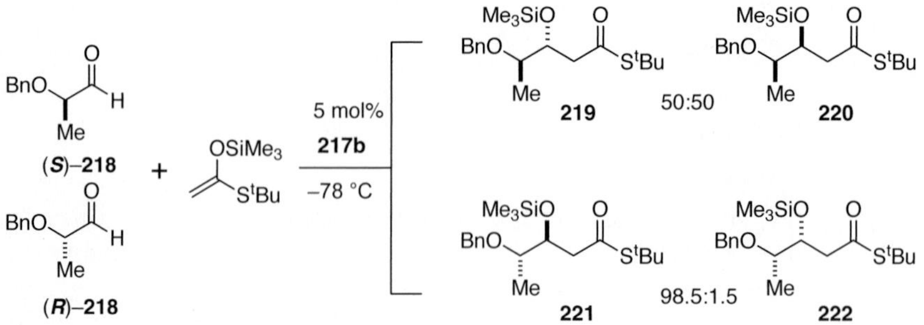

Scheme 18

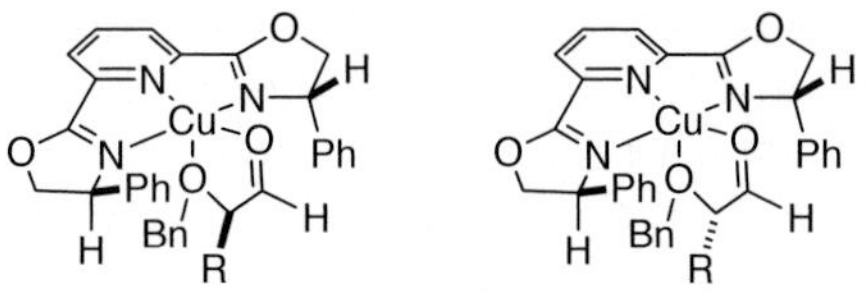

223   224

Fig. 20. Stereoanalysis of the metal-bound aldehyde complexes

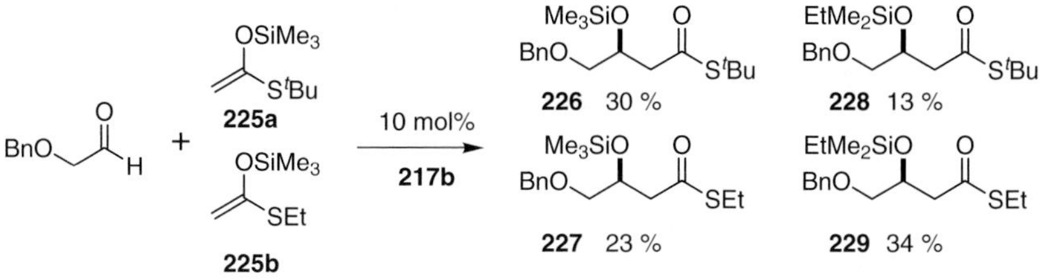

Scheme 19

under otherwise identical conditions afforded adducts as a 50:50 mixture of diastereomers. These experimental observations provide evidence for matched/mismatched substrate/catalyst pairing and further substantiate the presence of a chelated aldehyde in the activated complex. In this regard, analysis of the putative chelates formed from (R)-218 and (S)-218 reveal that the stereocontrolling features of the ligand operate in concert with only one of the two enantiomeric aldehydes (Fig. 20).

Studies employing doubly-labeled, sterically and electronically similar enol have revealed additional interesting mechanistic details on the nature of the silyl-transfer or turnover step [39a, 39b]. In the experiment, a 1:1 mixture of the two silyl ketene acetals 225a and 225b were allowed to react with α-benzyloxyacetaldehyde in the presence of 5 mol % 217b (Scheme 19). The reaction mixture yields a mixture of products in which the trialkylsilyl group has been scrambled, a result consistent with a turnover step in which silyl transfer occurs in an intermolecular fashion. The remarkable success of this system stems from the ability of a labile Cu(II) alkoxide to undergo silylation at a faster rate than the competing deleterious processes involving silyl-catalyzed aldol addition reaction.

4.5
Silver(I)

Yamamoto has pioneered the use of Ag(I) complexes as Lewis acids for aldehyde allylation [122] and aldol addition [123]. For the aldol addition process, ketone-derived tributyltin enolates have been employed as the nucleophilic component (Eq. 48). These enolates are readily prepared from the corresponding enolacetates upon treatment with Bu₃SnOMe. Importantly, although the resulting Bu₃Sn-enolates are known to exist as a mixture of C- and O-bound tautomers 230/231, this mixture can be used directly in the addition reaction. Control experiments had previously shown that tributylstannyl enolates undergo nucleophilic addition to aldehydes at ambient temperature over 14 h. Remarkably, however, Yamamoto has documented that the addition of Ag(I)·bisphosphine complex **232** substantially accelerate the addition reaction yielding adducts in up to 95% ee and 83% yield (Eq. 48).

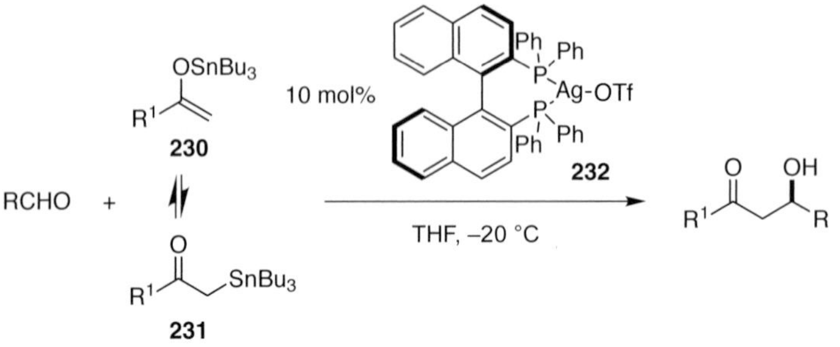

(48)

Entry	Nucleophile	Aldehyde	Yield	%ee
1	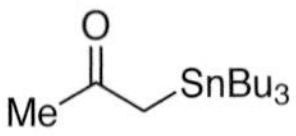 Me—C(O)CH₂SnBu₃ (230/231)	PhCHO	73%	77%
2		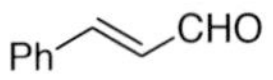 Ph—CH=CH—CHO	83%	53%
3		 Ph—CH₂CH₂—CHO	61%	59%
4	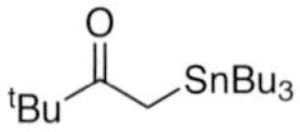 ᵗBu—C(O)CH₂SnBu₃	PhCHO	78%	95%
5		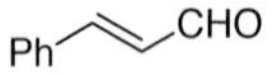 Ph—CH=CH—CHO	69%	86%
6		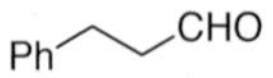 Ph—CH₂CH₂—CHO	75%	94%
7	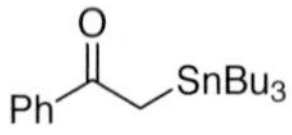 Ph—C(O)CH₂SnBu₃	PhCHO	57%	71%
8		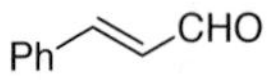 Ph—CH=CH—CHO	33%	41%
9		 Ph—CH₂CH₂—CHO	40%	50%

Yamamoto has also examined the reactions of substituted *E*- and *Z*-enol stannanes derived from cycloalkanones and acyclic *tert*-butylethyl and -propyl ketones (Eq. 49). The addition reactions of cyclic enolates afforded the adducts in excellent yields (92–96%) and 89/11 to 93/7 *anti/syn* diastereomeric ratio with the enantiopurity of the major diastereomer **234** uniformly high (92–95% ee). The use of acyclic *Z*-enol stannanes delivered the complementary *syn* adducts in superb diastereoselectivities (<1/99 *anti/syn*) and enantioselectivity (91–95% ee). The high degree of stereospecificity of the addition process with respect to the enolate component led the investigators to propose a closed transition state structure **236** wherein the chiral Ag(I) complex is exocyclic to the heterocyclic stannacyclohexane.

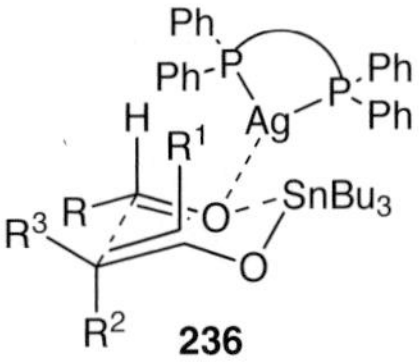

236

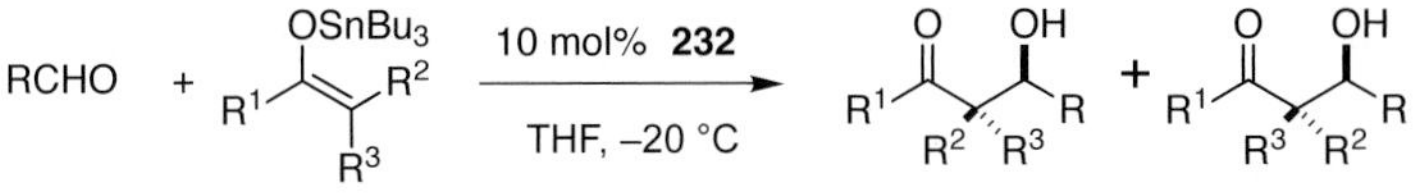

$$(49)$$

Entry	Nucleophile	Aldehyde	Yield	syn/anti	%ee
1	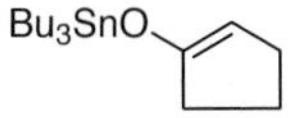 Bu₃SnO (cyclopentenyl)	PhCHO	92%	89/11	92%
2	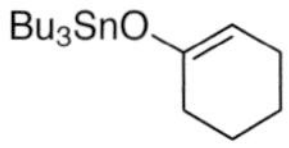 Bu₃SnO (cyclohexenyl)	PhCHO	94%	92/8	93%
3	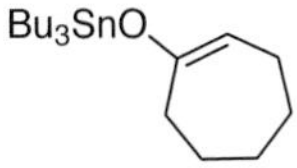 Bu₃SnO (cycloheptenyl)	PhCHO	90%	85/15	96%
4	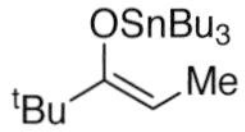 OSnBu₃ / ᵗBu / Me	PhCHO	81%	<1/99	95%
5		 Ph...CHO	77%	<1/99	95%
6	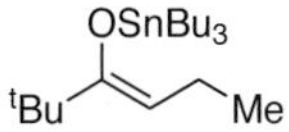 OSnBu₃ ᵗBu Me	PhCHO	98%	<1/99	91%

4.6
Carbocationic Lewis Acids

A series of reports by Mukaiyama and coworkers have highlighted the ability of triarylmethyl cations to function as promoters for the aldol addition reaction of enol silanes and aldehydes [27a, 27b, 27c, 27d, 27e, 27f, 27g, 90]. Subsequent studies by Denmark have provided the mechanistic and conceptual groundwork for the design of catalytic strategies utilizing 1-phenyldibenzosuberyl perchlorate **237** and triflate **238** salts as novel carbon-based Lewis acid catalysts for asymmetric aldol addition reactions [73].

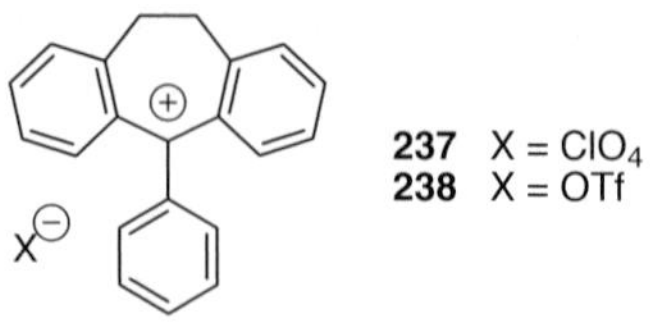

237 X = ClO$_4$
238 X = OTf

Recently, Chen has synthesized and resolved chiral suberylcarbenium ions and successfully demonstrated their use for Mukaiyama aldol addition reactions [74]. The asymmetric synthesis of the optically active triarylmethylcarbocation commences with the C_2-symmetric diol **239**, whose preparation had been reported by Platzke and Snatzke in an approach to a variety of anti-inflammatory agents [124] (Scheme 20). With the chiral Lewis acids in hand, Chen documented their use as asymmetric catalysts (Eq. 50); interestingly, both the yield and enantioselectivity of the process were shown to be sensitive to the counterion (ClO$_4^-$, PF$_6^-$, versus SbCl$_6^-$) of the triarylmethylcation.

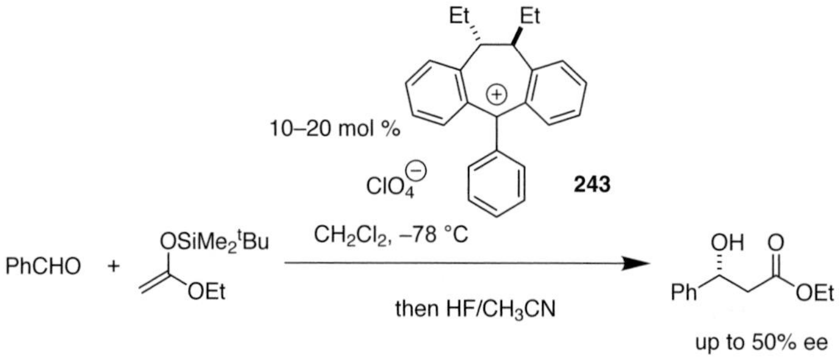

$$(50)$$

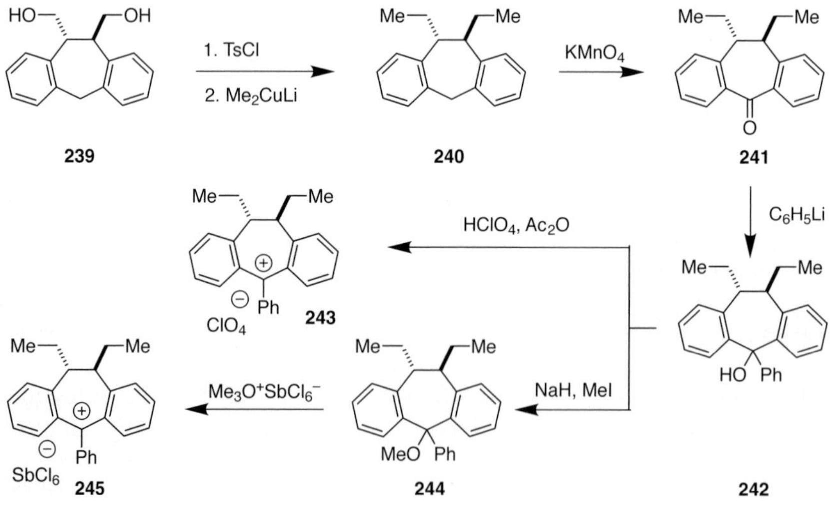

Scheme 20

5
Lewis Base-Mediated Aldol Addition Reactions

In a novel departure from the traditional approach to the asymmetric Mukaiyama aldol, Denmark has reported a Lewis base-catalyzed aldol addition reaction of enol trichlorosilanes and aldehydes. These unusual silyl ketene acetals are readily prepared by treatment of the tributylstannyl enolates **246** with $SiCl_4$ (Eq. 51). In the initial ground-breaking studies, the methyl acetate-derived trichlorosilyl ketene acetal **247** was shown to add rapidly to a broad range of aldehydes at $-80\,°C$ to give adducts (89–99% yield, Eq. 52).

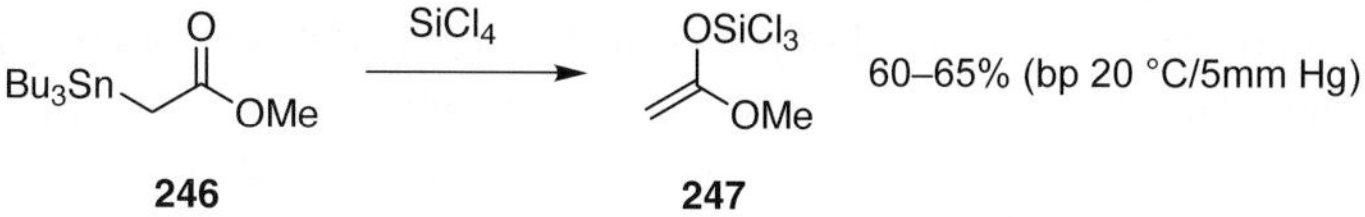

$$(51)$$

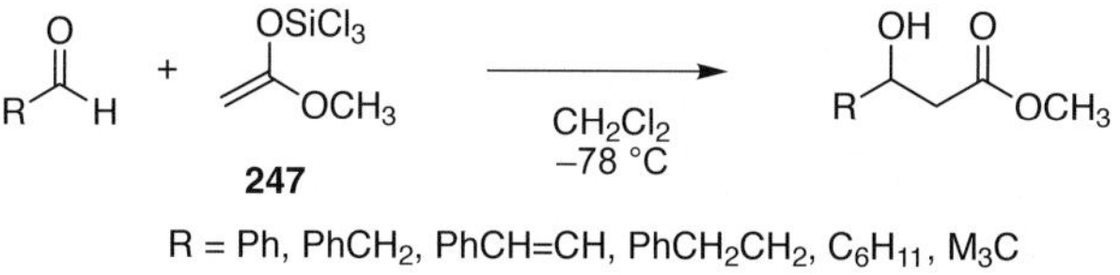

$$(52)$$

R = Ph, PhCH$_2$, PhCH=CH, PhCH$_2$CH$_2$, C$_6$H$_{11}$, M$_3$C

Despite the fact that the uncatalyzed reactions are reported to be very rapid at $-78\,°C$, Denmark has demonstrated that the addition reaction is dramatically accelerated in the presence of a catalytic amounts of Lewis-basic phosphoramides, such as hexamethylphosphoric triamide (HMPA). This remarkable observation coupled with mechanistic investigations has led to the successful development of chiral phosphoramides **248** to **250** as Lewis-base catalysts for enantioselective Mukaiyama aldol addition reactions. Initial investigations documented the superiority of phosphoramide **250** in delivering products of high optical purity; for example, the addition reaction of enol trichlorosilane derived from methyl acetate with trimethylacetaldehyde at $-78\,°C$ affords the aldol adduct in 62% ee and 78% yield (Eq. 53).

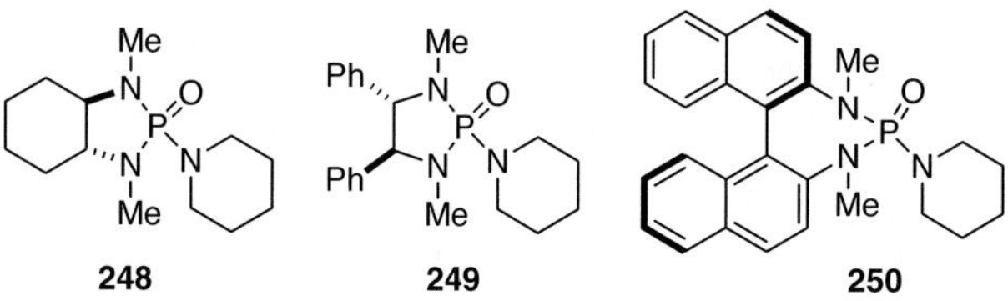

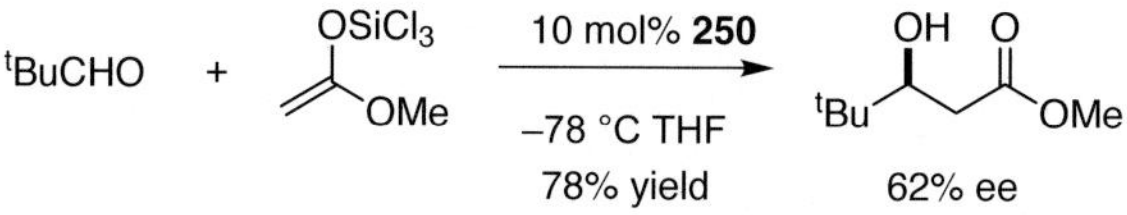

$$(53)$$

This system has been successfully applied to the addition reaction of substituted enol silanes to aldehydes, leading to the formation of adducts displaying
useful levels of simple and absolute stereocontrol (Eqs. 54 and 55). For example,
the addition of cyclohexanone-derived enol silane **251** to a broad range of aldehydes selectively affords the anti-adducts **252** in up to 95% enantiomeric excess.
It is particularly interesting that the catalyzed and uncatalyzed reactions have
divergent simple stereochemical outcomes. Thus, in contrast to reactions catalyzed by **250**, the reaction of **248** and benzaldehyde is predominantly *syn*-selec-
tive. The addition of acyclic Z-enolates **254** exhibited the opposite trend; thus,
while the uncatalyzed addition of propiophenone-derived trichlorosilyl enolate
with aldehydes has a modest preference for the *anti* adduct, the phosphoramide
catalyzed reaction displays *syn*-selectivity with all but one substrate (phenylpro-
pynal). The acyclic 1,2-*syn* products were formed in 84–96% ee and 89–97%
yields. The high selectivity of this process suggests a tight, well-defined transition state structure; Denmark has postulated a cyclic array organized about the
silyl moiety.

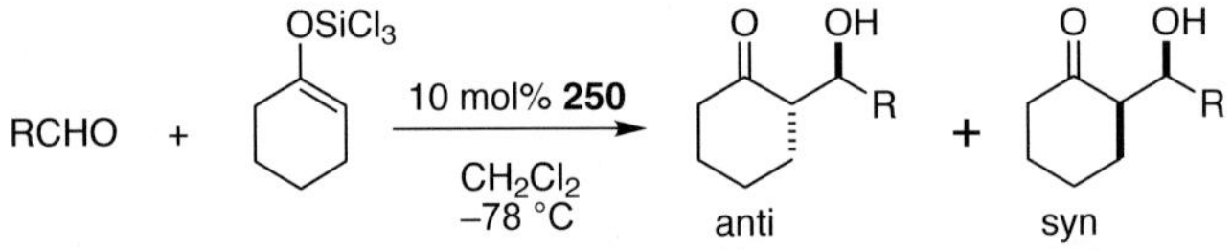

(54)

Aldehyde	Yield	syn/anti	% ee (syn)
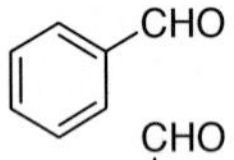 PhCHO	95%	1:1	93%
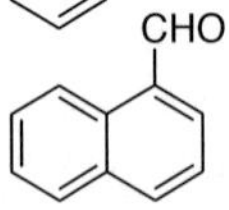 naphthyl-CHO	94%	1:99	97%
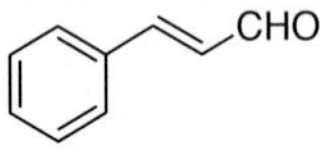 cinnamaldehyde-CHO	94%	1:99	88%
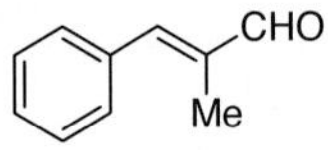 α-methyl cinnamaldehyde	98%	1:99	92%
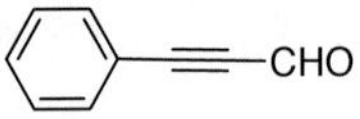 phenylpropynal-CHO	90%	1:5	82%

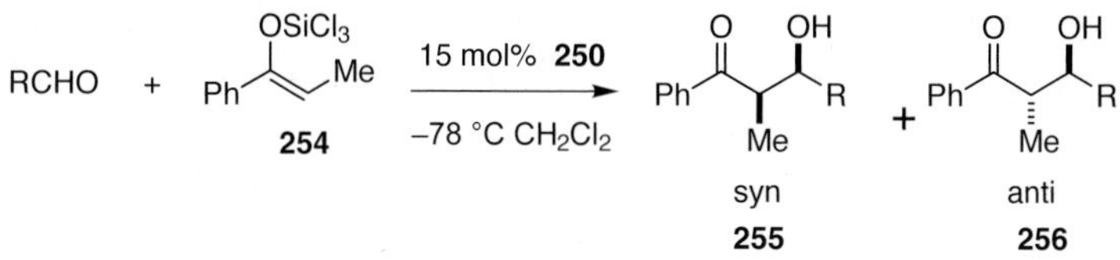

$$
RCHO \ + \ \underset{\mathbf{254}}{\overset{\overset{OSiCl_3}{\underset{Ph}{\bigwedge}}Me}{}} \quad \xrightarrow[-78\ ^{\circ}C\ CH_2Cl_2]{15\ mol\%\ \mathbf{250}} \quad \underset{\substack{Me \\ syn \\ \mathbf{255}}}{\overset{O\quad OH}{\underset{Ph}{\bigwedge}R}} \ + \ \underset{\substack{Me \\ anti \\ \mathbf{256}}}{\overset{O\quad OH}{\underset{Ph}{\bigwedge}R}} \qquad (55)
$$

Aldehyde	Yield	syn/anti	% ee (syn)
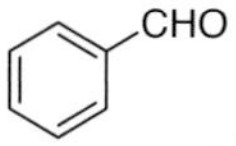 ⟨Ph⟩—CHO	95%	18:1	95%
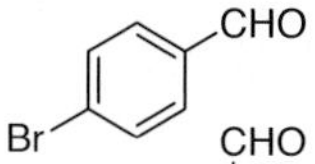 Br—⟨Ph⟩—CHO	89%	12:1	96%
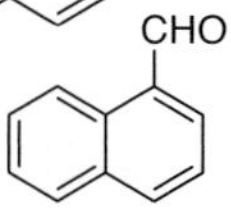 naphthyl-CHO	96%	3:1	84%
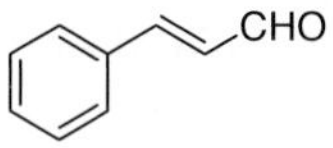 Ph—CH=CH—CHO	97%	9:1	92%
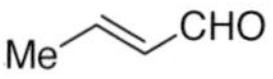 Me—CH=CH—CHO	94%	7:1	91%

6
Mukaiyama Additions via Metalloenolate Intermediates

6.1
Silver(I), Copper(II), Palladium(III)

Shortly after the discovery of the Lewis acid-mediated Mukaiyama aldol addition reaction of enol silanes the general mechanistic aspects of the reaction were intensely investigated [30a, 30b, 30c, 30d]. These processes are considered to proceed by electrophilic activation of the aldehyde towards addition by the nucleophilic enol silane. However, aldol addition processes that proceed by alternative mechanistic pathways have been documented and studied. It is worth considering those systems that have been developed for catalytic, enantioselective aldehyde addition reactions through metalloenolate intermediates.

Two general type of processes that proceed by way of a putative enolate-metal complex have been documented: (1) those in which the metalloenolate nucleophile is generated following deprotonation of C-H acid $\mathbf{257}{\rightarrow}\mathbf{258}$, and (2) those in which the metalloenolate is generated upon desilylative metallation of an enol silane $\mathbf{261}{\rightarrow}\mathbf{262}$ (Scheme 21). Examples of the former processes have been documented and utilize activated C-H acids with pK_a (H$_2$O)<20 such as isonitrile esters $\mathbf{259}$, nitroalkanes $\mathbf{260}$, and ketones. Fewer cases have been reported for the catalytic addition of enol silanes through a putative metalloenolate intermedi-

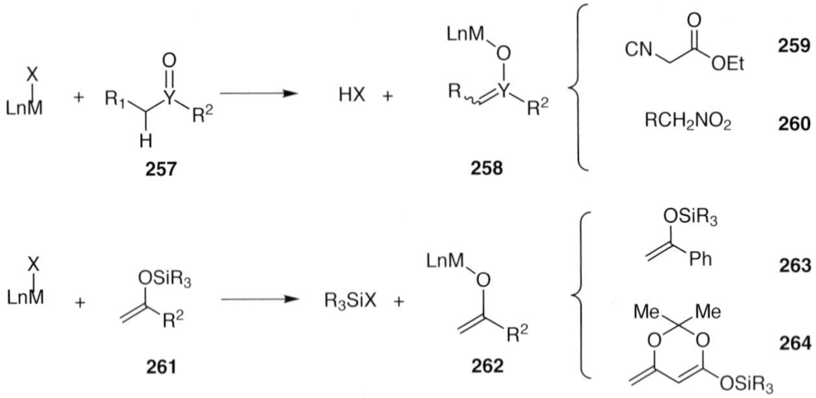

Scheme 21

ate; these include acetophenone and 2,2,3-dimethyldioxinone derived enol silanes **263** and **264**.

The pioneering work of Hayashi and Ito has set high standards for the class of
carbonyl addition reactions involving the first type of aldol addition reactions
(Eq. 56) [25]. These investigators documented the Au(I)-catalyzed addition of α-
isocyanocarboxylic acid esters to aldehydes to afford substituted oxazoline adducts. A family of chiral ferrocenylbisphosphines ligands and the corresponding
Au(I) and Ag(I) complexes were developed for this process and were the key to
the success of this remarkable process. The typical procedure prescribes the use
of 1 mol % of the complex in dichloromethane at 25 °C with reaction times ranging from 20 to 40 h. The product oxazolines are isolated in superb yields as a
mixture of trans/cis diastereomers (70/30 to >99/1) and 74–97% enantiomeric
excess. The synthetic chemistry of these oxazoline adducts has been studied and
developed extensively [125a, 125b].

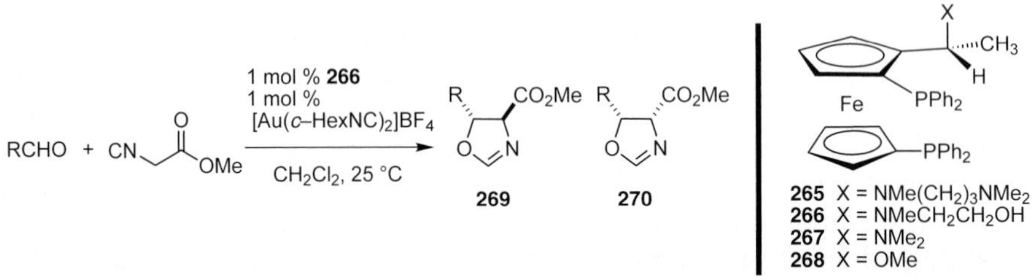

$$(56)$$

The efficiency and selectivity of this process has been explained on the basis
of several critical structural parameters that are synergistically operating in this
system. The coordination of an isonitrile to the soft metal center is proposed to
lead to a facile deprotonation of the enolate precursor by the pendant basic
alkylamino sidechain. The formation of a contact ion-pair between enolate and
ammonium groups should lead to a preferred orientation of the metal-bound

enolate wherein one of the two diastereomeric enolate faces is exposed. Coordination of the aldehyde to the cationic metal center completes the coordination sphere at Au(I) leading to stereoselective C-C bond formation.

Shibasaki and coworkers have reported an asymmetric, catalytic aldol addition reaction of ketone-derived enol silanes and aromatic aldehydes that is proposed to proceed through the intermediacy of a palladium enolate [126]. The active catalyst mixture is prepared upon treating a PdCl$_2$ and (R)-BINAP in the presence of 4 Å molecular sieves and AgOTf, with the optimal formulation also prescribing the addition of 2 equivalents of water. Aromatic aldehydes constitute the ideal substrates for the reaction giving adduct in up to 80% yield and 73% ee (Eq. 57). Shibasaki and co-workers have carried out some spectroscopic investigations that provide insight into the putative catalytic cycle and have suggested a model with a Pd-enolate **272** as the catalytically active intermediate.

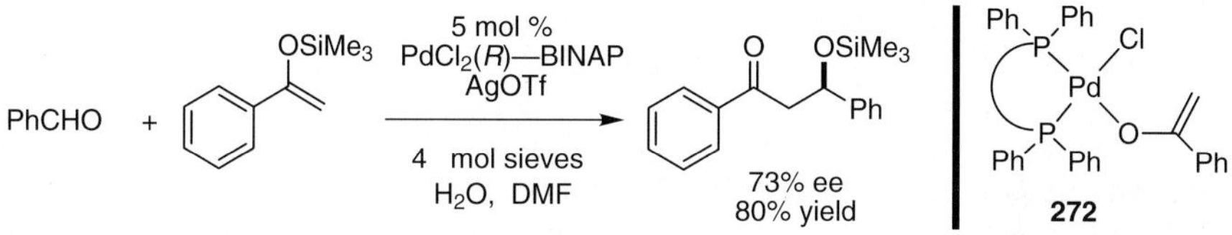

$$(57)$$

Carreira and co-workers have described a Cu–mediated process that effects the catalytic, enantioselective addition of silyl dienolates **87** to aldehydes [24]. The active complex that is believed to initiate the reaction is readily prepared in situ upon mixing optically active bisphosphine, Cu(OTf)$_2$ and (Bu$_4$)NPh$_3$SiF$_2$ in THF (Eq. 58). The addition reactions catalyzed by this system proceed with a broad range of aldehydes to afford adducts **88** in up to 95% ee and 98% yield. Moreover, the reaction may be conducted on a preparative multigram scale utilizing as little as 0.5 mol % of **273** without deleterious effects on the product enantiomeric excess or yields.

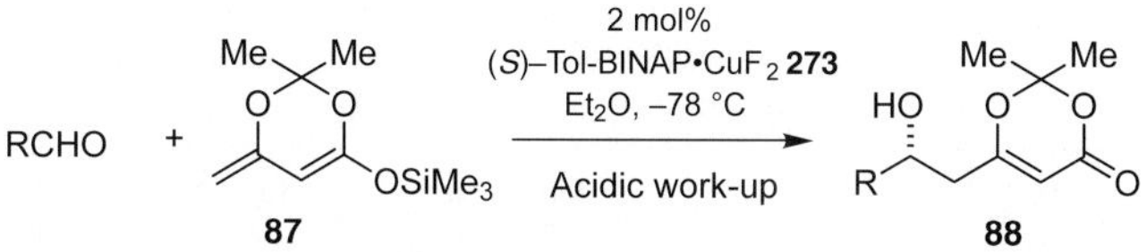

$$(58)$$

Entry	Aldehyde	Yield	% ee
1	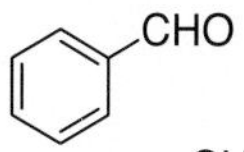	92%	94%
2	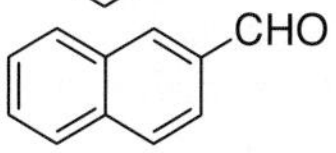	86%	93%
3	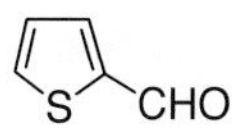	98%	95%

4	furfural CHO	91%	94%
5	CH₃O–C₆H₄–CHO	93%	94%
6	Ph–CH=CH–CHO	83%	85%
7	(2-OMe-phenyl)–CH=CH–CHO	82%	90%
8	Me–CH=CH–CHO	48%	91%
9	Me₂C=CH–CHO	81%	83%
10	Ph–CH=C(Me)–CHO	74%	65%

In analogy to the Au(I) and Pd(II) systems of Hayashi and Shibasaki, respectively, this process is proposed to proceed through a metalloenolate intermediate. The catalytically active metalloenolate species is generated upon desilylative metallation of the enol silane by the cupric fluoride complex. In support of the hypothesis that a soft-metal enolate is an intermediate in the reaction, the investigators have observed that the reaction can be successfully executed under conditions that directly promote transmetallation of the enol silane in the absence of fluoride (Scheme 22). When a solution of enol silane is successively treated with 10 mol % of either MeLi or (Bu₄N)Ph₃SiF₂ at 0 °C, followed by 5 mol % of (S)-BINAP·Cu(OTf)₂ at –78 °C and benzaldehyde, the aldol adduct was isolated

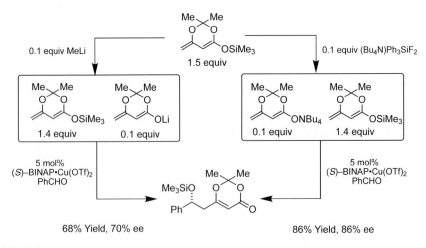

Scheme 22

in good yields and enantioselectivities. Thus, the fluoride counterion is only responsible for initiating the catalytic cycle by generation of the metalloenolate.

6.2
Lanthanides

Shibasaki has developed a family of heterobimetallic complexes **274** to **279** derived from the alkali metal diaryloxy salt of chiral binaphthols and lanthanide alkoxides; these complexes function as catalysts in a variety of useful enantioselective addition reactions [127, 128, 13j].

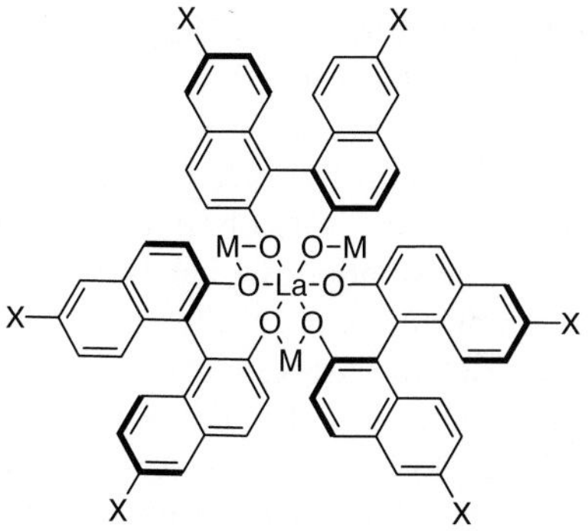

M = Na	M = Li
274 X = H	**277** X = H
275 X = Br	**278** X = Br
276 X = C≡CSiEt₃	**279** X = C≡CSiEt₃

The direct addition of enolizable ketones to aldehydes in the presence of 20 mol % of **277** furnished adducts in up to 94% ee and 81% yield (Eq. 59). Although additional examination and optimization of the system is warranted, it represents an important ground-breaking advance in the field of catalytic aldol addition methodology since it obviates the preparation and use of enol silanes.

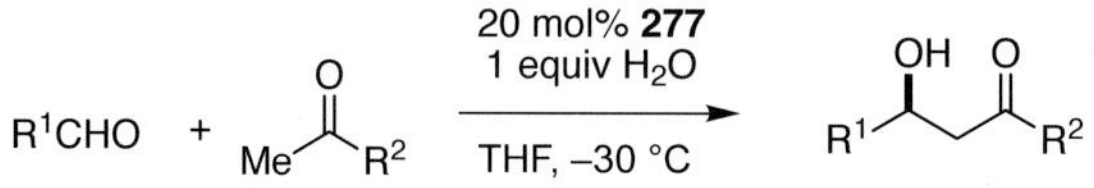

$$R^1CHO \ + \ \underset{Me}{\overset{O}{\|}}{R^2} \quad \xrightarrow[\text{THF, } -30\ °C]{\substack{20\ mol\%\ \textbf{277} \\ 1\ equiv\ H_2O}} \quad \underset{R^1}{\overset{OH}{\quad}}\overset{O}{\|}{R^2} \tag{59}$$

Entry	Aldehyde	R²	Yield	%ee
1	Me₃CCHO	Ph	81%	91%
2		CH₃	53%	73%
3		1-napthyl	55%	76%
4		CH₃CH2	71%	94%
5	c–C₆H₁₁CHO	Ph	72%	44%
6	Me₂HCCHO	Ph	59%	54%
7	Ph(CH₂)₂CHO	Ph	28%	52%

In-depth investigation of these systems has provided mechanistic and structural details of this remarkable catalytic system. X-ray crystal structures of the rare earth/Na/BINOL complexes have been obtained which form the basis of the structural models of the active species that have been proposed. Moreover, laser-desorption/ionization time-of-flight mass spectrometry data corroborates the solid state structural data and substantiates the proposal of a 3:3:1 ligand, alkali metal:lanthanide complex as the active species. Shibasaki and co-workers have proposed that the lanthanide alkoxide functions as a Brønsted base and, following deprotonation of the nitroalkane, furnishes a metalloenolate. The nature of the alkali metal in the lanthanide complex is important to the successful production of nitro aldol adducts in high enantiomeric excesses. Thus, while the Li-derived complex furnishes adducts in excellent enantioselectivities, the sodium-derived complex gives racemic product. Shibasaki has proposed that the nitroenolate is bound to the active catalytst by coordination to the lithium site. The nature of the lanthanide metal has also been shown to be important; thus the optimal catalyst for the aldol addition with optimal selectivites obtained for La(III) (97% ee) while the complexes prepared from the smaller rare earth metals give products with diminished induction. The lanthanide center is thus postulated to function as the Lewis acid site to which the aldehyde coordinates. The resulting active complex constitutes the lithium nitronate and La-bound aldehyde complex that leads to product formation. The Shibasaki catalyst represents a remarkable example of a self-assembled bimetallic system for which it is possible to fine tune structure and function by subtle variations in three variables: alkali metal, lanthanide, and substituted binaphthol. In this regard, Shibasaki and coworkers have elegantly documented other permutations of these variables that lead to impressive catalysts for asymmetric Michael additions, hydrophosphonylation of imines, and enoate epoxidation.

7
Conclusions

The successful development of an asymmetric, catalytic reaction process is a multidimensional problem at the interfaces of inorganic, organometallic, and organic chemistry that demands the consideration and integration of a multitude of parameters such as reaction mechanism as well as catalyst design and synthesis. Additionally, environmental concerns along with the constraints of the marketplace require that newer processes be invented with attention to experimental practicality. The numerous asymmetric catalytic aldol addition processes that have been detailed in this chapter highlight the intensity of interest in the generation of practical enantioselective carbonyl addition reactions. The diversity of catalysts, reaction conditions, and mechanistic possibilities underscore the intellectual richness that this general area provides for discovery and innovation in chemistry. The breathtaking pace of developments will surely continue unabated, guaranteeing the continued evolution of this field.

References

1. For general reviews, see: (a) Heathcock CH (1981) Science 214:395
1b. Mukaiyama T (1982) Organic Reactions. Wiley, New York
1c. Evans DA (1982) Aldrichimica Acta 15:23
1d. Evans DA, Nelson JV, Taber TR (1982) Top in Stereochem 13:1
1e. Heathcock CH (1984) The aldol addition reaction. In: Morrison JD (ed) Asymmetric Synthesis. Academic, New York, vol 3, chap 2
1f. Masamune S, Choy W, Petersen JS, Sita LR (1985) Angew Chem Int Ed Engl 24:1
1g. Heathcock CH (1990) Aldrichimica Acta 23:99
1h. Gennari C (1991) In: Trost BM, Fleming I, Heathcock CH (eds) Comprehensive Organic Synthesis: Additions to C-X π-Bonds, Pergamon Press: New York, chap 2.4, p 629
1i. Yamamoto H, Maruoka K (1993) In: Ojima I (ed) Catalytic Asymmetric Synthesis. VCH: New York, chap 9; p 413
1j. Ito Y, Sawamura M (1993) In: Ojima I (ed) Catalytic Asymmetric Synthesis. VCH: New York, chap 7; p 367
1k. Braun M, Sacha H (1993) J Prakt Chem 335:653
1l. Franklin AS, Paterson I (1994) Contemporary Organic Synthesis 317
1m. Bach T (1994) Angew Chem Int Ed Engl 33:417
1n. Noyori R (1994) Asymmetric Catalysis in Organic Synthesis; Wiley: New York; Santelli M, Pons JM (1995) Lewis Acids and Selectivity in Organic Synthesis; CRC: Boca Raton
2. Corey EJ, Cheng X-M (1989) The Logic of Chemical Synthesis, Wiley, New York
3a. Henderson I, Sharpless KB, Wong CH (1994) J Am Chem Soc 116:558
3b. Wong CH, Halcomb RL, Ichikawa Y, Kajimoto T (1995) Angew Chem Int Ed Engl 34:412
3c. Gijsen HMJ, Qiao L, Fitz W, Wong CH (1996) Chem Rev 96:443
3d. Gijsen HJM, Wong CH (1995) J Am Chem Soc 117:7585
4a. Reymond JL, Chen YW (1995) TetrahedronLett 36:2575
4b. Wagner J, Lerner RA, Barbas CF III (1995) Science 270:1797
4c. Zhing G, Hoffman T, Lerner RA, Danishefsky S, Barbas III CF (1997) J Am Chem Soc 119:8131
4d. Babas III CF, Heine A, Zhing G, Hoffman T, Gramatikova S, Björnestedt R, List B, Anderson J, Stura EA, Wilson EA, Lerner RA (1997) J Organomet Chem 278:2085
5a. Kobayashi S, Onozawa S, Mukaiyama T (1992) Chem Lett 2419
5b. Kobayashi S, Hayashi T, Iwamoto S, Furuta T, Matsumura M (1996) Synlett 672
5c. Rychnovsky SD, Khire UR, Yang G (1997) J Am Chem Soc 119:2058
5d. Li KW, Wu J, Xing WN, Simon JA (1996) J Am Chem Soc 118:7237
5e. Kobayashi S, Matsumura M, Furuta T, Hayashi T, Iwamoto S (1997) Synlett 301
5f. Kim Y, Singer RA, Carreira EM (1998) Angew Chem Int Ed Engl (in press)
6a. Mukaiyama T, Banno K, Narasaka K (1974) J Am Chem Soc 96:7503
6b. Mukaiyama T (1977) Angew Chem Int Ed Engl 16:817
7. Ewing WR, Harris BD, Li WR, Joullie MM (1989) Tetrahedron Lett 30:3757
8a. Iwasawa N, Mukaiyama T (1987) Chem Lett 463
8b. Mukaiyama T, Kobayashi S, Tamura M, Sagawa Y (1987) Chem Lett 491
8c. Kobayashi S, Tamura M, Mukaiyama T (1988) Chem Lett 91
8d. Mukaiyama T, Shimpuku T, Takashima T, Kobayashi S (1989) Chem Lett 145
8e. Kobayashi S, Mukaiyama T (1989) Chem Lett 297
8f. Mukaiyama T, Uchiro H, Kobayashi S (1989) Chem Lett 1757
8g. Mukaiyama T, Uchiro H, Kobayashi S (1990) Chem Lett 1147
8h. Mukaiyama T, Kobayashi S, Uchiro H, Shiina I (1990) Chem Lett 129
8i. Mukaiyama T, Shiina, I, Kobayashi S (1991) Chem Lett 1901
8j. Kobayashi S, Furuya M, Ohtsubo A, Mukaiyama T (1991) Tetrahedron Asymmetry 2:635
8k. Mukaiyama T, Asanuma H, Hachiya I, Harada T, Kobayashi S (1991) Chem Lett 1209
8l. Mukaiyama T, Furuya M, Ohtsubo A, Kobayashi S (1991) Chem Lett 989
8m. Kobayashi S, Hachiya I (1992) J Org Chem 57:1324

8n. Kobayashi S, Kawasuji T (1993) Synlett 911
8o. Kobayashi S, Uchiro H, Shiina I, Mukaiyama T (1993) Tetrahedron 49:1761
8p. Kobayashi S, Kawasuji T, Mori N (1994) Chem Lett 217
8q. Akiyama T, Ishikawa K, Ozaki S (1994) Synlett 275
8r. Kobayashi S, Hayashi T, Kawasuji T (1994) Tetrahedron Lett 35:9573
8s. Kobayashi S, Horibe M, Saito Y (1994) Tetrahedron 50:9629
8t. Mukaiyama T, Shiina I, Uchiro H, Kobayashi S (1994) Bull Chem Soc Jpn 67:1708
8u. Kobayashi S, Kawasuji T (1994) Tetrahedron Lett 35:3329
8v. Kobayashi S, Horibe M (1994) Synlett 147
8w. Mukaiyama T, Anan H, Shiina I, Kobayashi S (1993) Bull Soc Chim Fr 130:388
8x. Kobayashi S, Horibe M, Matsumura M (1995) Synlett 675
8y. Kobayashi S, Horibe M, Hachiya I (1995) Tetrahedron Lett 36:3173
8z. Kobayashi S, Uchiro H, Fujishita Y, Shiina I, Mukaiyama T (1991) J Am Chem Soc 113:4247
9.   Evans DA, Gage JR (1990) Tetrahedron Lett 31,6129
10a. Kita Y, Yasuda H, Tamura O, Itoh F, Ke YY, Tamura Y (1985) Tetrahedron Lett 26:5777
10b. Hayashi M, Inubushi A, Mukaiyama T (1987) Chem Lett 1975
10c. Mukaiyama T, Takashima T, Kusaka H, Shimpuku T (1990) Chem Lett 1777
10d. Chini M, Crotti P, Gardelli C, Minutolo F, Pineschi M (1993) Gaz Chim Ital 123:673
11a. Reetz MT, Fox DNA (1993) Tetrahedron Lett 34:1119
11b. Ipaktschi J, Heydari A (1993) Chem Ber 126:1905
11c. Charleux B, Pichot C (1993) Polymer 34:195
12a. Ohki H, Wada M, Akiba KY (1988) Tetrahedron Lett 29:4719
12b. Wada M, Takeichi E, Matsumoto T (1991) Bull Chem Soc Jpn 64:990
12c. Le Roux C, Maraval M, Borredon ME, Gaspard-Iloughmane H, Dubac J (1992) Tetrahedron Lett 33:1053
12d. Le Roux C, Gaspard-Iloughmane H, Dubac J, Jaud J, Vignaux P (1993) J Org Chem 58:1835
13. Mukaiyama T, Ohno T, Han JS, Kobayashi S (1991) Chem Lett 949
14a. Vougioukas AE, Kagan HB (1987) TetrahedronLett 28:5513
14b. Gong L, Streitwieser A (1990) J Org Chem 55:6235
14c. Kobayashi S (1991) Chem Lett 2187
14d. Mikami K, Terada M, Nakai T (1991) Tetrahedron Asymmetry 2:993
14e. Mikami K, Terada M, Nakai T (1991) J Org Chem 56:5456
14f. Gu JH, Terada M, Mikami K, Nakai T (1992) TetrahedronLett 33:1465
14g. Terada M, Gu JH, Deka DC, Mikami K, Nakai T (1992) Chem Lett 29
14h. Kobayashi S, Hachiya I (1992) Tetrahedron Lett 33:1625
14i. Van de Weghe PV, Collin J (1993) Tetrahedron Lett 34:3881
14j. Makioka Y, Nakagawa I, Taniguchi Y, Takaki K, Fujiwara Y (1993) J Org Chem 58:4771
14k. Kobayashi S, Hachiya I, Ishitani H, Araki M (1993) Synlett 472
14l. Kobayashi S, Hachiya I, Takahori T (1993) Synthesis 371
14m. Fukuzawa S, Tsuchimoto T, Kanai T (1994) Bull Chem Soc Jpn 67:2227
14n. Kobayashi S, (1994) Synlett 689
14o. Kobayashi S, Hachiya I (1994) J Org Chem 59:3590
15. Sodeoka M, Ohrai K, Shibasaki M (1995) J Org Chem 60:2648
16a. Gennari C, Colombo L, Bertolini G, Schimperna G (1987) J Org Chem 52:2754
16b. Hagiwara H, Kimura K, Uda H (1986) J Chem Soc Chem Commun 860
16c. Shirai F, Gu JH, Nakai T (1990) Chem Lett 10:1931
16d. Panyachotipun C, Thornton, ER (1990) Tetrahedron Lett 31:6001
16e. Vasconcellos ML, Desmaele D, Costa PRR, Dangelo J (1992) Tetrahedron Lett 33:4921
16f. Gennari C, Moresca D, Vieth S, Vulpetti A (1993) Angew Chem Int Ed Engl 32:1618
16g. Hagiwara H, Kimura K, Uda H (1992) J Chem Soc Perkin Trans 1 693
17a. Hollis TK, Robinson NP, Bosnich B (1992) Tetrahedron Lett 33:6423
17b. Hong Y, Norris DJ, Collins S (1993) J Org Chem 58:3591
18a. Odenkirk W, Whelan J, Bosnich B (1992) Tetrahedron Lett 33:5729

18b. Hollis TK, Odenkirk W, Robinson NP, Whelan J, Bosnich B (1993) Tetrahedron 49:5415
19a. Sato S, Matsuda I, Izumi Y (1986) Tetrahedron Lett 27:5517
19b. Reetz MT, Vougioukas AE (1987) Tetrahedron Lett 28:793
19c. Sato S, Matsuda I, Izumi Y (1987) Tetrahedron Lett 28:6657
20a. Seebach D, Imwinkelried R, Stucky G (1987) Helv Chim Acta 70:448
20b. Homma K, Takenoshita H, Mukaiyama T (1990) Bull Chem Soc Jpn 63:1898
20c. Bach T, Fox DNA, Reetz MT (1992) J Chem Soc Chem Commun 1634
21a. Reetz MT, Kunisch F, Heitmann P (1986) Tetrahedron Lett 27:4721
21b. Furuta K, Maruyama T, Yamamoto H (1991) Synlett 439
21c. Parmee ER, Hong YP, Tempkin O, Masamune S (1992) Tetrahedron Lett 33:1729
21d. Watanabe M, Kobayashi H, Yoneda Y (1995) Chem Lett 163
22a. Reetz MT, Kyung SH, Bolm C, Zierke T (1986) Chem Ind 824
22b. Kobayashi S, Horibe M (1993) Synlett 855
22c. Naruse Y, Ukai J, Ikeda N, Yamamoto H (1995) Chem Lett 1451
23a. Evans DA, Murry, JA, Kozlowski MC (1996) J Am Chem Soc 118:5814
23b. Evans DA, Kozlowski MC, Burgey CS, MacMillan DWC (1997) J Am Chem Soc 119:7893
24. Krüger J, Carreira EM (1998) J Am Chem Soc 120:837
25. Ito Y, Sawamura M, Hayashi T (1986) J Am Chem Soc 108:6405
26a. Murata S, Suzuki M, Noyori R (1980) J Am Chem Soc 102:3248
26b. Sakurai H, Sasaki K, Hosomi A (1983) Bull Chem Soc Jpn 56:3195
26c. Iwasawa N, Mukaiyama T (1987) Chem Lett 463
26d. Yura T, Iwasawa N, Narasaka K, Mukaiyama T (1988) Chem Lett 1025
26e. Myers AG, Widdowson KL (1990) J Am Chem Soc 112:9672
26f. Denmark SE, Griedel BD, Coe DM, Schnute ME (1994) J Am Chem Soc 116:7026
27a. Mukaiyama T, Kobayashi S, Murakami M (1984) Chem Lett 1759
27b. Mukaiyama T, Kobayashi S, Murakami, M (1985) Chem Lett 447
27c. Kobayashi S, Murakami M, Mukaiyama T (1985) Chem Lett 1535
27d. Mukaiyama T, Kobayashi S, Tamura M, Sagawa Y (1987) Chem Lett 491
27e. Murakami M, Minamikawa H, Mukaiyama T (1987) Chem Lett 1051
27f. Mukaiyama T, Sugumi H, Uchiro H, Kobayashi S (1988) Chem Lett 1291
27g. Kobayashi S, Matsui S, Mukaiyama T (1988) Chem Lett 1491
28. Otera J, Wakahara Y, Kamei H, Sato T, Nozaki H, Fukuzumi S (1991) TetrahedronLett 32:2405
29. Kawai M, Onaka M, Izumi Y (1986) Chem Lett 1581
30a. Noyori R, Yokoyama K, Sakata J, Kuwajima I, Nakamura E, Shimizu M (1977) J Am Chem Soc 99:1265
30b. Nakamura E, Shimizu M, Kuwajima I, Sakata J, Yokoyama K, Noyori R (1983) J Org Chem 48:932
30c. Boyer J, Corriu R, JP, Perz R, Reye C (1984) J Orgmet Chem 184:157
30d. Nakamura E, Yamago S, Machii D, Kuwajima I (1988) TetrahedronLett 29:2207
31a. Denmark SE, Winter SBD, Su XP, Wong KT (1996) J Am Chem Soc 118:7404
31b. Denmark SE, Wong KT, Stavenger RA (1997) J Am Chem Soc 119:2333
32. For example, see Evans, D. A.; Gage, J. R.; Leighton, J. L. (1992) J Am Chem Soc114:9434
33. For an insightful discussion, see: Evans DA, Dart MJ, Duffy JL, Yang MG (1996) J Am Chem Soc 118:4322
34a. Corey EJ, Lee DH (1993) Tetrahedron Lett 1737
34b. Kobayashi S, Horibe M, Hachiya I (1995) Tetrahedron Lett 36:3173
35a. Evans DA, McGee LR (1980) Tetrahedron Lett 21:3975
35b. Yamamoto Y, Maruyama K (1980) TetrahedronLett 4607
36a. Kariharan PC, Pople JA (1972) Chem Phys Lett 66:217
36b. Reetz MT, Raguse B, Marth CF, Hugel HM, Bach T, Fox DNA (1992) Tetrahedron 48:5731
36c. Myers AG, Kephart SE, Chen H (1992) J Am Chem Soc 114:7922
36d. Myers AG, Widdowson KL, Kukkola PJ (1992) J Am Chem Soc 114:2765
36e. Denmark SE, Lee W (1994) J Org Chem 59:707
36f. Gung BW, Zhu ZH, Fouch RA (1995) J Org Chem 60:2860

37.   For a selection of theoretical studies involving enolborinates and lithium enolates, see:
      (a) Hoffmann RW, Ditrich K, Froech S, Cremer D (1985) Tetrahedron 41:5517
37b. Goodman JM, Kahn SD, Paterson I (1990) J Org Chem 55:3295
37c. Li Y, Paddon-Row MN, Houk KN (1990) J Org Chem 55:481
37d. Paterson I, Goodman JM, Lister MA, Schumann RC, McClure CK, Norcross RD (1990)
      Tetrahedron 46:4663
37e. Bernardi A, Cassinari A, Comotti A, Gardner M, Gennari C, Goodman JM, Paterson I
      (1992) Tetrahedron 48:4183
38a. Heathcock CH, Hug KT, Flippin LA (1984) Tetrahedron Lett 25:5973
38b. Heathcock CH, Davidsen SK, Hug KT, Flippin LA (1986) J Org Chem 51:3027
39a. Mikami K, Matsukawa S (1994) J Am Chem Soc 116:4077
39b. Yanagisawa A, Matsumoto Y, Nakashima H, Asakawa K, Yamamoto H (1997) J Am Chem
      Soc 119:9319
40a. Yamamoto Y, Maruyama K (1980) Tetrahedron Lett 4607
40b. Gennari C, Beretta MG, Bernardi A, Mero G, Scolastico C, Todeschini R (1986) Tetrahe-
      dron 42:893
41a. Denmark SE, Henke BR (1991) J Am Chem Soc 113:2177
41b. Denmark SE, Almstead NG (1992) Tetrahedron 48:5565
41c. Denmark SE, Almstead NG (1993) J Am Chem Soc 115:3133
42a. Fleming I (1991) Chemtracts 4:21
42b. Keck GE, Savin KA, Cressman ENK, Abbott DE (1994) J Org Chem 59:7889
42c. Keck GE, Dougherty SM, Savin KA (1995) J Am Chem Soc 117:6210
43.   Hoffman RW (1989) Chem Rev 89:1841
44.   Yamamoto Y (1987) Acc Chem Res 20:243
45.   Wilcox CS, Babston RE (1984) J Org Chem 49:1451
46.   Babston RE, Lynch V, Wilcox CS (1989) Tetrahedron Lett 30:447
47a. Burzlaff H, Voll U, Bestman HJ (1974) Chem Ber 107:1949
47b. Seebach D, Amstutz R, Laube T, Schweizer WB, Dunitz JD (1985) J Am Chem Soc
      107:5403
48.   Seebach D, Golinski J (1981) Helv Chim Acta 64:1413
49a. Bürgi HB, Dunitz JD, Shefter EJ, (1973) J Am Chem Soc 95:5065
49b. Bürgi HB, Dunitz JD, Lehn JM, Wipff G (1974) Tetrahedron 30:1563
49c. Bürgi HB, Lehn JM, Wipff G (1974) J Am Chem Soc 96:1956
50a. Roush WR (1991) J Org Chem 56:4151
50b. Gustin DJ, VanNieuwenhze MS, Roush WR (1995) Tetrahedron Lett 36:3443
50c. Gustin DJ, VanNieuwenhze, MS, Roush WR (1995)Tetrahedron Lett 36:3447
51.   Anh NT, Eisenstein O (1977) Nouv J Chim 1:61
52a. 4322 Evans DA, Dart MJ, Duffy JL, Yang MG, Livingston AB (1995) J Am Chem Soc
      117:6619
52b. Evans DA, Dart MJ, Duffy JL, Yang MG (1996) J Am Chem Soc 118:4322
53a. Chan TH, Aida T, Lau PWK, Gorys V, Harpp DN (1979) Tetrahedron Lett 4029
53b. Helmchen G, Leikauf U, Taufer-Knöpfel I (1985) Angew Chem Int Ed Engl 24:874
53c. Trost BM, Urabe H (1990) J Org Chem 55:3982
54.   Denmark has referred to these as bridging halometallates, see Ref 6b
55.   Trost BM, Urabe H (1990) J Org Chem 55:3982
56a. Aben RW, Scheeren JW (1978) Synthesis 400
56b. Scheeren JW (1986) Recl Trav Chim Pays-Bas 105:71
56c. Sugimura H, Osumi K (1989) Tetrahedron Lett 30:1571
56d. Vasudevan S, Brock CP, Watt DS, Morita H (1994) J Org Chem 59:4677
56e. Bach T (1994) TetrahedronLett 35:5845
57.   Abe M, Ikeda M, Shirodai Y, Nojima M (1996) Tetrahedron Lett 37:5901
58.   Shambayati S, Crowe WE, Schreiber SL (1990) Angew Chem Int Ed Engl 29:256
59a. Keck G, Castellino S (1986) J Am Chem Soc 108:3847
59b. Denmark SE, Henke BR, Weber E (1987) J Am Chem Soc 109:2512
59c. LePage TJ, Wiberg KB (1988) J Am Chem Soc 110:6642

59d. Corcoran RC, Ma J (1991) J Am Chem Soc 113:8973
59e. Delbecq F, Sautet P (1992) J Am Chem Soc 114:2446
60. Power MB, Bott SG, Atwood JL, Barron AR (1990) J Am Chem Soc 112:3446
61a. Corey EJ, Loh TP, Roper TD, Azimioara MD, Noe MC (1992) J Am Chem Soc 114:8290
61b. Hawkins JM, Loren S, Nambu M (1994) J Am Chem Soc 116:1657
62. Hawkins JM, Loren S, Nambu M (1994) J Am Chem Soc 116:1657
63a. Corey EJ, Rohde JJ, Fischer A, Azimioara MD (1997) Tetrahedron Lett 38:33
63b. Corey EJ, Rohde JJ (1997) Tetrahedron Lett 38:37
63c. Corey EJ, Barnes-Seeman D, Lee TW (1997) Tetrahedron Lett 38:1699
63d. Corey EJ, Barnes-Seeman D, Lee TW (1997) Tetrahedron Lett 38:4351
63e. Corey EJ, Barnes-Seeman D, Lee TW, Goodman SN (1997) Tetrahedron Lett. 38:6513
64a. Sogah DY (1986) Polym. Prep. (Am Chem Soc Div Polym Chem) 27:163
64b. Sogah DY, Webster OW (1986) Macromolecules 19:1775
64c. Sumi H, Ishikawa K, Inai Y, Hirabayashi T, Yokota K (1995) Polym J 27:34
65. Chan TH, Brook MA (1985) TetrahedronLett 26:2943
66a. Kuwajima I, Nakamura E (1985) Acc Chem Res 18:181
66b. Harrison CR (1987) Tetrahedron Lett 28:4135
67a. Heathcock CH, Doney JJ, Bergman RG (1985) Pure & Appl Chem 57:1789
67b. Burkhardt ER, Doney JJ, Bergman RG, Heathcock CH (1987) J Am Chem Soc 109:2022
67c. Nurkhardt ER, Doney JJ, Slough GA, Stack JM, Heathcock CH, Bergman RG (1988) Pure & Appl Chem 60:1
67d. Slough GA, Bergman RG, Heathcock CH (1989) J Am Chem Soc 111:938
67e. Burkhardt ER, Bergman RG, Heathcock CH (1990) 9:30
68. Carreira EM, Singer RA (1994) Tetrahedron Lett 35:4323
69a. Hollis TK, Bosnich B (1995) J Am Chem Soc 117:4570
69b. Ellis WW, Bosnich B (1996) Mechanism of the Catalyzed Mukaiyama CrossAldol Reaction In: Helmchen G, Dibo J, Flubacher D, Wiese B (eds) Organic Synthesis via Organometallics OSM 5 proceedings of the fifth symposium Vieweg, p 209
70a. Denmark SE, Almstead NG (1992) Tetrahedron 48:5565
70b. see also, Denmark SE, Weber EJ, Wilson TM, Willson TM (1989) Tetrahedron 45:1053
71. Ellis WW, Bosnich B (1998) J Chem Soc Chem Commun 193
72a. Masamune S, Sato T, Kim BM, Wollmann TA (1986) J Am Chem Soc 108:8279
72b. Parmee ER, Tempkin O, Masamune S, Abiko A (1991) J Am Chem Soc 113:9365
72c. Parmee ER, Hong YP, Tempkin O, Masamune S (1992) Tetrahedron Lett 33:1729
73. Denmark SE, Chen CT (1994) Tetrahedron Lett 35:4327
74. Chen CT, Chao SD, Yen KC, Chen CH, Chou IC, Hon SW (1997) J Am Chem Soc 119:11341
75a. Mukaiyama T, Uchiro H, Kobayashi S (1989) Chem Lett 1001
75b. Kobayashi S, Sano T, Mukaiyama T (1989) Chem Lett 1319
75c. Kobayashi S, Fujishita Y, Mukaiyama T (1989) Chem Lett 2069
76. Mukaiyama T, Kobayashi S (1990) J Organomet Chem 382:39
77a. Mukaiyama T, Iwasawa N, Stevens RW, Haga T (1984) Tetrahedron 40:1381
77b. Mukaiyama T, Kobayashi S, Sano T (1990) Tetrahedron 46: 4653
78a. Kobayashi S, Horibe M (1994) J Am Chem Soc 116:9805
78b. Kobayashi S, Hayashi T (1995) J Org Chem 60:1098
78c. Kobayashi S, Horibe M (1995) Chem Lett 1029
78d. Kobayashi S, Horibe M (1995) Tetrahedron Asymmetry 6:2565
78e. Kobayashi S., Horibe M (1996) Tetrahedron 52:7277
78f. Kobayashi S, Horibe M (1997) Chem Eur 3:1472
79. Kobayashi S, Fujishita Y, Mukaiyama T (1990) Chem Lett 1455
80. Mukaiyama T, Uchiro H, Kobayashi S (1990) Chem Lett 1147
81. Kobayashi S, Ohtsubo A, Mukaiyama T (1991) Chem Lett 831
82. For a study on the stoichiometric use of the chiral diamines to control *syn* and *anti* diastereoselectivity, see: Kobayashi S, Hayashi T (1995) J Org Chem 60:1098
83. Evans DA, MacMillan DWC, Campos KR (1997) J Am Chem Soc 119:10859

84. Nishiyama H, Kondo M, Nakamura T, Itoh K (1991) Organometallics 10:500
85. Denmark SE, Nakajima N, Nicaise OJC, Faucher AM, Edwards JP (1995) J Org Chem 60:4884
86. For additions to ketones utilizing stoichiometric Sn·diamine complexes, see: Kobayashi S, Hachiya I (1992) J Org Chem 57: 1324
87. The Ti(IV) reagent utilized as starting material is suggested to be (^{i}PrO)$_2$Ti=O, prepared in a similar fashion as (EtO)$_2$Ti=O. It is important to note that the identity of (EtO)$_2$Ti=O has not been rigorously established: calculated MW=154, found: 1730, see: Bradley BC, Gaze R, Wardlaw W (1955) J Chem Soc 721. The complex (acac)$_2$Ti=O is known to be dimeric with a Ti(μ-O)$_2$Ti core, see: Smith GD, Caughan CN, Campbell JA (1972) Inorg Chem 11:2989
88. Mukaiyama T, Inubushi A, Suda S, Hara R, Kobayashi S (1990) Chem Lett 1015
89. Substituted aromatic solvents have been shown to have a dramamtic effect on the enantioselectivity of Ti·TADDOL-catalyzed Diels Alder addition reactions, see: Narasaka K, Iwasawa N, Inoue M, Yamada T, Nakashima M, Sugimori J (1989) J Am Chem Soc 111:5340
90. Mikami K, Matsukawa S (1993) J Am Chem Soc 115:7039
91. Cai DW, Hughes DL, Verhoeven TR, Reider PJ (1995) TetrahedronLett 36:7991
92. Mikami K, Yajima T, Takasaki T, Matsukawa S, Terada M, Maruta M (1996) Tetrahedron52:85
93. Mikami K, Matsukawa S, Sawa E, Harada A, Koga N (1997) Tetrahedron Lett 38:1951
94. Sato T, Wakahara Y, Otera J, Nozaki H, Fukuzumi S (1991) J Am Chem Soc 113:4028
95a. Keck GE, Tarbet KH, Geraci LS (1993) J Am Chem Soc 115:8467,
95b. Keck GE, Geraci LS (1993) TetrahedronLett 34:7827
95c. Keck GE, Krishnamurthy D, Grier MC (1993) J Org Chem 58:6543
96. Keck GE, Krishnamurthy D (1995) J Am Chem Soc 117:2363
97. The enantiomeric excess were determined by[1]H-NMR analysis using the chiral shift reagent Eu(hfc)$_3$
98a. Smrcina M, Polakova J, Vyskocil S, Kocovsky P (1993) J Org Chem 58:4534
98b. Smrcina M, Vyskocil S, Polivkova J, Polakova J, Kocovsky P (1996) Collect Czech Chem Commun 61:1520
99. For a discussion of electronic tuning, see: Jacobsen, E. N.; Zhang, W.; Guler, M. L. (1991) J Am Chem Soc 113:6703
100. Carreira EM, Singer RA, Lee WS (1994) J Am Chem Soc 116:8837
101. Li KW, Wu J, Xing WN, Simon JA (1996) J Am Chem Soc 118:7237
102. Rychnovsky SD, Khire UR, Yang G (1997) J Am Chem Soc 119:2058
103. *Syn* reduction: Mori, Y.; Kuhara, M.; Takeuchi, A.; Suzuki, M. (1988) TetrahedronLett 29:5419
104. Anti reduction: Evans DA, Hoveyda AH (1990) J Am Chem Soc 112:6447
105. For the use of boryl complexes in imine additions, see: Ishihara K, Miyata M, Hattori K, Tada T, Yamamoto H (1994) J Am Chem Soc 116:10520
106. Furuta K, Maruyama T, Yamamoto H (1991) J Am Chem Soc 113:1041
107. Furuta K, Miwa Y, Iwanaga K, Yamamoto H (1988) J Am Chem Soc 110:6254
108. Gao QZ, Maruyama T, Mouri M, Yamamoto H (1992) J Org Chem 57:1951
109. Furuta K, Maruyama T, Yamamoto H (1991) Synlett 439
110. Ishihara K, Gao QZ, Yamamoto H (1993) J Am Chem Soc 115:10412
111. Kiyooka S, Kaneko Y, Kume K (1992) TetrahedronLett 33:4927
112. Kaneko Y, Matsuo T, Kiyooka S (1994) TetrahedronLett 35:4107
113. Application to the synthesis of insect attractant: Kiyooka S, Kaneko Y, Harada Y, Matsuo T (1995) Tetrahedron Lett 36:2821
114. For a polymer bound version of the catalyst, see: Kiyooka S, Kido Y, Kaneko Y (1994) Tetrahedron Lett 35:5243
115. Kiyooka S, Kaneko Y, Komura M, Matsuo H, Nakano M (1991) J Org Chem 56:2276
116a. Corey EJ, Loh TP (1991) J Am Chem Soc 113:8966
116b. Corey EJ, Loh TP, Roper TD, Azimioara MD, Noe MC (1992) J Am Chem Soc 114:8290

117. For the addition of difluoroketene silyl acetals, see: Iseki K, Kuroki Y, Asada D, Takahashi M, Kishimoto S, Kobayashi Y (1997) Tetrahedron53:10271
118. Corey EJ, Cywin CL, Roper TD. (1992) TetrahedronLett 33:6907
119. For an enantioselective addition of enol silanes to acetals mediated by a boryl complex incorporating a chiral indolyl group, see: Kinugasa M, Harada T, Oku A (1996) J Org Chem 61:6772
120a. Evans DA, Murry JA, Kozlowski MC (1996) J Am Chem Soc 118:5814
120b. Evans DA, Kozlowski MC, Burgey CS, MacMillan DWC (1997) J Am Chem Soc 119:7893
121. For the use of Cu·bisoxazoline complexes in the realted reaction involving heteroDiels-Alder cycloaddition of Danishefsky's diene, see: (a) Ghosh AK, Mathivanan P, Cappiello J, Krishnan K (1996) Tetrahedron Asymmetry 7:2165
121b. Ghosh AK, Mathivanan P, Cappiello J (1997) Tetrahedron Lett 38:2427
122. Yanagisawa A, Nakashima H, Ishiba A, Yamamoto H (1996) J Am Chem Soc 118:4723
123. Yanagisawa A, Matsumoto Y, Nakashima H, Asakawa K, Yamamoto H (1997) J Am Chem Soc 119:9319
124. Platzek J, Snatzke G (1987) Tetrahedron43:4947
125a. Soloshonok VA, Kacharov AD, Avilov DV, Hayashi T (1996) TetrahedronLett 37:7845
125b. Soloshonok VA, Kacharov AD, Avilov DV, Ishikawa K, Nagashima N, Hayashi T (1997) J Org Chem 62:3470
126. Sodeoka M, Ohrai K, Shibasaki M (1995) J Org Chem 60:2648
127. Shibasaki M, Sasai H, Arai T (1997) Angew Chem Int Ed Engl 36:1236
128. Yamada Y M A, Yoshikawa N, Sasai H, Shibasaki M (1997) Angew Chem Int Ed Engl 36:1871

Chapter 29.2
Addition of Isocyanocarboxylates to Aldehydes

Ryoichi Kuwano · Yoshihiko Ito

Department of Synthetic Chemistry and Biological Chemistry, Graduate School
of Engineering, Kyoto University, Sakyo-ku, Kyoto 606–8501, Japan
e-mail: kuwano@sbchem.kyoto-u.ac.jp; yoshi@sbchem.kyoto-u.ac.jp

Keywords: Isocyanocarboxylate, Aldol reaction, Gold, Silver, Ferrocenylphosphine, β-Hydroxy-α-amino acid

1
Introduction

α-Isocyanocarboxylates are useful as precursors of α-amino acid enolates, which are readily generated with mild bases such as tertiary amines or K_2CO_3. The enolates react with an electrophile to yield a variety of α-amino acid derivatives. In 1985, Ito developed an aldol-type reaction with ethyl isocyanoacetate catalyzed by CuCl together with Et_3N, which yields useful 5-alkyl-2-oxazoline-4-carboxylates as synthetic intermediates of β-hydroxy-α-amino acids [1]. In the catalytic reaction, the copper salts act as Lewis acids to activate the α-hydrogen of the isocyano group, and one of the activated α-hydrogens is abstracted by the Et_3N to generate the ammonium enolate of isocyanoacetate, which reacts with aldehyde to form the aldol adduct. Therefore, the ammonium enolate coordinated to the metal atom is expected to be a key intermediate in the stereocontrol of the aldol reaction. In 1986, Ito and Hayashi reported that chiral ferrocenylphosphine-gold(I) complexes are effective for asymmetric aldol reactions of isocyanoacetate [2]. This chapter presents an overview of gold(I)-catalyzed asymmetric aldol reactions of isocyanocarboxylates and their analogs.

2
Asymmetric Aldol Reaction of Isocyanoacetates with Aldehydes

Gold(I) complexes prepared in situ from bis(cyclohexyl isocyanide)gold(I) tetrafluoroborate (**1**) [3] and chiral ferrocenylphosphine ligands (*R*)-(*S*)-**2a–e** [4, 5] bearing a 2-(dialkylamino)ethyl side chain are effective chiral catalysts for asymmetric aldol reactions of methyl isocyanoacetate (**3**) with aldehydes, which give *trans*-(4*S*,5*R*)-5-alkyl-2-oxazoline-4-carboxylates (*trans*-**4**) with high enantiomeric excess (Scheme 1, Table 1) [2, 6, 7, 8, 9, 10]. The corresponding chiral copper(I) and silver(I) catalysts are less stereoselective. Both the enantio- and *trans*-selectivity are affected by the terminal amino group on the side chain of ligands **2a–e**. Especially, enantioselectivity of the reaction with acetaldehyde is significantly improved by the modification of the terminal amino group, and six-membered ring amino groups such as piperidino (**2c**) and morpholino (**2d**) groups are superior in general (entries 1–5,10,11) [11, 12]. The 2-(dialkylamino)ethyl side chain is essential for the high degree of stereocontrol. Ferrocenylphosphine **2f** bearing a 3-(dimethylamino)propyl side chain is much less enantioselective, and **2g** without a pendant side chain gives almost racemic oxazolines with low *trans/cis* ratio (entries 12,13) [2].

Secondary and tertiary alkyl aldehydes give the corresponding *trans*-oxazolines almost exclusively with high enantioselectivity (entries 6,7). The reaction with α,β-unsaturated aldehydes catalyzed by the chiral gold complex is free of any products resulting from 1,4-additions, and proceeds with high stereoselectivity (entries 8,9). Various functional groups on aromatic aldehydes are acceptable for the highly enantio- and diastereoselective aldol reaction with **3** (entries 14–16). The reactions of highly fluorinated benzaldehydes, such as C_6F_5CHO

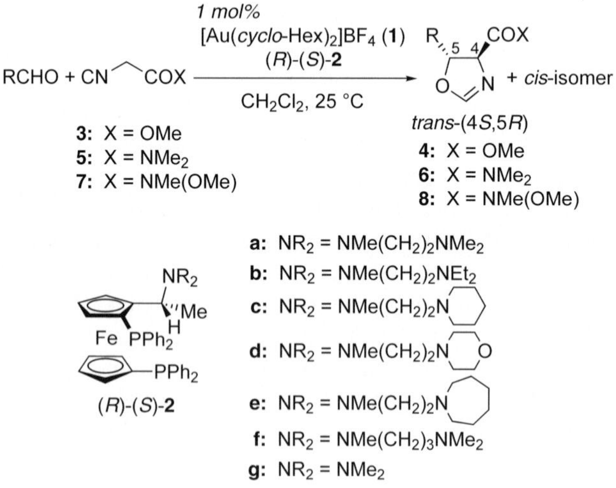

Scheme 1

Table 1. Gold(I)-catalyzed asymmetric aldol reaction of **3**

Entry	R	Ligand	Yield [%]	*trans/cis*	ee of *trans*-**4** [%]
1	Me	**2a**	94	78/22	37
2	Me	**2b**	100	84/16	72
3	Me	**2c**	100	85/15	85
4	Me	**2d**	99	89/11	89
5	Me	**2e**	100	86/14	80
6	*i*-Pr	**2c**	99	99/1	94
7	*t*-Bu	**2d**	94	100/0	97
8	(E)-*n*-PrCH=CH	**2d**	85	87/13	92
9	(E)-MeCH=C(Me)	**2a**	89	91/9	95
10	Ph	**2a**	91	90/10	91
11	Ph	**2d**	93	95/5	95
12	Ph	**2f**	99	89/11	23
13	Ph	**2g**	80	68/32	0
14	2-MeOC$_6$H$_4$	**2d**	98	92/8	92
15	4-ClC$_6$H$_4$	**2d**	97	94/6	94
16	4-NO$_2$C$_6$H$_4$	**2d**	80	83/17	86
17	2-thienyl	**2a**	90	95/5	33
18	2-furyl	**2a**	62	68/32	32
19	2-pyridyl	**2a**	45	75/25	6

and 2,3,5,6-F$_4$-C$_6$HCHO, form preferentially the corresponding *cis*-oxazolines with high enantiomeric excesses (86% and 90% ee, respectively), however, the enantiopurity of *trans*-oxazolines is fairly low (36% and 33% ee, respectively) and the ratios of *trans*- to *cis*-**4** are about 4/6 [13, 14]. Low enantioselectivities for the *trans*-oxazoline were observed in the aldol reactions of 2-heteroaromatic aldehydes, such as 2-thiophene-, 2-furan-, and 2-pyridinecarboxaldehyde (entries 17–19) [15].

The *trans*-oxazolines with high enantiomeric excess can readily be converted to optically active *threo*-β-hydroxy-α-amino acids without epimerization by acid hydrolysis. Moreover, the aldol reaction was applied to the total synthesis of Cyclosporin's unusual amino acid MeBmt [16], and to the asymmetric synthesis of D-*threo*- and D-*erythro*-sphingosine, important membrane components [17]. The [substrate]/[catalyst] ratio can be raised to 10,000/1 without significant loss of the stereoselectivity in the reaction of **3** with 3,4-methylenedioxybenzaldehyde (91% ee, *trans/cis*=91/9), indicating that the gold-catalyzed aldol reaction may provide a practical process to produce optically active *threo*-β-hydroxy-α-amino acids [6].

Use of isocyanoacetamide **5** instead of isocyanoacetate **3** improves the enantioselectivity of the aldol reaction with acetaldehyde and primary alkyl aldehydes (R=Me: 99% ee, *trans/cis*=91/9, R=Et: 96% ee, *trans/cis*=95/5, R=*i*-Bu:

97% ee, *trans/cis*=94/6) [18]. A remarkable improvement in stereoselectivity attained by the use of **5** is observed for the aldol reactions with highly fluorinated benzaldehydes, which give the corresponding *trans*-oxazolines **6** with high enantio- and diastereoselectivity (80–91% ee, *trans/cis*=77/23 to 85/15) [14, 19]. Methyl 2-oxopropanoate also is a good electrophile for the asymmetric aldol reaction of **5**, producing 90% ee of *cis*-(4S,5S)-oxazoline (*cis/trans*=88/12) [20].

The asymmetric aldol reaction of the α-isocyano Weinreb amide **7** also proceeds with high enantio- and diastereoselectivity, yielding *trans*-oxazoline **8** [R=Me: 97% ee (*trans/cis*=95/5), R=*i*-Pr: 97% ee (*trans/cis*=98/2), R=Ph: 97% ee (*trans/cis*=98/2), R=(E)-MeCH=CH: 99% ee (*trans/cis*=97/3)] [21]. The oxazolines **8** can be transformed to optically active *N,O*-protected β-hydroxy-α-aminoaldehydes and ketones in high yield, which are useful chiral building blocks for the synthesis of highly functionalized unusual amino acids, amino polyols, and peptide mimics.

IR studies of the coordination chemistry of gold(I) and silver(I) coordinated with (R)-(S)-**2a** in the presence of **3** revealed a significant difference between these metals in the coordination number of the isocyanide to metal [22]. The tricoordinated gold(I) complex **9** coordinated with one isocyanide was observed without the formation of tetracoordinated species bearing two isocyanides, even in the presence of a large excess of isocyanide at 25 °C, while the silver complex is in equilibrium between the tricoordinated complex **10** and the tetracoordinated complex **11**, in the presence of one equivalent of **3** (Scheme 2). Consequently, it may be expected that the tricoordinated complexes **9** and **10** coordinated with two phosphorous atoms of the ligand and one isocyanide may be key intermediates for high enantioselectivity. Actually, slow addition, over 1 h, of isocyanide **3** to prevent the formation of species **11** enables the asymmetric aldol reaction with the AgClO$_4$/(R)-(S)-**2c** catalyst to give *trans*-oxazoline **4** with high enantiomeric excess (R=Ph: 80% ee, R=*i*-Pr: 90% ee), although the aldol reaction with **3** added in one portion gives only 37% ee (R=Ph) of *trans*-**4**.

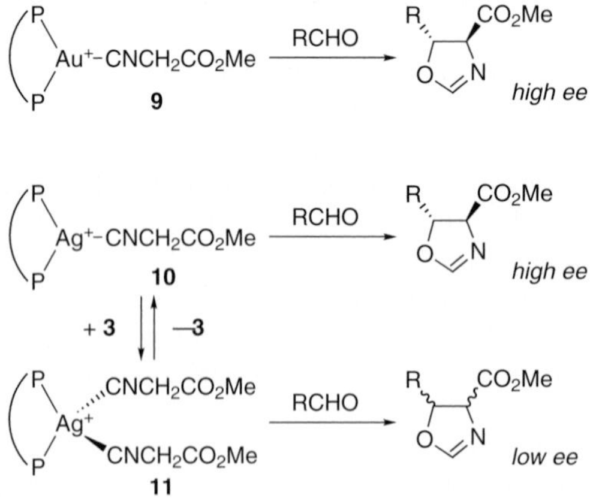

Scheme 2

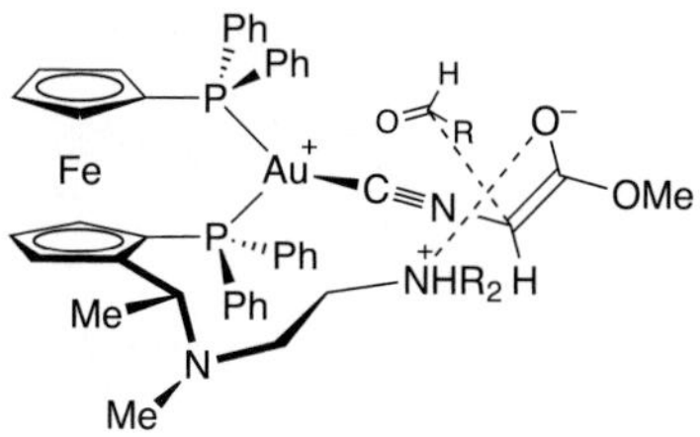

Fig. 1

The high efficiency of the gold catalyst can be explained by a postulated transition state as shown in Fig. 1, where the terminal amino group of (R)-(S)-2 abstracts one of the α-methylene protons of 3 activated by coordination to the gold cation, forming the ammonium enolate of the isocyanoacetate [2]. Ionic interaction between the enolate anion and the ammonium cation seems to control the enantioface of the enolate reacting with aldehyde [23, 24]. The distance between the terminal amino group and the ferrocene moiety of 2a–e will be crucially important for such conformational control. The pendant side chain tethered to 2 shields the re-face of the enolate, therefore, aldehydes approach the si-face preferentially. Such a conformation of the side chain was demonstrated by the structure of the AgOTf/(R)-(S)-2a complex coordinated with two molecules of 3 in a solution, which was determined by ^{1}H{^{1}H} NOE experiments [25].

Togni and Pastor reported that the combination of the carbon central chirality at ferrocenylmethyl position and the ferrocene planar chirality was also an important factor for the high enantioselectivity [26, 27]. Ferrocenylphosphine (R)-(S)-2a, which has R-central chirality along with S-planar chirality, attains high enantioselectivity for the aldol reaction with benzaldehyde to provide trans-(4S,5R)-4, while the diastereomeric (S)-(S)-2a gives much lower ee of trans-4 (41% ee, trans/cis=84/16) with the R-configuration at the 4-position. NMR studies on (R)-(S)- and (S)-(S)-2a suggest that the directions of the aminoethyl side chains differ. The inversion of the central chirality of (R)-(S)-2a may bring about a conformational change from the transition state as shown in Fig. 1, so that aldehydes attack preferentially the re-face of the ammonium enolate of 3.

3
Catalytic Asymmetric Aldol Reaction of α-Isocyanocarboxylates and Their Analogues

Gold(I)/ferrocenylphosphine 2a–d complexes are applicable to asymmetric aldol reactions of α-alkyl substituted α-isocyanoacetates 12. Although the dependency of stereoselectivity on the structures of the substrates is fairly large, some combinations of 12 and aldehydes show high enantio- and diastereoselectivity (Scheme 3). The reaction with paraformaldehyde yields (S)-4-alkyl-2-oxazoline-4-carboxylates in 64 to 81% ee, which can be readily transformed to the

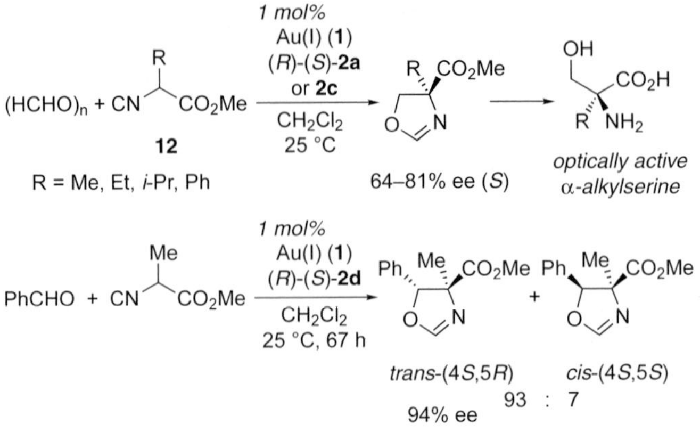

Scheme 3

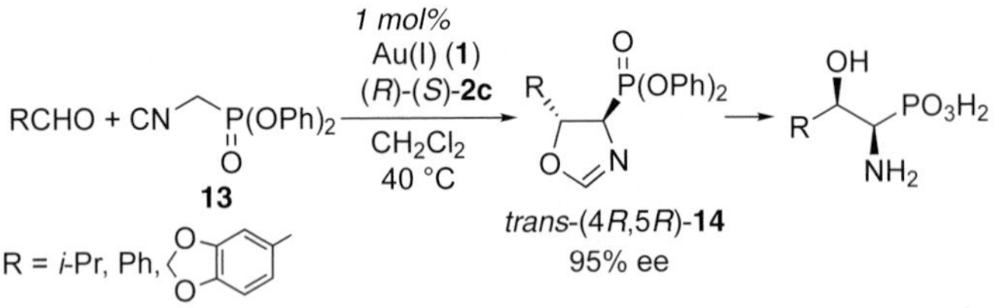

Scheme 4

corresponding optically active α-alkylserines, a class of biologically interesting compounds [23]. The combination of **12** with acetaldehyde or benzaldehyde gives the corresponding *trans*-oxazolines with high enantioselectivity, although the *trans*-selectivity is very low, except for the reaction of α-isocyanopropionate with benzaldehyde [24].

(Isocyanomethyl)phosphonate **13** reacts with aldehydes at 40 °C in the presence of gold(I)/(R)-(S)-**2c** to give 95% ee of *trans*-(4R,5R)-5-alkyl-2-oxazoline-4-phosphonates **14**, with no *cis*-isomer detectable by ^{1}H-NMR (Scheme 4) [28, 29]. The asymmetric reaction provides useful access to optically active (1-aminoalkyl)phosphonic acids, which are a class of biologically interesting phosphorous analogs of α-amino acids.

Interestingly, the aldol-type condensation of tosylmethyl isocyanide **15** with aldehydes is catalyzed by the silver(I)/ferrocenylphosphine **2c** or **2e** catalyst more selectively than it is catalyzed by the chiral gold catalyst (about 20% ee) under the standard reaction conditions (Scheme 5) [30]. Oxazoline **16** can be converted to optically active α-alkyl-β-(N-metylamino)ethanols by reduction with LiAlH$_4$.

Recently, a rhodium(I)-catalyzed, highly enantioselective aldol reaction of the 2-cyanopropionate **17** has been achieved by the use of the *trans*-chelating

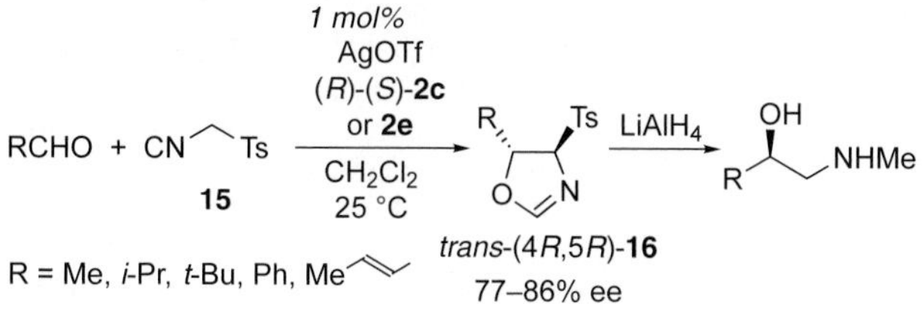

Scheme 5

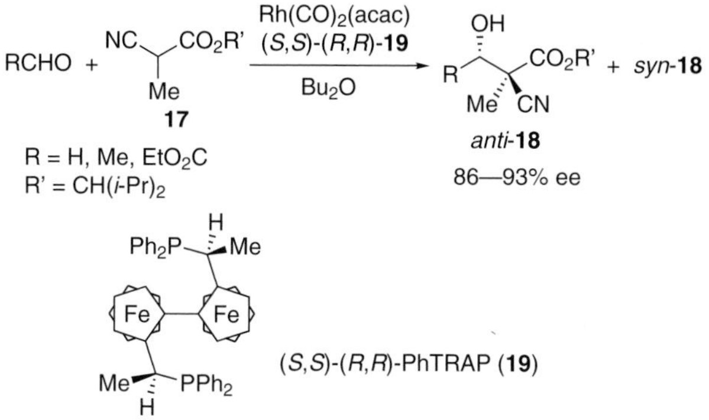

Scheme 6

chiral diphosphine ligand (*S*,*S*)-(*R*,*R*)-PhTRAP (**19**), yielding the corresponding optically active *anti*-aldol **18** with 2*S*-configuration (Scheme 6) [31].

References

1. Ito Y, Matsuura T, Saegusa T (1985)Tetrahedron Lett 26:5781
2. Ito Y, Sawamura M, Hayashi T (1986) J Am Chem Soc 108:6405
3. Bonati F, Minghetti G (1973) Gazz Chim Ital 103:373
4. Hayashi T, Mise T, Fukushima M, Kagotani M, Nagashima N, Hamada Y, Matsumoto A, Kawakami S, Konishi M, Yamamoto K, Kumada M (1980) Bull Chem Soc Jpn 53:1138
5. Hayashi T, Yamazaki A (1991) J Organomet Chem 413:295
6. Sawamura M, Ito Y (1993) Asymmetric Aldol Reactions. In: Ojima I (ed) Catalytic Asymmetric Synthesis. VCH, New York, p 367
7. Hayashi T (1995) Asymmetric Catalysis with Chiral Ferrocenylphosphine Ligands. In:Togni A, Hayashi T (eds) Ferrocenes. VCH, Weinheim, p105
8. Pastor SD, Togni A (1990) Tetrahedron Lett 31:839
9. Togni A, Pastor SD (1990) J Organmomet Chem 381:C21
10. Pastor SD, Togni A (1991) Helv Chim Acta 74:905
11. Ito Y, Sawamura M, Hayashi T (1987) Tetrahedron Lett 28:6215
12. Hayashi T, Sawamura M, Ito Y (1992) Tetrahedron 48:1999
13. Soloshonok VA, Hayashi T (1994) Tetrahedron Lett 35:2713
14. Soloshonok VA, Kacharov AD, Hayashi T (1996) Tetrahedron 52:245
15. Togni A, Pastor SD (1989) Helv Chim Acta 72:1038
16. Togni A, Pastor SD, Rihs G (1989) Helv Chim Acta 72:1471

17. Ito Y, Sawamura M, Hayashi T (1988) Tetrahedron Lett 29:239
18. Ito Y, Sawamura M, Kobayashi M, Hayashi T (1988) Tetrahedron Lett 29:6321
19. Soloshonok VA, Hayashi T (1994) Tetrahedron Asymmetry 5:1091
20. Ito Y, Sawamura M, Hamashima H, Emura T, Hayashi T (1989) Tetrahedron Lett 30:4681
21. Sawamura M, Nakayama Y, Kato T, Ito Y (1995) J Org Chem 60:1727
22. Hayashi T, Uozumi Y, Yamazaki A, Sawamura M, Hamashima H, Ito Y (1991) Tetrahedron Lett 32:2799
23. Ito Y, Sawamura M, Shirakawa E, Hayashizaki K, Hayashi T (1988) Tetrahedron Lett 29:235
24. Ito Y, Sawamura M, Shirakawa E, Hayashizaki K, Hayashi T (1988) Tetrahedron 44:5253
25. Sawamura M, Ito Y, Hayashi T (1990) Tetrahedron Lett 31:2723
26. Pastor SD, Togni A (1989) J Am Chem Soc 111:2333
27. Togni A, Pastor SD (1990) J Org Chem 55:1649
28. Sawamura M, Ito Y, Hayashi T (1989) Tetrahedron Lett 30:2247
29. Togni A, Pastor SD (1989) Tetrahedron Lett 30:1071
30. Sawamura M, Hamashima H, Ito Y (1990) J Org Chem 55:5935
31. Kuwano R, Miyazaki H, Ito Y (1998) Chem Commun 71

Chapter 29.3
Nitroaldol Reaction

Masakatsu Shibasaki · Harald Gröger

Graduate School of Pharmaceutical Sciences, The University of Tokyo, Hongo 7-3-1,
Bunkyo-ku, Tokyo 113, Japan
e-mail: mshibasa@mol.f.u-tokyo.ac.jp

Keywords: Heterobimetallic lanthanoid catalysts, Asymmetric nitroaldol reaction, Henry reaction, Enantioselective and diastereoselective reactions, syn-Selectivity

1
Introduction

Just over 100 years ago, Henry et al. discovered the addition reaction of nitroalkanes to aldehydes with formation of compounds containing a β-nitroalcohol framework [1]. During the history of modern organic chemistry, the importance of these nitroalcohols as versatile intermediates in the synthesis of natural products and many other useful compounds has increased rapidly [2, 3, 4]. This was especially due to the easy transformation of the nitro group into other functional groups, e.g., amine derivatives. Consequently, up to now the nitroaldol reaction (Henry reaction) has developed to one of the most classical C-C bond forming processes [4]. However, in contrast to the high interest in the Henry reaction as a method to produce a wide pool of useful compounds, the lack of a synthetic tool for an enantioselective design prevented the application of the reaction to the challenging field of asymmetric synthesis for a long time.

First encouraging results for a stereoselective synthesis in general were reported by Seebach in 1982, who investigated the *syn/anti*-diastereoselectivity starting from achiral aldehydes and nitroalkanes [4, 5]. Barrett et al. examined the influence of nonchiral Lewis acids on the *syn/anti* diastereoselectivity [6]. Stoichiometric amounts of an enantiomerically pure aldehyde were used in a diastereoselective reaction with 3-nitropropionate by Hanessian et al. [7]. However, an approach to enantioselective synthesis of nitroalcohols via the route of the asymmetric Henry reaction could not be carried out until almost one hundred years after the discovery of the nitroaldol reaction.

In 1992, Shibasaki et al. [8] reported for the first time on the use of recently developed chiral heterobimetallic lanthanoid complexes (LnLB) as chiral catalysts in the catalytic asymmetric Henry reaction (Scheme 1). In the following sections, this efficient concept of an asymmetric nitroaldol reaction, its scope and limitations, and its applications to complex stereoselective synthetic topics are described.

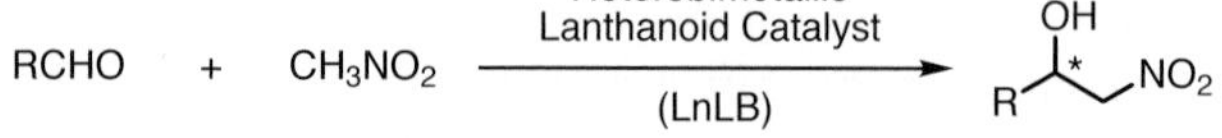

RCHO + CH₃NO₂ → (Heterobimetallic Lanthanoid Catalyst (LnLB)) → R-CH(OH)-CH₂NO₂

Scheme 1. Catalytic asymmetric nitroaldol reaction promoted by heterobimetallic lanthanoid catalysis LnLB

2
Mechanism

The proposed mechanism for the asymmetric nitroaldol reaction catalyzed by heterobimetallic lanthanoid complexes is shown in Scheme 2 [9]. In the initial step, the nitroalkane component is deprotonated and the resulting lithium nitronate coordinates to the lanthanoid complex under formation of the intermediate **I** [10]. Subsequent addition of the aldehyde by coordination of the C=O double bond to the lanthanoid(III) ionic center leads to intermediate **II**, in which the carbonyl function should be attacked by the nitronate via a six-membered transition state (in an asymmetric environment). A proton exchange reaction step will then generate the desired optically active nitroalkanol adduct with regeneration of the "free" rare earth complex LnLB.

The same basic principle of this catalytic cycle with slight modifications concerning the structure of several intermediates can also be proposed when using the improved, second generation catalysts of the LnLB type [11]. The Henry reaction with this type of catalysts together with detailed mechanistic considerations will be described in section 6.

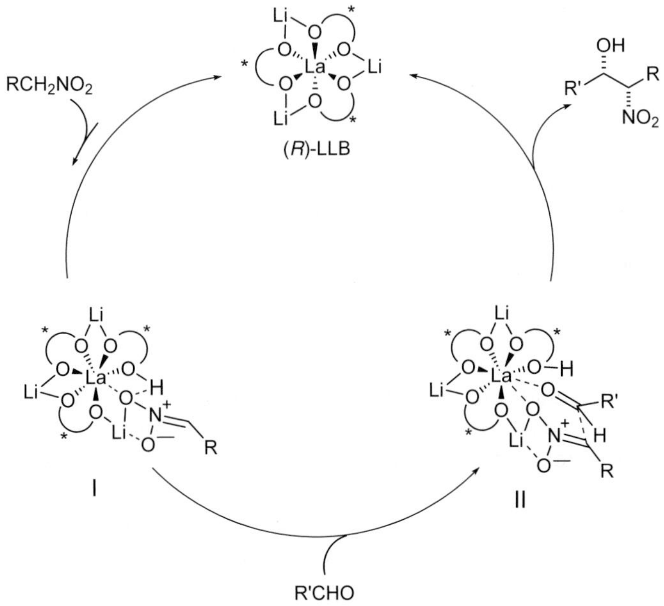

Scheme 2. A possible mechanism for catalytic asymmetric nitroaldol reactions

3
The Catalytic Concept: Catalyst Design and Development of an Efficient Catalysis in Model Reactions

3.1
The First Steps and Applications in Model Reactions

The first promising investigations on the asymmetric nitroaldol reaction showed that this reaction proceeded efficiently in the presence of a catalytic amount of a chiral rare earth metal complex using optically active BINOL as a bidentate asymmetric ligand [8]. In contrast, the use of less acidic bidentate asymmetric diols as ligands led to a lack of asymmetric induction due to an undesired exchange of the asymmetric ligand for (acidic) nitromethane. The chiral BINOL based catalyst was prepared starting from anhydrous $LaCl_3$ and an equimolar amount of the dialkali metal salt of BINOL in the presence of a small amount of water [12].

By using the catalyst prepared as described above, the first example of a catalytic asymmetric nitroaldol reaction was realized. The results are summarized in Scheme 3. Starting from prochiral aldehydes 1 to 3, the desired products 4 to 6 were obtained in good chemical yields and with enantioselectivities up to 90% ee [8]. The amount of the catalyst is not shown in Scheme 3 due to the unknown structure of the catalyst (at this time).

3.2
Structural Requirements for an Efficient Catalysis

Investigations concerning the influence of the rare earth metal component showed pronounced differences both in the reactivity and in the enantioselectivity among the various rare earth metals used [13]. When benzaldehyde and nitromethane were used as starting materials, the corresponding Eu complex gave 7 in 72% ee (91%) in contrast to 37% ee (81%) in the case of the La complex (−40 °C, 40 h). The unique relationship between the ionic radii of rare earth metals and the enantioselectivities of several nitroaldols 4, 6, 7 is depicted in Fig. 1.

Consequently, small changes in the structure of the catalyst (ca. 0.1 Å in ionic radius of the rare earth cation) cause drastic changes in the optical purity of the produced nitroaldols. Although in general nitroaldol reactions are regarded as

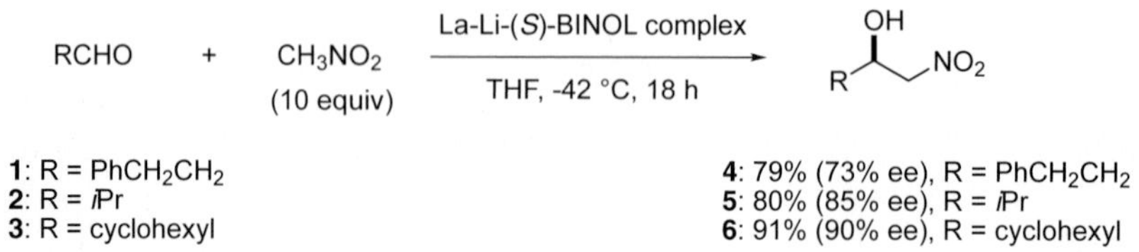

$$RCHO \quad + \quad CH_3NO_2 \quad \xrightarrow[\text{THF, -42 °C, 18 h}]{\text{La-Li-(S)-BINOL complex}} \quad \underset{R}{\overset{OH}{\diagup}}\diagdown NO_2$$

(10 equiv)

1: R = PhCH$_2$CH$_2$ 4: 79% (73% ee), R = PhCH$_2$CH$_2$
2: R = iPr   5: 80% (85% ee), R = iPr
3: R = cyclohexyl   6: 91% (90% ee), R = cyclohexyl

Scheme 3. The first catalytic asymmetric nitroaldol reaction catalyzed by chiral lanthanoid complexes

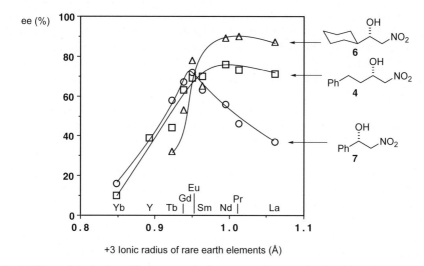

Fig. 1. Effects of the ionic radii of rare earth elements on the enantioselectivity

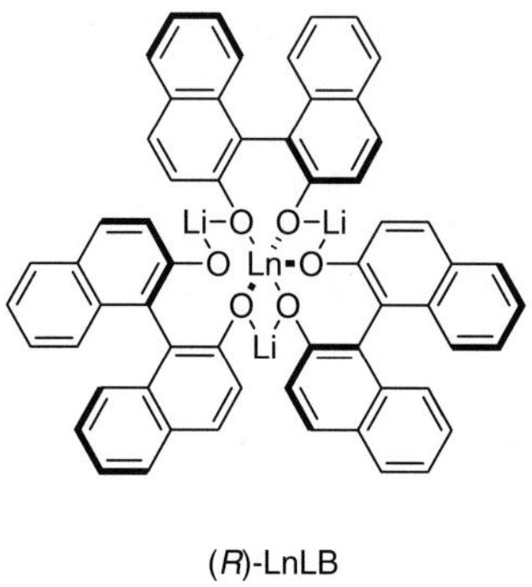

(R)-LnLB

Fig. 2. The structure of (R)-LnLB

equilibrium processes, in the Ln-BINOL complex catalyzed asymmetric nitroaldol reactions no detectable retro-nitroaldol reaction was observed.

A breakthrough in the catalyst design and development of improved catalytic systems for the Henry reaction was achieved by the clarification of the catalyst structure. According to LDI-mass spectrometric investigations, the structure of the heterobimetallic complex LnLB consists of one lanthanoid, three lithiums, and three BINOL moieties (Fig. 2) [14, 15, 16]. The oligomeric structure of the catalyst in the reaction mixture was supported by a slightly positive asymmetric amplification [12, 17, 18, 19]. Thus, the LnLi$_3$tris(binaphthoxide) complex (LnLB) appeared to be an effective asymmetric catalyst for nitroaldol reactions. Having succeeded in determining the structure of the LnLB complex, the conditions for the preparation of LLB [the abbreviation LLB is used in case of lanthanum (La) as the lanthanoid component (Ln)] were further optimized with establishment of two efficient procedures for preparation, starting from $LaCl_3 \cdot 7H_2O$ and La(O-i-Pr)$_3$, respectively [20]. On preparing the catalyst from La(O-i-Pr)$_3$, it is interesting to note that the addition of 1 equiv of water to LLB was found to improve the catalyst's activity [14].

In addition to the early results of the general and effective catalytic asymmetric nitroaldol reaction (Scheme 3), which proceeds efficiently in the presence of 3.3 mol % of LLB, the knowledge of the structure of the LnLB complexes led to an extension of this catalytic method to a wide range of further applications.

However, structural modification of the BINOL ligand system also plays an important role with regard to stereoselection in the asymmetric Henry reaction. Improved enantioselectivites were obtained using a number of (R)-BINOL derivatives 8 (3 mol equiv) in which the 6,6'-positions were substituted [21]. Their utility as asymmetric catalysts was assessed using the nitroaldol reaction of nitromethane with hydrocinnamaldehyde 1. Enantioselectivities up to 88% ee accompanied by chemical yields up to 85% were obtained using 3.3 mol % of various catalysts 9 and 10 equiv of nitromethane (–40 °C, 91 h) (Scheme 4).

In conclusion, surprisingly the substitution at the 6,6'-positions of BINOL proved to be effective in obtaining superior asymmetric catalysts, whereas the use of complexes derived from 3,3'-disubstituted BINOL derivatives [22, 23] gave racemic 4 while the BIPOL derived catalyst [24] gave 4 in only 39% ee. The reason for the positive effect of 6,6'-substituents on BINOL might be that the introduction of 6,6'-bis(trialkylsilyl)ethynyl substituents completely suppresses undesired ligand exchange between nitroalkane and BINOL, whereas this appears to occur in the case of LLB (albeit in only small amounts) [9, 25].

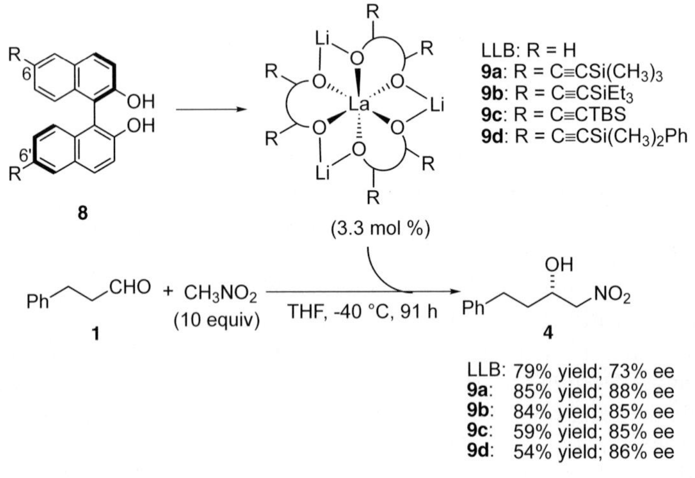

Scheme 4. Catalytic asymmetric nitroaldol reactions promoted by various LLB type complexes

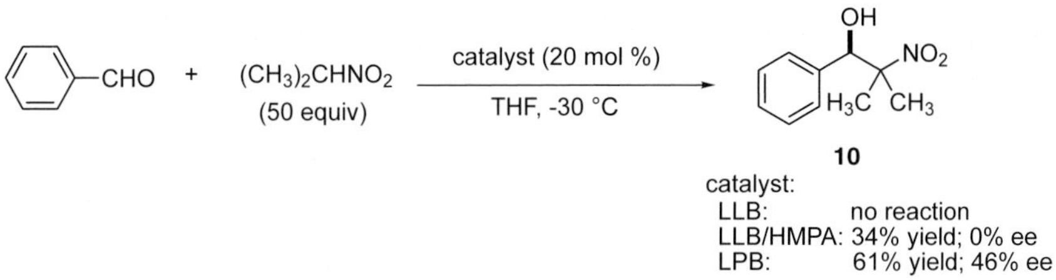

Scheme 5. Catalytic asymmetric nitroaldol reactions using 2-nitropropanol

Recently, Okamoto et al. showed that the reactivity and selectivity also depends on the alkali metal component in the heterobimetallic catalysts [26]. Using the bulkier 2-nitropropane as starting material in a model reaction with benzaldehyde, almost no reaction occurred at −30 °C in the presence of the lithium containing catalyst LLB, whereas higher temperatures as well as the use of HMPA as a co-solvent led to racemic product **10**. However, in the presence of the corresponding potassium containing catalyst LPB, which has been developed and applied previously by Shibasaki et al. [27, 28], the desired reaction to **10** proceeded with 46% ee in 61% yield (Scheme 5).

In contrast, the use of LLB was connected with superior enanioselectivity and chemical yield (compared to LPB) when replacing 2-nitropropane by the less bulkier nitromethane (LLB: 91% yield; 48% ee; LPB: 71% yield; 6% ee) [26].

4
Application of LnLB Catalysis I: Enantioselective Construction of Nitroaldol Adducts with One Stereogenic Center

4.1
Asymmetric Synthesis of β-Blockers

A first example of an efficient application of the LnLB catalyzed nitroaldol reaction as key step in a multi-step syntheses was presented by Shibasaki et al. in the asymmetric approach to three kinds of optically active β-blockers **13, 16,** and **19** (Scheme 6) [13, 29, 30, 31].

Using **17** and 10 mol equiv of nitromethane at −50 °C in the presence of 3.3 mol % of (R)-LLB catalyst, a 76% yield of nitroaldol **18** in 92% ee was obtained. Reductive alkylation of the nitroaldol **18** to **19** was accomplished in 88% yield by a PtO_2 catalyzed hydrogenation in the presence of 5 mol equiv of acetone in methanol. Thus, (S)-(−)-pindolol **19** was synthesized in only four steps from 4-hydroxyindole [30]. Interestingly, the nitroaldols **12, 15,** and **18** were found to have (S)-configuration when (R)-LLB was used. The nitronates thus appear to react preferentially with the *Si* face of the aldehydes, in the opposite sense to the enantiofacial selectivity which might have been expected on the basis of the previous results (cf. Scheme 3). These results suggested that the presence of

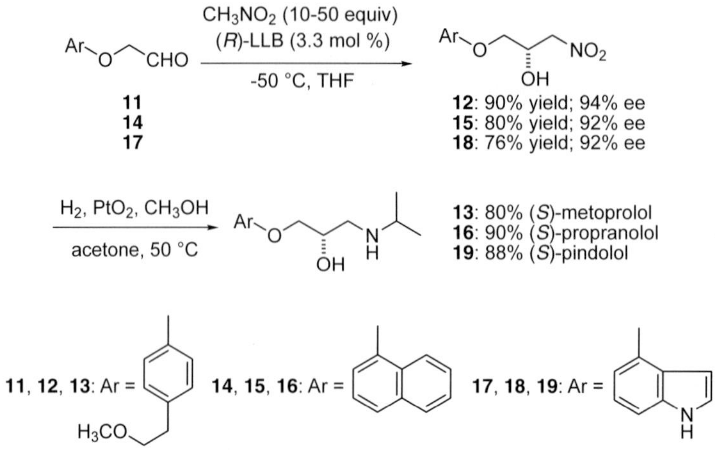

Scheme 6. Catalytic asymmetric synthesis of β-blockers using (R)-LLB as a catalyst

an oxygen atom at the β-position had a pronounced influence on the enantiofacial selectivity.

4.2
Asymmetric Synthesis of Fluorine-Containing Nitroaldol Adducts

The LLB type catalysts were also successfully applied in the asymmetric nitroaldol reaction of the quite unreactive α,α-difluoro aldehydes. In general, catalytic asymmetric syntheses of fluorine-containing compounds are rather difficult [32]. However, catalytic asymmetric nitroaldol reaction of a broad variety of α,α-difluoro aldehydes **20, 22, 24, 26, 28,** and **30** proceeded satisfactorily when using the heterobimetallic asymmetric catalysts with modified, 6,6'-disubstituted BINOL ligands [33] (Scheme 7). The best results were obtained with the samarium(III) complex (5 mol %) generated from 6, 6'-bis{(triethylsilyl)ethynyl}BINOL with enantioselectivities up to 95% ee.

The (S)-configuration of the nitroaldol adduct **21** showed that the nitronate reacted preferentially on the *Si* face of aldehyde in the presence of (R)-LLB (20 mol %; 74% yield; 55% ee). In previous examples (R)-LLB generally caused the attack of the nitronate with *Re* face preference on aldehydes. Therefore, it is noteworthy that the enantiotopic face selection for α,α-difluoro aldehydes is the reverse to that for nonfluorinated aldehydes. The stereoselectivity for α,α-difluoro aldehydes is identical with that of β-oxa-aldehydes, suggesting that the fluorine atoms at the α-position have a great influence on enantioface selection.

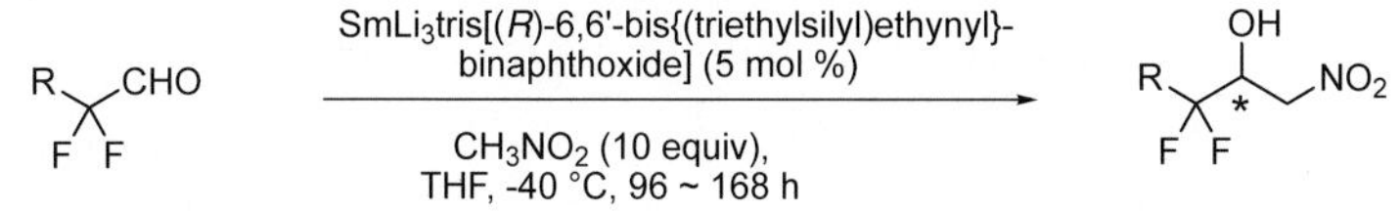

SmLi$_3$tris[(R)-6,6'-bis{(triethylsilyl)ethynyl}-binaphthoxide] (5 mol %)

R-CHO (with F, F at alpha carbon) → R—C(OH)(*)—CH$_2$NO$_2$ (with F, F)

CH$_3$NO$_2$ (10 equiv), THF, -40 °C, 96 ~ 168 h

20: R = PhCH$_2$CH$_2$CH$_2$
22: R = CH$_3$(CH$_2$)$_6$
24: R = PhCH$_2$O(CH$_2$)$_2$
26: R = iPrSCH$_2$
28: R = 4-(CH$_3$OC$_2$H$_4$)C$_6$H$_4$O
30: R = c-C$_6$H$_{11}$

21: R = PhCH$_2$CH$_2$CH$_2$:   55% yield; 92% ee
23: R = CH$_3$(CH$_2$)$_6$:   73% yield; 70% ee
25: R = PhCH$_2$O(CH$_2$)$_2$:   52% yield; 80% ee
27: R = iPrSCH$_2$:   55% yield; 85% ee
29: R = 4-(CH$_3$OC$_2$H$_4$)C$_6$H$_4$O :   52% yield; 77% ee
31: R = c-C$_6$H$_{11}$:   58% yield; 95% ee

Scheme 7. Catalytic asymmetric nitroaldol reactions of α,α-difluoro aldehydes

5
Application of LnLB Catalysis II: Enantioselective Construction of Nitroaldol Adducts with Two or More Stereogenic Centers

5.1
Diastereoselective Catalytic Nitroaldol Reaction Starting from Chiral Aldehydes

The diastereoselective catalytic nitroaldol reaction has been investigated starting from optically active α-amino aldehydes, e.g., **32**. The adducts of type **33** are attractive intermediates for the synthesis of unnatural *erythro*-amino-2-hydroxy acids, which are important components of several biologically active compounds. For example, the promising HIV-protease inhibitor KNI-272 [34, 35] contains (2S,3S)-3-amino-2-hydroxy-4-phenylbutanoic acid (erythro-AHPA, **34**) as a subunit. A conventional diastereoselective synthesis in the presence of achiral bases led to limited internal induction with *erythro/threo* ratios of **33** in the range between 62:38 and 74:26. The use of the achiral complex La(O-i-Pr)$_3$ gave the product **33** in an 89:11 *erythro/threo* ratio [36]. However, this limitation of diastereoselection has been overcome by using catalytic amounts of lithium-containing heterobimetallic complexes LnLB (Scheme 8).

In the presence of (R)-LLB (3.3 mol %), the treatment of N-phthaloyl-L-phenylalanal **32** with nitromethane at –40 °C gave practically a single stereoisomer of (2R,3S)-2-hydroxy-4-phenyl-3-phthaloylamino-1-nitrobutane **33** in 92% yield (>99:1 *erythro*-selectivity) [36]. The enantiofacial selectivity for the C-2 hydroxy group of **33** agreed with results previously observed in enantioselective nitroaldol reactions for non-β-oxa-aldehydes using LLB. Interestingly, reaction of the (S)-aldehyde **32** with nitromethane, using the (S)-LLB complex as a catalyst, led to a reduced diastereo- and enantioselectivity (96% yield; *erythro/threo* 74:26; 90% ee (*erythro*)). The conversion of the nitroaldol adduct **33** into **34** was achieved in a one pot process (80% yield). Investigations to study use of this type of diastereoselective asymmetric nitroaldol reaction on an industrial scale are now in progress [37].

A further example of a diastereoselective nitroaldol reaction using heterobimetallic lanthanoid complexes as catalysts was recently reported by Okamoto et

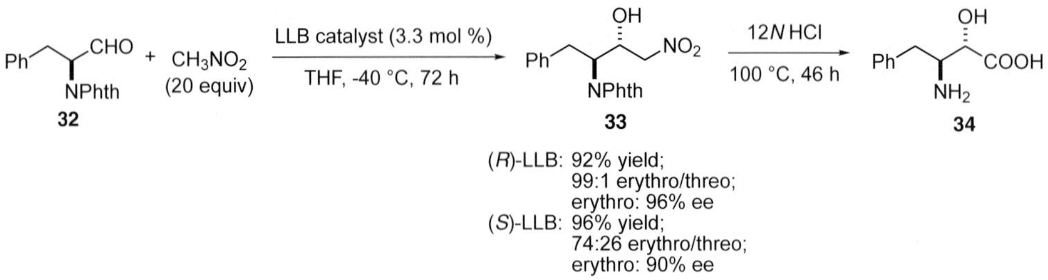

Scheme 8. Diastereoselective nitroaldol reaction as key step in the synthesis of *erythro*-AHPA **34**

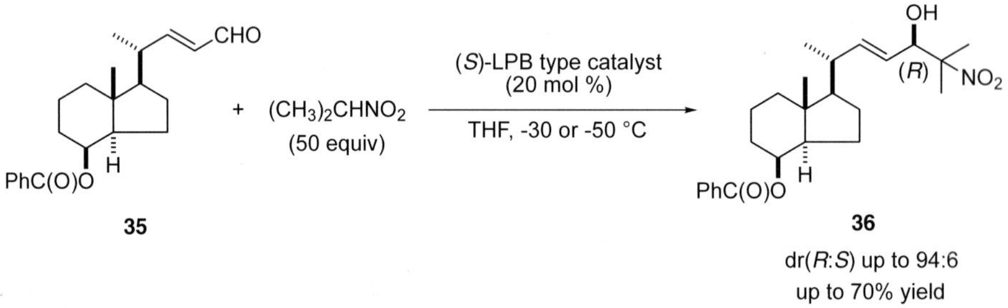

dr(*R:S*) up to 94:6
up to 70% yield

Scheme 9. Catalytic diastereoselective nitroaldol reaction promoted by the LPB type catalyst

al. [26] in connection with a novel approach to 1α,24(*R*)-dihydroxyvitamin D$_3$ [38], which is an active analogue of vitamin D$_3$ and induces keratinocyte differentiation [39, 40]. Here, several rare earth metal complexes were used to catalyze the nitroaldol reaction of the C/D-ring 24-aldehyde precursor **35** with 2-nitropropane (Scheme 9). In accordance with the results of the corresponding model reaction with benzaldehyde, when using 2-nitropropane as starting material (see Chapter 3.2) the best results were achieved in the presence of the potassium containing lanthanoid complex of type LPB with (*S*)-6,6'-{(triethylsilyl)ethynyl}BINOL as ligand. The desired nitroaldol adduct **36** was formed in yields up to 71% and with diastereomeric ratios (dr) up to 94:6 (Scheme 9). It is noteworthy that a conjugate double bond in the aldehyde component was needed for good asymmetric induction.

The obtained nitroaldol adduct **36** was easily converted into a synthetic intermediate of 1α,24(*R*)-dihydroxyvitamin D$_3$ by a denitration reaction using 2,2'-azobisisobutyronitrile (AIBN) and Bu$_3$SnH [26].

5.2
Diastereoselective and Enantioselective Synthesis: Asymmetric Construction of Two New Stereogenic Centers Starting from Prochiral Compounds

LnLB type catalysts are also able to promote diastereoselective and enantioselective nitroaldol reactions starting from prochiral materials. In preliminary work, LLB gave unsatisfactory results in terms of both diastereoselectivity (*syn/anti* ratio 63:37 to 77:23) and enantioselectivity (<78% ee) [29]. However, an effective asymmetric induction was obtained in the presence of LLB type catalysts **9** containing 6,6'-substituted BINOL.

The application of the catalysts of type **9** (3.3 mol %) to diastereoselective nitroaldol reactions led to high *syn*-selectivity and enantioselectivity [41]. In all cases, much higher *syn*-selectivity (*syn/anti* ratio up to 94:6) and enantioselectivity (up to 97% ee) were obtained using the catalysts with 6,6'-substituted BINOL instead of LLB (representative results are given in Scheme 10). The optical purities of the minor *anti*-adducts **41**, **43**, and **45** were lower than those of the *syn*-adducts **40**, **42**, and **44**, indicating that the former were not generated by epimerization of the nitro group. In fact, treatment of the *syn*-adducts with catalysts such as LLB and its derivatives resulted in near-quantitative recovery of the starting materials with unchanged optical purities.

The *syn*-selective asymmetric nitroaldol reaction was successfully applied to the catalytic asymmetric synthesis of *threo*-dihydrosphingosine **46**, which elicits a variety of cellular responses by inhibiting protein kinase C (Scheme 11) [42]. Nitroaldol reaction of hexadecanal **47** with 3 equiv of nitroethanol catalyzed by **9b** gave the corresponding nitroaldol adduct **48** in high *syn*-selectivity (91:9) and 78% yield, with the *syn*-adduct **48** being obtained with up to 97% ee [41]. In this case, under similar conditions the LLB-catalyzed reaction proceeded only

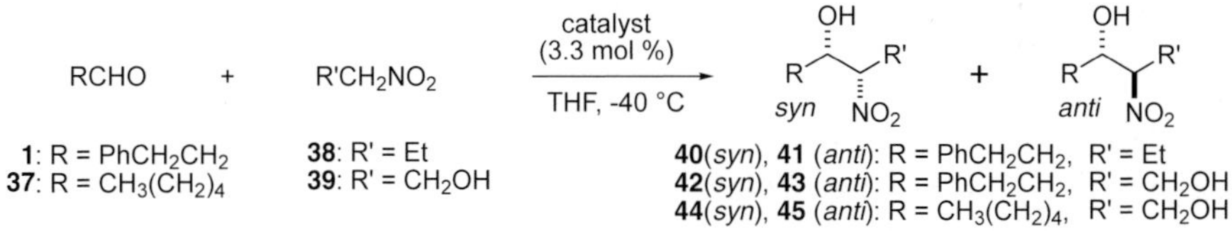

RCHO + R'CH$_2$NO$_2$ → (catalyst (3.3 mol %), THF, -40 °C) → syn + anti

1: R = PhCH$_2$CH$_2$
37: R = CH$_3$(CH$_2$)$_4$

38: R' = Et
39: R' = CH$_2$OH

40(*syn*), 41 (*anti*): R = PhCH$_2$CH$_2$, R' = Et
42(*syn*), 43 (*anti*): R = PhCH$_2$CH$_2$, R' = CH$_2$OH
44(*syn*), 45 (*anti*): R = CH$_3$(CH$_2$)$_4$, R' = CH$_2$OH

Reagent	Catalyst	Time [h]	Product	Yield [%]	syn/anti	ee (syn) [%]
1+38	LLB	138	40+41	89	85:15	87
1+38	9b	138	40+41	89	93:7	95
1+39	LLB	111	42+43	62	84:16	66
1+39	9b	111	42+43	97	92:8	97
37+39	LLB	93	44+45	79	87:13	78
37+39	9b	93	44+45	96	92:8	95

Scheme 10. Diastereoselective and enantioselective nitroaldol reaction

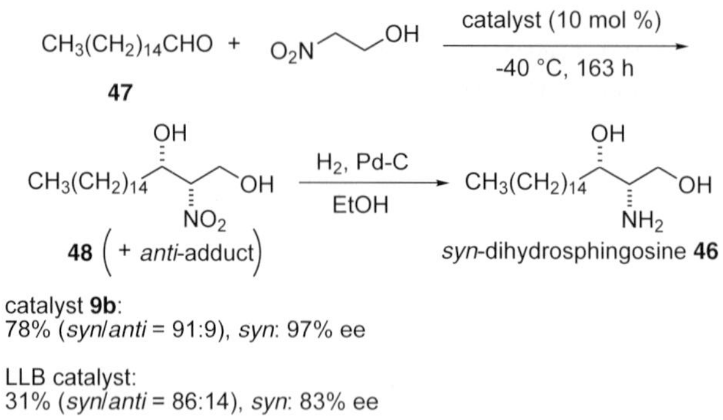

$CH_3(CH_2)_{14}CHO$ + O_2N⌒⌒OH $\xrightarrow[\text{-40 °C, 163 h}]{\text{catalyst (10 mol %)}}$

47

$CH_3(CH_2)_{14}$—$\overset{OH}{\underset{NO_2}{\diagdown}}$—$OH$ $\xrightarrow[\text{EtOH}]{H_2,\ Pd\text{-}C}$ $CH_3(CH_2)_{14}$—$\overset{OH}{\underset{NH_2}{\diagdown}}$—$OH$

48 (+ *anti*-adduct)   *syn*-dihydrosphingosine **46**

catalyst **9b**:
78% (*syn/anti* = 91:9), *syn*: 97% ee

LLB catalyst:
31% (*syn/anti* = 86:14), *syn*: 83% ee

Scheme 11. Catalytic asymmetric synthesis of *syn*-dihydrosphingosine

slowly to give a 86:14 ratio of the *syn* and *anti*-adducts in 31% yield (with lower optical purity: 83% ee). The hydrogenation of **48** in the presence of 10% Pd on charcoal afforded *threo*-dihydrosphingosine **46** in 71% yield.

5.3
Tandem Inter-Intramolecular Catalytic Asymmetric Nitroaldol Reaction

The asymmetric catalytic nitroaldol reaction was also successfully extended to the field of asymmetric tandem reactions [43]. Tandem reactions are especially useful to construct compounds with several chiral centers in a one-pot process starting from simple achiral components in the presence of a chiral catalyst. The first tandem inter-intramolecular catalytic asymmetric nitroaldol reaction was realized in the reaction of the cyclopentanedione derivative **49** with nitromethane using a catalytic amount of LnLB according to Scheme 12 [43].

In addition to temperature effects, the optical purity of the product **51b** strongly depends on the lanthanoid center ion. In the presence of the (*R*)-PrLB complex (5 mol %) as the most efficient catalyst, the hexahydro-1-indanone derivative **51b** was formed with enantioselectivities up to 65% ee [for comparison: LLB (10 mol %; –20 °C): 39% ee; YbLB (10 mol %; –20 °C): 7% ee] [43]. After crystallization, **51b** was isolated with up to 79% ee and 41% yield.

6
Recent Extensions and Improvements of the Catalytic Concept: The Second Generation of LnLB Catalysts

The catalytic asymmetric nitroaldol reactions promoted by LLB or its derivatives require at least 3.3 mol % of asymmetric catalysts for efficient conversion. However, even in the case of 3.3 mol % of catalyst, reactions are rather slow. Attempts were made to reduce the required mol % of asymmetric catalysts and accelerate the reactions, which led to a second-generation of heterobimetallic lan-

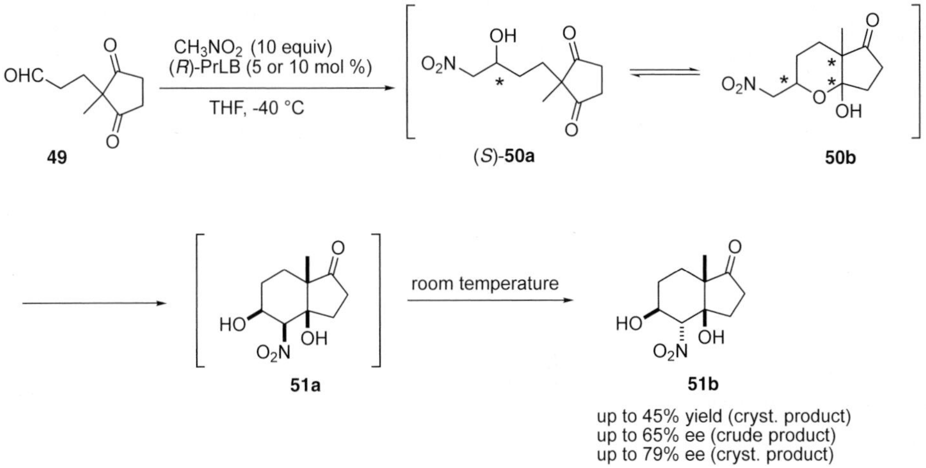

OHC
49
CH_3NO_2 (10 equiv)
(*R*)-PrLB (5 or 10 mol %)
THF, -40 °C
O_2N
OH
(*S*)-50a
50b
room temperature
HO
OH
51a
51b
up to 45% yield (cryst. product)
up to 65% ee (crude product)
up to 79% ee (cryst. product)

Scheme 12. Tandem inter-intramolecular catalytic asymmetric nitroaldol reaction

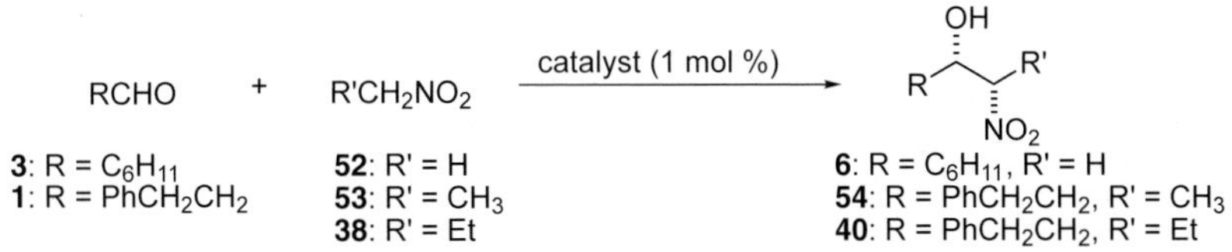

RCHO
+
R'CH2NO2
catalyst (1 mol %)
3: R = C_6H_{11}
1: R = $PhCH_2CH_2$
52: R' = H
53: R' = CH_3
38: R' = Et
6: R = C_6H_{11}, R' = H
54: R = $PhCH_2CH_2$, R' = CH_3
40: R = $PhCH_2CH_2$, R' = Et

Reagent	Catalyst	Time [h]	Temp. [°C]	Product	Yield [%]	syn/anti	ee (syn) [%]
3+**52**	LLB	24	–50	**6**	5.6	-	88
3+**52**	LLB-II	24	–50	**6**	73	-	89
1+**53**	9b	113	–30	**54**	25	70:30	62
1+**53**	**9b**-II	113	–30	**54**	83	89:11	94
1+38	9b	166	–40	40	Trace	-	-
1+**38**	**9b**-II	166	–40	40	84	95:5	95

Scheme 13. Comparison of the catalytic activity of LLB and second-generation LLB (LLB-II) or **9b** and **9b**-II. [LLB-II: LLB+H2O (1 mol equiv)+BuLi (0.9 mol equiv); **9b**-II: **9b**+H_2O (1 mol equiv)+BuLi (0.9 mol equiv)

thanoid catalysts (LLB-II), prepared from LLB, 1 mol equiv of H_2O, and 0.9 mol equiv of butyllithium. The use of only 1 mol % of LLB-II efficiently promoted catalytic asymmetric nitroaldol reactions and additionally LLB-II (3.3 mol %) accelerated the catalytic asymmetric nitroaldol reaction [11]. A comparison of the efficiency of LLB and the second-generation LLB catalysts is given in Scheme 13.

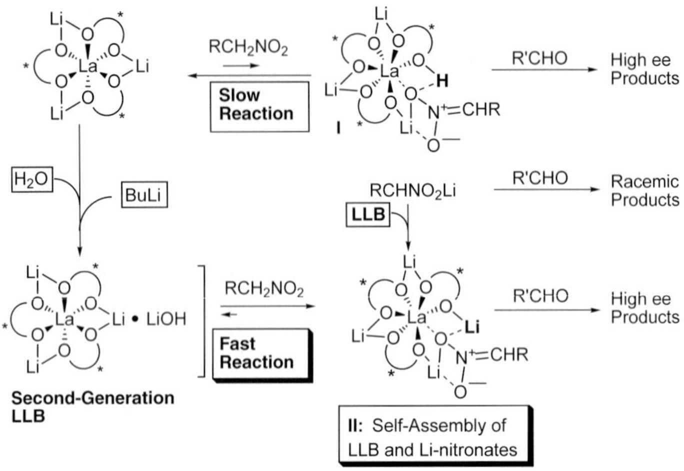

Scheme 14. Proposed mechanism of catalytic asymmetric nitroaldol reaction promoted by LLB, LLB-II or LLB-Li-nitronate

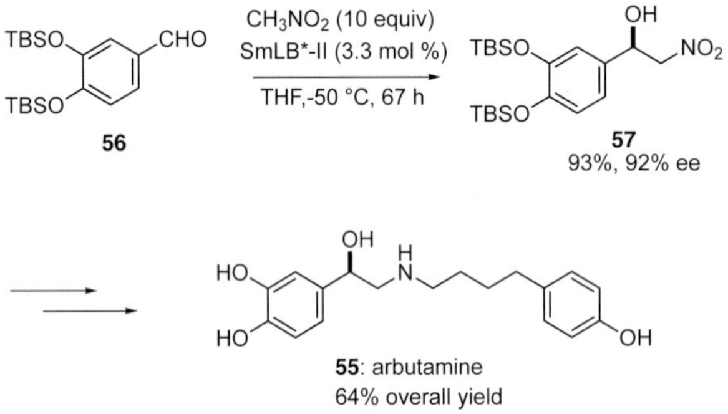

Scheme 15. A catalytic asymmetric synthesis of arbutamine. [SmLB*-II=SmLi$_3$tris((R)-6,6'-bis(trimethylsilylethynyl)binaphthoxide)+H$_2$O (1.0 mol equiv to Sm)+BuLi (0.6 mol equiv to Sm)]

The structure of LLB-II has not yet been unequivocally determined. However, it appears that it is a complex of LLB and LiOH. A proposed reaction course for the improved catalytic asymmetric nitroaldol reaction is shown in Scheme 14.

It is also noteworthy that treatment of the lithium nitronate (0.9 mol %), generated from nitropropane and butyllithium, with **9b** (1 mol %), **1**, and nitropropane **38** under similar conditions as described above gave comparable results (59% yield, *syn/anti* ratio 94:6, 94% ee), suggesting the presence of a heteropolymetallic intermediate **II** as shown in Scheme 14.

Using a second-generation LnLB catalyst consisting of 6,6'-bis{(trimethylsilyl)ethynyl}BINOL and Sm, an efficient catalytic asymmetric synthesis of arbutamine **55**, a useful β-agonist [44, 45], was achieved (Scheme 15) [46]. In the key step, the nitroaldol adduct **57** was formed in 93% yield and with 92% ee [46].

7
Summary and Outlook

From its recent beginnings in 1992, Shibasaki et al. have developed the catalytic asymmetric nitroaldol reaction (Henry reaction) into a highly efficient synthetic method for the stereoselective synthesis of nitroalkanols [9]. Alkali metal-containing heterobimetallic lanthanoid complexes were applied as catalysts. Using these catalysts, a broad variety of nitroalkanol derivatives containing one, two, or more stereogenic centers has been constructed in a highly stereoselective manner. In the mean time, this new and innovative catalytic concept has been applied to the synthesis of several biologically active and pharmaceutically interesting compounds (or their precursors). Very recently, the asymmetric nitroaldol reaction catalyzed by heterobimetallic complexes was successfully carried out in 100 g scale (with 90% ee) [47]. Further extension to an industrial scale (up to 15 kg) is now in progress [47].

Concerning future work, an improvement might be the extension of the asymmetric catalytic nitroaldol reaction to the field of solid-support synthesis. An interesting contribution concerning the use of dendritic catalysts in the Henry reaction was reported very recently by Cossio et al. [48], who demonstrated that dendrimers based on achiral triethanolamine exhibit catalytic properties. Several nitroalkanols were synthesized with *syn/anti* ratio up to 2:1 (racemic *syn*- and *anti*-products). The design of enantiomerically pure dendrimers and their application to the field of asymmetric catalytic nitroaldol reaction should be of high interest.

References

1. Henry L, (1895) Hebd CR Seances Acad Sci 120:1265
2. Seebach D, Colvin EW, Lehr F, Weller T (1979) Chimia 33:1
3. Rosini G (1991) In: Heathcock CH (ed) Comprehensive Organic Synthesis, vol 2. Pergamon, Oxford, p 220
4. Seebach D, Beck AK, Mukhopadhyay T, Thomas E (1982) Helv Chim Acta 65:1101
5. Eyer M, Seebach D (1985) J Am Chem Soc 107:3601
6. Barrett AGM, Robyr C, Spilling CD (1989) J Org Chem 54:1234
7. Hanessian S, Kloss J (1985) Tetrahedron Lett 26:1261
8. Sasai H, Suzuki T, Arai S, Arai T, Shibasaki M (1992) J Am Chem Soc 114:4418
9. Sibasaki M, Sasai H, Arai T (1997) Angew Chem 109:1290; Angew Chem Int Ed Engl 36:1236
10. Although the lithium nitronate is generated first, there also appears to be a significant possibility that the aldehyde coordinates to La first
11. Arai T, Yamada YMA, Yamamoto N, Sasai H, Shibasaki M (1996) Eur J Chem 2:1368
12. Sasai H, Suzuki T, Itoh N, Shibasaki M (1993) Tetrahedron Lett 34:851
13. Sasai H, Suzuki T, Itoh N, Arai S, Shibasaki M (1993) Tetrahedron Lett 34:2657
14. Sasai H, Suzuki T, Itoh N, Tanaka K, Date T, Okamura K, Shibasaki M (1993) J Am Chem Soc 115:10372
15. Additionally, the proposed structure of the LnLB complexes was confirmed by X-ray analytical data of the corresponding sodium complexes, described in, e.g., [14] and [16]; the successful conversion of crystalline LnSB to LnLB on treatment with LiCl also supported the structure of LnLB (cf. [14} and [16])

16. Sasai H, Arai T, Satow Y, Houk KN, Shibasaki M (1995) J Am Chem Soc 117:6194
17. Puchot C, Samuel O, Dunach E, Zhao S, Agami C, Kagan HB (1986) J Am Chem Soc 108:2353
18. Noyori R, Kitamura M (1991) Angew Chem 103:34; Angew Chem Int Ed Engl 30:49
19. Terada M, Mikami K, Nakai T (1990) J Chem Soc Chem Commun 1623
20. Sasai H, Watanabe S, Shibasaki M (1997) Enantiomer 2:267
21. A 6,6'-disubstituted BINOL derivative was also used in the asymmetric Diels-Alder reaction by: Terada M, Motoyama Y, Mikami K (1994) Tetrahedron Lett 35:6693
22. Lingfelter DS, Hegelson RC, Cram DJ (1981) J Org Chem 46:393
23. Maruoka K, Itoh T, Araki Y, Shirasaka T, Yamamoto H (1988) Bull Chem Soc Jpn 61:2975
24. Yamamoto K, Noda K, Okamoto Y (1985) J Chem Soc Chem Commun 1065
25. Sasai H, Watanabe T, Suzuki T, Shibasaki M (1997) Org Synth submitted
26. Oshida JI, Okamoto M, Azuma S, Tanaka T (1997) Tetrahedron: Asymmetry 8:2579
27. Sasai H, Arai S, Tahara Y, Shibasaki M (1995) J Org Chem 60:6656
28. Gröger H, Saida Y, Arai S, Martens J, Sasai H, Shibasaki M (1996) Tetrahedron Lett 37:9291
29. Sasai H, Itoh N, Suzuki T, Shibasaki M (1993) Tetrahedron Lett 34:855
30. Sasai H, Yamada YMA, Suzuki T, Shibasaki M (1994) Tetrahedron 50:12313
31. Sasai H, Suzuki T, Itoh N, Shibasaki M (1995) Appl Organomet Chem 9:421
32. Seebach D (1990) Angew Chem 102:1363; Angew Chem Int Ed Engl 29:1320
33. Iseki K, Oishi S, Sasai H, Shibasaki M (1996) Tetrahedron Lett 37:9081
34. Mimoto T, Imai J, Kisanuki S, Enomoto H, Hattori N, Akaji K, Kiso Y (1992) Chem Pharm Bull 40:2251
35. Kageyama S, Mitsuto T, Murakawa Y, Nomizu M, Ford H Jr, Shirasaka T, Gulnik S, Erickson J, Takada K, Hayashi H, Broder S, Kiso Y, Mitsuya H (1993) Antimicrob Agents Chemother 37:810
36. Sasai H, Kim WS, Suzuki T, Shibasaki M, Mitsuda M, Hasegawa J, Ohashi T (1994) Tetrahedron Lett 35:6123
37. Mitsuda M (1997) Kaneka Corporation. Personal communication
38. Morisaki M, Koizumi N, Ikekawa N (1975) J Chem Soc Perkin Trans I 1421
39. Tanaka H, Abe E, Miyaura C, Kurihashi T, Konno K, Nishii Y, Suda T (1982) Biochem J 204:713
40. Kobayashi T, Okumura H, Azuma Y, Kiyoki M, Matsumoto K, Hashimoto K, Yoshikawa K (1990) J Dermatol 17:707
41. Sasai H, Tokunaga T, Watanabe S, Suzuki T, Itoh N, Shibasaki M (1995) J Org Chem 60:7388
42. Schwartz GK, Jiang J, Kelsen D, Albino AP (1993) J Natl Cancer Inst 85:402
43. Sasai H, Hiroi M, Yamada YMA, Shibasaki M (1997) Tetrahedron Lett 38:6031
44. Young M, Pan W, Wiesner J, Bullough D, Browne G, Balow G, Potter S, Metzner K, Mullane K (1994) Drug Dev Res 32:19
45. Hammond HK, McKirnan MD (1994) J Am Coll Cardiol 23:475
46. Takaoka E, Yoshikawa N, Yamada YMA, Sasai H, Shibasaki M (1997) Heterocycles 46:157
47. Urata Y (1997) Chisso Corporation. Personal communication
48. Morao I, Cossio FP (1997) Tetrahedron Lett 38:6461

Chapter 30

Addition of Acyl Carbanion Equivalents to Carbonyl Groups and Enones

Chapter 30
Addition of Acyl Carbanion Equivalents to Carbonyl Groups and Enones

Dieter Enders · Klaus Breuer

Institut für Organische Chemie, Rheinisch-Westfälische Technische Hochschule, Professor-Pirlet-Straße 1, D-52074 Aachen, Germany
e-mail: enders@rwth-aachen.de

Keywords: Nucleophilic acylation, triazolium salts, thiazolium salts, benzoin reaction, acyloin reaction, Stetter reaction

Introduction

In the course of evolution, nature has devised a multitude of enzymes which are capable of stereoselectively forming C–C bonds in vivo. Depending on the kind of substrate employed, nucleophilic acylation reactions are accomplished in nature by means of different lyases such as transketolases and pyruvate decarboxylases, which all require thiamine (1) as coenzyme [1, 2, 3, 4, 5, 6]. For the synthetic organic chemist, asymmetric catalytic C–C bond formation reactions, in

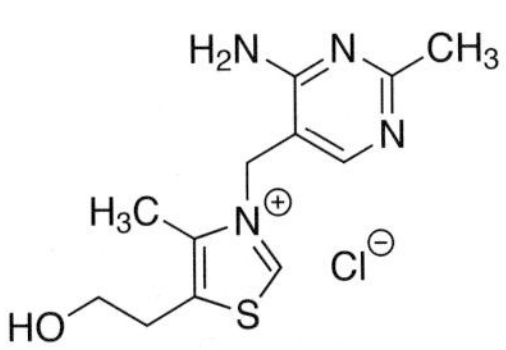

Fig. 1

general, constitute tools of prime importance for the efficient and straightforward stereoselective synthesis of chiral building blocks [7, 8, 9, 10, 11, 12]. However, only a few methods have been published that allow the asymmetric catalytic nucleophilic acylation of carbonyl compounds or activated double bonds leading to the corresponding enantiomerically pure or enriched 1,2- and 1,4-bifunctionalized compounds. All approaches described so far mimic the catalytic systems involving the naturally occurring coenzyme thiamine (1), Fig. 1.

2
The Mechanism of Heterazolium Catalysis

The basic mechanism of heterazolium catalysis was elucidated by Breslow et al. [13] in 1958 in the course of their pioneering work (Scheme 1). First the heterazolium ion 2 used as the catalyst is deprotonated affording the corresponding nucleophilic carbene 3 which represents the actual catalytically active species. This species attacks the carbonyl function of an aldehyde leading to the hydroxy-enamine-type Breslow intermediate 4, which subsequently functions as the nucleophilic acylation reagent (d^1-synthon). Reaction with an electrophilic substrate such as a second aldehyde molecule or Michael acceptors yields the bifunctional products 5 or 6, respectively, and the original carbene catalyst 3.

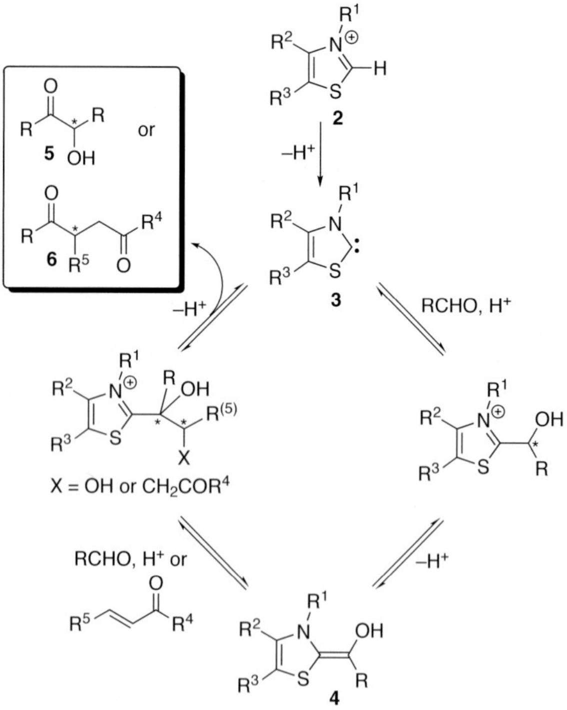

Scheme 1

The catalytic cycle was recently challenged by López-Calahorra et al. [14, 15, 16, 17, 18, 19, 20], who claimed that the catalysis proceeds via the corresponding dimers of the aforementioned nucleophilic carbenes leading to dimeric intermediates. However, recent investigations by Breslow et al. [21, 22] and others [23] seem to confirm the original proposal by Breslow.

3
Nucleophilic Acylation of Aldehydes

The nucleophilic acylation of aldehydes is a short and efficient pathway to 1,2-bifunctionalized building blocks. This reaction type is extremely valuable, since it allows the catalytic formation of α-hydroxy ketones, which are important synthetic intermediates in organic chemistry.

In general, the starting materials, i.e. the aldehydes, are readily accessible or even commercially available. Despite this synthetic significance, only a few catalytic asymmetric variants for the nucleophilic acylation of aldehydes have been reported.

3.1
Thiazolium Catalysts in the Nucleophilic Acylation of Aldehydes

The first investigations in the field of heterazolium-catalyzed asymmetric nucleophilic acylation go back to Sheehan et al. (Scheme 2) [24], who tried to devise an asymmetric variant of the benzoin reaction, which was known to be catalyzed by thiazolium salts from the work of Ukai et al. [25, 26] and of Mizuhara et al. [27].

Sheehan et al. [28] developed several chiral thiazolium salts, which were shown to catalyze the formation of benzoin with low to moderate enantiomeric excesses, up to 52% in the case of a 1-naphthylethyl-substituted catalyst. However, the yields were very low (6%), leading to the consumption of a stoichiometric amount of the catalyst (TTN < 1) which limited the applicability of the reaction. Tagaki et al. [29] reported chiral menthyl-substituted thiazolium salts, the best of which catalyzed the formation of benzoin with enantiomeric excesses up to 35% and slightly improved yields of 20% (TTN 4) by carrying out the reaction in a micellar two-phase system. Zhao et al. [30] combined the superior catalyst concept of Sheehan et al. with the favorable micellar reaction conditions of

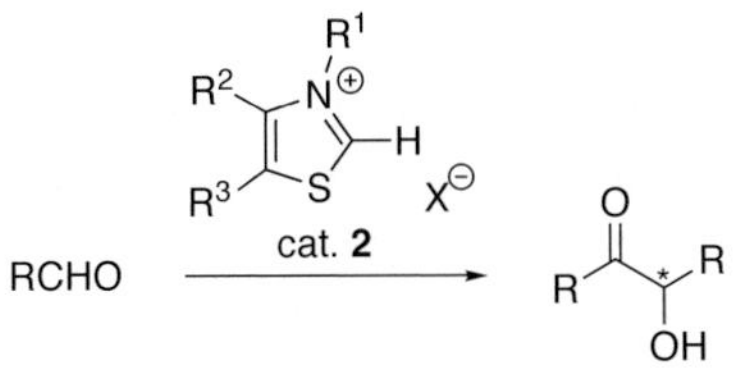

Scheme 2

Tagaki et al. and obtained benzoin with enantiomeric excesses of up to 57% and yields ranging from 20 to 30% using 1-naphthylethyl-based catalysts. For some catalysts, a critical micelle concentration could be observed in several buffer solutions, which was understood as evidence of micelle formation being involved in the catalytic process. López-Calahorra et al. [31] introduced bridged *bis*thiazolium salts to the catalytic benzoin reaction, however, the enantiomeric excesses found were quite low (ee up to 26%, yields up to 21%).

In general, the unsatisfactory yields obtained in the asymmetric variants of the thiazolium-catalyzed benzoin reaction are probably caused by the low inherent activity of the thiazolium salts, which is aggravated by the steric bulk accumulated in the neighborhood of the active site.

3.2
Triazolium Catalysts in the Nucleophilic Acylation of Aldehydes

Recent work by Teles et al. [32] has shown that triazolium salts are highly active catalysts for the condensation of formaldehyde affording glycolaldehyde (formoin reaction). In terms of activity, these catalysts proved considerably superior to the thiazolium salts previously used for this transformation. Investigations into the mechanism have shown that the catalytic cycle corresponds to the mechanism proposed by Breslow [32].

Accordingly, our research group [33] synthesized a variety of chiral triazolium salts and examined their ability to catalyze the benzoin reaction. However, the enantiomeric excesses and catalytic activities proved to vary strongly with slight structural changes in the substitution pattern of the triazolium system.

The catalyst 7, which is accessible via bisformylation of phenylhydrazine [34] and cyclization to the oxadiazolium salt [35] with subsequent ring opening ring closure substitution [36] using (4S,5S)-5-amino-2,2-dimethyl-4-phenyl-1,3-dioxan, yielded benzoin with 75% ee and a satisfactory yield of 66% using a significantly reduced catalyst amount of 1.25 mol % allowing total turnover numbers over 50 (Scheme 3). This represents an increase in activity of almost two orders of magnitude compared with those results obtained with chiral thiazolium salts. We subsequently extended the applicability of this new catalyst type to other aromatic substrates to give the respective aromatic α-hydroxy ketones. This illustrates that electron-rich aldehydes generally furnish the respective benzoins in moderate to good enantiomeric excesses up to 86%, whereas the asymmetric inductions achieved with electron-deficient aldehydes are significantly lower. Apparently, deactivation of the aldehyde function leads to lower catalytic activity but higher enantioselectivities. Accordingly, the highest enantioselectivities of 86% (and the lowest yields) were obtained with p-methoxybenzaldehyde, in which the carbonyl function is considerably deactivated due to the +M effect of the methoxy group. All in all, this procedure has been routinely used to prepare different benzoins on a multigram scale. However, despite the significant improvement of the enantioselectivities and catalyst activities achieved upon use

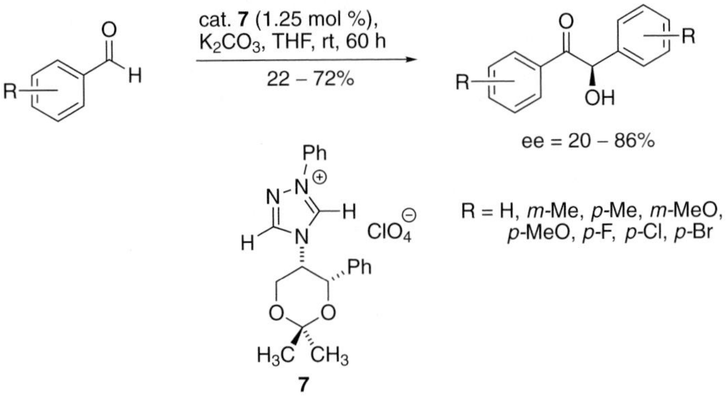

Scheme 3

R^1	R^2	Yield [%]	ee [%]
H	H	66	75
MeO	H	41	66
H	MeO	22	86
Me	H	54	76
H	Me	46	82
H	F	48	44
H	Cl	51	29
H	Br	72	20

Scheme 3

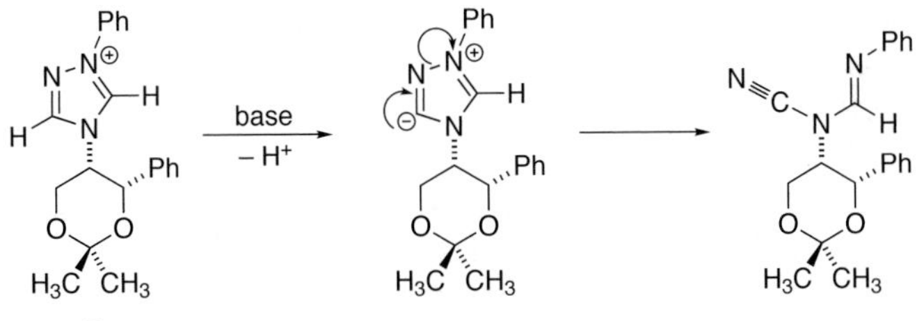

Scheme 4

of triazolium salt **7**, the method does not yet present a fully satisfacory approach to the asymmetric synthesis of α-hydroxy ketones.

The major deactivation pathway of the catalyst **7** in the course of the catalytic cycle proceeds via deprotonation at C-3, since the proton at C-3 is almost as acidic as that one at C-5. After the removal of the proton, the triazol ring opens to the respective N-cyanobenzamidine, leading to the irreversible destruction of the catalyst (Scheme 4) [33].

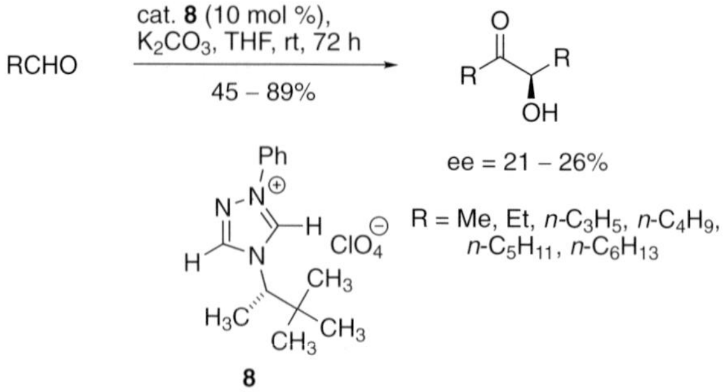

Scheme 5

R	Yield [%]	ee [%]
Me	45	23
Et	65	23
n-C$_3$H$_5$	84	21
n-C$_4$H$_9$	84	26
n-C$_5$H$_{11}$	89	21
n-C$_6$H$_{13}$	71	21

The reaction has to be carried out in the absence of oxygen and water, otherwise the intermediately formed nucleophilic carbene is oxidized to the triazolinone or suffers hydrolysis with subsequent aminal-type ring opening. Attempts to apply catalyst 7 to the synthesis of aliphatic acyloins gave very low yields and low enantioselectivities. Optimization of the catalyst structure with regard to activity and enantioselectivity yielded triazolium salt 8 as the best-suited system for the condensation of aliphatic aldehydes (Scheme 5) [37]. However, the enantiomeric excesses obtained, only up to 26%, and the rather low total turnover numbers, ranging from 4 to 8, are modest and a search for new, more active catalyst systems is highly desirable.

An asymmetric variant of a mixed acyloin or benzoin condensation using heterazolium catalysts has not yet been reported. However, enzymes have been shown to catalyze a number of mixed acyloin condensations efficiently [38, 39, 40].

4
Nucleophilic Acylation of Enones and Enoates (Stetter Reaction)

The Stetter reaction is an extremely useful Umpolung procedure for the synthesis of 1,4-dicarbonyl compounds [41, 42, 43]. Since its discovery in 1973, it has found widespread application in the preparation of key organic intermediates and in natural product synthesis. However, despite the importance and useful-

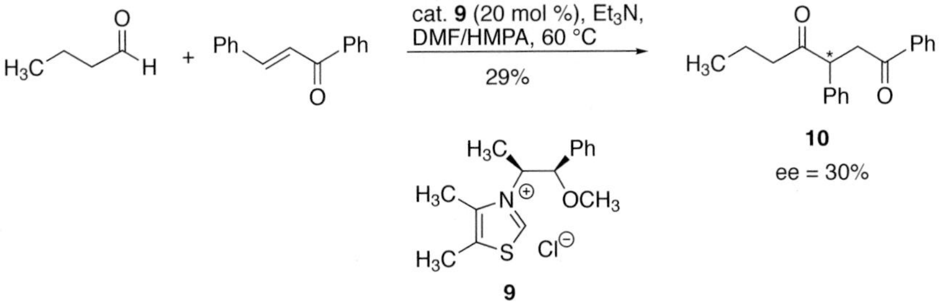

Scheme 6

ness of this protocol, very few asymmetric approaches have been reported and those only recently.

The first attempts to develop a heterazolium-catalyzed asymmetric variant of the Stetter reaction were carried out by our group [44, 45, 46], employing the chiral thiazolium salt **9** to catalyze the addition of butanal to chalcone. The resultant 1,4-dicarbonyl compound **10** was obtained in 29% yield with enantiomeric excesses up to 30% (Scheme 6).

Attempts to use triazolium catalysts instead of thiazolium salts proved to be unsuccessful, although the activity of triazolium salts in the non-enantioselective Stetter reaction had been previously reported [41]. However, some triazol-5-ylidenes have been shown to give stable adducts with several Michael acceptors and this could be the reason for their failure to react [47, 48].

We used triazolium salt **7** as the catalyst for the intramolecular Stetter reaction of 2-formylphenoxycrotonates **11** affording the corresponding 4-chromanones **12** [49], since these were known to be highly active substrates in the non-enantioselective thiazolium-catalyzed Stetter reaction [50] (Scheme 7). Apparently, the entropically favorable proximity of the reacting functionalities leads to a strong enhancement of the reactivity.

As already observed in the triazolium-catalyzed benzoin reaction, electron-donating substituents (e.g. methoxy) lead to moderate to good enantiomeric excesses, up to 71%, whereas electron-withdrawing groups cause a strong decrease in enantioselectivity. Again, deactivation of the aldehyde function leads to lower catalytic activity but higher enantioselectivities. The chromanones accessible via this method are useful intermediates for the synthesis of pterocarpans [51, 52].

In general, however, the activity of the triazolium salts in this asymmetric Stetter protocol is quite low, i.e. the total turnover numbers obtained ranged from 0.5 to 8. The development of more active catalysts which are also suitable for the intermolecular Stetter reaction is desirable.

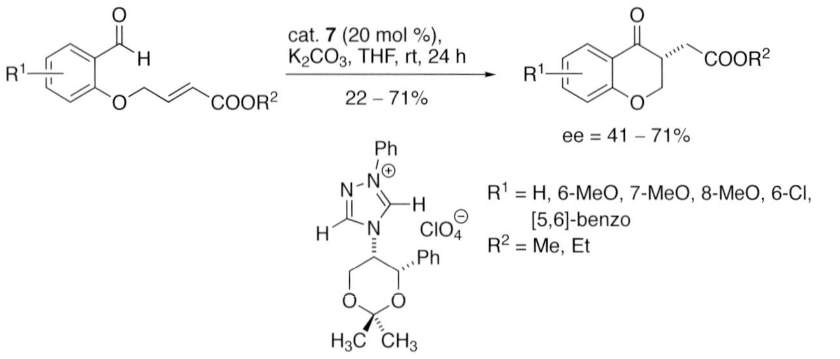

Scheme 7 reaction scheme showing:

Starting material: aryl aldehyde with R^1 substituent and O-CH=CH-$COOR^2$ chain

Conditions: cat. 7 (20 mol %), K_2CO_3, THF, rt, 24 h, 22 – 71%

Product: chromanone with $COOR^2$ substituent, ee = 41 – 71%

Catalyst 7: triazolium salt with Ph groups, $ClO_4^\ominus$, dioxolane ring with H_3C CH_3

R^1 = H, 6-MeO, 7-MeO, 8-MeO, 6-Cl, [5,6]-benzo
R^2 = Me, Et

R^1	R^2	R^3	R^4	R^5	Yield [%]	ee [%]
H	H	H	H	Me	73	60
H	H	H	H	Et	69	56
H	H	H	MeO	Me	44	68
H	H	H	MeO	Et	69	62
H	H	MeO	H	Me	22	71
H	MeO	H	H	Me	56	61
H	Cl	H	H	Me	50	41
–CH=CH-CH=CH–	H	H	Me	51	65

Scheme 7

5
Principal Alternatives

Highly enantiomerically enriched α-hydroxy ketones are readily accessible via various stoichiometric methods, e.g. carbamoylation reactions [52], nucleophilic acylation using metalated aminonitriles [54] or the α-hydroxylation of chiral hydrazones [55] and α-silyl ketones, as well as the α-hydroxylation of ketones using chiral oxaziridines [56]. A number of catalytic approaches are based on the asymmetric dihydroxylation of enol ethers [57] and the enantioselective reduction of diketones using enzymes [58], or via catalytic hydrogenation [59]. As was mentioned earlier, a highly enantioselective enzymatic variant of the mixed acyloin condensation with aromatic aldehydes and aliphatic α-oxocarboxylic acids yielding the mixed acyloins with high enantiomeric excesses has been described [38, 39, 40].

The nucleophilic acylation of Michael acceptors can be accomplished by a number of stoichiometric methods, i.e. via the addition of metalated α-aminonitriles [60, 61] or under neutral conditions with formaldehyde-SAMP-hydrazone acting as an aza-enamine [62, 63]. Enzymatic approaches to an asymmetric nucleophilic acylation of Michael acceptors have not been published so far.

References

1. Shin W, Oh D-G, Chae C-H, Yoon T-S (1993) J Am Chem Soc 115: 12238
2. Sax M, Pulsinelli P, Pletcher J (1974) J Am Chem Soc 96: 155
3. Kluger R, Gish G (1988) Stereochemical aspects of thiamin catalysis. In: Schellenberg A, Schowen RL (eds) Thiamin pyrophosphate biochemistry. CRC Press, Boca Raton, Florida, chap 1
4. Lobell M, Crout DHG (1996) J Am Chem Soc 118: 1867
5. Effenberger F, Null V, Ziegler T (1992) Tetrahedron Lett 33: 5157
6. Kobori Y, Myles DC, Whitesides GM (1992) J Org Chem 57: 5899
7. Morrison JD (1985) Asymmetric synthesis. Academic Press, New York
8. Eliel EL, Wilen SH (1994) Stereochemistry of organic compounds. Wiley, New York
9. Helmchen G, Hoffmann RW, Mulzer J, Schaumann E (eds) (1995) Houben-Weyl, 4th edn. Stereoselective synthesis, vol E21. Thieme, Stuttgart
10. Bosnich B (1986) Asymmetric catalysis. Martinus Nijhoff, Dordrecht
11. Ojima I (1993) Catalytic asymmetric synthesis. VCH, Weinheim
12. Noyori R (1994) Asymmetric catalysis in organic synthesis. Wiley, New York
13. Breslow R (1958) J Am Chem Soc 80: 3719
14. Castells J, López-Calahorra F, Geijo F, Pérez-Dolz R, Bassedas M (1986) J Heterocyclic Chem 23: 715
15. Castells J, López-Calahorra F, Domingo L (1988) J Org Chem 53: 4433
16. Castells J, Domingo L, López-Calahorra F, Martí J (1993) Tetrahedron Lett 34: 517
17. López-Calahorra F, Rubires R (1995) Tetrahedron 51: 9713
18. López-Calahorra F, Castro E, Ochoa A, Martí J (1996) Tetrahedron Lett 37: 5019
19. López-Calahorra F, Castells J, Domingo L, Martí J, Bofill JM (1994) Heterocycles 37: 1579
20. Martí J, López-Calahorra F, Bofill JM (1995) J Mol Struct (Theochem) 339: 179
21. Breslow R, Kim R (1994) Tetrahedron Lett 35: 699
22. Breslow R, Schmuck C (1996) Tetrahedron Lett 37: 8241
23. Chen Y-T, Barletta GL, Haghjoo K, Cheng JT, Jordan F (1994) J Org Chem 59: 7714
24. Sheehan J, Hunneman DH (1966) J Am Chem Soc 88: 3666
25. Ugai T, Tanaka S, Dokawa S (1943) J Pharm Soc Jpn 63: 269
26. Ugai T, Dokawa S, Tsubokawa S (1943) J Pharm Soc Jpn 63: 269
27. Mizuhara S, Tamura R, Arata H (1951) Proc Jpn Acad 87: 302
28. Sheehan J, Hara T (1974) J Org Chem 39: 1196
29. Tagaki W, Tamura Y, Yano Y (1980) Bull Chem Soc Jpn 53: 478
30. Zhao C, Chen S, Wuz P, Wen Z (1988) Huaxue Xuebao 46: 784
31. Martí J, Castells J, López-Calahorra F (1993) Tetrahedron Lett 34: 521
32. Teles JH, Melder J-P, Ebel K, Schneider R, Gehrer E, Harder W, Brode S, Enders D, Breuer K, Raabe G (1996) Helv Chim Acta 79: 61
33. Enders D, Breuer K, Teles JH (1996) Helv Chim Acta 79: 1899
34. Freund M, Horst G (1895) Ber Dtsch Chem Ges 28: 944
35. Boyd GV, Dando SR (1970) J Chem Soc C 1397
36. Boyd GV, Summers AJH (1971) J Chem Soc C 409
37. Breuer K (1997) PhD thesis, Technical University of Aachen
38. Fessner WD, Walter C (1996) Top Curr Chem 184: 97
39. Servi S (1990) Synthesis 1
40. Csuk R, Glänzer BI (1991) Chem Rev 91: 49
41. Stetter H, Kuhlmann H (1991) Org React 40: 407
42. Stetter H (1976) Angew Chem 88: 695; Angew Chem Int Ed Engl 15: 639
43 Kreiser W (1981) Nachr Chem Tech Lab 29: 172

44. Tiebes J (1991) Diploma work, Technical University of Aachen
45. Enders D (1994) Enzymemimetic C-C and C-N bond formations. In: Ottow E, Schöllkopf K, Schulz BG (eds) Stereoselective synthesis. Springer, Berlin Heidelberg New York, p 63
46. Enders D, Bockstiegel B, Dyker H, Jegelka U, Kipphardt H, Kownatka D, Kuhlmann H, Mannes D, Tiebes J, Papadopoulos K (1993) Dechema-Monographien, vol 129. VCH, Weinheim, p 209
47. Enders D, Breuer K, Raabe G, Runsink J, Teles JH, Melder J-P, Ebel K, Brode S (1995) Angew Chem 107: 1119; Angew Chem Int Ed Engl 34: 1021
48. Enders D, Breuer K, Runsink J, Teles JH (1996) Liebigs Ann Chem 2019
49. Enders D, Breuer K, Runsink J, Teles JH (1996) Helv Chim Acta 79: 1899
50. Ciganek E (1995) Synthesis 1311
51. Ozaki Y, Mochida K, Kim S-W (1989) J Chem Soc Perkin Trans 1: 1219
52. Ozaki Y, Kim S-W (1985) Synth Commun 15: 1171
53. Enders D, Lotter H (1981) Angew Chem 93: 831; Angew Chem Int Ed Engl 20: 795
54. Enders D, Lotter H, Maigrot N, Mazaleyrat J-P, Welvart Z (1984) Nouv J Chim 8: 747
55. Enders D, Bhushan V (1988) Tetrahedron Lett 29: 2437
56. Davis FA, Weismiller MC, Murphy CK, Reddy RT, Chen BC (1992) J Org Chem 57: 7274 and literature cited therein
57. Hashiyama T, Morikawa K, Sharpless KB (1992) J Org Chem 57: 5067
58. Gu JX, Li ZY, Lin GQ (1995) Chin Chem Lett 6: 457
59. Takeuchi S, Ohgo Y (1988) Chem Lett 403
60. Enders D, Gerdes P, Kipphardt H (1990) Angew Chem 102: 226; Angew Chem Int Ed Engl 29: 179
61. Enders D, Kirchhoff J, Mannes D, Raabe G (1995) Synthesis 659 and literature cited therein
62. Enders D, Syrig R, Raabe G, Fernández R, Gasch C, Lassaletta J-M, Llera J-M (1996) Synthesis 48
63. Lassaletta J-M, Fernández R, Martín-Zamora E, Díez E (1996) J Am Chem Soc 51: 7002

Note added in proof

Some more conformationally restricted thiazolium catalysts have been recently published by Leeper et al. [1, 2] and Rawal et al. [3].

Using a chiral, bicyclic Thiazolium salt Leeper et al. [1] obtained enantiometric excesses of up to 20.5% (c.y. 50%, 10 mol % catalyst, TTN ca. 2) in the benzoin reaction..A similar bicyclic catalyst gave ee's of up to 33% when used for the synthesic of butyroin (c.y. 75%, 20 mol % catalyst, TTN ca. 4).

A novel thiazolium catalyst with a norbornane backbone gave benzoin with enantiomeric excesses of up to 26% (c.y. 100%, 5 mol % catalyst, TTN = 20) [2].

Rawal et al. [3] achieved enantiomeric excesses up to 52% for the formation of benzoin with improved yields of up to 48%, using the best thiazoilim catalyst previously descrived by Sheehan et al. (10 mol % catalyst, TTN = 5).

1. Knight RL, Leeper FJ (1997) Tetrahedron Lett. 38: 3611-3614
2. Gerhards AV, Leeper FJ (1997) Tetrahedron Lett. 38: 3615-3618
3. Dvorak CA, Rawal VH (1998) Tetrahedron Lett. 39: 2925-2928

Chapter 31   Conjugate Addition Reactions

Chapter 31.1
Conjugate Addition of Organometallic Reagents

Kiyoshi Tomioka · Yasuo Nagaoka

Graduate School of Pharmaceutical Sciences, Kyoto University, Yoshida, Sakyo-ku,
Kyoto 606-8501, Japan
e-mail: tomioka@pharm.kyoto-u.ac.jp

Keywords: Ligand, Ether, Amine, Nucleophilic addition, Organolithium, Grignard reagent, Organozinc, Copper, Phosphorus, Conjugate addition, Alkylation, Phosphine, Activated olefin, β-Substituted carbonyl compound

1
Introduction

Conjugate additions of organometallic reagents to electrophilically activated olefins constitute one of the versatile methodologies for forming carbon-carbon bonds [1]. The products are the corresponding β-substituted carbonyl compounds. Because of the usefulness of the reaction as well as the products, many approaches to asymmetric conjugate addition reactions and successful achievements have been reported, especially using chirally modified olefins [2, 3]. However, the approach towards enantioselective conjugate addition reaction is currently a developing area [4, 5, 6, 7, 8, 9, 10]. In this chapter the recent progress in the enantioselective conjugate addition reactions of organolithium and organocopper reagents with achiral activated olefins under the control of an external chiral ligand or chiral catalysts is summarized. The Michael reaction of active methylene compounds is not included in this chapter.

2
Reaction of Organolithium Reagents Using External Chiral Ligands

Organolithium compounds are highly reactive species and are used in a variety of organic transformations. Asymmetric conjugate addition of organolithium reagents to α,β-unsaturated carbonyl compounds has recently reached a useful level with the use of external chiral ligands, especially the chiral DME modification, 1, and the chiral diamine, (–)-sparteine 2 [11]. Chiral diethers are used as ligands for lithium, and the use of a stoichiometric amount of diether 1 has shown the greatest efficiency for the asymmetric addition of organolithium reagents to α,β-unsaturated N-cyclohexylimines (Scheme 1) [12, 13, 14]. After hydrolysis, β-substituted aldehydes are obtained with an excellent selectivity. The ligand 1 is recoverable for reuse in high yield. The use of a poor coordinating solvent such as toluene or ether is essential for high enantioselectivity, probably because of the formation of the tight lithium-ligand chelated complex.

The diether 1 was readily prepared by dimethylation of the chiral stilbene diol which was prepared by an AD-mix reaction of stilbene in high yield.

The observed enantiofacial selection has been interpreted in terms of lithium-coordinated complex formation between the organolithium, the imine and the chiral diether 1 (Fig. 1). The R group of the organolithium is then transferred from the favored complex to the less hindered face of the double bond of the unsaturated imine.

The regioselectivity, that is 1,4- vs. 1,2-addition, is directed mainly by the larger LUMO coefficient of the corresponding reaction site [15]. A change of the cyclohexyl group of the imine moiety to an aromatic group leads to larger coefficients at the imine carbon and results in selective 1,2-conjugate addition. A cat-

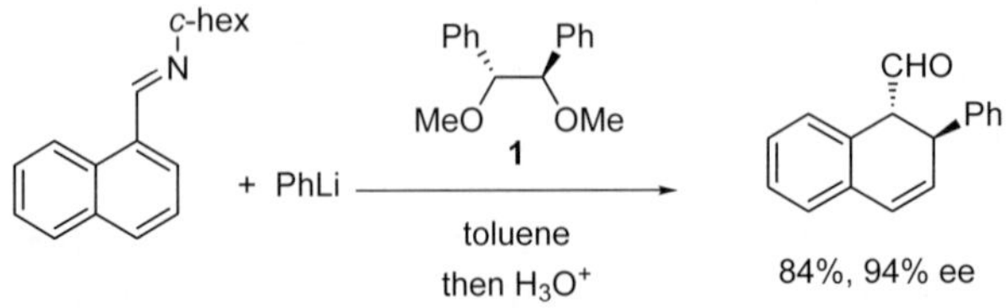

Scheme 1

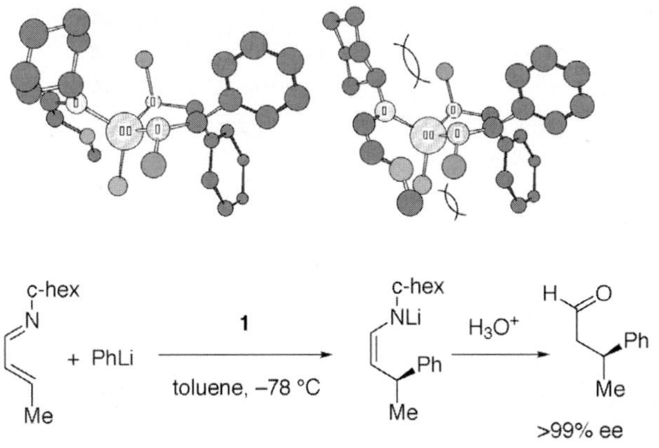

Fig. 1.

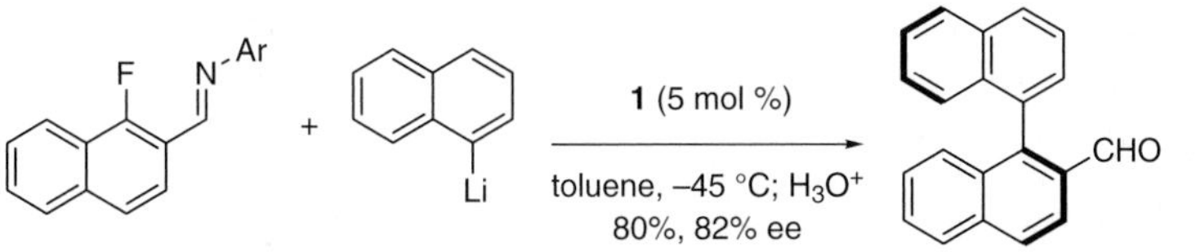

Scheme 2

alytic asymmetric 1,2-addition reaction to provide a chiral amine is also possible [16, 17].

The first prominent catalytic asymmetric addition of an organolithium reagent was realized in the reaction of 1-naphthyllithium with 1-fluoro-2-naphthylaldehyde imine in the presence of the chiral diether **1** to afford chiral binaphthyls in over 82% ee (Scheme 2). Merely a catalytic amount of **1** (5 mol %) is required to effect the reaction, in which an enantioselective conjugate addition-elimination mechanism is operative [18].

The same ligand for organolithium has been used to achieve a high level of catalytic asymmetric conjugate addition to hindered α,β-unsaturated and naphthyl esters as shown in Scheme 3 [19, 20]. The chiral diether **1** shows high efficiency for aryllithium reagents and (−)-sparteine for alkyllithium compounds. The catalytic turnover of (−)-sparteine is superior to that of **1** [21].

The chromium complex of benzaldehyde imine is also good substrate for addition of organolithium reagents mediated by a stoichiometric amount of the chiral diether **1** in toluene to give the corresponding addition products in up to 93% ee (Scheme 4)[22].

(−)-Sparteine **2** is also an excellent chiral diamine ligand in stoichiometric amounts as shown by the ligand-directed conjugate addition of chirally fixed organolithium species. The choice of ligand for lithium can provide control of 1,2- vs 1,4-addition of organolithium species to α,β-unsaturated carbonyl substrates

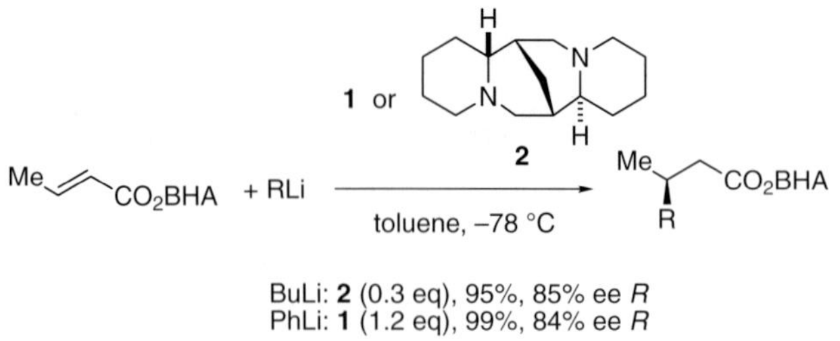

Scheme 3

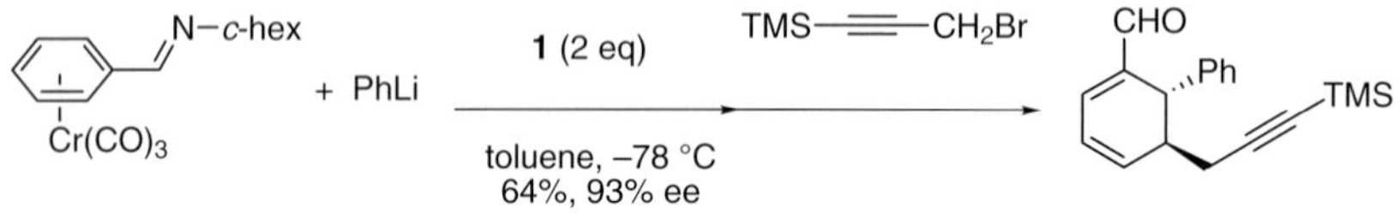

Scheme 4

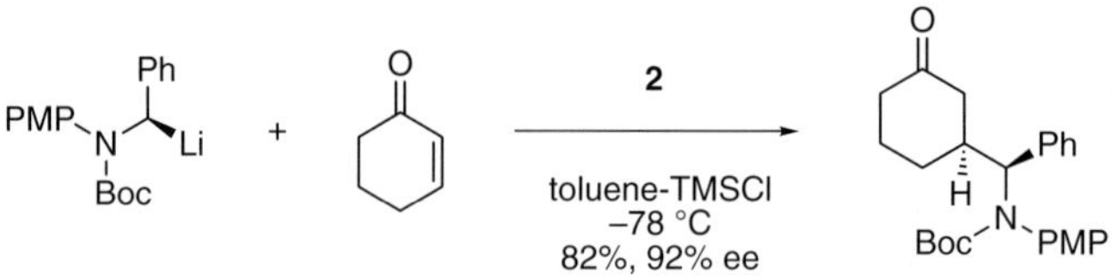

Scheme 5

(Scheme 5) [23]. Furthermore in these addition reactions, two stereocenters are constructed with high diastereo- and enantioselectivities.

3
Reaction of Heteroorganocuprates Prepared from Organolithium and Grignard Reagents

Organocopper reagents are the most reliable species for conjugate additions and a number of approaches towards chiral cuprates has been developed. The approaches are classified into two categories; one is the chiral heterocuprate obtained by treatment with chiral alcohols, amines, sulfonamides, and thiols. The other involves organocopper compounds coordinated by chiral external ligands such as phosphines, sulfides, and oxazolines.

3.1
Chiral Alkoxycuprates

Chirally modified alkoxycuprates can be generated from organolithium or Grignard reagents and copper(I) salt in the presence of a lithium or magnesium alcoholate of a chiral alcohol **3**. Although in the early attempts the enantioselectivity was not high [24, 25], use of *N*-methylprolinol **4** as a chiral alkoxide source opens a new route for the asymmetric conjugate addition reaction of the Grignard reagent, methylmagnesium bromide, with chalcone to provide a relatively good enantioselectivity of 68% [26]. The reaction was optimized to afford the addition product in 88% ee [27, 28]. A breakthrough was achieved by using the ephedrine-derived chiral amino alcohol **5** to effect conjugate addition of organolithium reagents in over 90% enantioselectivity [29]. Even a small amount of alkoxide impurity in the alkyllithium solution was found to be deleterious for the enantioselectivity. The relationships between the cluster structure and the enantiofacial selection are a matter of discussion [30]. The observed enantiofacial selection was interpreted in terms of the model **6** (Scheme 6).

By use of an amino alcohol having a bornane skeleton (**7**) the conjugate addition of methyllithium to cyclic alkenones to afford the corresponding methyl adduct in excellently high ee was realized (Scheme 7). Although these reactions need a stoichiometric amount of the chiral alcohol, batch process techniques are applicable in the reaction [31, 32].

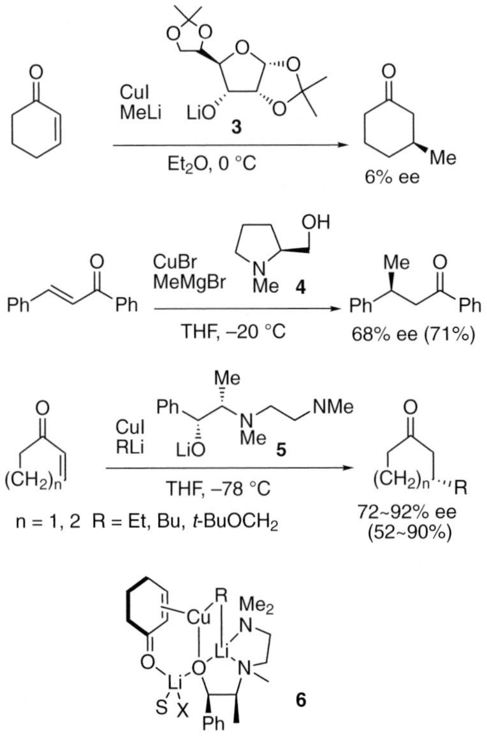

Scheme 6

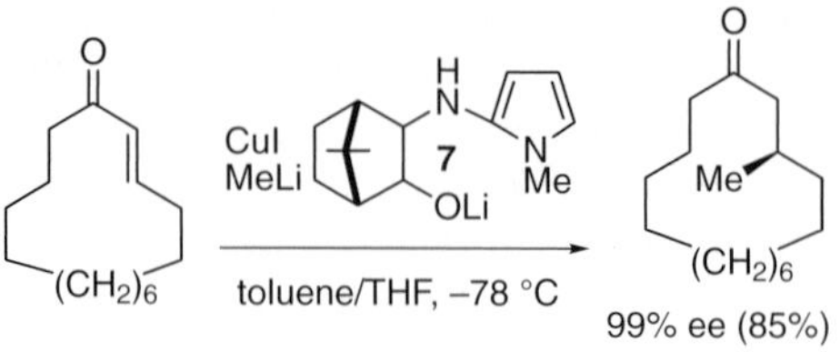

Scheme 7

3.2
Chiral Amidocuprates

Chirally modified amidocuprates can be generated from organolithium or Grignard reagents and copper(I) salt in the presence of a lithium or magnesium amide of a chiral amine. A relatively high enantioselectivity was first reported by Bertz in which a chiral amide **8** and copper iodide were used to effect the reaction of phenyllithium to afford the adduct in 50% ee (Scheme 8) [33]. The more simple prolinol derived lithium amide **9** (Scheme 8) is interesting in that it affords either enantiomers by choosing bromide or thiocyanate as a copper source in over 82% ee [34]. The linear lithium amide **10** (Scheme 8) was also introduced to effect the addition to provide the adducts in up to 97% ee [35, 36]. The observed enantiofacial selection was interpreted by means of a model assuming a dimeric structure in which the presence of the phenyl group blocks the bottom face and leads to top face reaction (Fig. 2). The dimeric structure was supported by the observation of amplification effect.

Application of the prolinol-derived lithium amide **11** (Scheme 9)in the asymmetric synthesis of (+)-confertin was successful with the addition of isopropenyllithium to 2-methylcyclopentenone being a key step [37].

The first epoch-making catalytic process was developed using a chiral copper amide in 1990. In the presence of 3 mol % of aminotroponeimine-copper **12** (Scheme 9) butylmagnesium chloride reacted with cyclohexenone to afford the corresponding adduct in 74% ee. A weakly basic nitrogen-copper structure is proposed as the reason for the success [38, 39].

3.3
Chiral Thiocuprates

Chirally modified thiocuprates are used mostly in the catalytic process, probably because of the high affinity of sulfur atom to copper and their good stability. Thiocuprates can be generated from organolithium or Grignard reagents and a copper(I) salt in the presence of the lithium or magnesium thiolate of a chiral thiol.

The reaction of the Grignard reagents is catalyzed by a catalytic amount of chiral copper thiolates **13–15** (Scheme 10) to afford the corresponding adduct in relatively high ee [40, 41, 42].

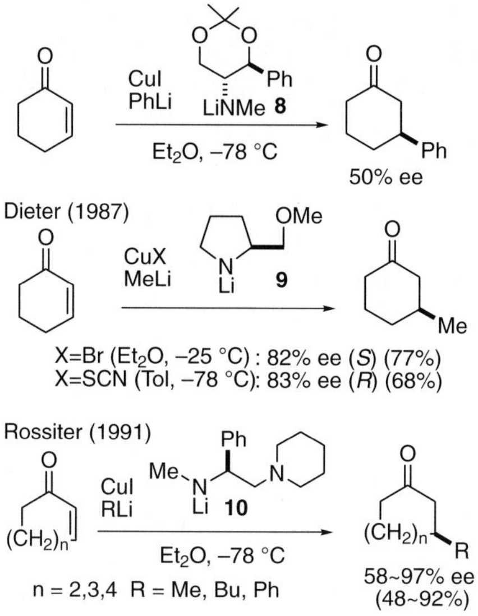

Scheme 8

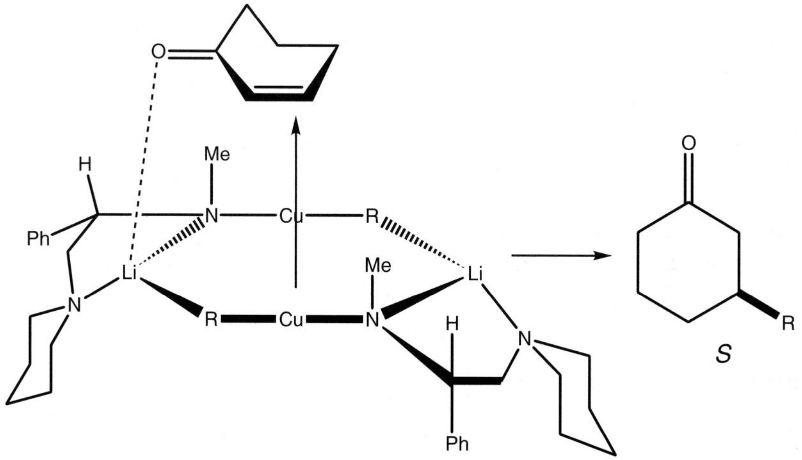

Fig. 2

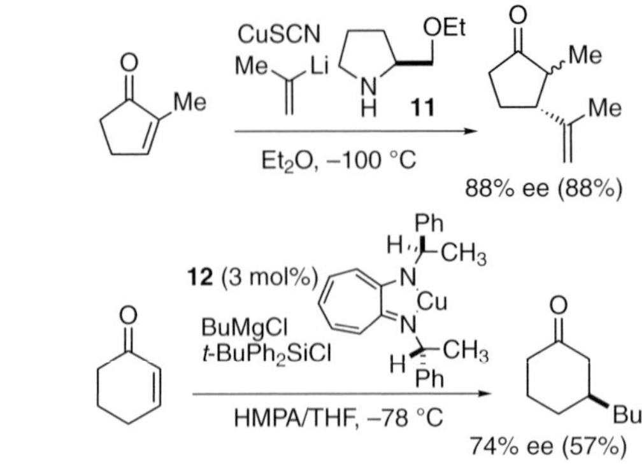

Scheme 9

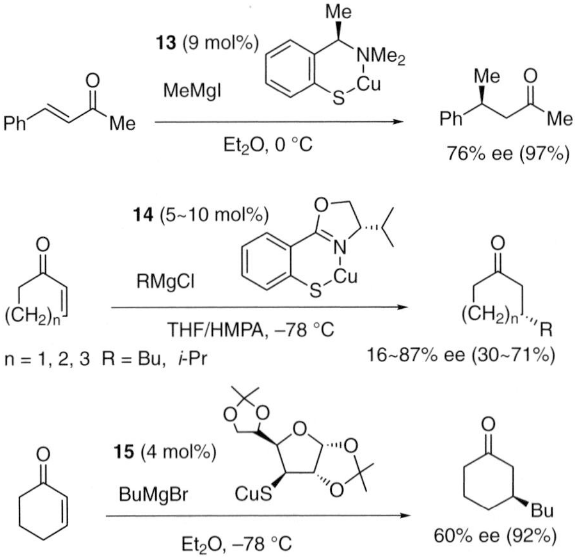

Scheme 10

4
Reaction of Homoorganocopper Reagents Prepared from Organolithium Reagents and Grignard Reagents with External Chiral Ligands

Organocopper reagents, prepared from a copper(I) salt and organometallic species such as organolithium or Grignard reagents, contain two different metals in the cluster. A chiral modification requires a chiral ligand, the heteroatoms of which coordinate to copper and other metal. Kretchmer was the first to use the chiral diamine (–)-sparteine 2 as a ligand for methylcopper in 1972 [43]. However, the reaction with cyclohexenone gave the addition product in only 6% ee.

The breakthrough in the stoichiometric reaction was realized by Leyendecker in 1983 with the use of the hydroxyprolinol-derived sulfide **16** (Scheme 11) with three coordinating sites as shown in Fig. 3 [44, 45]. The reaction of dimethylcopper lithium with chalcone gave the product in 94% ee. In 1991, Alexakis introduced chiral phosphines as ligands **17** (Scheme 11)in the reaction of a medium-order cuprate with cycloalkenones in the presence of lithium bromide to afford the products in 76–95% ee [46]. Unfortunately, the catalytic process with an organolithium reagent was described as being unsuccessful.

The proline-derived bidentate amidophosphines **18–20** (Scheme 12) were developed by us on the basis of the concept of metal-differentiating coordination. The carbonyl oxygen and phosphorus atoms of the ligand selectively coordinate to lithium and copper of organocopper species, which discriminates the reaction face of the complex (Fig. 4). In fact the reaction of dimethylcopperlithium with chalcone gave the adduct in 84% ee [47]. Enantioselectivity was later improved to 90% with more bulky amidophosphine **20** based on the model shown [48, 49]. The metal selective coordination was supported by NMR studies [50]. The reaction with cycloalkenone was also highly efficient to give the adducts in up to 95% ee by the reaction of lithium cyanocuprate in the presence of lithium bromide [51]. However, the catalytic version of the reaction with the lithium cyanocuprate was unsuccessful. On the other hand, magnesium cyanocuprate prepared from the corresponding Grignard reagent was highly effective to afford the products in up to 98% ee. It is noteworthy that the same chiral ligand gave the products with the reversed absolute configuration on replacing lithium by

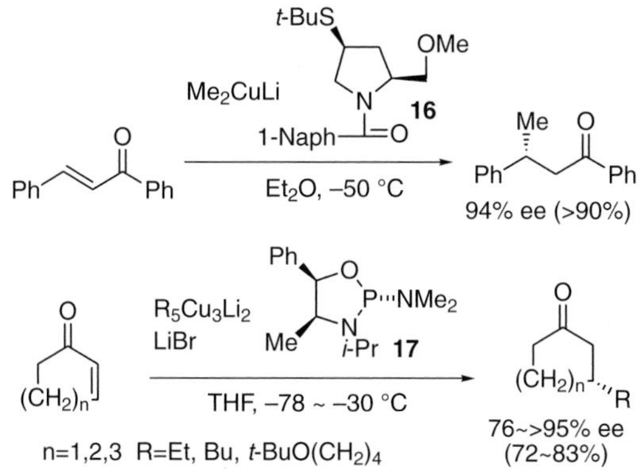

Scheme 11

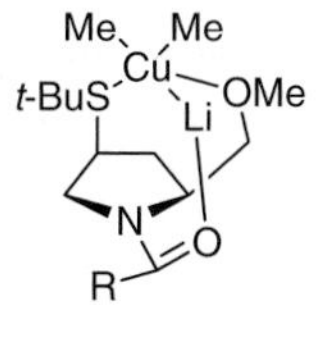

Fig. 3.

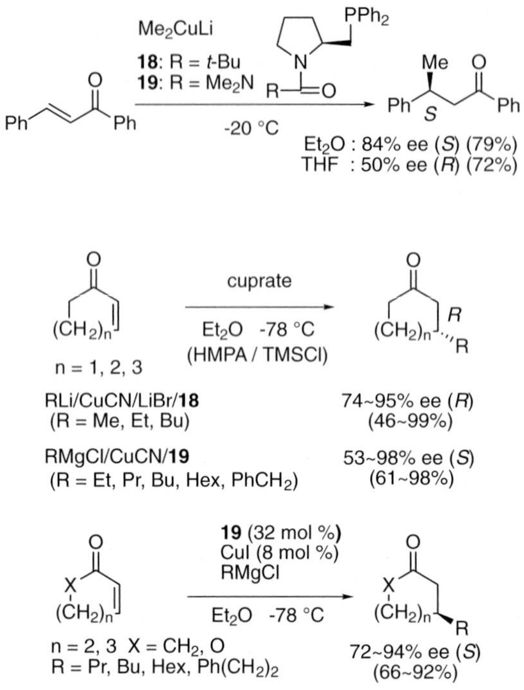

Me₂CuLi

18: R = t-Bu
19: R = Me₂N

Ph—CH=CH—C(=O)—Ph

-20 °C

$$\text{Ph}\overset{Me}{\underset{S}{\frown}}\overset{O}{\frown}\text{Ph}$$

Et₂O : 84% ee (S) (79%)
THF : 50% ee (R) (72%)

cuprate

Et₂O -78 °C
(HMPA / TMSCl)

n = 1, 2, 3

RLi/CuCN/LiBr/**18**   74~95% ee (R)
(R = Me, Et, Bu)   (46~99%)

RMgCl/CuCN/**19**   53~98% ee (S)
(R = Et, Pr, Bu, Hex, PhCH₂) (61~98%)

19 (32 mol %)
CuI (8 mol %)
RMgCl

Et₂O -78 °C

n = 2, 3 X = CH₂, O
R = Pr, Bu, Hex, Ph(CH₂)₂

72~94% ee (S)
(66~92%)

Scheme 12

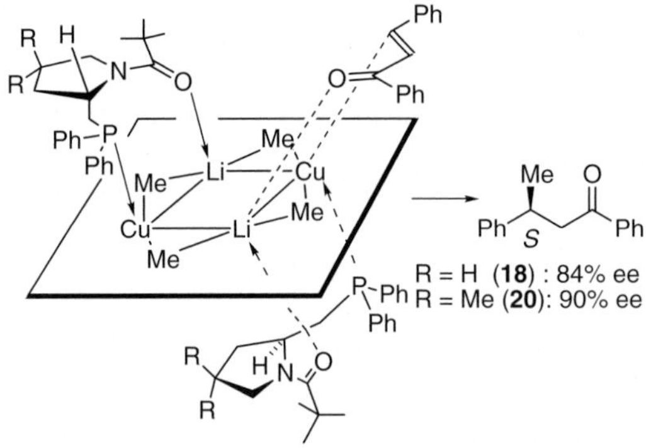

R = H (**18**) : 84% ee
R = Me (**20**): 90% ee

Fig. 4.

magnesium [52]. Catalytic asymmetric conjugate addition was realized by using 8 mol % of copper iodide and 32 mol % of the chiral amidophosphine **19** to afford the products in 72–94% ee [53]. The amidophosphine is recoverable for reuse in high yield.

The chiral ferrocenylphosphine oxazoline **21** (Scheme 13) was also introduced as a chiral ligand for use in catalytic amounts (12 mol %) in the reaction

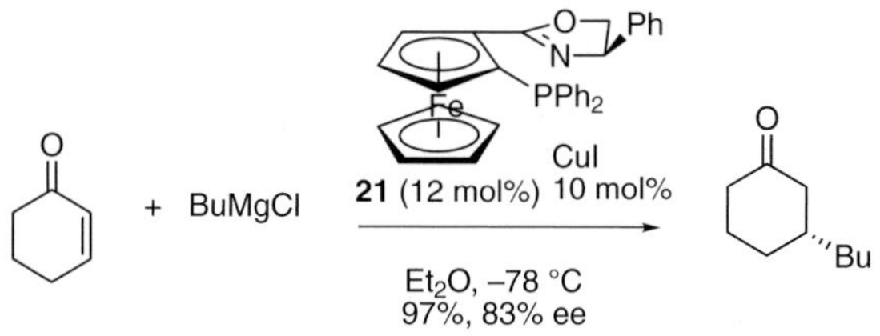

Scheme 13

of Grignard reagents and 10 mol % of copper iodide with cyclohexenone to afford the product in 83% ee [54].

5
Reaction of Organocopper Reagents Prepared from Organozinc Reagents with External Chiral Ligands

Asymmetric conjugate additions of organozinc reagents to enones in the presence of chiral ligands are a rapidly developing and exciting new field in conjugate addition chemistry. Using **17**, Alexakis discovered with the copper-catalyzed asymmetric conjugate addition of diethylzinc to cyclohexenone, giving the product in 32% ee. The binaphthol-based phosphorus amidite **24** was developed by Feringa to afford the product in 63% ee [55]. Later this ligand was greatly improved (**25**) to afford the ethylcyclohexenone adduct in over 98% ee (Scheme 14) [56]. However, high enantioselectivity is limited to cyclohexenone and rather poor selectivities were observed with cyclopentenone (10% ee) and cycloheptenone (53% ee).

The success is attributed to the use of copper triflate as the copper source. The ligand catalyzes not only the reaction of cycloalkenone but also that of acyclic enones. The chiral thiazolidinone **23** (Scheme 14) was also developed as a chiral ligand to afford the product in 63% ee [57].

Since the bisphosphine or monophosphine greatly accelerates the copper-catalyzed reaction [58], a survey of the known diphosphines was carried out and revealed that 0.5% of copper(II) triflate and 0.5% of phosphine are sufficient, although enantioselectivity was at the most 44% [59]. The chiral phosphite ligand based on tartrate **22** (Scheme 14) was also observed to exert the same ligand acceleration [60], but the ee was not so satisfactory [61].

The symmetrical aminophosphine ligand **26** (Scheme 14) was synthesized and examined in the reaction with cyclohexenone in the presence of 5 mol % of copper triflate to afford the product in 55% ee [62].

Based on Noyori's finding that N-monosubstituted sulfonamide and copper(I) catalyze the addition of diorganozinc reagents to cycloalkenone [63], the effect of the chiral sulfonamide **27** (Scheme 14) was examined by Sewald; it was found that catalytic amounts of both sulfonamide and copper(I) are necessarily to effect the reaction, but the ee was at the most 32% [64].

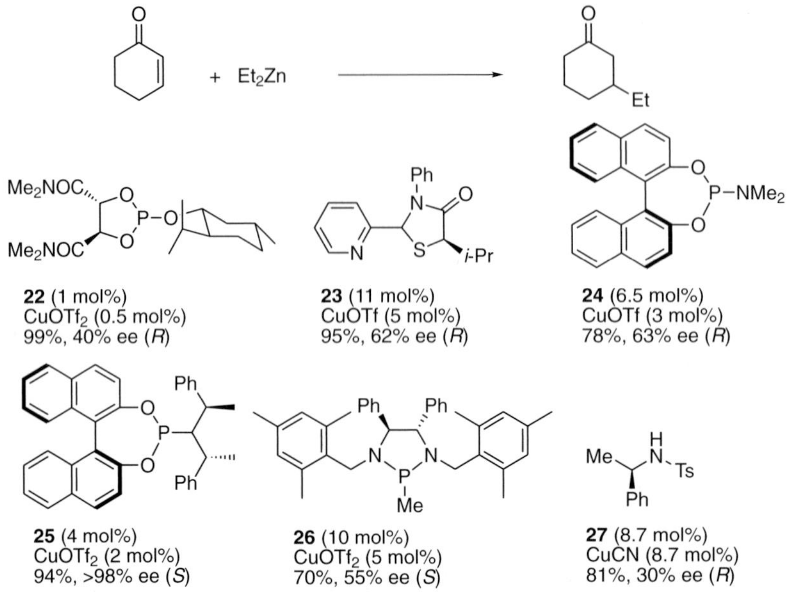
+ Et2Zn
Et
22 (1 mol%)
CuOTf2 (0.5 mol%)
99%, 40% ee (R)
23 (11 mol%)
CuOTf (5 mol%)
95%, 62% ee (R)
24 (6.5 mol%)
CuOTf (3 mol%)
78%, 63% ee (R)
25 (4 mol%)
CuOTf2 (2 mol%)
94%, >98% ee (S)
26 (10 mol%)
CuOTf2 (5 mol%)
70%, 55% ee (S)
27 (8.7 mol%)
CuCN (8.7 mol%)
81%, 30% ee (R)

Scheme 14

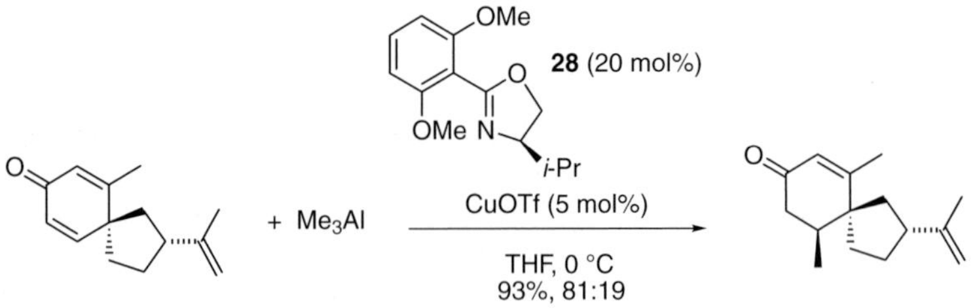
28 (20 mol%)
+ Me3Al
CuOTf (5 mol%)
THF, 0 °C
93%, 81:19

Scheme 15

The reaction of trimethylaluminum [65, 66] with cyclohexadienone was also catalyzed by the combination of the oxazoline ligand **28** and copper(I) triflate to afford relatively high selectivity (Scheme 15) [67]. The process was successfully applied to the asymmetric total synthesis of (–)-solavetivone [68].

6
Reaction of Organometallic Reagents with External Chiral Ligands

The chiral diamine-zinc(II) complex **29** catalyzes the addition of Grignard reagent to cyclohexenone, though the ee was poor (Scheme 16) [69].

A catalytic process was achieved by using a combination of 17 mol % of amino alcohol **30** and nickel acetylacetonate in the reaction of diethylzinc and chal-

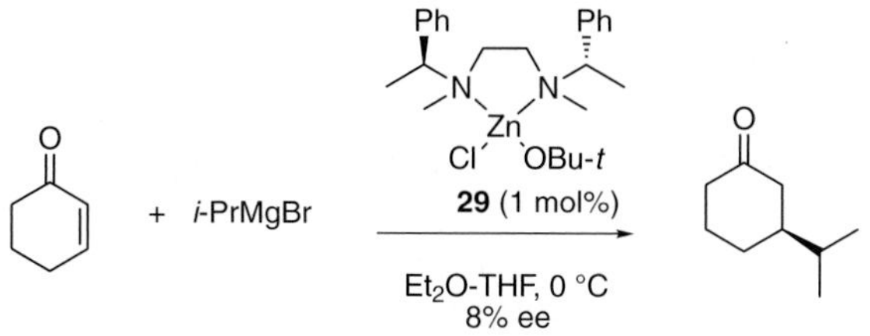

Scheme 16

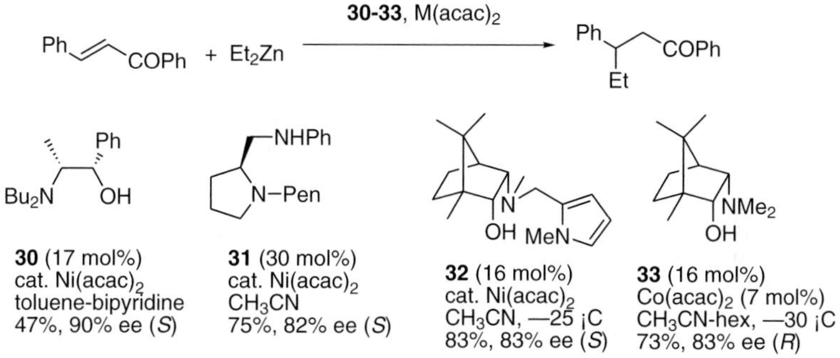

Scheme 17

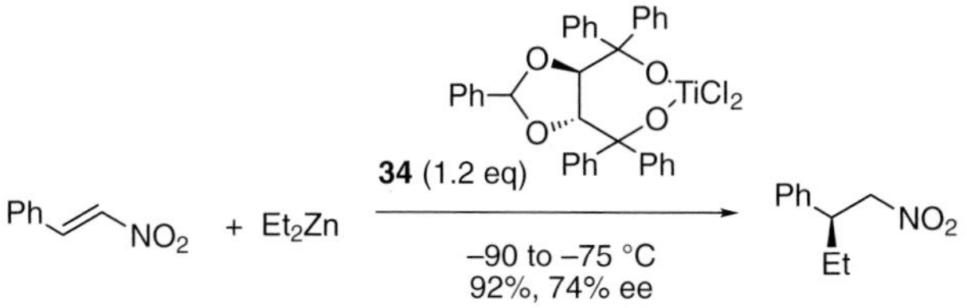

Scheme 18

cone to provide the product in 90% ee (Scheme 17)[70, 71, 72, 73]. The proline derived chiral diamine **31** was also effective to give 82% ee [74].

The camphor-derived tridentate amino alcohol **32** (Scheme 17) also catalyzes the conjugate addition reaction of diethylzinc in the presence of nickel acetylacetonate to afford the product in 83% ee [75]. Similarly the ligand **33**-cobalt acetylacetonate complex catalyzes the reaction to afford the product in 83% ee [76].

The combination of titanium-TADDOL **34** mediates the reaction of diethylzinc with nitroolefins to afford the products in relatively high ees (Scheme 18) [77].

7
Principal Alternatives

β-Substituted carbonyl compounds are readily accessible via conjugate addition of organometallic reagents to α,β-unsaturated carbonyl compounds [7]. Although a stoichiometric amount of chiral auxiliary is necessary, the asymmetric addition to the olefin bonded covalently by a chiral activating group has been well documented to give the adduct with high level of diastereoselectivity [78, 79, 80, 81, 82, 83, 84, 85, 86]. Removal of the chiral auxiliary provides the chiral β-substituted carbonyl compounds.

Catalytic asymmetric hydrogenation is a well-established method for the conversion of the stereochemically defined α,β-unsaturated carbonyl compounds to the chiral β-substituted carbonyl compounds in high enantioselectivity [87, 88].

Rhodium(I)-catalyzed isomerization of allylic amines is also good route to β-substituted aldehydes [89].

References

1. Perlmutter P (1992) Conjugate Addition Reactions in Organic Synthesis, Tetrahedron Organic Chemistry Series, vol 9 Pergamon Press, Oxford
2. Tomioka K, Koga K (1983) Asymmetric Synthesis, vol 2, Academic Press, New York, chap 7; Posner GH, ibid. chap 8; Lutomoski KA, Meyers AI (1984) Asymmetric Synthesis, vol 3, Academic Press, New York chap 3
3. Oppolzer W (1987) Tetrahedron 43:1969
4. Tomioka K (1990) Synthesis 541
5. Rossiter BE, Swingle NM (1992) Chem Rev 92:771
6. Noyori R (1994) Asymmetric Catalysis in Organic Synthesis, John Wiley, New York
7. Seyden-Penne J (1995) Chiral Auxiliaries and Ligands in Asymmetric Synthesis. John Wiley, New York
8. Kanai M, Nakagawa Y, Tomioka K (1996) J Syn Org Chem Jpn 54:474
9. Krause N (1998) Angew Chem Int Ed Engl 37:283
10. Hayashi T, Tomioka K, Yonemitsu O (eds) (1998) Asymmetric Synthesis, Graphical Abstracts and Experimental Methods, Kodansya and Gordon and Breach Science Publishers, Tokyo
11. Tomioka K, Sudani M, Shinmi Y, Koga K (1985) Chemistry Lett 329
12. Tomioka K, Shindo M, Koga K (1989) J Am Chem Soc 111:8266
13. Inoue I, Shindo M, Koga K, Kanai M, Tomioka K (1995) Tetrahedron Asymmetry 6:2527
14. Fujieda H, Kanai M, Kambara T, Iida A, Tomioka K (1997) J Am Chem Soc 119:2060
15. Tomioka K, Okamoto T, Kanai M, Yamataka H (1994) Tetrahedron Lett 35:1891
16. Denmark SE, Nicaise O J-C (1996) J Chem Soc Chem Commun 999
17. Inoue I, Shindo M, Koga K, Tomioka K (1994) Tetrahedron 50:4429 and references cited therein
18. Shindo M, Koga K, Tomioka K (1992) J Am Chem Soc 114:8732
19. Tomioka K, Shindo M, Koga K (1993) Tetrahedron Lett 34:681
20. Asano Y, Iida A, Tomioka K (1997) Tetrahedron Lett 38:8973
21. Asano Y, Iida A, Tomioka K (1998) Chem Pharm Bull 46:184
22. Amurrio D, Khan K, Kundig EP (1996) J Org Chem 61:2258
23. Park YS, Weisenburger GA, Beak P (1997) J Am Chem Soc 119:10537
24. Zweig JS, Luche JL, Barreiro E, Crabbé P (1975) Tetrahedron Lett 2355
25. Huché M, Berlan J, Pourcelot G, Cresson P (1981) Tetrahedron Lett 22:1329
26. Mukaiyama T, Imamoto T (1980) Chem Lett 45

27. Leyendecker F, Laucher D (1983) Tetrahedron Lett 24:3517
28. Leyendecker F, Laucher D (1985) Nouv J Chim 9:13
29. Corey EJ, Naef R, Hannon FJ (1986) J Am Chem Soc 108:7114
30. Dieter RK, Lagu B, Deo N, Dieter JW (1990) Tetrahedron Lett 31:4105
31. Tanaka K, Ushio H, Suzuki H (1990) J Chem Soc Chem Commun 795
32. Tanaka K, Ushio H, Kawabata Y, Suzuki H (1991) J Chem Soc Perkin Trans I 1445
33. Bertz SH, Dabbagh G, Sundararajan G (1986) J Org Chem 51:4953
34. Dieter RK, Tokles M (1987) J Am Chem Soc 109:2040
35. Swingle NM, Reddy KV, Rossiter BE (1994) Tetrahedron 50:4455
36. Miano G, Rossiter BE (1995) J Org Chem 60:8424
37. Quinkert G, Müller T, Königer A, Schultheis O, Sickenberger B, Düner G (1992) Tetrahedron Lett 33:3469
38. Villacorta GM, Rao CP, Lippard SJ (1988) J Am Chem Soc 110:3175
39. Ahn K-H, Klassen RB, Lippard SJ (1990) Organometallics 9:3178
40. van Klaveran M, Lambert F, Eijkelkamp DJFM, Grove DM, van Koten G (1994) Tetrahedron Lett 35:6135
41. Spescha M, Rihs G (1993) Helv Chim Acta 76:1219
42. Zhou Q-L, Pfaltz A (1994) Tetrahedron 50:4467 and references cited therein
43. Kretchmer RA (1972) J Org Chem 37:2744
44. Leyendecker F, Laucher D (1983) Tetrahedron Lett 24:3517
45. Leyendecker F, Laucher D (1985) Nouv J Chim 9:13
46. Alexakis A, Mutti S, Normant JF (1991) J Am Chem Soc 113:6332
47. Kanai M, Koga K, Tomioka K (1992) Tetrahedron Lett 33:7193
48. Nakagawa Y, Kanai M, Nagaoka Y, Tomioka K (1996) Tetrahedron Lett 37:7805
49. Nakagawa Y, Kanai M, Nagaoka Y, Tomioka K (1998) Tetrahedron in press
50. Kanai M, Koga K, Tomioka K (1993) J Chem Soc Chem Commun 1248
51. Kanai M, Tomioka K (1994) Tetrahedron Lett 35:895
52. Kanai M, Tomioka K (1995) Tetrahedron Lett 36:4273
53. Kanai M, Tomioka K (1995) Tetrahedron Lett 36:4275
54. Stangeland EL, Sammakia T (1997) Tetrahedron 53:16503
55. de Vries AHM, Meetsma A, Feringa BL (1996) Angew Chem Int Ed Engl 35:2374
56. Feringa BL, Pineschi M, Arnold LA, Imbos R, de Vries AHM (1997) Angew Chem Int Ed Engl 36:2620
57. de Vries AHM, Hof RP, Staal D, Kellogg RM, Feringa BL (1997) Tetrahedron Asymmetry 8:1539
58. Alexakis A, Vastra J, Mangeney P (1997) Tetrahedron Asymmetry 8:7745
59. Alexakis A, Burton J, Vastra J, Mangeney P (1997) Tetrahedron Asymmetry 8:3987
60. Berrisford DJ, Bolm C, Sharpless KB (1995) Angew Chem Int Ed Engl 34:1050
61. Alexakis A, Vastra J, Burton J, Mangeney P (1997) Tetrahedron Asymmetry 8:3193
62. Mori T, Tomioka K, unpublished result
63. Kitamura M, Miki T, Nakano K, Noyori R (1996) Tetrahedron Lett 37:5141
64. Wendish V, Sewald N (1997) Tetrahedron Asymmetry 8:1253
65. Westermann J, Nickisch K (1993) Angew Chem Int Ed Engl 32:1368
66. Kabbara J, Fleming S, Nickisch K, Neh H, Westermann J (1995) Tetrahedron 51:743
67. Takemoto Y, Kuraoka S, Hamaue N, Aoe K, Hiramatsu H, Iwata C (1996) Tetrahedron 52:14177
68. Takemoto Y, Kuraoka Ohra T, Yonetoku Y, Iwata C (1996) J Chem Soc Chem Commun 1655
69. Jansen JFGA, Feringa BL (1989) J Chem Soc Chem Commun 741
70. Soai K, Hayasaka T, Ugajin S, Yokoyama S (1989) J Chem Soc Chem Commun 516
71. Soai K, Hayasaka T, Ugajin S, Yokoyama S (1988) Chem Lett 1571
72. Soai K, Yokoyama S, Hayasaka T, Ebihara K (1988) J Org Chem 53:4149
73. Soai K, Okudo M, Okamoto M (1991) Tetrahedron Lett 32:95
74. Asami M, Usui K, Higuchi S, Inoue S (1994) Chem. Lett 297
75. de Vries AHM, Imbos R, Feringa BL (1997) Tetrahedron Asymmetry 8:1467
76. de Vries AHM, Feringa BL (1997) Tetrahedron Asymmetry 8:1377

77. Schäfer H, Seebach D (1995) Tetrahedron 51:2305
78. Oppolzer W, Poli G, Kingma AJ, Starkemann C, Bernardinelli G (1987) Helv Chim Acta 70:2201
79. Rück K, Kunz H (1991) Angew Chem Int Ed Engl 30:694
80. Fang C-L, Ogawa T, Suemune H, Sakai K (1991) Tetrahedron Asymmetry 2:389
81. Stephan E, Rocher R, Aubouet J, Pourcelot G, Cresson P (1993) Tetrahedron Asymmetry 5:41
82. Wu M-J, Yeh J-Y (1994) Tetrahedron 50:1073
83. Tsuge H, Takumi K, Nagai T, Okano T, Eguchi S, Kimoto H (1997) Tetrahedron 53:823
84. Tomioka K, Suenaga T, Koga K (1986) Tetrahedron Lett 27:369
85. Fleming I, Lawrence NJ (1990) Tetrahedron Lett 31:3645
86. Didiuk MT, Johannes CW, Morken JP, Hoveyda AH (1995) J Am Chem Soc 117:7097
87. Ohta T, Takaya H, Kitamura M, Nagai K, Noyori R (1987) J Org Chem 52:3174
88. Saburi M, Shao L, Sakurai T, Uchida Y (1992) Tetrahedron Lett 33:7877
89. Noyori R, Takaya H (1990) Acc Chem Res 23:345

Chapter 31.2
Conjugate Addition of Stabilized Carbanions

Masahiko Yamaguchi

Graduate School of Pharmaceutical Sciences, Tohoku University, Aoba, Sendai 980-8578, Japan
e-mail: yama@mail.pharm.tohoku.ac.jp

Keywords: Michael addition, Amine, Phase transfer catalyst, Alkoxide, Crown ether, Transition metal complex, Lewis acid

1
Introduction

Conjugate addition reactions are some of the most fundamental C-C bond-forming reactions in organic synthesis, and their asymmetric versions have been studied extensively [1]. Treated in this chapter is the catalytic conjugate addition of stabilized carbanions, especially enolate derivatives, for which the term "Michael addition and/or reaction" is used. The asymmetric Michael reactions can be categorized into two groups (Fig. 1):
(i)   enantioselective addition of prochiral donor (enolate) to acceptor; and
(ii) Enantioselective addition of donor (enolate) to prochiral acceptor [2].

The former reaction discriminates the enantiofaces of the donor, and the asymmetric center is formed on the donor carbon atom. The latter reaction proceeds *via* enantioface discrimination of the Michael acceptor generating a chiral carbon center on the acceptor. Although both reactions are known, their mecha-

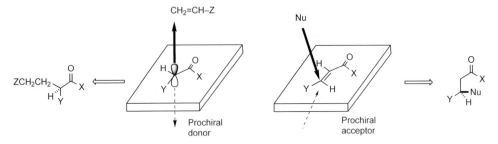

Fig. 1

nisms for asymmetric induction appear to differ. Distinct catalyst systems are generally used for the two groups of the reaction.

The mechanism of the former reactions, at least in principle, is fairly easy to understand. The anionic donor (enolate) interacts with chiral cationic species such as ammonium salts or metal cations, which differentiate the enantiofaces of the donor. In contrast, the latter reaction appears to be more complicated since the chiral enolate complex recognizes the enantiofaces of the Michael acceptor. The following discussions may be helpful in some cases for the better understanding of the prochiral acceptor reaction [2, 3].

The first aspect is the concept of the enantiofaces of the Michael acceptors. Fig. 2 shows how the _re_- and _si_-enantiofaces of the Michael acceptors can be defined. Here, the first priority is always given to the C= group irrespective of the other substituents. This definition is relatively insensitive to changes in the substituents compared to the conventional _re/si_-face definitions. Since the Michael acceptors possess two prochiral centers, the α-carbon and β-carbon, an enantioface can be described as, for example, _re_(α), _si_(β), _si_(α)/_si_(β), or _re_(α)/_si_(β). The enantiofaces _si_(α) or _re_(β) indicate the _si_-face with regard to the α-carbon atom and the _re_-face with regard to the β-carbon atom, respectively.

The prochiral acceptor reactions can then be classified into the α-enantioface-discriminating reaction and β-enantioface-discriminating reaction (Fig. 3). The absolute configurations of the adducts derived from (_E_)- and (_Z_)-acceptors provide the criteria. If both isomers give _si_(α) attack products, the α-enantiofaces are discriminated. The nucleophile recognizes the chiral environments in the vicinity of the acceptor α-carbon atom rather than the β-carbon. Asymmetric Michael addition reactions utilizing chiral auxiliaries show α-enantioface discrimination [4]. Since the amide or ester auxiliaries are located in the vicinity of the α-carbon atom, the asymmetrical reaction reasonably proceeds via this mechanism. β-Enantioface-discriminating reactions would give, for example, _si_(β) attack products from (_E_)- and (_Z_)-acceptors. Reaction of the _re_(β) attack is known for a Grignard addition reaction [5].

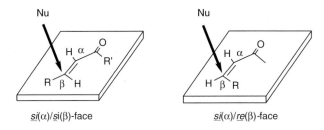

Fig. 2

α-Enantioface discriminating mechanism   β-Enantioface discriminating mechanism

Fig. 3

2
Amine Catalysts

Early studies on the catalytic asymmetric Michael reactions were conducted with readily available amines of natural origin as listed in Fig. 4. (–)-Quinine (**1**) and (+)-quinidine (**2**) are pseudo-enantiomeric concerning the aza[2.2.2]bicyclooctane and quinoline moiety, and generally give the antipodes. The same situation holds for (+)-cinchonine (**3**) and (–)-cinchonidine (**4**) which are demethoxylated derivatives of **1** and **2**.

The catalytic asymmetric Michael addition using chiral amine was first reported by Långström and Bergson [6]. Treatment of 2-methoxycarbonyl-1-indanone (**6**) and acrolein with 0.03 mol % of partially resolved (R)-2-(hydroxymethyl)quinuclidine (**5**, 57% ee) in benzene at room temperature gave optically active adduct, $[\alpha]_{546}^{21.0}$ +8.83° (c 6.53, CCl$_4$). It clearly indicated that the cluster formed from the base and the enolate reacted with the acceptor. The enantiomeric excess and the absolute configuration of this compound, however, have not yet been determined.

Wynberg studied the catalysis by *Cinchona* alkaloids [7, 8]. Use of **1** in the addition of **6** to 3-buten-2-one (**7**) gave the optically active adduct (S)-**8** in 76% ee (Scheme 1). As shown below, this prochiral donor reaction has become a standard to evaluate the efficiency of various catalysts. Several features of the reaction deserve comment:

(i)   The configuration at C(8) and C(9) of the catalyst is mainly responsible for the stereochemical outcome. The amine **1** with the (8S,9R)-configuration

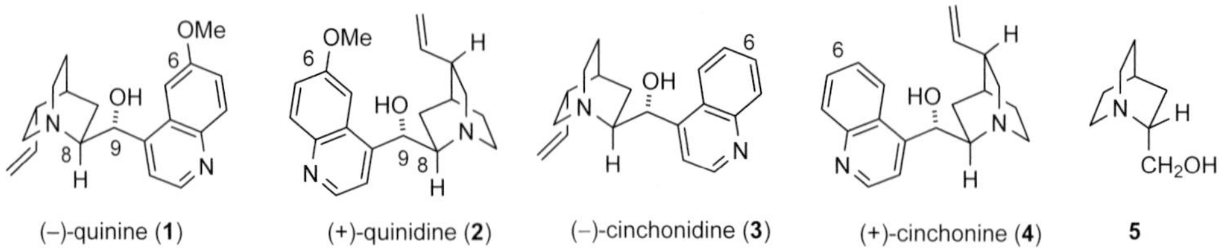

(–)-quinine (**1**) (+)-quinidine (**2**) (–)-cinchonidine (**3**) (+)-cinchonine (**4**) **5**

Fig. 4

99%, 76% ee

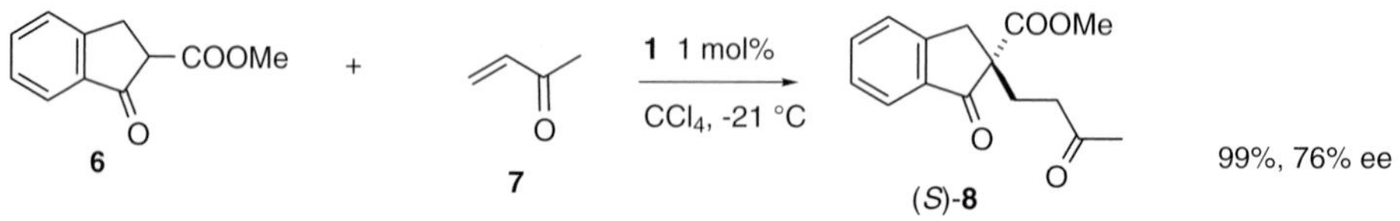

Scheme 1

and **2** with the (8*R*,9*S*)-configuration gave the antipodes with comparable optical purities.

(ii) The hydroxy group at the C(9) position probably participates in the transition state *via* hydrogen bonding. The reaction rate and the optical purity decreased when the corresponding acetate was employed. Addition of even a small amount of ethanol reduced the stereoselectivity.

(iii) Use of non-polar solvents such as toluene and CCl_4 gave better results than polar or protic solvents.

(iv) The activation by the amines is restricted to relatively acidic Michael donors such as **6**, and the less acidic 2-alkoxycarboxycyclohexanone was inert under these conditions.

The results of Långström were considerably improved by this alkaloid methodology as judged from the optical rotation of the same adduct, $[\alpha]_{546}^{rt}$ –61.1° (*c* 3.46, CCl_4).

The *Cinchona* alkaloids were used in other reactions of acidic prochiral donors (Scheme 2). Optically active spiro-compounds were synthesized by the double Michael addition of 1,3-cyclohexanedione to dibenzylideneacetone [9]. Trost reported a novel intramolecular addition of acetylenic ester **9** giving bicyclic **10** in 30% ee [10]. A series of 2-nitrocycloalkanones were added to **7** in the presence of **4** (100 mol %) [11]. Depending on the ring size, the enantiomeric excess and the absolute configuration differed. For example, the eight-membered ring compound **11** gave (*S*)-**12** (60% ee), while the twelve-membered ring compound gave the (*R*)-adduct (25% ee). Example of the kinetic resolution have also been reported [12].

In contrast to the above prochiral donor reactions, only a limited number of prochiral acceptor reactions have been reported (Scheme 3) [13]. Nitromethane did not add to chalcone (**13**) in the presence of **1** in aprotic solvents. Although

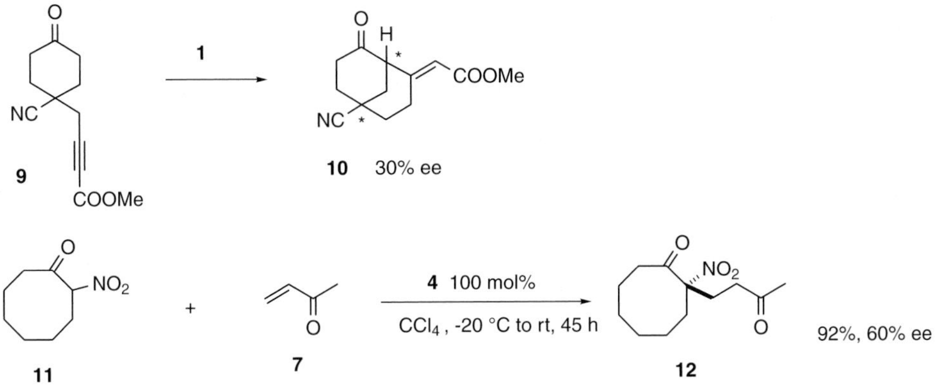

Scheme 2

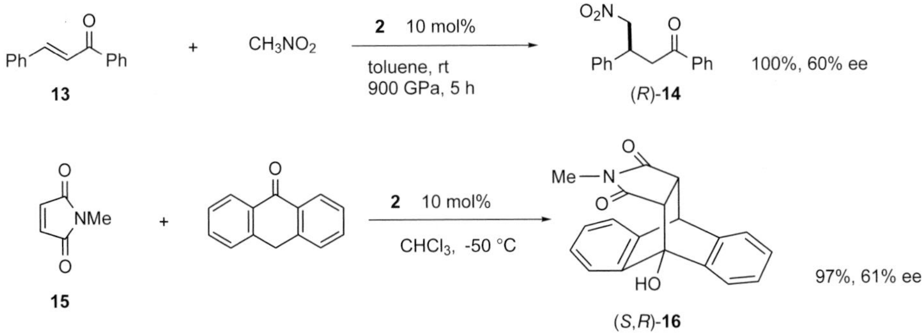

Scheme 3

the reaction took place in methanol, the product was racemic. Matsumoto found that nitromethane added to **13** under a high pressure of 900 GPa (Scheme 3) [14, 15]. The amine **1** and **3** gave (S)-**14**, while **2** and **4** gave (R)-**14**. The enantiomeric excesses were relatively insensitive to pressures between 400 and 900 GPa when **2** was used as the catalyst. Brucine and strychnine which lack the β-hydroxy amine moiety showed no asymmetric induction.

Kagan studied the base-catalyzed cycloaddition reaction of anthrone and N-methylmaleimide (**15**) [16]. In contrast to the above reactions, **2** and **3** gave (S,R)-**16**, while **1** and **4** gave (R,S)-**16**. The oxygen functionality at C(6) of the catalysts might be participating in the asymmetric induction. Besides the alkaloids, (S)-prolinol also gave (R,S)-**16** in 47% ee.

Polymer-supported chiral amine reagents represent an attractive extension of this methodology, since these catalysts can readily be recovered (Fig. 5). Succinated polystyrene-divinylbenzene (**17**) attached to **1** promoted the addition of **6** to **7** at a slower rate than the homogeneous reaction [17]. Although the absolute configuration of the adduct (S)-**8** was identical, the enantiomeric excess was low-

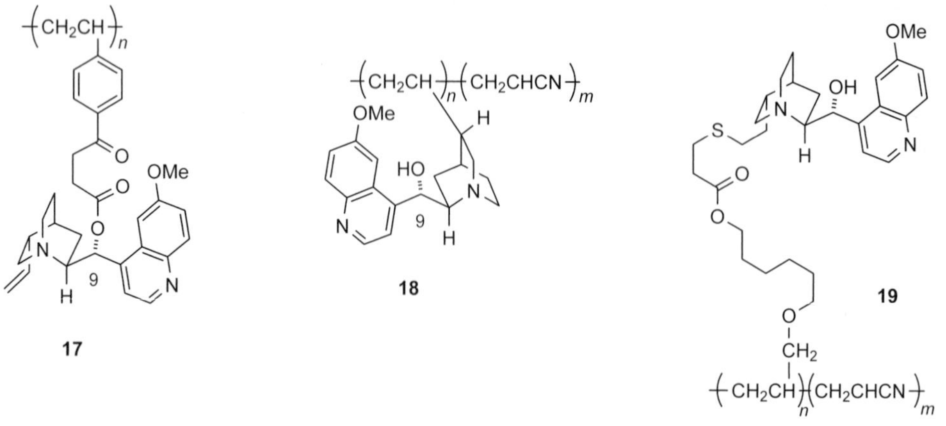

Fig. 5

er, a maximum of 11% ee. Later, copolymers were prepared by Kobayashi from the alkaloids and acrylonitrile by olefin polymerization, which exhibited much improved enantioselectivity [18, 19, 20]. The catalyst **18** gave (*R*)-**8** in 42% ee. This could be ascribed to the presence of the free C(9) hydroxy group. The catalysts were recovered and reused. Introduction of spacers between the polymer backbone and the chiral amine, as shown in **19**, further enhanced the stereoselectivity, up to 65% ee [21]. The value was close to that obtained by the homogeneous reaction.

In order to broaden the scope of the amine-catalyzed Michael addition, Yamaguchi examined the system of amine and alkali metal salt [2]. Although amine did not promote the addition of malonate to enones, the $LiClO_4$-Et_3N catalyst turned out to be effective. Optically active amines, however, gave racemic adducts. As an extension, the (*S*)-proline rubidium salt, (*S*)-**21**, was developed, which possessed a cation and an amine moiety in the same molecule [2, 22]. The catalyst (*S*)-**21** in chloroform promoted the asymmetric addition of malonate to a wide range of enones and enals as exemplified by the reaction of 2-cycloheptenone (**20**) giving (*R*)-**22** (Scheme 4). The role of the counter cation was important and (*S*)-**24** was obtained from 2-cyclohexenone (**23**) when (*S*)-proline lithium and tetrabutylammonium salts were employed. Taguchi used (*S*)-pyrrolidylalkylammonium hydroxide derived from (*S*)-proline and obtained (*S*)-**24** [23]. Changes in the side chain and cation structure dramatically varied the reaction course. Higher asymmetric induction was attained by reacting di(*t*-butyl) malonate in the presence of (*S*)-**21** and CsF, and (*S*)-**26** was obtained in 88% ee from acyclic (*E*)-enone **25**.

The amino acid salt (*S*)-**21** (5–10 mol %) catalyzed the asymmetric Michael addition of nitroalkanes (Scheme 5) [24, 25]. Substituted nitromethanes exhibited higher enantiomeric excesses, and addition of 2-nitropropane to cycloheptenone **20** gave (*R*)-**27** in 73% ee. In case of primary nitroalkane, two diastereomers of **28** were formed in comparable amounts, both of which possessed the

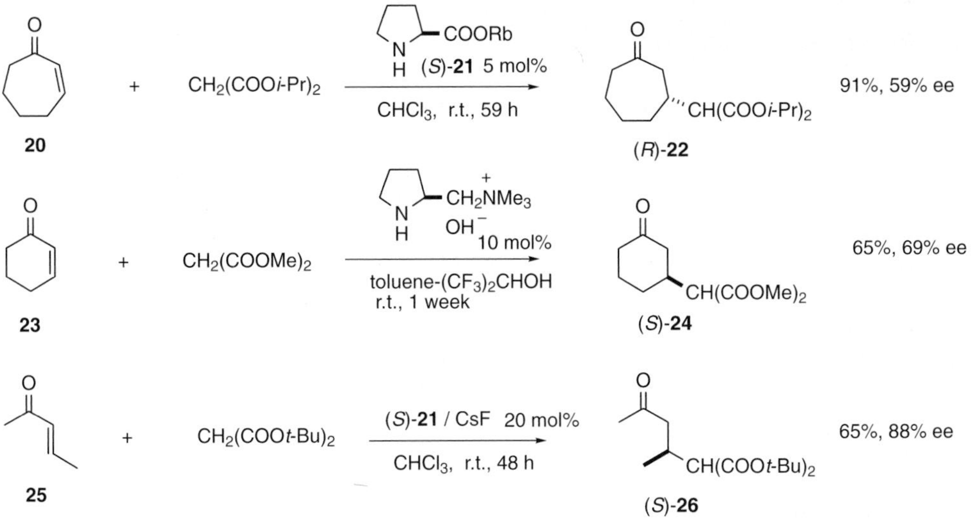

Scheme 4

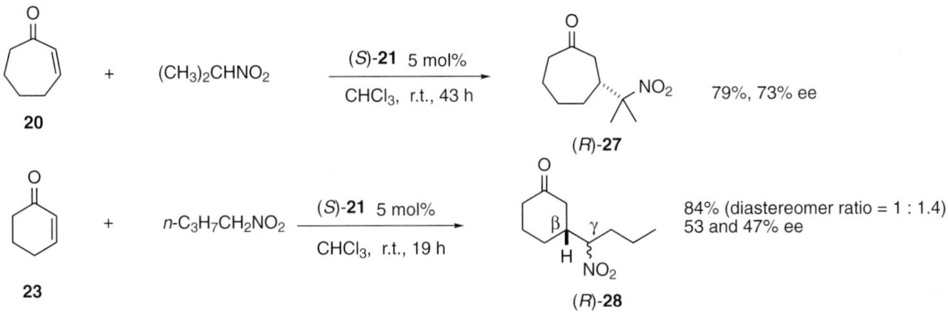

Scheme 5

(R)-configuration at the β-carbon atom. This implied that (S)-**21** controlled the stereochemistry of the C–C bond forming β-carbon atom, and marginally affected that of the γ-carbon atom.

These malonate and nitroalkane reactions gave the adducts with the predicted absolute configurations: (R)-adducts were obtained from cyclic (Z)-enones and (S)-adducts from acyclic (E)-enones when (S)-**21** was employed. The stereochemical outcome can be summarized as $\underline{si}(\alpha)$-attack. The involvement of the α-enantioface-discriminating mechanism suggests that the chiral catalysts are located in the vicinity of the enone carbonyl group at the transition state. The reaction of the primary nitroalkane mentioned above also supports this explanation.

3
Phase Transfer Catalysts

Chiral phase transfer catalysis involves a similar concept as chiral amine catalysis in the sense that the enolate forms an ion pair with chiral ammonium cation.
The former, however, has an advantage of being a stronger base compared to the
latter [12, 26, 27, 28]. Although the addition of nitromethane to **13** did not take
place with **1**, it proceeded with ammonium fluoride generated *in situ* by treating
N-benzyl-*N*-methylephedrinium bromide, (1*S*,2*S*)-**29**, with excess KF (Scheme 6).
Nitroketone (*S*)-**14** was obtained in 26% ee. When nitromethane was added to
(*E*)- and (*Z*)-2-propenylsulfone (**30**), the antipodes **31** were obtained. This phase
transfer reaction appears to proceed *via* the α-enantioface-discriminating mechanism, and the enolate-ammonium cation ion pair probably is interacting with
the sulfonyl group at the transition state. Tetraalkylammonium hydroxide was
also used for the asymmetric addition of **32** to **7** giving (*S*)-**33** [8, 29, 30].

Conn reported that *N*-(*p*-trifluoromethylbenzyl)cinchoninium bromide (**35**)
catalyzed the addition of 2-propylindanone **34** to **7** exhibiting (*S*)-selectivity in
80% ee (Scheme 7) [31]. The antipode (*R*)-**36** was obtained in 52% ee when *N*-
benzylcinchonidium bromide was used. Although the salt is not soluble in toluene, a homogeneous toluene solution obtained by partition between the organic
solvent and the aqueous base exhibited the activity. Related reactions were conducted with **37**, and asymmetric inductions of higher than 80% ee were attained
in the synthesis of (*R*)-**38** [32]. The *p*-trifluoromethylbenzyl group was considered to participate in a π-π interaction with the substrate aromatic ring. Hydro-

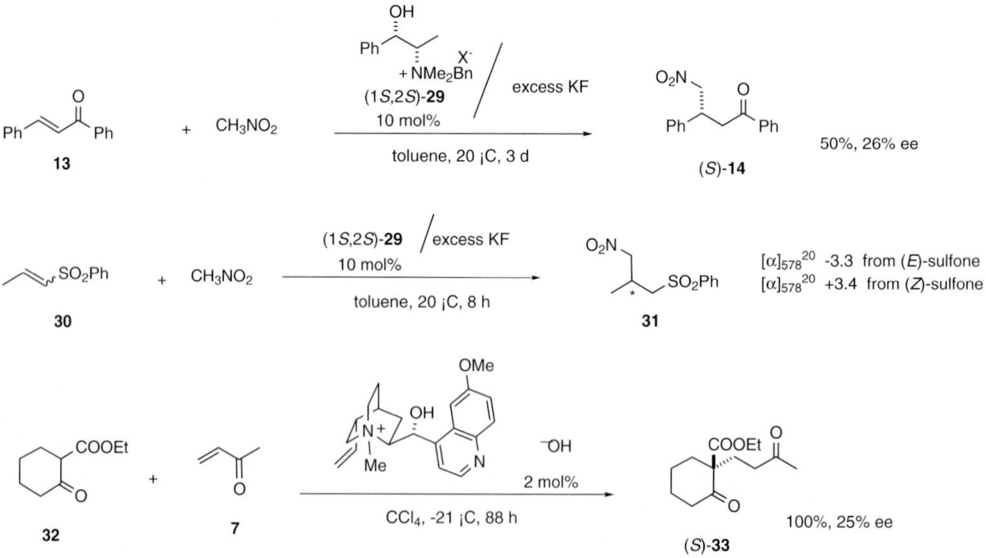

Scheme 6

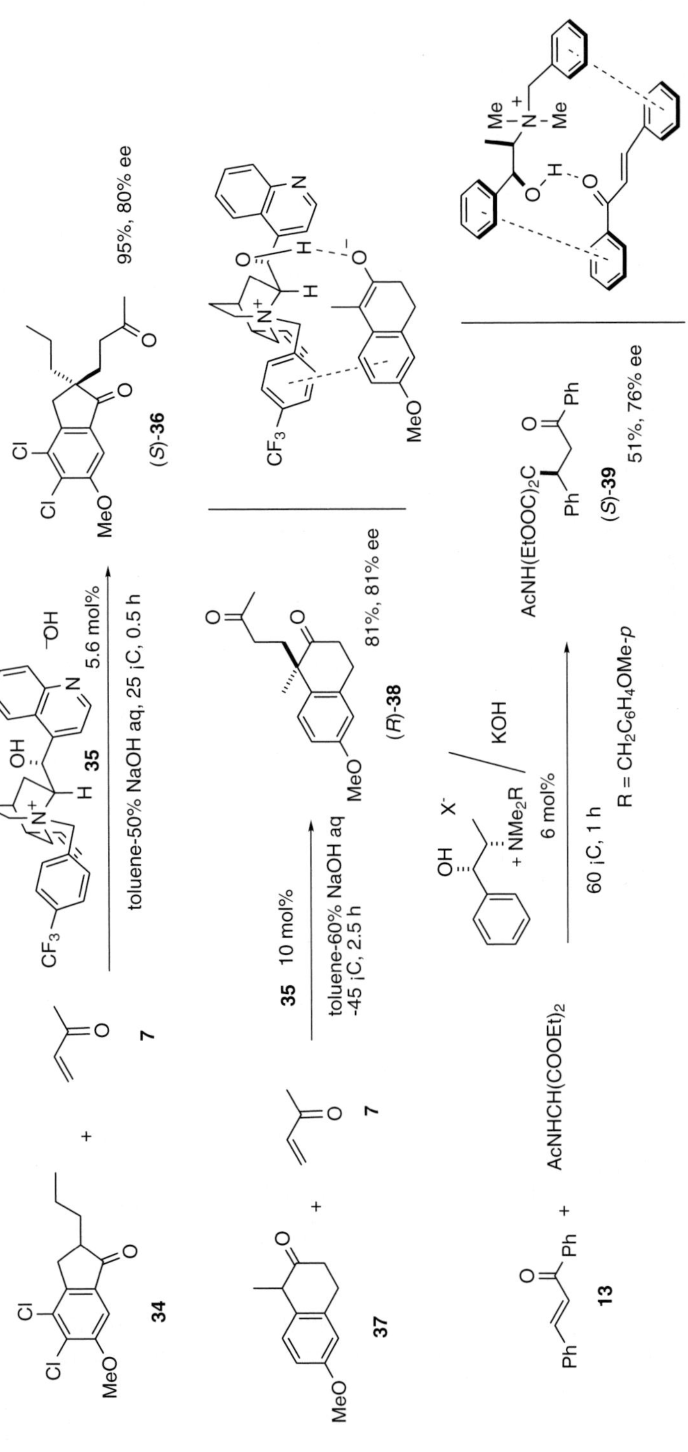

95%, 80% ee

(S)-36

CF₃

N

MeO

MeO

Cl

Cl

MeO

OH

OH

N

CF₃

35 5.6 mol%

toluene–50% NaOH aq, 25 ¡C, 0.5 h

7

+

34

Cl

Cl

MeO

(R)-38

81%, 81% ee

MeO

O

O

35 10 mol%

toluene–60% NaOH aq
–45 ¡C, 2.5 h

7

+

37

MeO

Me
N–Me

O–H···O

(S)-39

51%, 76% ee

AcNH(EtOOC)₂C

Ph

Ph

O

OH X⁻

+ NMe₂R / KOH

6 mol%

60 ¡C, 1 h

R = CH₂C₆H₄OMe-p

AcNHCH(COOEt)₂ +

Ph

O

Ph

13

Scheme 7

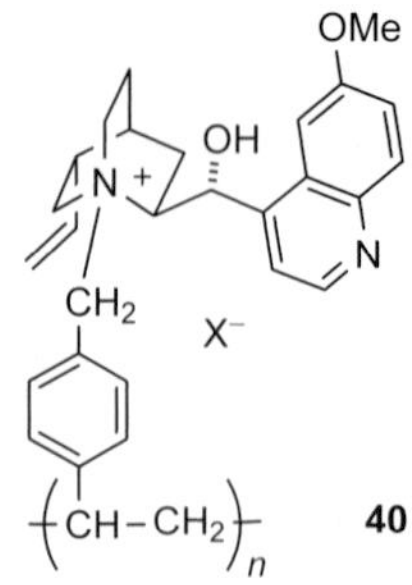

Fig. 6

gen bonding between the catalyst hydroxy group and enolate oxygen was also shown to be critical.

Solid-liquid phase transfer without solvent was reported for a prochiral acceptor reaction. In the presence of N-(p-methoxyphenylmethyl)ephedrinium salt, aminomalonate underwent addition to **13** giving (S)-**39** in 76% ee [33, 34, 35]. The selectivity was higher in the absence of solvent than in toluene or chloroform. Introduction of the electron-donating group at the N-benzyl arene moiety enhanced the selectivity. A π-π interaction between **13** and the aromatic ring of the catalyst was suggested, since the enantiomeric excesses correlated with the Hammett's factor.

The cationic polymer **40**, prepared from chloromethylated polystyrene and **1**, was subjected to ion-exchange to give the hydroxide or fluoride derivatives, and used in the asymmetric addition of **6** to **7** giving (S)-**8** in 27% ee (Fig. 6) [36].

4
Alkoxide and Phenoxide Catalysts

In 1953, chemists in Russia reported the asymmetric induction in the Michael addition of cyclohexanone and 2-methylcyclohexanone to acrylonitrile in the presence of optically active quartz coated with EtOLi, EtOK, or EtONa [37]. Maximum rotations of 0.07 and 0.157 were obtained, respectively.

Chiral metal alkoxides are apparently attractive candidates for the asymmetric catalyst since they are readily available. A very low level of asymmetric induction, however, was observed when lithium (S)-1-phenylethoxide was used in the reaction of phenylacetate **41** and acrylate **42**. Koga showed that a stoichiometric amount of the lithium alkoxide derived from an amino alcohol promoted the asymmetric addition giving (S)-**43** in 84% ee [38]. Use of sodium or potassium salts resulted in racemic **43**. A catalytic reaction (10 mol %) still exhibited a selectivity of 41% ee (Scheme 8).

As indicated by Miyano, alkali metal phenoxides derived from (R)-1,1'-bi-2-naphthol (BINOL) possessing an oligoether moiety promoted the asymmetric addition of ketoesters to **7** [39]. Simple BINOL or its monomethyl derivative gave the racemic product, and the appropriate length of the polyoxyethylene chain

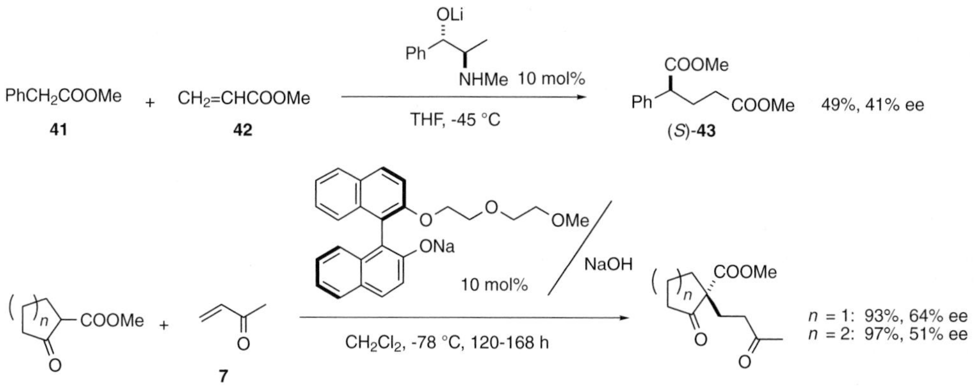

Scheme 8

was required. Sodium and potassium salts gave higher asymmetric induction than lithium, rubidium, or cesium salts, probably reflecting the affinity of the oligoether chain to the metal cation.

The most successful example of this approach has been provided by Shibasaki and Sasai using optically active phenoxides derived from BINOL (Scheme 9) [40, 41, 42, 43, 44, 45]. Four catalyst systems consisting of group 3 or group 13 elements were developed for the asymmetric Michael addition of malonates to prochiral enones. Three of them were associated with alkali metal cations, which also played an important role in the effective asymmetric induction. The oligomeric structure formed from BINOL and two metal components was named "heterobimetallic catalyst". Their initial work was conducted with lithium free lanthanoid complex prepared from La(Oi-Pr)$_3$ and (S)-BINOL, and addition of dibenzyl malonate to **23** with 10 mol % of the (S)-catalyst at –10 °C gave (S)-**44** in 92% ee. The composition of the complex was La:BINOL=2:3. Addition of t-BuONa to this complex generated lanthanum-sodium-(R)-BINOL, (R)-LSB catalyst, whose structure was determined by X-ray analysis as LaNa$_3$(BINOL)$_3$·6THF·H$_2$O. Of the possible combinations of lanthanoids and alkali metals, the lanthanum/sodium system gave the best results in terms of the stereoselectivity. NMR and computational studies suggested that the si-face of **23** was effectively shielded by the catalyst, which resulted in the formation of (R)-**44**. Both the basic and the Lewis acid nature of the catalyst is important. The LSB catalyst can be used for the reaction of **13** giving (S)-**45** provided that it is conducted at –50 °C in toluene. The LSB catalysts also promoted the asymmetric reaction of prochiral donor **32** and **7**.

An aluminum-lithium catalyst, (R)-ALB, prepared from (R)-BINOL, and lithium aluminium hydride promoted the addition of malonate to **23** giving (R)-**44** in 99% ee. X-ray analysis of the ALB catalyst showed an aluminum ate complex structure with Li coordination to the oxygen atom. The asymmetric tandem Michael-aldol reaction of **46** was conducted with this catalyst giving a single isomer **47** containing three asymmetric centers. The aluminum enolate under-

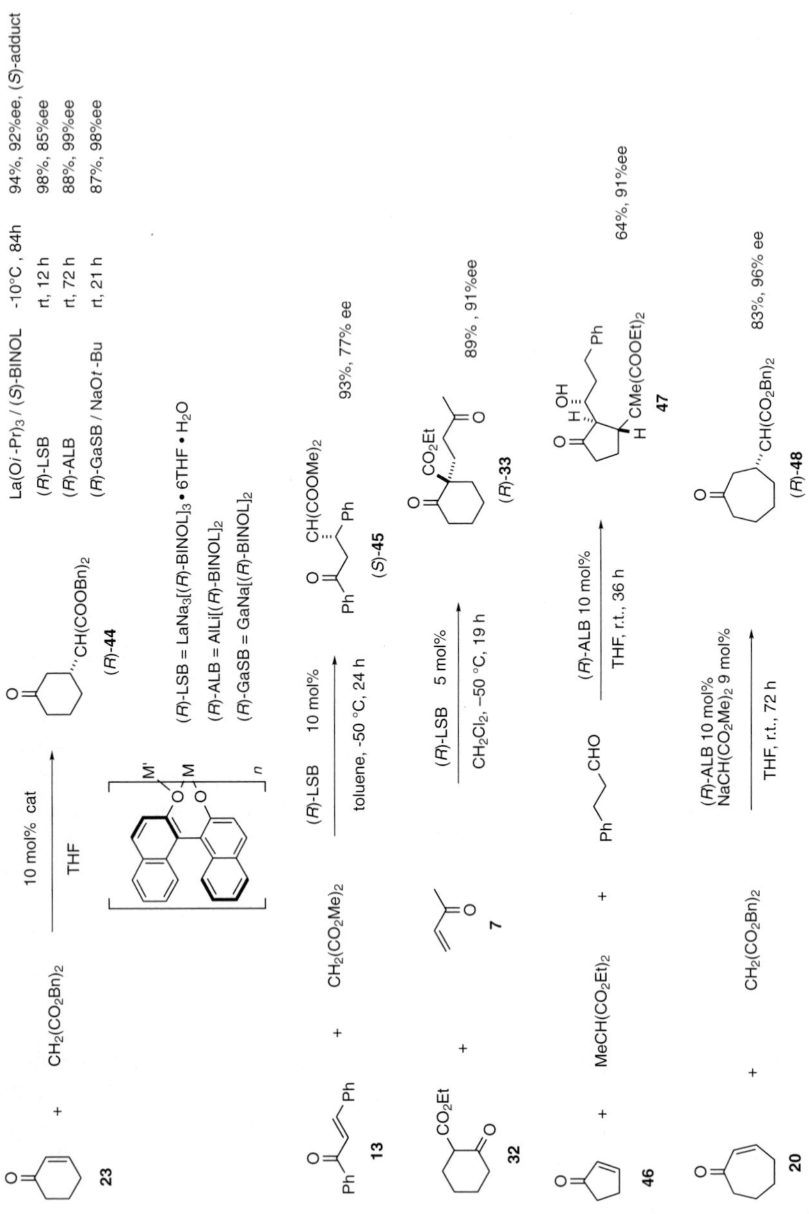
23
CH2(CO2Bn)2
10 mol% cat
THF
(R)-44
CH(COOBn)2
La(Oi-Pr)3 / (S)-BINOL -10°C, 84h 94%, 92%ee, (S)-adduct
(R)-LSB rt, 12 h 98%, 85%ee
(R)-ALB rt, 72 h 88%, 99%ee
(R)-GaSB / NaOt-Bu rt, 21 h 87%, 98%ee
M'
M
n
(R)-LSB = LaNa3[(R)-BINOL]3 • 6THF • H2O
(R)-ALB = AlLi[(R)-BINOL]2
(R)-GaSB = GaNa[(R)-BINOL]2
13
CH2(CO2Me)2
(R)-LSB 10 mol%
toluene, -50 °C, 24 h
(S)-45
CH(COOMe)2
93%, 77% ee
CO2Et
32
7
(R)-LSB 5 mol%
CH2Cl2, –50 °C, 19 h
(R)-33
89%, 91%ee
46
MeCH(CO2Et)2
CHO
(R)-ALB 10 mol%
THF, r.t., 36 h
47
CMe(COOEt)2
64%, 91%ee
20
CH2(CO2Bn)2
(R)-ALB 10 mol%
NaCH(CO2Me)2 9 mol%
THF, r.t., 72 h
(R)-48
CH(CO2Bn)2
83%, 96% ee

Scheme 9

went aldol reaction more rapidly than protonation. Another catalyst, (R)-GaNa-BINOL generated from $GaCl_3$, NaOt-Bu (4 equivalents), and (R)-BINOL (2 equivalents), was less active than (R)-ALB. However, addition of one mol equivalent of t-BuONa dramatically accelerated the reaction. Since the optical activity was not reduced by the presence of the added base, the sodium salt of malonate bound to the catalyst much more rapidly than it reacted with enone. A related method using malonate sodium salt was applied to the less reactive **20** giving (R)-**48** in 96% ee.

5
Crown Ether/Alkali Metal Base Catalysts

Cram found that chiral crown ethers in the presence of alkali metal bases catalyzed the asymmetric Michael addition [46]. Ketoester **6** underwent addition to **7** in more than 99% ee in the presence of (S,S)-**49** and KOt-Bu (4 mol %). Another crown ether, (R)-**50**, and KNH_2 promoted the addition of **41** to **42** giving (S)-**43** in 60% ee. Since then, this reaction was examined using various optically active crown ethers [47, 48, 49, 50, 51, 52, 53, 54, 55, 56], which are summarized in Scheme 10 showing the configuration and enantiomeric excess of **43**. Slight changes in the structure of the crown ethers drastically affected the stereochemistry of the reaction. A brief structure-activity relationships may be presented.

A simple derivative, (S,S)-1,2-dimethyl-18-crown-6 (**51**), and KOt-Bu gave (S)-**43** with a considerable level of asymmetric induction, 79% ee [49]. It is presumed that vicinal dimethyl group occupies a diaxial arrangement thus constructing an effective chiral environment. (R,R)-15-Crown-5 **52** gave (S)-**43** when NaOt-Bu was used as the base, and a very low ee was obtained with the potassium salt [50]. It may be due to a weaker interaction between the cation and the crown ether in the latter. Notably, the (S,S,S,S)-1,2,7,8-tetramethyl and (R,R,R,R,R,R)-hexaphenyl derivatives, **53** and **54**, gave the antipode (R)-**43** [49, 50, 51]. Crown ethers containing a disaccharide moiety such as **55** and **56** were also developed [52, 53, 54]. Deracemization phenomena were observed in which (±)-**43** was converted to (S)-**43** in the presence of KOt-Bu and **55**. The reaction took place under thermodynamic control. In the reactions of the 25-membered ring polyether **57** and the 27-membered compound **58**, the protecting group played an important role [55]. The structure of the ion pair consisting of **59**, enolate, and metal cation was discussed on the basis of theoretical calculations [56].

Asymmetric Michael additions of the prochiral acceptors using crown ethers are rare. The reaction of **60** and **46** using chiral crown ethers **62**, **63**, **64**, etc., was reported by Yamamoto and other researchers (Scheme 11) [57, 58]. The phenylthio group could be removed under radical conditions giving **61**.

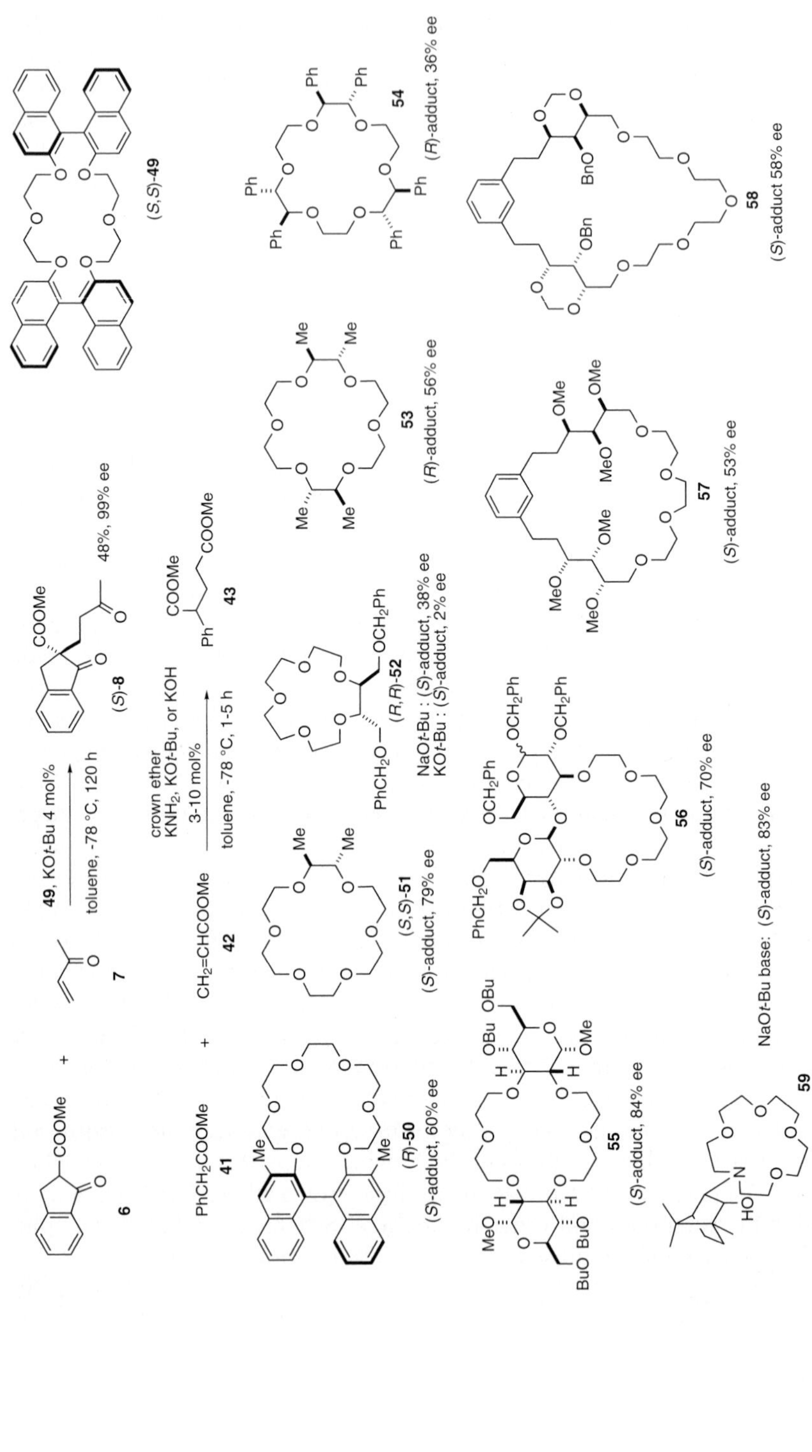

6
7
49, KOt-Bu 4 mol%
toluene, -78 °C, 120 h
(S)-8
48%, 99% ee
(S,S)-49
PhCH2COOMe
41
CH2=CHCOOMe
42
crown ether
KNH2, KOt-Bu, or KOH
3-10 mol%
toluene, -78 °C, 1-5 h
43
(R)-50
(S)-adduct, 60% ee
(S,S)-51
(S)-adduct, 79% ee
(R,R)-52
NaOt-Bu : (S)-adduct, 38% ee
KOt-Bu : (S)-adduct, 2% ee
53
(R)-adduct, 56% ee
54
(R)-adduct, 36% ee
55
(S)-adduct, 84% ee
56
(S)-adduct, 70% ee
57
(S)-adduct, 53% ee
58
(S)-adduct 58% ee
59
NaOt-Bu base: (S)-adduct, 83% ee

Scheme 10

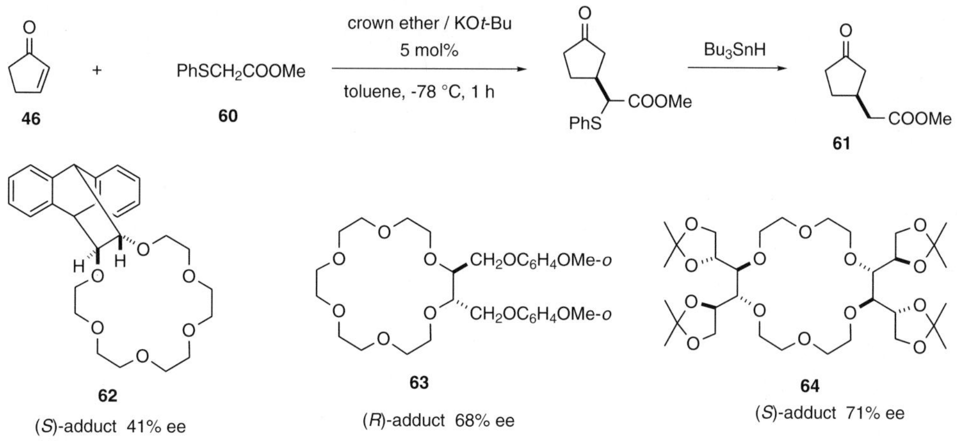

Scheme 11

6
Transition Metal Complexes

The transition metal-catalyzed asymmetric Michael addition reaction was first reported by Brunner employing the complex of Co(acac)$_2$ and (S,S)-1,2-diphenyl-1,2-ethylenediamine (Scheme 12) [59, 60]. An enantiomeric excess of 66% was attained in the reaction of **6** and **7** giving (R)-**8** at –50 °C in toluene. The dimeric copper complex **65** derived from salicylaldehyde and optically active (S)-hydroxyamines also promoted the reaction giving (S)-**8** in 75% ee [61, 62, 63]. The second hydroxy group is considered to occupy the axial position of the monomeric intermediate.

Based on the finding that ruthenium complexes catalyzed the Michael addition of cyanoesters, Ito developed a system of RhH(CO)(PPh$_3$)$_3$ and chiral bidentated phosphine, (S,S)-(R,R)-TRAP. The catalyst promoted the asymmetric addition of **66** to **7** giving (R)-**67** [64, 65, 66]. In the case of a reactive acceptor, acrolein, even 0.1 mol % of the complex effectively catalyzed the reaction. An enantiomeric excess of up to 93% was attained with the diisopropylmethyl ester. Since BINAP, DIOP, CHIRAPHOS, etc., did not induce such high stereoselectivities, the *trans*-coordinated structure constructed by the TRAP was considered to be critical. The structure of the ruthenium complex obtained by X-ray analysis indicated the interaction of the metal with the nitrile nitrogen atom. The *trans*-coordinated ligand might be required to affect the remote reaction site.

The combination of (S)-prolinamide and Ni(acac)$_2$ promoted the asymmetric addition of nitromethane to **13** in the selectivity up to 61% ee [67, 68, 69].

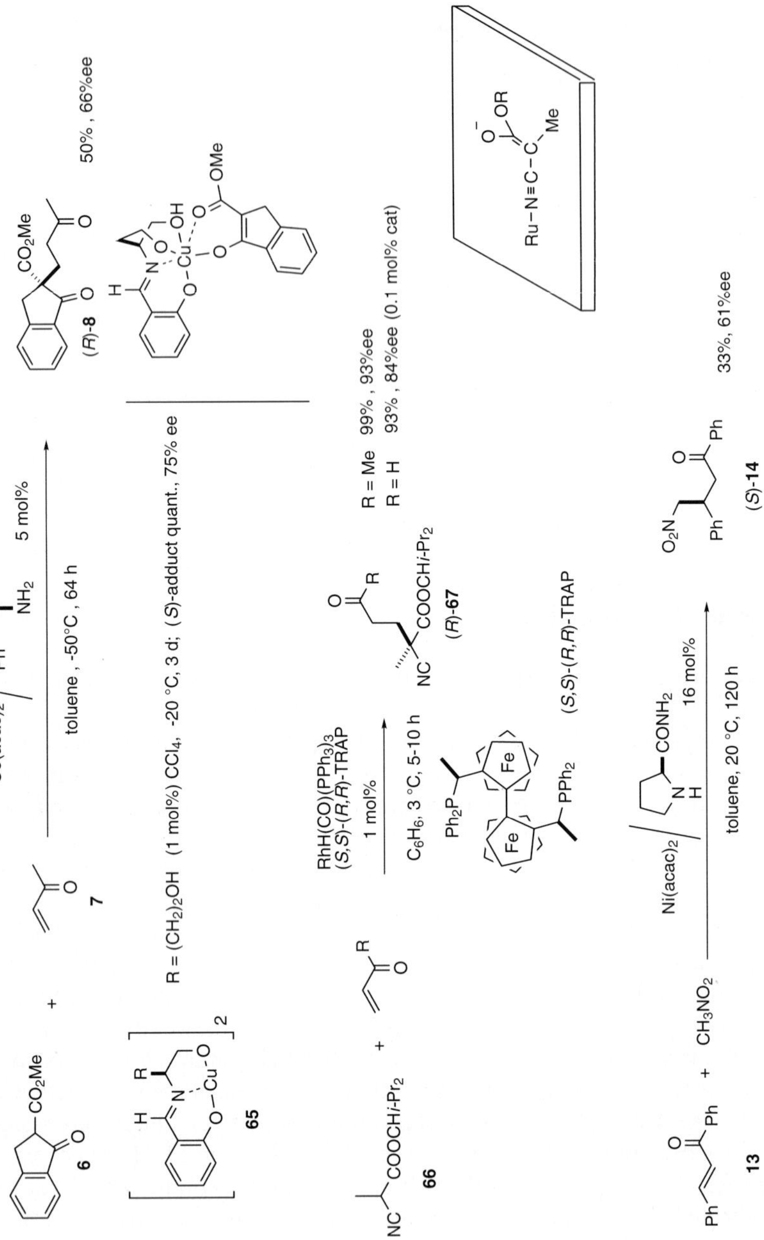

6
7
Co(acac)2
5 mol%
toluene, -50°C, 64 h
(R)-8
50%, 66%ee
65
R = (CH2)2OH (1 mol%) CCl4, -20 °C, 3 d; (S)-adduct quant., 75% ee
66
RhH(CO)(PPh3)3
(S,S)-(R,R)-TRAP
1 mol%
C6H6, 3 °C, 5-10 h
(R)-67
R = Me 99%, 93%ee
R = H 93%, 84%ee (0.1 mol% cat)
(S,S)-(R,R)-TRAP
Ru-N≡C-C
13
Ni(acac)2
16 mol%
toluene, 20 °C, 120 h
(S)-14
33%, 61%ee

Scheme 12

7
Metal Complexes with Lewis Acid Properties

Metal compounds possessing Lewis acid character are often used in the Michael addition reaction, and the methodology is reasonably applied to the asymmetric reaction in the presence of chiral ligands. The mechanism could involve either purely Lewis acidic activation of the Michael acceptor or generation of new organometallic species by the transmetalation or C-H activation, although they were not clear in many cases. The system of $Sn(OTf)_2$ and chiral (S)-diamine developed by Mukaiyama promoted the asymmetric addition of trimethylsilyl enethiolate **68** to β-arylvinyl ketones (Scheme 13) [70, 71]. The diamine-coordinated tin enolate was considered to be involved, and slow addition of **68** was essential to inhibit the racemate formation process.

Narasaka reported that the titanium compound generated from $TiCl_2(OiPr)_2$ and an optically active diol in the presence of 4 Å molecular sieve promoted the asymmetric addition of enamine **69** to the activated fumarate **70** [72]. Cyclobutane derivatives were formed when β,β-disubstituted enamines were employed. Titanium oxide derived from (R)-BINOL and $(iPrO)_2Ti=O$ catalyzed the asymmetric addition of silyl thioenol ether **71** to enones [73]. The sulfur derivative ex-

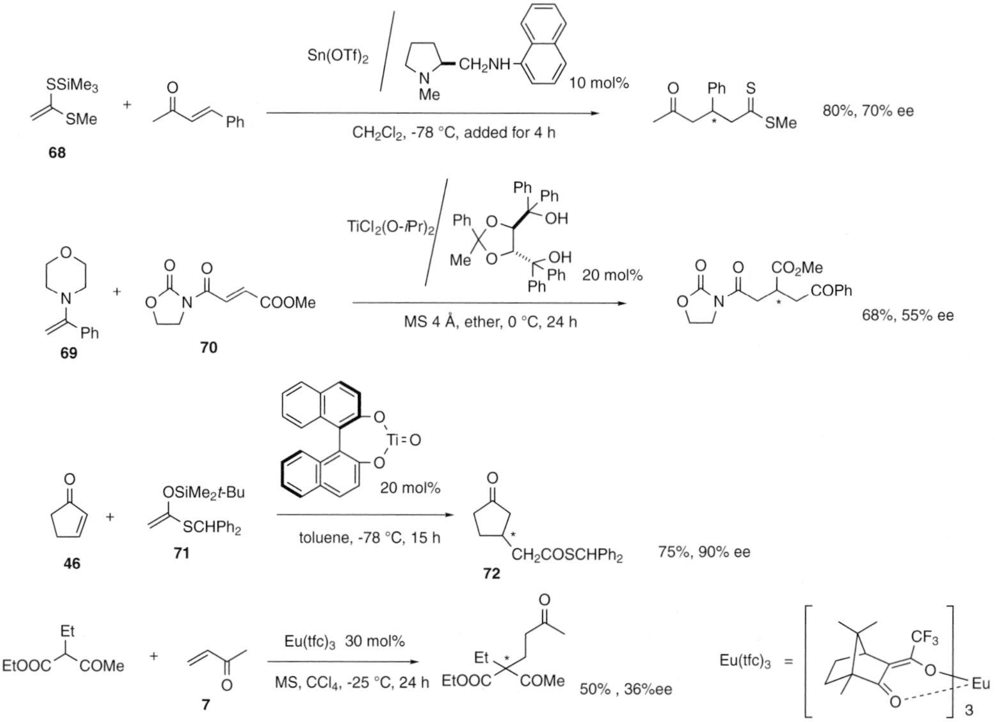

Scheme 13

hibited a much higher enantiomeric excess than simple ester, and addition to **47** gave **72** in 90% ee. Reaction of the lanthanoid shift reagent, europium tris[3-(trifluoromethylhydroxymethylene)-*d*-camphorato]europium(III) was also reported [74].

Several successful examples appeared for the catalytic asymmetric Michael addition reaction. It may be apparent, however, that they are not yet quite satisfactory in terms of stereoselectivity, catalyst efficiency, and applicability. Development of new methods is still required, which would also deepen the fundamental understanding of the Michael addition reaction, a very important reaction in organic synthesis.

Acknowledgments: I would like to thank Ms. Mie Tomizawa for the preparation of drawings.

References

1. Perlmutter P (1992) Conjugate Addition Reactions in Organic Synthesis, Pergamon Press, Oxford
2. Yamaguchi M, Shiraishi T, Hirama M (1996) J Org Chem 61:3520
3. Duhamel P, Duhamel L (1995) C R Acad Sci Paris IIb 320:689
4. For example; Oppolzer W, Löher HJ (1981) Helv Chim Acta 64:2808
5. Kretchmer RA (1972) J Org Chem 37:2744
6. Långström B, Bergson G (1973) Acta Chem Scand 27:3118
7. Wynberg H, Helder R (1975) Tetrahedron Lett 4057
8. Hermann K, Wynberg H (1979) J Org Chem 44:2238
9. ten Hoeve W, Wynberg H (1979) J Org Chem 44:1508
10. Trost BM, Shuey CD, DiNinno Jr F, McElvain SS (1979) J Am Chem Soc 101:1284
11. Latvala A, Stanchev S, Linden A, Hesse M (1993) Tetrahedron Asymmetry 4:173
12. Annunziata R, Cinquini M, Colonna S (1980) J Chem Soc Perkin Trans I 2422
13. Juliá S, Guixer J, Masana J, Rocas J, Colonna S, Annunziata R, Molinari H (1982) J Chem Soc Perkin Trans I 1317
14. Matsumoto K, Uchida T (1981) Chem Lett 1673
15. Sera A, Takagi K, Katayama H, Yamada H, Matsumoto K (1988) J Org Chem 53:1157
16. Riant O, Kagan HB, Ricard L (1994) Tetrahedron 50:4543
17. Hermann K, Wynberg H (1977) Helv Chim Acta 60:2208
18. Kobayashi N, Iwai K (1978) J Am Chem Soc 100:7071
19. Kobayashi N, Iwai K (1980) J Polym Sci Polym Chem Ed 18:923
20. Kobayashi N, Iwai K (1982) J Polym Sci Polym Chem Ed 20:85
21. Inagaki M, Hiratake J, Yamamoto Y, Oda J (1987) Bull Chem Soc Jpn 60:4121
22. Yamaguchi M, Shiraishi T, Hirama M (1993) Angew Chem Int Ed Engl 32:1176
23. Kawara A, Taguchi T (1994) Tetrahedron Lett 35:8805
24. Yamaguchi M, Shiraishi T, Igarashi Y, Hirama M (1994) Tetrahedron Lett 35:8233
25. Yamaguchi M, Igarashi Y, Reddy RS, Shiraishi T, Hirama M (1997) Tetrahedron 53:11223
26. Colonna S, Hiemstra H, Wynberg H (1978) J Chem Soc Chem Commun 238
27. Annunziata R, Cinquini M, Colonna S (1980) Chem Ind 238
28. Banfi S, Cinquini M, Colonna S (1981) Bull Chem Soc Jpn 54:1841
29. Wynberg H, Greijdanus B (1978) J Chem Soc Chem Commun 427
30. Brunner H, Zintl H (1991) Monatsh Chem 122:841
31. Conn RSE, Lovelle AV, Karady S, Weinstock LM (1986) J Org Chem 51:4710
32. Nedrickx W, Vandewalle M (1990) Tetrahedron Asymmetry 1:265
33. Loupy A, Sansoulet J, Zaparucha A, Merienne C (1989) Tetrahedron Lett 30:333
34. Loupy A, Zaparucha A (1993) Tetrahedron Lett 34:473
35. Colonna S, Re A, Wynberg H (1981) J Chem Soc Perkin Trans I 547

36. Hodge P, Khoshdel E, Waterhouse J (1983) J Chem Soc Perkin Trans I 2205
37. Terent'ev AP, Klabunovskii EI, Budovskii EI (1953) Sbornik Statei Obshchei Khim 2 1612; (1955) Chem Abstr 49:5263b
38. Kumamoto T, Aoki S, Nakajima M, Koga K (1994) Tetrahedron Asymmetry 5:1431
39. Tamai Y, Kamifuku A, Koshiishi E, Miyano S (1995) Chem Lett 957
40. Sasai H, Arai T, Shibasaki M (1994) J Am Chem Soc 116:1571
41. Sasai H, Arai T, Satow Y, Houk K N, Shibasaki M (1995) J Am Chem Soc 117:6194
42. Sasai H, Emori E, Arai T, Shibasaki M (1996) Tetrahedron Lett 37:5561
43. Arai T, Sasai H, Aoe K, Okamura K, Date T, Shibasaki M (1996) Angew Chem Int Ed Engl 35:104
44. Arai T, Yamada YMA, Yamamoto N, Sasai H, Shibasaki M (1996) Chem Eur J 2:1368
45. Shibasaki M, Sasai H, Arai T (1997) Angew Chem Int Ed Engl 36:1237
46. Cram DJ, Sogah GDY (1981) J Chem Soc Chem Commun 625
47. Raguse B, Ridley DD (1984) Aust J Chem 37:2059
48. Dehmlow EV, Sauerbier C (1989) Liebigs Ann Chem 181
49. Aoki S, Sasaki S, Koga K (1989) Tetrahedron Lett 30:7229
50. Dehmlow EV, Knufinke V (1992) Liebigs Ann Chem 283
51. Crosby J, Stoddart JF, Sun X, Venner MRW (1993) Synthesis 141
52. Alonso-López M, Martín-Lomas M, Penadés S (1986) Tetrahedron Lett 27:3551
53. Alonso-López M, Jimenez-Barbero J, Martín-Lomas M, Penadés S (1988) Tetrahedron 44:1535
54. Töke L, Fenichel L, Albert M (1995) Tetrahedron Lett 36:5951
55. van Maarschalkerwaart DAH, Willard NP, Pandit UK (1992) Tetrahedron 48:8825
56. Brunet E, Poveda AM, Rabasco D, Oreja E, Font LM, Batra MS, Rodríguez-Ubis JC (1994) Tetrahedron Asymmetry 5:935
57. Takasu M, Wakabayashi H, Furuta K, Yamamoto H (1988) Tetrahedron Lett 29:6943
58. Aoki S, Sasaki S, Koga K (1992) Heterocycles 33:493
59. Brunner H, Hammer B (1984) Angew Chem Int Ed Engl 23:312
60. Brunner H, Kraus J (1989) J Mol Cat 49:133
61. Desimoni G, Quadrelli P, Righetti PP (1990) Tetrahedron 46:2927
62. Desimoni G, Faita G, Mellerio G, Righetti PP, Zanelli C (1992) Gazz Chim Ital 122:269
63. Desimoni G, Dusi G, Faita G, Quadrelli P, Righetti PP (1995) Tetrahedron 51:4131
64. Sawamura M, Hamashima H, Ito Y (1992) J Am Chem Soc 114:8295
65. Sawamura M, Hamashima H, Ito Y (1994) Tetrahedron 50:4439
66. Sawamura M, Hamashima H, Shinoto H, Ito Y (1995) Tetrahedron Lett 36:6479
67. Schionato A, Paganelli S, Botteghi C, Chelucci G (1989) J Mol Cat 50:11
68. Botteghi C, Paganelli S, Schionato A, Boga C, Fava A (1991) J Mol Cat 66:7
69. Basato M, Corin B, De Roni P, Favero G, Jaforte R (1987) J Mol Cat 42:115
70. Yura T, Iwasawa N, Narasaka K, Mukaiyama T (1988) Chem Lett 1025
71. Iwasawa N, Yura T, Mukaiyama T (1989) Tetrahedron 45:1197
72. Hayashi Y, Otaka K, Saito N, Narasaka K (1991) Bull Chem Soc Jpn 64:2122
73. Kobayashi S, Suda S, Yamada M, Mukaiyama T (1994) Chem Lett 97
74. Bonadies F, Lattanzi A, Orelli LR, Pesci S, Scettri A (1993) Tetrahedron Lett 34:7649

Chapter 32   Ene-Type Reactions

Chapter 32
Ene-Type Reactions

Koichi Mikami · Masahiro Terada

Department of Chemical Technology, Tokyo Institute of Technology, Meguro-ku, Tokyo 152, Japan
e-mail: kmikami@o.cc.titech.ac.jp

Keywords: Ene reaction, Hetero-Diels-Alder reaction, Ene cyclization, Desymmetrization, Kinetic resolution, Non-linear effect, Asymmetric activation, Metallo-ene, Carbonyl addition reaction, Aldol-type reaction, Titanium, Aluminum, Magnesium, Palladium, Copper, Lanthanides, Binaphthol, Bisoxazoline, Diphosphine, TADDOL, Schiff base.

List of Abbreviations

BINOL: 1,1'-bi-2-naphthol
TADDOL: $\alpha,\alpha,\alpha',\alpha'$-tetraaryl-1,3-dioxolane-4,5-dimethanol
MPM: 4-methoxyphenylmethyl

1
Introduction

Asymmetric catalysis of organic reaction to provide enantiomerically enriched products is of central importance for modern synthetic and pharmaceutical chemistry. In particular, enantioselective catalysis is an economical and environmentally benign process, since it achieves "multiplication of chirality" [1] thereby affording a large amount of the enantio-enriched product, while producing a small amount of waste material, due to the very small amount of chiral catalyst employed. Thus, the development of enantioselective catalysts is a most challenging and formidable endeavor for synthetic organic chemists [2, 3]. Highly promising candidates for such enantioselective catalysts are metal complexes bearing chiral organic ligands. The degree of enantioselectivity should be critically influenced by metal-ligand bond lengths, particularly metal-oxygen and -nitrogen bond lengths in the cases of metal alkoxide and amide complexes [4, 5], as well as the steric demand of the organic ligands. Therefore, boron and aluminum are the main group elements of choice, and titanium is one of the best early transition metals with hexa- and pentacoordination, lanthanides are similarly useful. The Lewis acidity of the metal complexes is generally proportional to the value of (charge density)×(ionic radius)$^{-3}$ [6].

2
Carbonyl-Ene Reaction

C-H bond activation [7] and C-C bond formation are the key issues in organic synthesis. In principle, the ene reaction is one of the simplest methods for C-C bond formation, which converts readily available olefins, via activation of an allylic C-H bond and allylic transposition of the C=C bond, into more functionalized products. The ene reaction encompasses a vast number of variants in terms of the enophile used. Comprehensive reviews on ene reactions are given in Refs. [8a, 8b, 8c, 8d].

The class of ene reactions involving a carbonyl compound as the enophile, which we refer to as the carbonyl-ene reaction [8c], constitutes a useful synthetic method for the stereocontrolled construction of carbon skeletons using a stoichiometric or catalytic amount of various Lewis acids (Scheme 1) [9, 10]. From the synthetic point of view, the carbonyl-ene reaction should, in principle, constitute a more efficient alternative to the carbonyl addition reaction of allylmetal

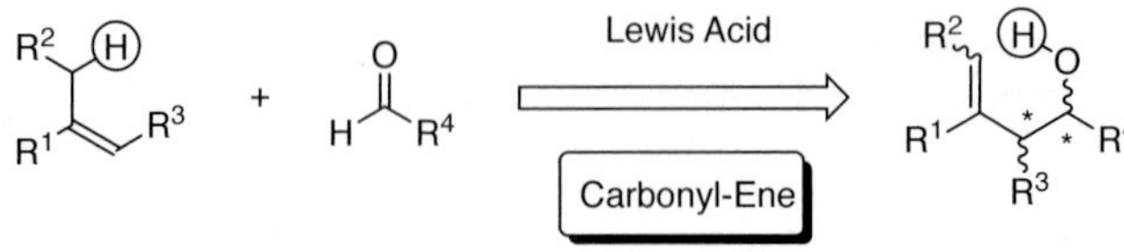

Scheme 1. Carbonyl-ene reaction catalyzed by chiral Lewis acids

species which has now become one of the most useful methods for stereocontrol [11a, 11b, 11c, 11d, 11e, 11f, 11g].

Yamamoto et al. have reported an asymmetric catalysis of carbonyl-ene reaction, which employs chloral as the enophile using an optically pure 3,3'-bissilylated binaphthol (BINOL) aluminum catalyst (Scheme 2) [12]. The 3,3'-diphenyl BINOL-derived aluminum catalyst provides the racemic product in low yield.

We have developed a chiral titanium catalyst for the glyoxylate-ene reaction which provides α-hydroxy esters of biological and synthetic importance [13] in an enantioselective fashion (Scheme 3) [14, 15a, 15b]. Various chiral titanium catalysts were screened [16]. The best result was obtained with the titanium catalyst

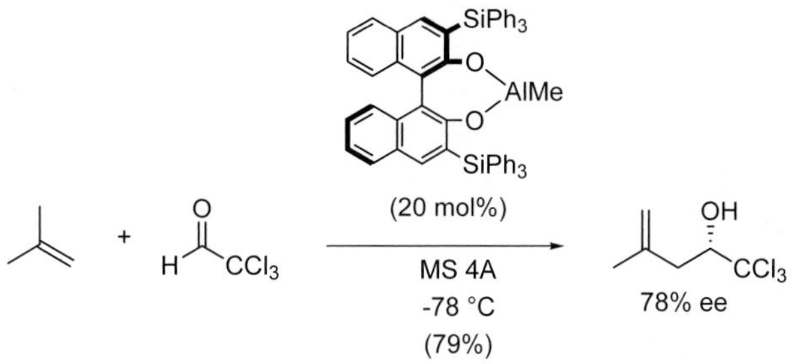

Scheme 2. Asymmetric carbonyl-ene reaction catalyzed by chiral Al complex

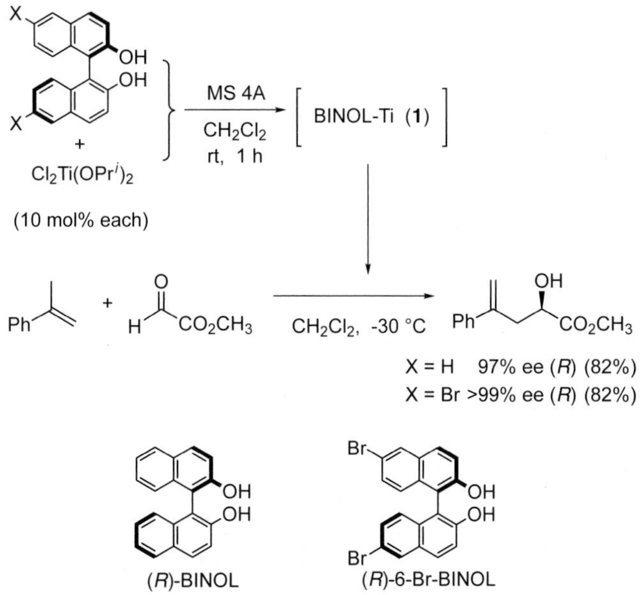

$$X = H \quad 97\% \text{ ee } (R) \ (82\%)$$
$$X = Br \quad >99\% \text{ ee } (R) \ (82\%)$$

Scheme 3. Asymmetric carbonyl-ene reaction catalyzed by BINOL-Ti complex

(1) prepared in situ in the presence of 4 Å molecular sieves (MS 4A) from diiso-propoxytitanium dihalides ($X_2Ti(O^iPr)_2$, X=Br [17] or Cl [18]) and optically pure BINOL or 6-Br-BINOL [19a, 19b, 19c, 19d]. (This ligand is now commercially available in either (R)- or (S)-form.) The remarkable levels of enantiose-lectivity and rate acceleration observed with these BINOL-Ti catalysts (1) [20] stem from the favorable influence of the inherent C_2 symmetry and the higher acidity of BINOLs compared to those of aliphatic diols. The reaction is applicable to a variety of 1,1-disubstituted olefins to provide the ene products in extremely high enantiomeric excess (Table 1).

In the reactions with mono- and 1,2-disubstituted olefins, however, no ene product was obtained. This limitation has been overcome by the use of vinylic sulfides and selenides instead of mono- and 1,2-disubstituted olefins. With these substrates, the ene products are formed with virtually complete enantioselectivity and high diastereoselectivity [21]. The synthetic utility of the vinylic sulfide and selenide approach is exemplified by the synthesis of enantiopure (R)-(−)-ipsdienol, an insect aggregation pheromone (Scheme 4), [22a, 22b, 22c].

We [23] and others [24] have also reported the lanthanide complex-catalyzed asymmetric glyoxylate-ene reaction (Scheme 5). Although the reaction of glyoxylate and α-methylstyrene proceeds catalytically under the influence of the lanthanide $Ln(NTf_2)_3$ or $Ln(OTf)_3$ [25] complexes with chiral ligands, the enantioselectivity is low-to-moderate.

The synthetic potential of the asymmetric catalytic carbonyl-ene reaction depends greatly on the functionality that is possible in the carbonyl enophile. How-

Table 1. Asymmetric catalytic glyoxylate-ene reaction with various olefins

Run	olefin	$X_2Ti(O^iPr)_2$ (X)	catalyst (mol %)	products	% yield	% ee
A		Cl	10	OH, CO_2CH_3	72	95
		Cl	1.0		78	93
		Br	10		87	94
B	Ph	Cl	1.0	Ph, OH, CO_2CH_3	97	97
		Br	1.0		98	95
C		Cl	10	OH, CO_2CH_3	82	97
		Br	5		89	98
D		Cl	10	OH, CO_2CH_3	87	88
		Br	5		92	89

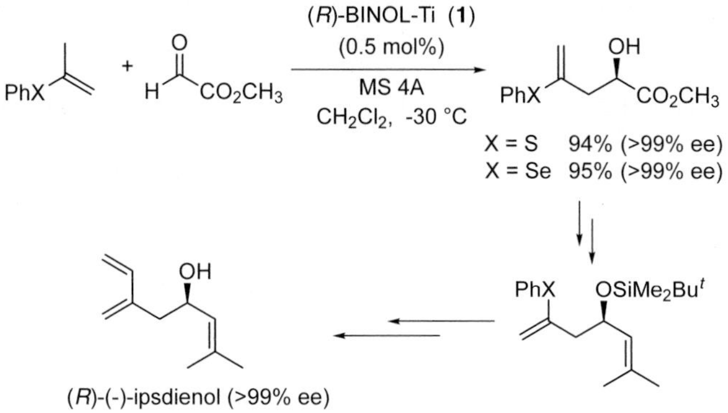

Scheme 4. Asymmetric carbonyl-ene reaction of vinylic sulfides and selenides

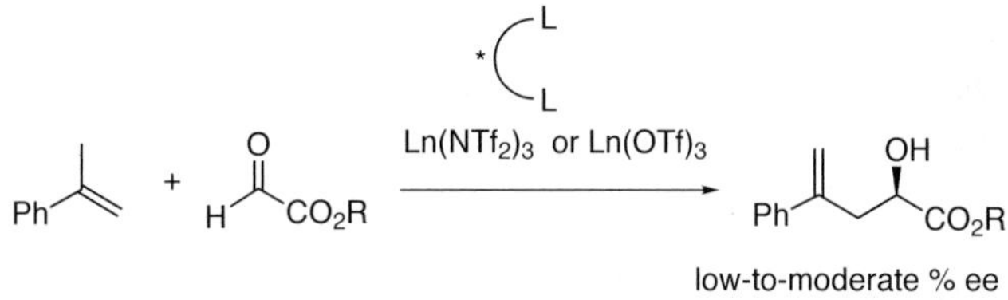

Scheme 5. Asymmetric carbonyl-ene reaction catalyzed by chiral lanthanide complexes

ever, the types of enophile that can be employed in the asymmetric catalytic ene reaction have previously been limited to aldehydes such as glyoxylate [15, 16, 26] and chloral [12, 27a, 27b]. Thus, it is highly desirable to develop other types of carbonyl enophiles to provide enantio-enriched molecules with a wider range of functionalities. We have developed an asymmetric catalytic fluoral-ene reaction [28], which provides an efficient approach for the asymmetric synthesis of some fluorine-containing compounds of biological and synthetic importance [29]. The reaction of fluoral with 1,1-disubstituted and trisubstituted olefins proceeds quite smoothly under catalysis by the BINOL-Ti complex (1) to provide the corresponding homoallylic alcohol with extremely high enantioselectivity (>95% ee) and *syn*-diastereoselectivity (>90%) (Scheme 6). The sense of asymmetric induction in the fluoral-ene reaction is exactly the same as that observed for the glyoxylate-ene reaction; (R)-BINOL-Ti (1) provides the (R)-α-CF$_3$ alcohol. The *syn*-diastereomers of α-trifluoromethyl-β-methyl-substituted compounds thus synthesized with *two stereogenic centers* show anti-ferroelectric properties preferentially to the *anti*-diastereomers [30a, 30b, 30c, 30d].

The BINOL-Ti catalyst can also be used for the carbonyl-ene reaction with formaldehyde or vinyl and alkynyl analogues of glyoxylates in an asymmetric catalytic desymmetrization (vide infra) approach to the asymmetric synthesis of isocarbacycline analogues (Scheme 7) [31a, 31b].

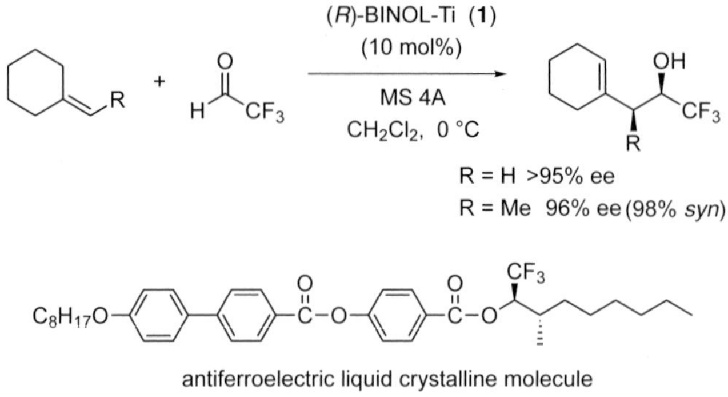

Scheme 6. Asymmetric carbonyl-ene reaction of fluoral

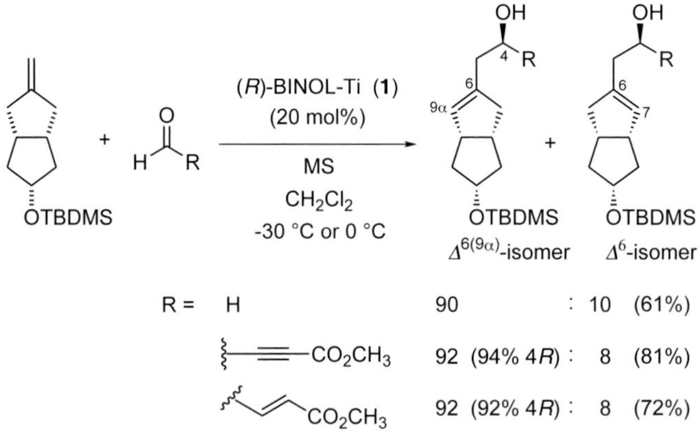

R =	H	90	:	10	(61%)
	$\equiv$—CO_2CH_3	92 (94% 4R) :		8	(81%)
	$\diagdown$$CO_2CH_3$	92 (92% 4R) :		8	(72%)

Scheme 7. Asymmetric carbonyl-ene reaction of formaldehyde or vinyl and alkynyl analogues of glyoxylate

2.1
Ene vs Hetero-Diels-Alder Reaction

In the reaction of a carbonyl compound with a conjugated diene having an al-lylic hydrogen, such as isoprene, there is the problem of the so-called periselec-tivity, arising from the formation of both the ene product and the hetero-Diels-Alder (HDA) product. In the reaction of glyoxylate with isoprene under catalysis of the BINOL-Ti complex (1) the ratio of ene/HDA product is dependent not only on the solvent employed but also on the chiral ligand of the titanium complex and further on the steric bulkiness of alkyl group (R) in glyoxylate (Scheme 8, Table 2) [19c].

The more polar solvent CH_2Cl_2 is more favorable than toluene for the forma-tion of the ene product (Run 1 vs 2). The modification of the BINOL-ligand, 6-

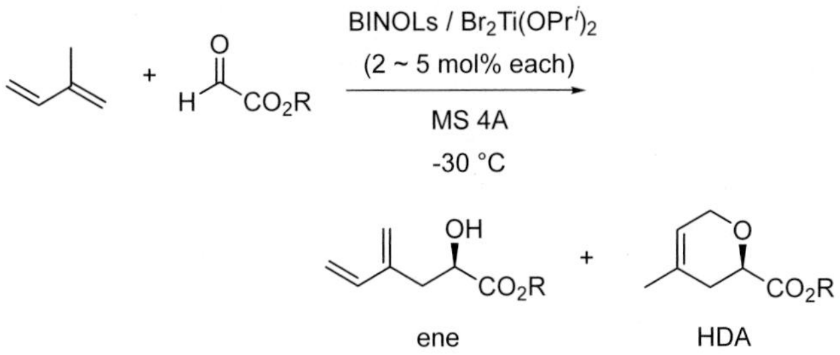

Scheme 8. Ene vs hetero Diels-Alder reaction catalyzed by BINOL-Ti complex

Table 2. The reaction of glyoxylate with isoprene

Run	R	BINOLs (mol %)	% yield	ene (% ee)/HDA (% ee)
1	Me	BINOL (2)	94	79 (97):21 (97)
2^a	Me	BINOL (2)	85	74 (98):26 (–)
3	Me	6-Br-BINOL (2)	95	83 (99):17 (–)
4	^nBu	6-Br-BINOL (2)	86	85 (>99):15 (–)
5	^iPr	6-Br-BINOL (5)	61	90 (92):10 (–)
6	CH_2CF_3	6-Br-BINOL (2)	95	92 (>99):8 (–)

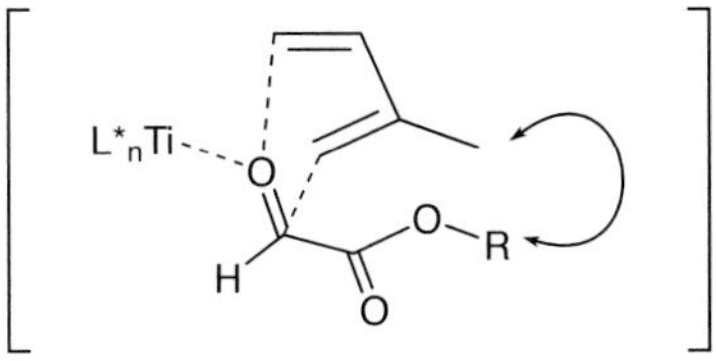

A

Fig. 1. Endo transition state of hetero-Diels-Alder reaction

Br-BINOL, is quite effective for enhancement of both the ene-selectivity and enantioselectivity as compared with those by the parent BINOL-Ti catalyst (1) (Run 1 vs 3). An increase in the steric bulkiness of the alkyl group (R) [32] in the glyoxylate leads to a substantial increase in the periselectivity for the ene product (Runs 3 to 6): With the more bulky alkyl group (R), the *endo* orientation of the ester moiety becomes less favorable through repulsive interaction between the alkyl group (R) and the methyl substituent of isoprene in the transition state (**A**) for the HDA reaction (Fig. 1), resulting, in turn, in the predominant formation of the ene products. Thus, the periselectivity for the ene reaction is increased up to 92%, particularly with the trifluoroethyl glyoxylate, accompanied by a high chemical yield (84%) and again complete enantioselectivity (Run 6) [33]. The

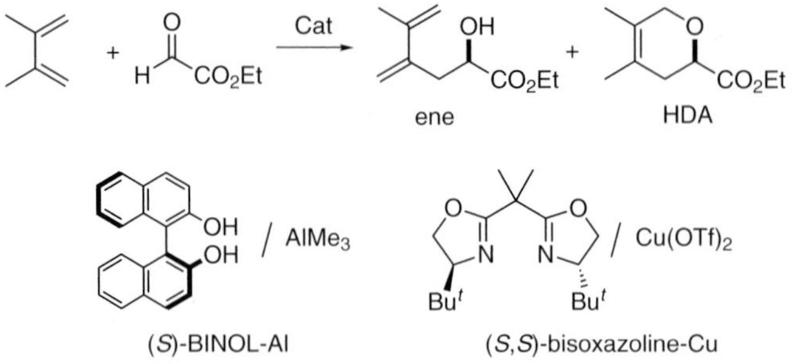

Cat	solvent	% yield	ene	HDA
(S)-BINOL-Al	CH$_2$Cl$_2$	82	11 (88% ee S)	: 89 (97% ee S)
(S,S)-bisoxazoline-Cu	CH$_2$Cl$_2$	56	64 (83% ee R)	: 36 (85% ee R)
(S,S)-bisoxazoline-Cu	CH$_3$NO$_2$	68	44 (78% ee R)	: 56 (90% ee R)

Scheme 9. Ene vs hetero Diels-Alder reaction catalyzed by BINOL-Al or bisoxazoline-Cu complex

enhanced ene-selectivity is presumably due not only to the steric but also to the electronic effect of the electron-withdrawing CF$_3$ group [32, 34].

A dramatic changeover is observed not only in the ene/HDA product ratio but also in the absolute stereochemistry when the central metal is changed from Ti to Al. Jørgensen and coworkers thus reported the HDA selective reaction of ethyl glyoxylate with 2,3-dimethyl-1,3-butadiene catalyzed by a BINOL-derived Al complex [35], where the HDA product was obtained in up to 89% periselectivity with high ee (Scheme 9). The absolute configuration was opposite to that observed when using the BINOL-Ti catalyst.

They have also reported on a solvent effect in the reaction of ethyl glyoxylate and 2,3-dimethyl-1,3-butadiene catalyzed by the cationic bisoxazoline-Cu complex (Scheme 9) [36]. In the less polar solvent CH$_2$Cl$_2$, the ene product is obtained predominantly. In contrast, the reaction in the more polar solvent CH$_3$NO$_2$ leads to a preference for the HDA product over the ene product.

2.2
Asymmetric Catalytic Desymmetrization

Desymmetrization of an achiral, symmetrical molecule through a catalytic process is a potentially powerful but relatively unexplored concept for asymmetric synthesis. While the ability of enzymes to differentiate between enantiotopic functional groups is well known [37a, 37b, 37c, 37d], little is known about a similar ability of non-enzymatic catalysts, particularly for carbon-carbon bond forming processes. Desymmetrization by the catalytic glyoxylate-ene reaction of prochiral ene substrates with the planar symmetry provides an efficient access to remote [38] and internal [39] asymmetric induction which is otherwise difficult to attain (Scheme 10) [40]. The (2R,5S)-syn-product is obtained in >99% ee along

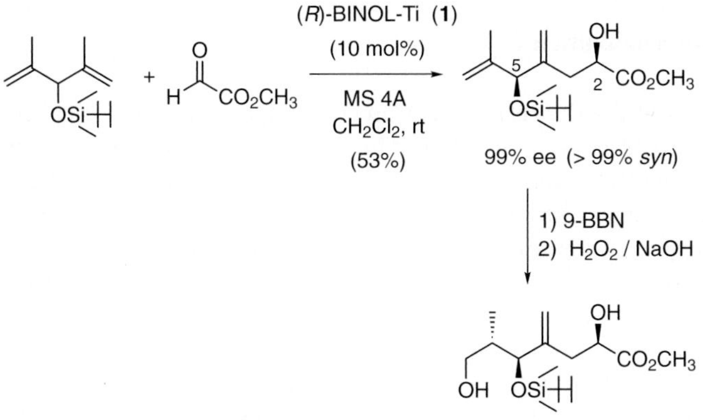

Scheme 10. Asymmetric desymmetrization in carbonyl-ene reaction catalyzed by BINOL-Ti complex

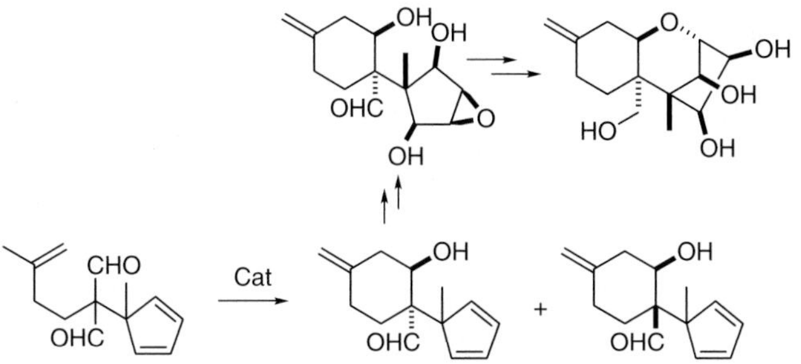

(+)-Eu(hfc)₃	5 (20% ee)	:	1
(+)-Eu(dppm)₃	4.5 (31% ee)	:	1
(S)-BINOL-TiCl₂	4.5 (38% ee)	:	1

Scheme 11. Asymmetric desymmetrization in asymmetric ene cyclization

with more than 99% diastereoselectivity. The diene thus obtained can be transformed to a more functionalized compound in a regioselective and diastereoselective manner.

Ziegler and Sobolov have reported an asymmetric desymmetrization approach to the synthesis of the tricothecene, anguidine, via an ene cyclization (Scheme 11) [41]. The (2,4) ene cyclization (vide infra) of the prochiral aldehyde on silica gel gives a 1:1 diastereomeric mixture. Cyclization with purified $Eu(fod)_3$ as Lewis acid catalyst gives an 8:1 mixture. The major isomer is a potential intermediate for the synthesis of anguidine. However, use of (+)-Eu(hfc)₃, (+)-Eu(dppm)₃, or (S)-BINOL-TiCl₂ complex as chiral Lewis acid affords only 20~38% ee.

2.3
Kinetic Optical Resolution

On the basis of the desymmetrization concept, the kinetic optical resolution of a racemic substrate [42a, 42b] might be recognized as an intermolecular version of the desymmetrization. The kinetic resolution of a racemic allylic ether by the glyoxylate-ene reaction also provides an efficient access to remote but relative [40] asymmetric induction. The reaction of allylic ethers catalyzed by the (R)-BINOL-derived complex (1) provides the 2R,5S-*syn*-products with >99% diastereoselectivity along with more than 95% ee (Scheme 12). The high diastereoselectivity, coupled with the high ee, strongly suggests that the catalyst/glyoxylate complex efficiently discriminates between the two enantiomeric substrates to accomplish effective kinetic resolution. In fact, the relative rates between the

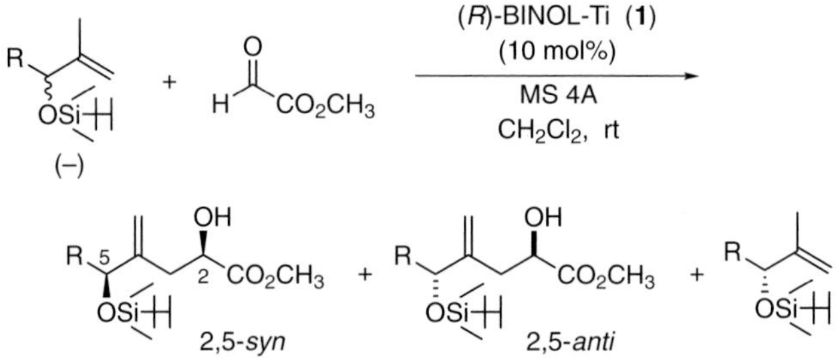

R	ene-product	recovered ene	relative rate (k_R/k_S)
i-Pr	99.6% ee (>99% *syn*)	37.8% ee	720
Me	96.2% ee (>99% *syn*)	22.0% ee	64

Scheme 12. Kinetic optical resolution in asymmetric carbonyl-ene reaction catalyzed by BINOL-Ti complex

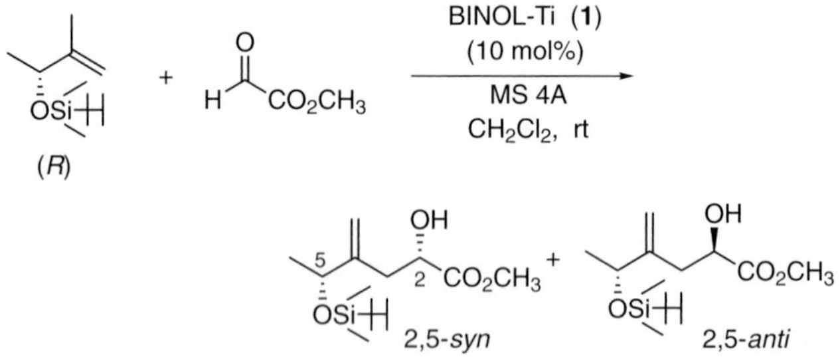

	2,5-*syn*	:	2,5-*anti*
(S)-BINOL-Ti (1)	>99	:	<1
(R)-BINOL-Ti (1)	50	:	50

Scheme 13. Double asymmetric induction in carbonyl-ene reaction catalyzed by BINOL-Ti complex

reactions of the either enantiomers, calculated by the equation $\ln[(1-c)(1-ee_{recov})] \times \{\ln[(1-c)(1+ee_{recov})]\}^{-1}$, $c=(ee_{recov}) \times (ee_{recov}+ee_{prod})^{-1}$, $0<c$, $ee<1$ where c is the fraction of consumption, were ca. 700 for R=i-Pr and 65 for R=Me. As expected, the double asymmetric induction [43a, 43b, 44] in the reaction of the (R)-ene component using the catalyst (S)-1 ("matched" catalytic system) leads to the complete (>99%) 2,5-*syn*-diastereoselectivity in high chemical yield, whereas the reaction of the (R)-ene using (R)-1 ("mis-matched" catalytic system) produces a diastereomeric mixture in quite low yield (Scheme 13).

2.4
Positive Non-Linear Effect of Non-Racemic Catalysts

A chiral catalyst is not necessarily in an enantiopure form. Deviations from the linear relationship, namely "non-linear effects" are sometimes observed between the enantiomeric purity of chiral catalysts and the optical yields of the products (Fig. 2). Among these, the convex deviation, which Kagan [45a] and Mikami [46] independently refer to as a positive non-linear effect, (abbreviated as (+)-NLE (asymmetric amplification [45c]) has attracted current attention by achieving a higher level of asymmetric induction than the enantio-purity of the non-racemic (partially resolved) catalysts [45a, 45b, 45c]. In turn, (−)-NLE stands for the opposite phenomenon of concave deviation, namely a negative non-linear effect.

We have observed a remarkable level of (+)-NLE in the catalytic ene reaction. For instance, in the glyoxylate-ene reaction, the use of a catalyst prepared from BINOL of 33.0% ee provides the ene product with 91.4% ee in 92% chemical yield (Scheme 14) [46]. The ee thus obtained is not only much higher than the ee of the BINOL employed, but also very close to the value (94.6% ee) obtained using enantiomerically pure BINOL (Fig. 3).

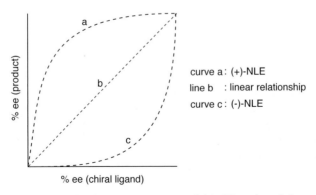

Fig. 2. Relationship between the enantiomeric purity of chiral ligands and the optical yield of products

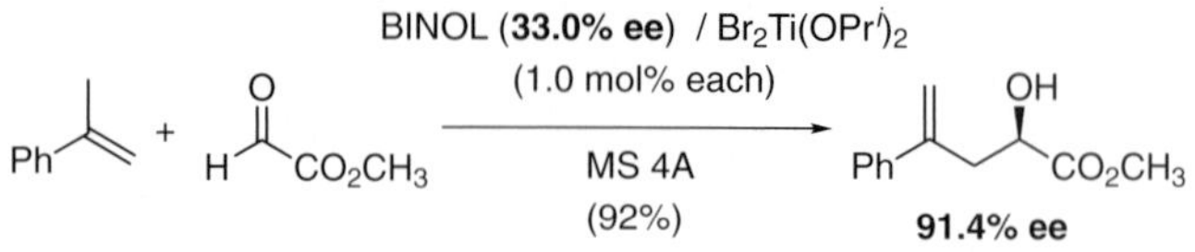

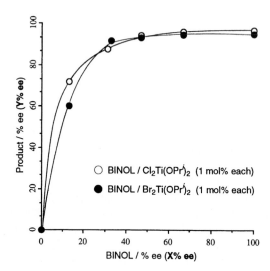

Scheme 14. Positive non-linear effect in asymmetric carbonyl-ene reaction

Fig. 3. Positive non-linear effect in asymmetric glyoxylate-ene reaction

2.5
Asymmetric Activation of Racemic Catalysts

While non-racemic catalysts can generate non-racemic products with or without NLE, racemic catalysts inherently produce only racemic products. A strategy whereby a racemic catalyst is enantiomer-selectively de-activated by a chiral molecule has been shown to yield non-racemic products [47, 48]. However, the level of asymmetric induction does not exceed the level attained by the enantiopure catalyst (Fig. 4a). Recently, "chiral poisoning" [49] has been named as such a *deactivating* strategy. In contrast, we have reported an alternative but conceptually opposite strategy to asymmetric catalysis by racemic catalysts. A *chiral activator* selectively activates one enantiomer of a racemic chiral catalyst. Higher enantioselectivity might be attained than that achieved by an enantiopure catalyst (% $ee_{act} >> $% $ee_{enantio-pure}$), in addition to a higher level of catalytic efficiency ($k_{act} >> k_{enantio-pure}$) (Fig. 4b).

Catalysis with racemic BINOL-Ti(O^{i}Pr)$_2$ (**2**) achieves extremely high enantioselectivity by adding another diol for the enantiomer-selective activation (Scheme 15, Table 3) [50a, 50b, 50c, 50d]. Significantly, a remarkably high enan-

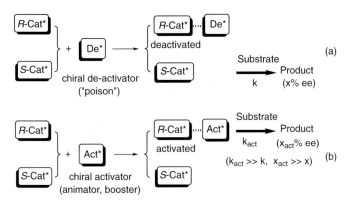

Fig. 4. Asymmetric activation vs de-activation

tioselectivity (90% ee, R) was achieved using just a half-molar amount (5 mol %) of (R)-BINOL activator added to a *racemic* (±)-BINOL-Ti(O^{i}Pr)$_2$ complex (**2**) (10 mol %).

The activation of the enantiopure (R)-BINOL-Ti(O^{i}Pr)$_2$ [51] catalyst (**2**) was investigated by further addition of (R)-BINOL (Scheme 16, Table 4). The reaction proceeded quite smoothly to provide the carbonyl-ene product in higher chemical yield (82%) and enantioselectivity (97% ee) (Run 3) than without additional BINOL (95% ee, 20%) (Run 1). Comparing the results of enantiomer-selective activation of the racemic catalyst (90% ee, R) (Table 3, Run 4) with those of the enantiopure catalyst [with (97% ee, R) or without activator (95% ee, R)], the reaction catalyzed by the (R)-BINOL-Ti(O^{i}Pr)$_2$/(R)-BINOL complex (**2'**) is calculated to be 27 times faster than that catalyzed by (S)-BINOL-Ti(O^{i}Pr)$_2$ (**2**) in the racemic case (Fig. 5a). Indeed, kinetic studies involving a rapid-quench GC analysis show that the reaction catalyzed by the (R)-BINOL-Ti(O^{i}Pr)$_2$/(R)-BINOL complex (**2'**) is 26 times faster than that catalyzed by the (R)-BINOL-Ti(O^{i}Pr)$_2$ (**2**). These results imply that the racemic (±)-BINOL-Ti(O^{i}Pr)$_2$ (**2**) and half-molar amount of (R)-BINOL assemble preferentially into the (R)-BINOL-Ti(O^{i}Pr)$_2$/(R)-BINOL complex (**2'**) and unchanged (S)-BINOL-Ti(O^{i}Pr)$_2$ (**2**). In contrast, the enantiomeric form of the additional chiral ligand ((S)-BINOL) activates the (R)-BINOL-Ti(O^{i}Pr)$_2$ (**2**) to a smaller degree (Run 6), thus providing the carbonyl-ene product in lower optical (86% ee, R) and chemical (48%) yields than (R)-BINOL does.

The great advantage of asymmetric activation of the racemic BINOL-Ti(O^{i}Pr)$_2$ complex (**2**) is highlighted in a catalytic version (Table 3, Run 5). High enantioselectivity (80.0% ee) is obtained by adding less than the stoichiometric amount (0.25 molar amount) of additional (R)-BINOL. A similar phenomenon of enantiomer-selective activation has been observed in aldol and (hetero-) Diels-Alder reactions catalyzed by a racemic BINOL-Ti(O^{i}Pr)$_2$ catalyst (**2**) [52].

Another possibility was explored using racemic BINOL as an activator. Racemic BINOL was added to the (R)-BINOL-Ti(O^{i}Pr)$_2$ (**2**) (Run 8), giving higher

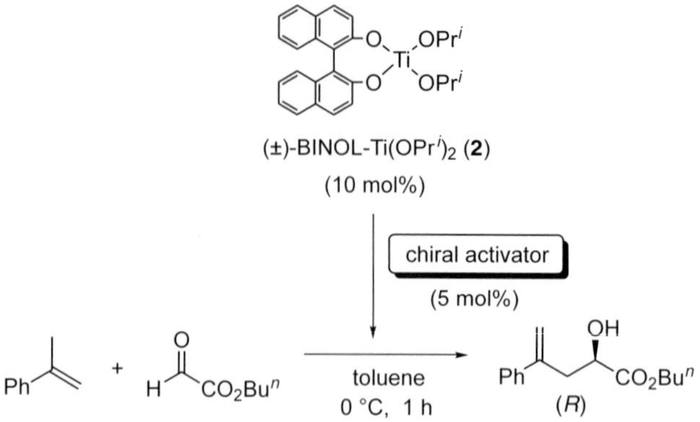

(±)-BINOL-Ti(OPr^i)$_2$ (**2**)

(10 mol%)

chiral activator

(5 mol%)

$$Ph \overset{}{\diagup\!\!\diagdown} + \overset{O}{\underset{H}{\diagdown}}CO_2Bu^n \xrightarrow[\text{0 °C, 1 h}]{\text{toluene}} Ph\diagup\!\!\diagdown\overset{OH}{\underset{(R)}{\diagdown}}CO_2Bu^n$$

Scheme 15. Enantiomer selective activation of racemic BINOL-Ti(O^{i}Pr)$_2$

Table 3. Enantiomer selective activation of racemic BINOL-Ti(O^{i}Pr)$_2$ (**2**)

Run	chiral activator	% yield	% ee
1	none	1.6	0
2	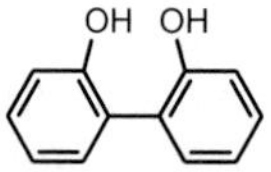	20	0
3	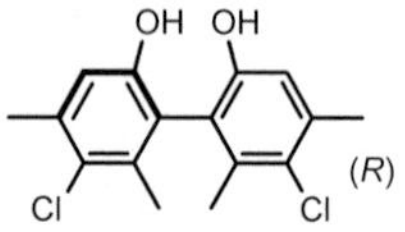	38	81
4	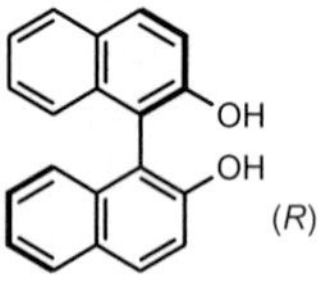	52	90
5[a]		35	80

[a] 2,5 mol% of (R)-BINOL as an activator

yield and enantioselectivity (96% ee, 69%) than those obtained by the original catalyst (R)-BINOL-Ti(O^{i}Pr)$_2$ (**2**) without additional BINOL (95% ee, 20%) (Run 1). Comparing the results (96% ee, R) with the racemic activator with those of enantiopure catalyst, (R)-BINOL-Ti(O^{i}Pr)$_2$/(R)-BINOL (**2'**) (97% ee, R) or (R)-BINOL-Ti(O^{i}Pr)$_2$/(S)-BINOL (86% ee, R), the reaction catalyzed by the (R)-BINOL-Ti(O^{i}Pr)$_2$ catalyst/(R)-BINOL complex (**2'**) is calculated to be 10 times faster than that catalyzed by the (R)-BINOL-Ti(O^{i}Pr)$_2$/(S)-BINOL (Fig. 5b). A rapid-quench GC analysis revealed the reaction catalyzed by the (R)-BINOL-Ti(O^{i}Pr)$_2$/(R)-BINOL complex (**2'**) to be 9.2 times faster than that catalyzed by the (R)-BINOL-Ti(O^{i}Pr)$_2$/(S)- BINOL.

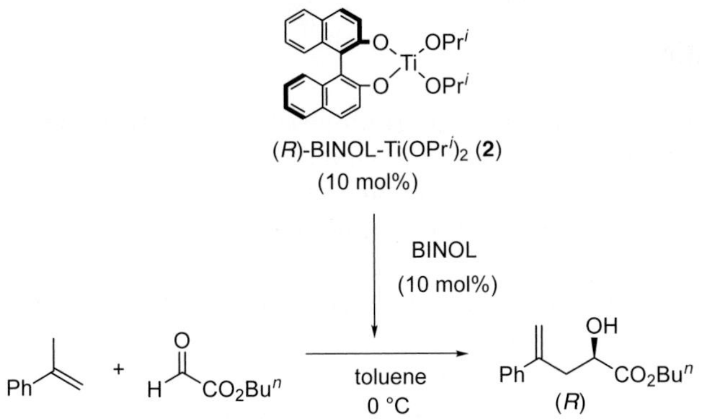

Scheme 16. Asymmetric activation of enantiopure BINOL-Ti(O^{i}Pr)$_2$

Table 4. Asymmetric activation of enantiopure (R)-BINOL-Ti(O^{i}Pr)$_2$ (2)

Run	BINOL	time (min)	% yield	% ee
1	none	60	20	95
2		1	1.6	95
3	(R)-BINOL	60	82	97
4		1	41	97
5		0.5	24	97
6	(S)-BINOL	60	48	86
7		0.5	2.6	86
8	(±)-BINOL	60	69	96

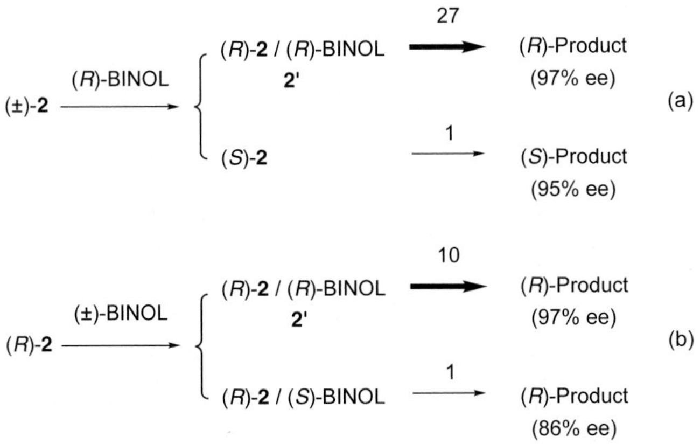

Fig. 5. Kinetic feature of asymmetric activation of BINOL-Ti(O^{i}Pr)$_2$

3
Ene Cyclization

Conceptually, intramolecular ene reactions [53a, 53b, 53c, 53d] (ene cyclizations) can be classified into six different modes (Fig. 6) [8a, 54]. In the ene cyclizations, the carbon numbers where the tether connects the [1,5]-hydrogen shift system, are expressed in (m,n) type. A numerical prefix stands for the forming ring size.

Asymmetric catalysis of ene reactions was initially explored in the intramolecular cases, since the intramolecular versions are much more facile than their intermolecular counterparts. The first example of an enantioselective 6-(3,4) carbonyl-ene cyclization was reported using a BINOL-derived zinc reagent [55]. However, this was successful only when using an excess of the zinc reagent (at least 3 equivalents). Recently, an enantioselective 6-(3,4) olefin-ene cyclization has been developed using a stoichiometric amount of a TADDOL-derived chiral titanium complex (Scheme 17) [56]. In this ene reaction, a hetero Diels-Alder product was also obtained, the ratio depending critically on the solvent system

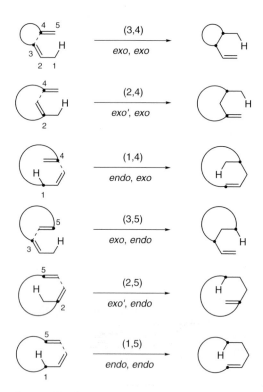

Fig. 6. Classification of ene cyclization

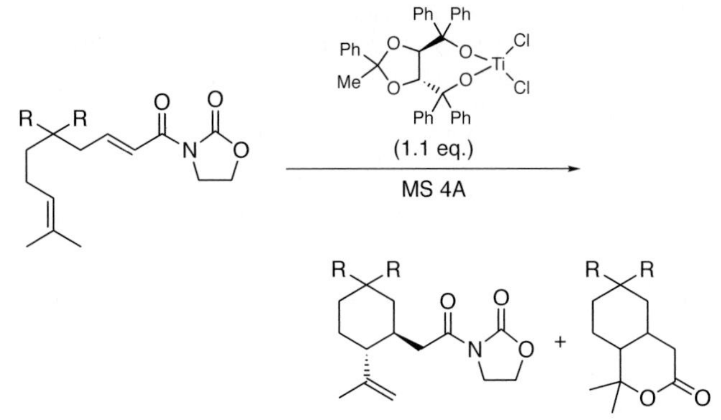

Scheme 17. Asymmetric ene cyclization catalyzed by TADDOL-Ti complex

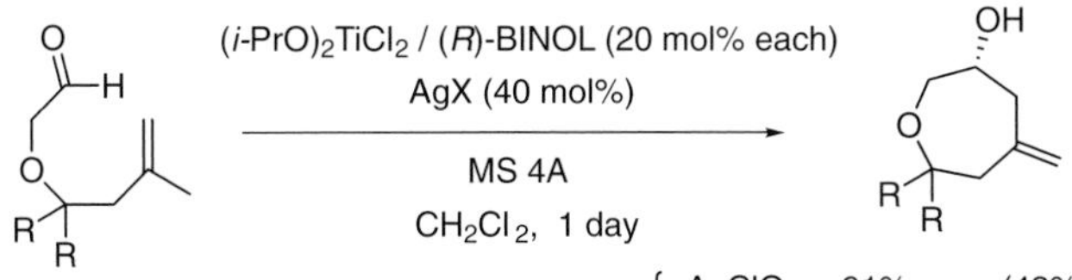

toluene	R = H	20 days	17%		37%	
	R = Me	4 days	39%	(82% ee)	36%	(92% ee)
mesitylene	R = Me	4 days	32%	(86% ee)	37%	(>98% ee)
$CFCl_2CF_2Cl$	R = Me	4 days	63%	(>98% ee)	25%	(-)

R = H	AgClO$_4$	91% ee	(43%)
	AgOTf	92% ee	(40%)
R = Me	AgClO$_4$	82% ee	(40%)

Scheme 18. Asymmetric 7-(2,4) carbonyl-ene cyclization catalyzed by BINOL-Ti complex

employed. In both cases, geminal disubstitution is required in order to obtain high ee's. However, neither reaction constitutes an example of a truly catalytic asymmetric ene cyclization.

We reported the first examples of asymmetric catalysis of intramolecular carbonyl-ene reactions of types (3,4) and (2,4), using the BINOL-derived titanium complex (1) [54, 57]. The catalytic 7-(2,4) carbonyl-ene cyclization gives the oxepane with high ee, and *gem*-dimethyl groups are not required (Scheme 18). In a similar catalytic 6-(3,4) ene cyclization, the *trans*-tetrahydropyran is preferentially obtained with high ee (Scheme 19). The sense of asymmetric induction is exactly the same as observed for the glyoxylate-ene reaction: the (R)-BINOL-Ti catalyst provides the (R)-cyclic alcohol. Therefore, the chiral BINOL-Ti catalyst works efficiently for both the chiral recognition of the enantioface of the aldehyde and the discrimination of the diastereotopic protons of the ene component in a truly catalytic fashion.

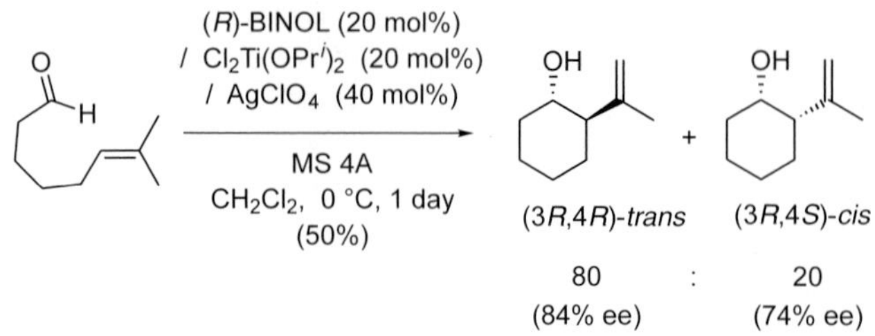

Scheme 19. Asymmetric 6-(3,4) carbonyl-ene cyclization catalyzed by BINOL-Ti complex

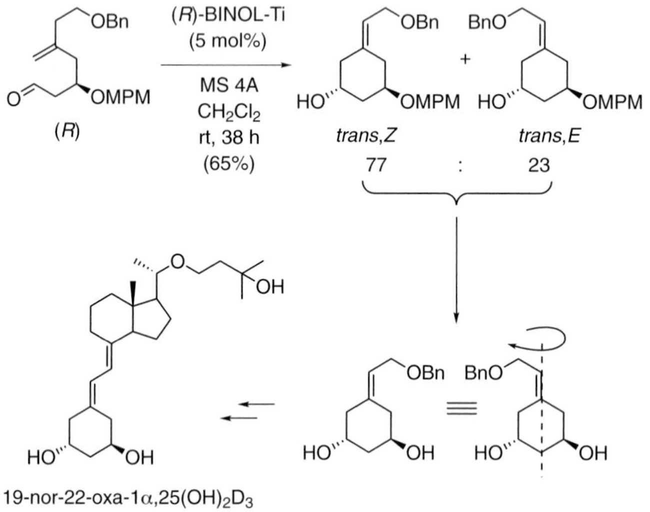

Scheme 20. "Symmetry"-assisted enantiospecific synthesis of A-ring of Vitamin D analogues based on asymmetric ene cyclization

Basic research on the synthesis of analogues of the biologically active form of vitamin D_3, $1\alpha,25$-dihydroxyvitamin D_3 $[1\alpha,25(OH)_2D_3]$ has brought about the development of an important new field in medicinal chemistry [58, 59]. We have reported "symmetry"-assisted enantiospecific synthesis of the A-ring of the vitamin D hybrid analogues, 19-nor-22-oxa-$1\alpha,25(OH)_2D_3$ (Scheme 20) [60]. It should be noted here that the "*gem*-dialkyl" substituents are not necessary in obtaining a high level of enantioselectivity.

Desimoni and coworkers reported on the enantioselective ene cyclization using a stoichiometric or catalytic amount of a bisoxazoline-$Mg(ClO_4)_2$ complex (Scheme 21) [61]. In all cases, although a hetero-Diels-Alder product was formed, an ene product was obtained predominantly in *trans,syn* fashion as the major kinetic product. The minor kinetic ene products, the *trans,anti* and *cis,anti* products, were readily isomerized at the C3 atom by silica gel to give the thermodynamically stable products, the *trans,syn* and *cis,syn* isomers, respectively. The

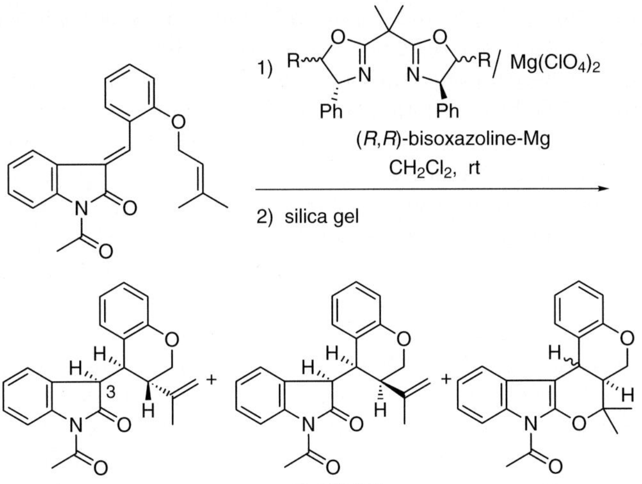

ligand	mol%	% yield	*trans,syn*-ene		*cis,syn*-ene		HDA
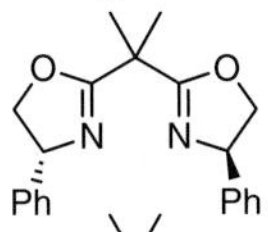	100	quant.	65 (30% ee)	:	16	:	19
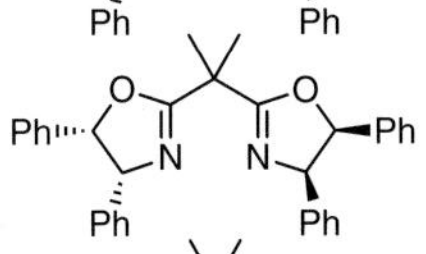	100	quant.	53 (51% ee)	:	19	:	28
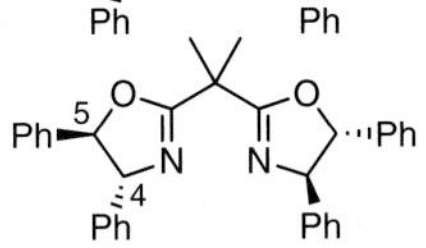	100	quant.	75 (88% ee)	:	15	:	10
	33	85	73 (81% ee)	:	15	:	12

Scheme 21. Asymmetric ene cyclization catalyzed by bisoxazoline-Mg complex

ratio of ene/HDA products and the enantioselectivity were critically dependent on the substituent at the 5 position of the chiral bisoxazoline ligand. A 4,5-*trans* disubstituent is important to obtain the ene product predominantly and in high % ee. When the reaction was performed with a reduced the amount of bisoxazoline-Mg complex, the product distribution and enantioselectivity were close to those of the corresponding stoichiometric reaction.

4
Metallo-Ene Cyclization

Metallo-ene cyclizations, intramolecular Pd(0), Ni(0), Rh(I), Zn(0) catalyzed alkene (or alkyne) allylations [62, 63, 64], have been recognized as a powerful tool in organic synthesis due to the synthetically useful functionalizations such as carbonylation or C-C coupling. The high regio- and stereoselectivities also allow the application of the metallo-ene cyclization to the syntheses of a variety of natural products. However, the enantioselective version of the catalytic metallo-ene cyclization, where the absolute stereochemistry of newly created stereogenic centers is controlled by chiral metal complexes, remains to be explored. Recently, Oppolzer and coworkers reported the enantioselective metallo-ene cyclization catalyzed by chiral palladium, nickel, and rhodium complexes (Scheme 22) [65]. Moderate enantioselectivity in terms of differentiation between the enantiotopic olefin functionality of prochiral ene substrate with planar symmetry was observed by using a chiral palladium complex. The use of a chiral bidentate ligand with a large bite angle [66] is effective for this type of enantioselective catalytic metallo-ene cyclizations.

Pd complexes also catalyze the cyclization of the 1,6-enyne system to provide carbo- or heterocyclic compounds. The catalytic enantioselective version was reported by Trost and coworkers using chiral binaphthyldicarboxylic acid ligands [67] or cyclohexyldiamine-derived diphospine [68] to provide moderate level of % ee (Schemes 23 and 24).

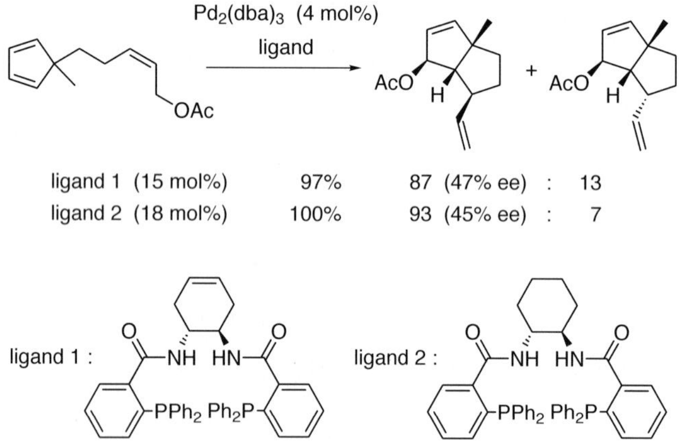

Scheme 22. Asymmetric metallo-ene cyclization catalyzed by chiral Pd complex

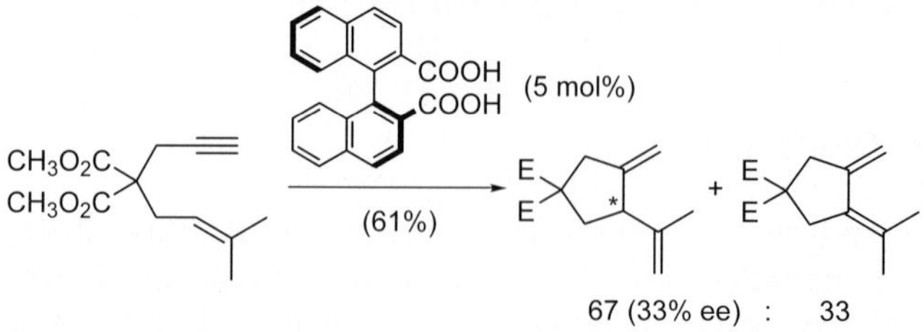

Scheme 23. Asymmetric cyclization of 1,6-enyne catalyzed by Pd/chiral acid

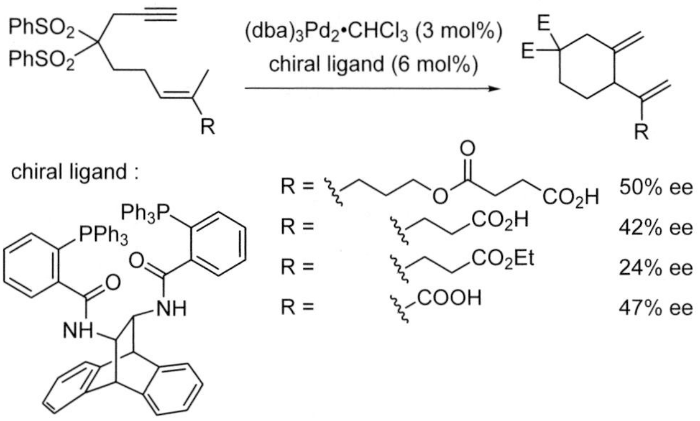

Scheme 24. Asymmetric cyclization of 1,7-enyne catalyzed by chiral Pd complex

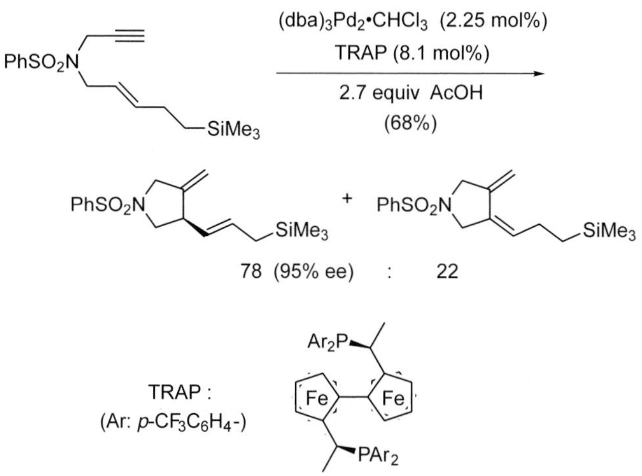

Scheme 25. Asymmetric cyclization of 1,6-enyne catalyzed by TRAP-Pd complex

Extremelly high enantioselectivity was attained by using the *trans* chelate di-phosphine ligand, TRAP (Scheme 25) [69]. There are strong implications in the catalytic efficiency between electron density at the phosphine atom and the enantioselectivity: Increasing the electron-withdrawing ability of the ligand *P*-aryl substituents resulted in higher enantioselectivity and also higher catalytic activity. The best enantioselectivity, up to 95% ee, was obtained in the reaction of the silicon-substituted 1,6-enyne with a *para*-CF$_3$-substituted arylphoshine as a chiral ligand.

Iron complexes can also be employed for ene cyclization of triene systems (Scheme 26). Though a chiral bisoxazoline complex exhibits not only higher 1,3-stereoinduction but also diastereoselectivity, no asymmetric induction was observed by the use of chiral bisoxazoline iron complex [70].

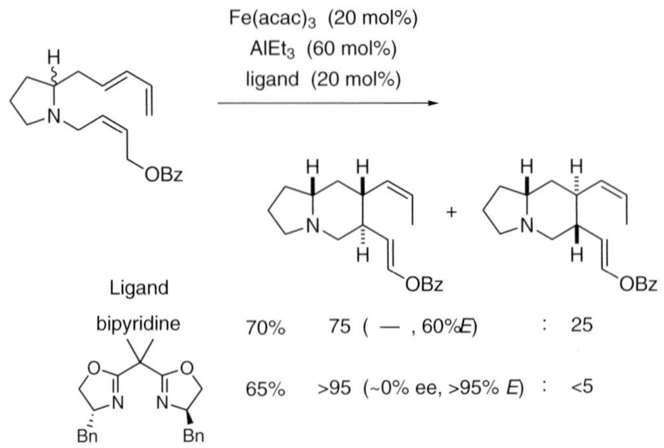

Fe(acac)₃ (20 mol%)
AlEt₃ (60 mol%)
ligand (20 mol%)

Ligand

bipyridine   70%   75 (— , 60%*E*)   :   25

Bn   Bn   65%   >95 (~0% ee, >95% E)   :   <5

Scheme 26. Ene-type cyclization of triene system catalyzed by iron complex

5
Carbonyl Addition Reaction

During the course of our research project on asymmetric catalysis of the carbonyl-ene reaction, we have found that the BINOL-Ti complexes (**1**) catalyze rather than promote stoichiometrically the carbonyl addition reaction of allylic silanes and stannanes [11, 71]. The addition reactions to glyoxylate of (*E*)-2-butenylsilane and -stannane proceed smoothly to afford the *syn*-products in high enantiomeric excess (Scheme 27) [71]. The *syn*-product thus obtained could readily be converted to the lactone portion of verrucaline A [72].

We have further found that BINOL-Ti (**1**) catalyzes the Sakurai-Hosomi reaction of methallylsilanes with glyoxylates (Scheme 28) [73]. Surprisingly, however, the products were obtained in the allylic silane (ene product) form with high enantioselectivity.

Asymmetric catalysis by BINOL-Ti complexes of the reaction of aliphatic and aromatic aldehydes with an allylstannane has been reported independently by Umani-Ronchi/Tagliavini [74] and Keck [75]. In Ronchi/Tagliavini's case [74], a new complex generated by reaction of the BINOL-Ti complex with allylstannane has been suggested to be the catalytic species which provides the remarkably high enantioselectivity (Scheme 29). Interestingly enough, no reaction occurs if 4 Å molecular sieves are not present during the preparation stage of the chiral catalyst, and 4 Å molecular sieves affect the subsequent allylation reaction. 4 Å molecular sieves dried for 12 h at 250 C and 0.1 torr were recommended. Keck reported that addition of CF₃CO₂H or CF₃SO₃H strongly accelerates the reactions catalyzed by the BINOL-Ti(O^{i}Pr)₂ complex (**2**) [75].

Scheme 27. Asymmetric carbonyl addition reaction catalyzed by BINOL-Ti

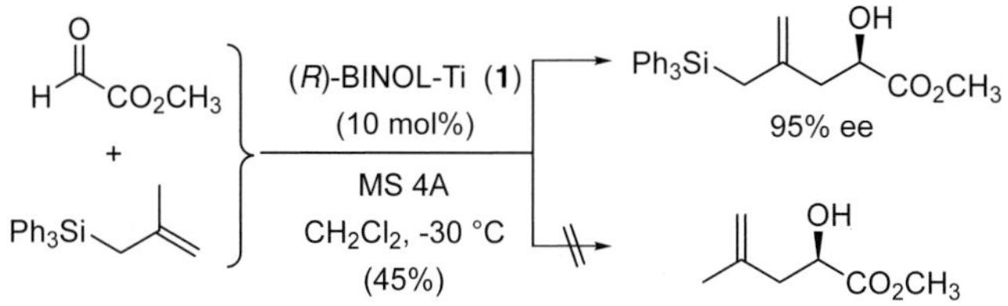

Scheme 28. Asymmetric ene-type reaction of methallylsilanes

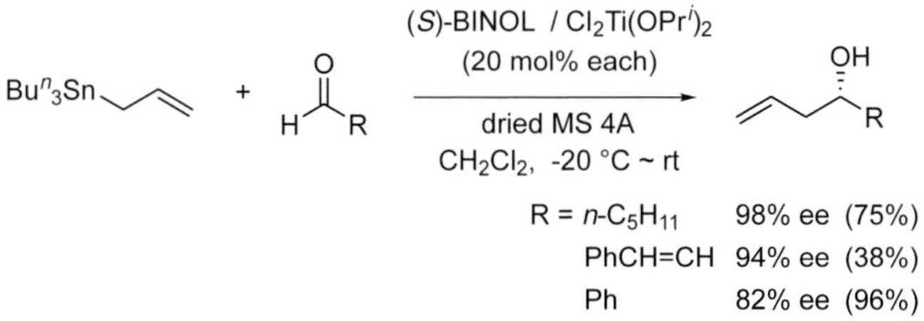

Scheme 29. Asymmetric carbonyl addition reaction of aliphatic and aromatic aldehydes

6
Aldol-Type Reaction

The aldol reaction constitutes one of the most fundamental bond construction processes in organic synthesis [76a, 76b]. Therefore, much attention has been focused on the development of asymmetric catalysts for aldol reactions using silyl enol ethers of ketones or esters as storable enolate components, the so-called Mukaiyama aldol condensation.

We have found that the BINOL-derived titanium complex serves as an efficient catalyst for the Mukaiyama-type aldol reaction of ketone silyl enol ethers with good control of both absolute and relative stereochemistry (Scheme 30) [77]. Surprisingly, however, the aldol products were obtained in the silyl enol

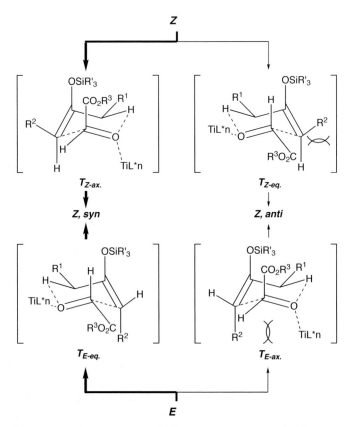

Scheme 30. Asymmetric aldol (prototropic ene-type) reaction

Fig. 7. Transition states of Mukaiyama-type aldol reaction of ketone silyl enol ether

ether (ene product) form, with high *syn*-diastereoselectivity from either geometrical isomer of the starting silyl enol ethers.

It appears likely that the reaction proceeds through an ene reaction pathway.
Such an ene reaction pathway has not been previously recognized as a possible
mechanism in the Mukaiyama aldol condensation. Usually, an acyclic antiperi-

planar transition state model has been used to explain the formation of the *syn*-diastereomer from either (*E*)- or (*Z*)-silyl enol ethers [78a, 78b]. However, the cyclic ene mechanism now provides another rationale for the *syn*-diastereose-lection irrespective of the enol silyl ether geometry (Fig. 7).

The aldol reaction of a silyl enol ether proceeds in a double and two-direc-tional fashion – upon addition of an excess amount of an aldehyde – to give the silyl enol ether in 77% isolated yield and in more than 99% ee and 99% de (Scheme 31) [79]. The present asymmetric catalytic aldol reaction is character-ized by a kinetic amplification phenomenon of the product chirality on going from the one-directional aldol intermediate to the two-directional product (Fig. 8). Further transformation of the *pseudo* C_2 symmetric product whilst still being protected as the silyl enol ether leads to a potent analogue of an HIV pro-tease inhibitor.

The silatropic ene pathway, that is, direct silyl transfer from an enol silyl ether to an aldehyde, may be involved as a possible mechanism in the Mukaiyama al-dol-type reaction. Indeed, *ab initio* calculations show the silatropic ene pathway involving the cyclic (boat and chair) transition states for the BH_3-promoted al-dol reaction of the trihydrosilyl enol ether derived from acetaldehyde with for-maldehyde to be favored [80]. Recently, we have reported the possible interven-tion of a silatropic ene pathway in the asymmetric catalytic aldol-type reaction of silyl enol ethers of thioesters [81]. The chloro and amino compounds thus obtained are useful intermediates for the synthesis of carnitine and GABOB (Scheme 32) [82a, 82b].

There is a dichotomy in the sense of *syn*- vs *anti*-diastereofacial preference, dictated by the bulkiness of the migrating group [80]. The sterically demanding silyl group shows *syn*-diastereofacial preference but the less demanding proton leads to *anti*-preference (Scheme 33). The *anti*-diastereoselectivity in carbonyl-

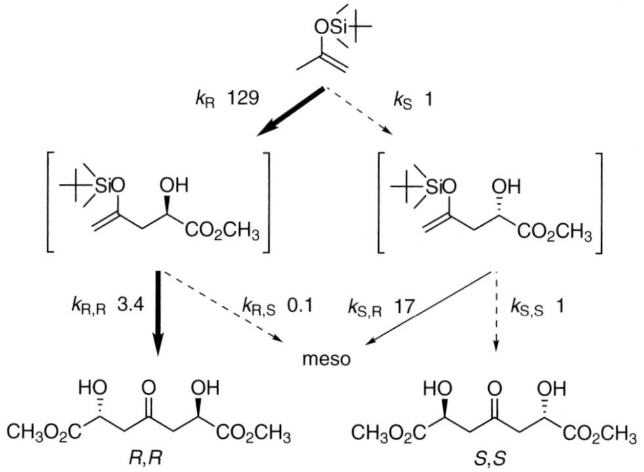

Fig. 8. Kinetic feature of two-directional aldol reaction

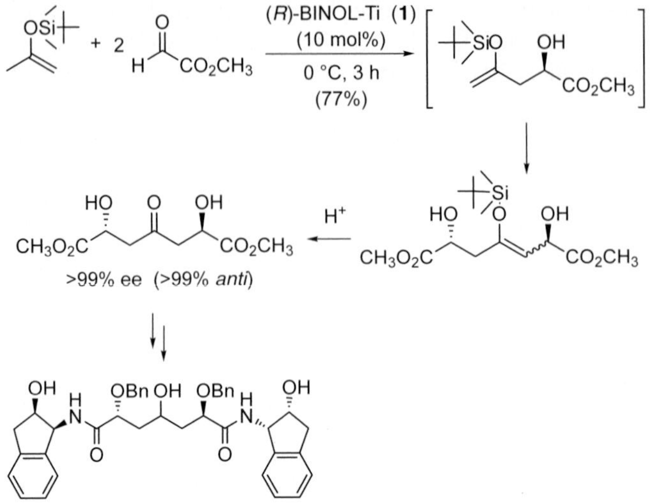

Scheme 31. Tandem and two-directional aldol-type reaction

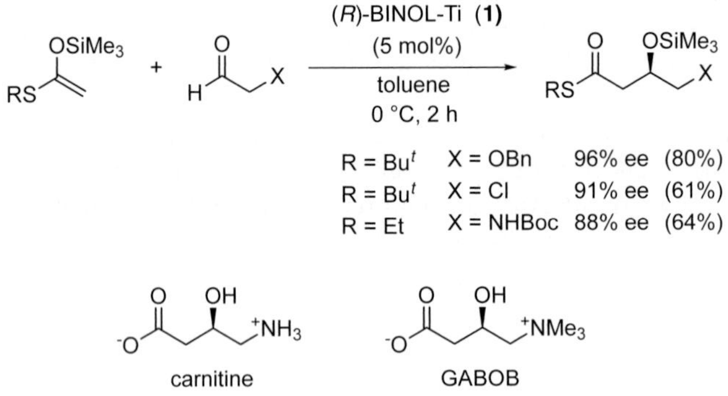

Scheme 32. Asymmetric aldol (metallotropic ene-type) reaction

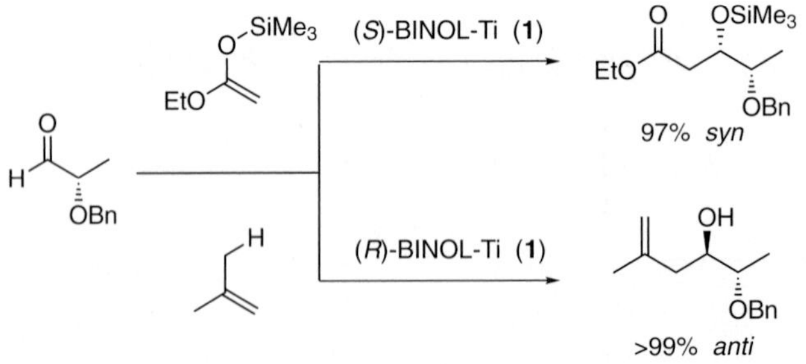

Scheme 33. Asymmetric carbonyl-ene vs aldol-type reaction of α-benzyloxy aldehyde

ene reactions can be explained by the Felkin-Anh-like cyclic transition state model (T_1) (Fig. 9). In the aldol reaction, by contrast, the inside-crowded transition state (T_1') is less favorable than T_2', because of the steric repulsion between the trimethylsilyl group and the inside methyl group of aldehyde (T_1'). Therefore, the *syn*-diastereofacial selectivity is visualized by the *anti*-Felkin-like cyclic transition state model (T_2').

Keck [83] and Carreira [84a, 84b] have independently reported catalytic asymmetric Mukaiyama aldol reactions. Keck et al. also reported an aldol reaction of α-benzyloxy aldehyde with Danishefsky's diene. The aldol product was transformed to the hetero-Diels-Alder type product through acid-catalyzed cyclization. In their method, the catalyst is prepared using 1:1 and 2:1 stoichiometry of BINOL and Ti(O^{i}Pr)$_4$ (Scheme 34) [85]. In their cases, oven dried 4 Å molecular sieves are used to generate the catalyst, which they reported to be of the BINOL-Ti(O^{i}Pr)$_2$ structure, under refluxing conditions.

Carreira employed a chiral BINOL-derived Schiff base-titanium complex as a catalyst for aldol reactions with acetate-derived ketene silyl acetals (Scheme 35) [84a]. The catalyst was prepared in toluene in the presence of salicylic acid, which was reported to be crucial to attain a high enantioselectivity. The similar Schiff base-titanium complex is also applicable to the carbonyl-ene type reac-

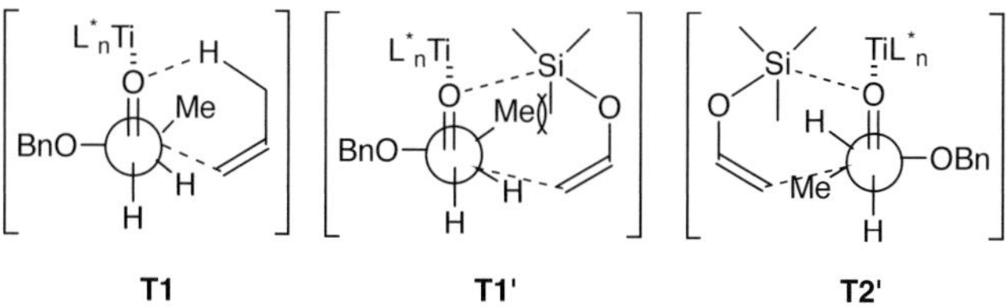

Fig. 9. Transition states of carbonyl-ene and aldol-type reaction of α-alkoxy aldehyde

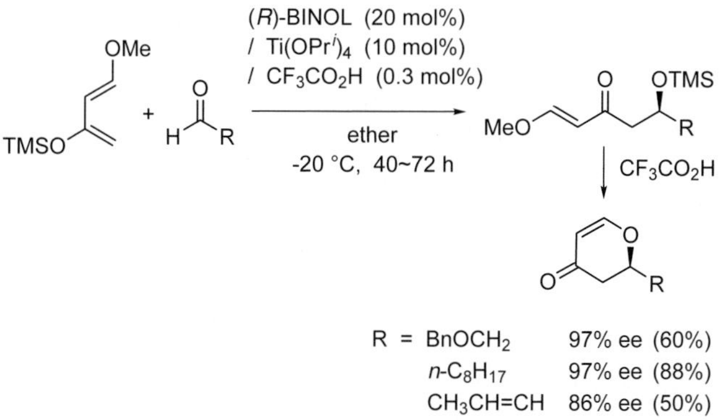

$$R = \text{BnOCH}_2 \quad 97\% \text{ ee } (60\%)$$
$$n\text{-C}_8\text{H}_{17} \quad 97\% \text{ ee } (88\%)$$
$$\text{CH}_3\text{CH=CH} \quad 86\% \text{ ee } (50\%)$$

Scheme 34. Asymmetric aldol-type reaction of Danishefsky's diene

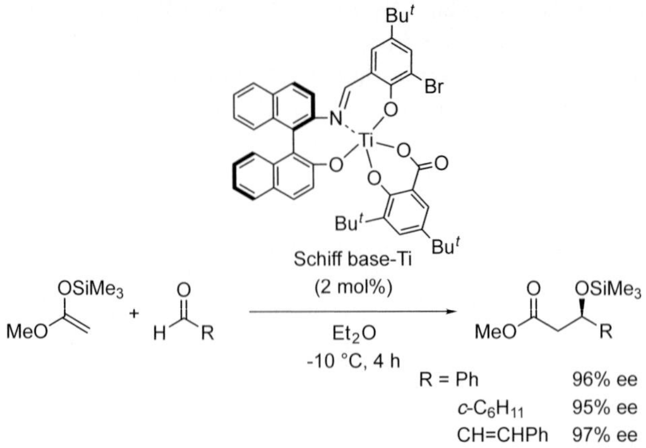

Scheme 35. Asymmetric aldol-type reaction catalyzed by chiral Schiff base-Ti

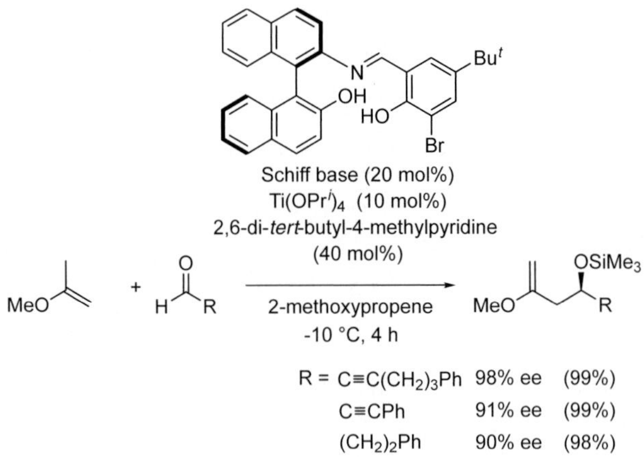

Scheme 36. Asymmetric ene-type reaction of 2-methoxypropene

tion with 2-methoxypropene [84b]. Although the reaction was conducted in the toluene or ether solution to provide no addition product, excellent chemical yield and enantioselectivity were achieved by the use of 2-methoxypropene as a solvent (Scheme 36).

7
Conclusion

The class of ene reactions, the carbonyl-ene reaction in particular, has gained a wide scope for synthetic applications and a mechanistic basis for the stereocontrolled construction of carbon skeletons by catalytic amounts of various chiral

Lewis acids. For practical applications, the development of more efficient catalysts is important, for which molecular design of asymmetric catalysts is the key in view of the structure-catalytic activity relationship. Any progress along this line is highly promising and worth the effort.

References

1.   Noyori R (1990) Science 248:1194
2.   Noyori R (1994) Asymmetric catalysis in organic synthesis. Wiley, New York
3.   Kagan HB (1997) Asymmetric synthesis. Thieme, Stuttgart
4.   Reetz MT (1986) Organotitanium reagents in organic synthesis. Springer, Berlin Heidelberg New York
5.   Seebach D (1990) Angew Chem Int Ed Engl 29:1320
6.   Also see the review on coordination numbers depending on ionic radii: Brown, ID (1988) Acta Cryst B44:545
7.   Reviews: Shilov AE, Shul'pin GB (1997) Chem Rev 97:2879; Schneider JJ (1996) Angew Chem Int Ed Engl 35:1068; Arndtsen BA, Bergman RG, Mobley TA, Peterson TH (1995) Acc Chem Res 28:154; Crabtree RH (1993) Angew Chem Int Ed Engl 32:789; Davies JA, Watson PL, Liebman JF, Greenberg A (1990) Selective hydrocarbon activation. Wiley, New York
8a. Mikami K, Shimizu M (1992) Chem Rev 92:1021
8b. Snider BB (1991) In: Trost BM, Fleming I (eds) Comprehensive organic synthesis. Pergamon, London, vol 2, p 527 and vol 5, p 1
8c. Mikami K, Terada M, Shimizu M, Nakai T (1990) J Synth Org Chem Jpn 48:292
8d. Hoffmann HMR (1969) Angew Chem Int Ed Engl 8:556
9.   Review: Mikami K (1995) Advances in asymmetric synthesis. JAI Press, Greenwich, Connecticut, vol 1, p 1
10. Mikami K, Loh T-P, Nakai T (1988) Tetrahedron Lett 29:6305; Mikami K, Loh T-P, Nakai T (1988) J Chem Soc Chem Commun 1430
11a. Hoffmann RW (1983) Angew Chem Int Ed Engl 22:489
11b. Weidmann B, Seebach D (1983) Angew Chem Int Ed Engl 22:31
11c. Yamamoto Y (1987) Acc Chem Res 20:243
11d. Hoffmann RW (1987) Angew Chem Int Ed Engl 26:489
11e. Roush WR (1991) In: Trost BM, Fleming I (eds) Comprehensive organic synthesis. Pergamon, London, vol 2, p 1
11f. Marshall JA (1992) Chemtracts 5:75
11g. Yamamoto Y, Asao N (1993) Chem Rev 93:2207
12. Maruoka K, Hoshino Y, Shirasaka T, Yamamoto H (1988) Annual Meeting of the Chemical Society of Japan, Tokyo, April 1–4, Abstract No. 1XIIB27; (1988) Tetrahedron Lett 29:3967
13. Omura S (1986) J Synth Org Chem Jpn 44:127; Hanessian S (1983) Total synthesis of natural products: the 'chiron' approach. Pergamon, Oxford; Mori K (1981) The total synthesis of natural products. Wiley, New York, vol. 4; Seebach D, Hungerbuhler E (1980) In: Scheffold R (ed) Modern synthetic methods. Otto Salle Verlag, Frankfurt am Main, vol. 2, p 91
14. Mikami K (1996) Pure Appl Chem 68:639; Mikami K, Terada M, Nakai T (1995) In: Doyle MP (ed) Advances in catalytic processes. JAI Press, London, vol. 1, p 123; Mikami K, Terada M, Narisawa S, Nakai T (1992) Synlett 255
15a. Mikami K, Terada M, Nakai T (1988) Annual Meeting of the Chemical Society of Japan, Tokyo, April 1–4, Abstract No. 1XIB43; (1989) J Am Chem Soc 111:1940
15b. Mikami K, Terada M, Nakai T (1990) J Am Chem Soc 112:3949
16. Mikami K, Terada M, Nakai T (1989) Chem Express 4:589

17. For the experimental details of the glyoxylate-ene reaction including the preparation of Br$_2$Ti(O^{i}Pr)$_2$: Mikami K, Terada M, Narisawa S, Nakai T (1992) Org Synth 71:14
18. Dijkgraff C, Rousseau JPG (1968) Spectrochim Acta 2:1213
19a. Mikami K, Motoyama Y, Terada M (1994) Inorg Chim Acta 222:71
19b. Terada M, Motoyama Y, Mikami K (1994) Tetrahedron Lett 35:6693
19c. Terada M, Mikami K (1995) J Chem Soc Chem Commun 2391
19d. Motoyama Y, Terada M, Mikami K (1995) Synlett 967
20. Mikami K (1995) In: Paquette LA (ed) Encyclopedia of reagents for organic synthesis. Wiley, Chichester, Vol. 1, p 403
21. Terada M, Matsukawa S, Mikami K (1993) J Chem Soc Chem Commun 327
22a. >96% ee: Mori K, Takigawa H (1991) Tetrahedron 47:2163
22b. 96% ee: Brown HC, Randad RS (1990) Tetrahedron 46:4463
22c. 91% ee: Ohloff G, Giersch W (1977) Helv Chim Acta 60:1496
23. Mikami K, Kotera O, Motoyama Y, Tanaka M, Maruta M, Sakaguchi H (1997) Annual Meeting of the Chemical Society of Japan, Tokyo, March 27–30, Abstract No. 1F108
24. Qian C, Huang T (1997) Tetrahedron Lett 38:6721
25. Kobayashi S, Hachiya I, Ishitani H, Araki M (1993) Synlett 472
26. Kabat MM, Lange M, Wovkulich PM, Uskokovic MR (1992) Tetrahedron Lett 33:7701; van der Meer FT, Feringa BL (1992) Tetrahedron Lett 33:6695; Kitamoto D, Imma H, Nakai T (1995) Tetrahedron Lett 36:1861
27a. Faller JW, Liu X (1996) Tetrahedron Lett 37:3449
27b. Akhmedov IM, Tanyeli C, Akhmedov MA, Mohammadi M, Demir AS (1994) Synth Commun 24:137
28. Mikami K, Yajima T, Terada M, Uchimaru T (1993) Tetrahedron Lett 34:7591; Mikami K, Yajima T, Terada M, Kato E, Maruta M (1994) Tetrahedron Asymmetry 5:1087; Mikami K, Yajima T, Takasaki T, Matsukawa S, Terada M, Uchimaru T, Maruta M (1996) Tetrahedron 52:85
29. Review: Welch JT, Eswarakrishnan S (1990) Fluorine in bioorganic chemistry. Wiley, New York
30a. Mikami K, Siree N, Yajima T, Terada M, Suzuki Y (1995) Annual Meeting of the Chemical Society of Japan, Tokyo, March 28–31, Abstract No. 3H218
30b. Mikami K, Yajima T, Siree N, Terada M, Suzuki Y, Kobayashi I (1996) Synlett 837
30c. Mikami K, Yajima T, Terada M, Kawauchi S, Suzuki Y, Kobayashi I (1996) Chem Lett 861
30d. Mikami K, Yajima T, Terada M, Suzuki Y, Kobayashi I (1997) J Chem Soc Chem Commun 57
31a. Mikami K, Yoshida A (1995) Synlett 29
31b. Mikami K, Yoshida A, Matsumoto Y (1996) Tetrahedron Lett 37:8515
32. Whereas the steric size of the CF$_3$ group is estimated to be as bulky as the isopropyl group by dynamic NMR measurements of 2,2'-disubstituted biphenyls, CF$_3$ group tends to act as the more bulky group like *tert*-butyl group in the reactions such as asymmetric borane reductions: Ramachandran PV, Teodorovic AV, Brown HC (1993) Tetrahedron 49:1725; see also: Leslie GB, Field D, Sternhell S (1980) J Am Chem Soc 102:5618
33. Significant decrease in the enantioselectivity was observed with the carbonyl-ene reaction of isopropyl glyoxylate catalyzed by the parent BINOL-Ti complex (1). See ref. 15b.
34. Acrylate derivatives of fluoroalkyl or fluorophenyl esters as activated dienophiles in Diels-Alder reactions: Kan T, Ohfune Y (1995) Tetrahedron Lett 36:943
35. Graven A, Johannsen M, Jørgensen KA (1996) J Chem Soc Chem Commun 2373
36. Johannsen M, Jørgensen KA (1996) Tetrahedron 52:7321; (1995) J Org Chem 60:5757
37a. Ward RS (1990) Chem Soc Rev 19:1
37b. Baba SE, Sartor K, Poulin J-C, Kagan HB (1994) Bull Soc Chim Fr 131:525; Rautenstrauch V (1994) Bull Soc Chim Fr 131:515
37c. Ward DE, How D, Liu Y (1997) J Am Chem Soc 119:1884
37d. Schreiber SL, Schreiber TS, Smith DB (1987) J. Am Chem Soc 109:1525
38. Review: Mikami K, Shimizu M. (1993) J Synth Org Chem Jpn 51:21

39. Bartlett PA (1980) Tetrahedron 36:2
40. Mikami K, Narisawa S, Shimizu M, Terada M (1992) J Am Chem Soc 114:6566; (1992) J Am Chem Soc 114:9242
41. Ziegler FE, Sobolov SB (1990) J Am Chem Soc 112:2749
42a. Kagan HB, Fiaud JC (1988) Topics in stereochemistry. Interscience, New York, vol 18
42b. Brown JM (1988) Chem Ind (London) 612
43a. Masamune S, Choy W, Peterson J, Sita LR (1985) Angew Chem Int Ed Engl 24:1
43b. Heathcock CH (1985) In: Morrison JD (ed) Asymmetric synthesis. Academic Press, New York, vol 3, p 111
44. Double asymmetric induction with terpenes as chiral ene components: Terada M, Sayo N, Mikami K (1995) Synlett 411
45a. Guillaneux D, Zhao S-H, Samuel O, Rainford D, Kagan HB (1995) J Am Chem Soc 116:9430; Puchot C, Samuel O, Dunach E, Zhao S, Agami C, Kagan HB (1986) J Am Chem Soc 108:2353; Girard C, Kagan HB (1998) Angew Chem Int Ed 37:2922
45b. Noyori R, Kitamura M. (1991) Angew Chem Int Ed Engl 30:49; Kitamura M, Suga S, Niwa M, Noyori R (1995) J Am Chem Soc 117:4832
45c. Oguni N, Matsuda Y, Kaneko T (1988) J Am Chem Soc 110:7877
46. Mikami K, Terada M (1992) Tetrahedron 48:5671; Terada M, Mikami K, Nakai T (1990) J Chem Soc Chem Commun 1623; see also: Terada M, Mikami, K (1994) J Chem Soc Chem Commun 833
47. Alcock NW, Brown JM, Maddox PJ (1986) J Chem Soc Chem Commun 1532
48. Maruoka K, Yamamoto H (1988) J Am Chem Soc 111:789
49. Faller JW, Parr J (1993) J Am Chem Soc 115:804; see also ref. 27a
50a. Mikami K, Matsukawa S (1997) Nature 385:613
50b. Matsukawa S, Mikami K (1995) Tetrahedron Asymmetry 6:2571
50c. (1996) Enantiomer 1:69
50d. (1997) Tetrahedron Asymmetry 8:815
51. Mikami K (1995) In: Paquette LA (ed) Encyclopedia of reagents for organic synthesis. Wiley, Chichester, Vvol 1, p 407
52. Matsukawa S, Mikami K (1997) Tetrahedron Asymmetry 8:815
53a. Taber DF (1984) Intramolecular Diels-Alder and Alder ene reactions. Springer Verlag, Berlin
53b. Fujita Y, Suzuki S, Kanehira K (1983) J Synth Org Chem Jpn 41:1152
53c. Oppolzer W, Snieckus V (1978) Angew Chem Int Ed Engl 17:476
53d. Conia JM, Le Perchec P (1975) Synthesis 1
54. Mikami K, Sawa E, Terada M (1991) Tetrahedron Asymmetry 2:1403
55. Sakane S, Maruoka K, Yamamoto H (1986) Tetrahedron 42:2203; (1985) Tetrahedron Lett 26:5535
56. Narasaka K, Hayashi Y, Shimada S (1988) Chem Lett 1609; Narasaka K, Hayashi Y, Shimada S, Yamada J (1991) Isr J Chem 31:261
57. Mikami K, Terada M, Sawa E, Nakai T (1991) Tetrahedron Lett 32:6571
58. Reviews on the synthesis and structure-function relationships of vitamine D analogues: Bouillon R, Okamura WH, Norman AW (1995) Endocrine Reviews 16:200; Dai H, Posner GH (1994) Synthesis 1383
59. Norman AW, Schaefer K, Grigoleit HG, Herrath D (ed) (1985) Vitamin D: a chemical, biochemical, and clinical update. Walter de Gruyter, Berlin
60. Mikami K, Osawa A, Isaka A, Sawa E, Shimizu M, Terada M, Kubodera N, Nakagawa K, Tsugawa N, Okano T (1998) Tetrahedron Lett 39:3359; Okano T, Nakagawa K, Ozono K, Kubodera N, Osawa A, Terada M, Mikami K (1998) Biol Pharm Bull 21:1300
61. Desimoni G, Faita G, Righetti PP, Sardone N (1996) Tetrahedron 52:12019
62. Oppolzer W (1995) In: Abel EW, Stone FGA, Wilkinson G (eds) Comprehensive organometallic chemistry, vol 12, p 905; Oppolzer W (1990) Pure Appl Chem 62:1941, and references cited therein
63. Ojima I, Tzamarioudaki M, Li Z, Donovan RJ (1996) Chem Rev 96:635

64. Yamamoto K, Terakado M, Murai K, Miyazawa M, Tsuji J, Takahashi K, Mikami K (1989) Chem Lett 955
65. Oppolzer W, Kuo DL, Hutzinger MW, Leger R, Durand L-O, Leslie C (1997) Tetrahedron Lett 38:6213
66. Trost BM, Van Vranken DL, Bingel C (1992) J Am Chem Soc 114:9836; Trost BM, Van Vranken DL (1992) Angew Chem In Ed Engl 31:228
67. Trost BM, Czeskis BA (1994) Tetrahedron Lett 35:211
68. Trost BM, Lee DC, Rise F (1989) Tetrahedron Lett 30:651
69. Goeke A, Sawamura M, Kuwano R, Ito Y (1996) Angew Chem Int Ed Engl 35:662
70. Takacs JM, Weidner JJ, Takacs BE (1993) Tetrahedron Lett 34:6219
71. Aoki S, Mikami K, Terada M, Nakai T (1993) Tetrahedron 49:1783
72. For the synthesis, see: Roush WR, Blizzad TA (1982) Tetrahedron Lett 23:2331; Still WC, Ohmizu H (1981) J Org Chem 46:5242
73. Mikami K, Matsukawa S (1994) Tetrahedron Lett 35:3133
74. Costa AL, Piazza MG, Tagliavini E, Trombini C, Umani-Ronchi A (1993) J Am Chem Soc 115:7001
75. Keck GE, Tarbet KH, Geraci LS (1993) J Am Chem Soc 115:8467; Keck GE, Krishnamurthy D, Grier MC (1993) J Org Chem 58:6543; see also: Weigand S, Brückner R (1996) Chem Eur J 2:1077
76a. Evans DA, Nelson JV, Taber TR (1982) Topics in stereochemistry. Interscience, New York, vol. 13
76b. Mukaiyama T (1982) Org React 28:203
77. Mikami K, Matsukawa S (1993) J Am Chem Soc 115:7039; Mikami K, Yajima T, Takasaki T, Matsukawa S, Terada M, Uchimaru T, Maruta M (1996) Tetrahedron 52:85; Mikami K, Takasaki T, Matsukawa S, Maruta M (1995) Synlett 1057. Also see: Mukaiyama T, Inubushi A, Suda S, Hara R, Kobayashi S (1990) Chem Lett 1015
78a. Murata S, Suzuki M, Noyori R (1980) J Am Chem Soc102:3248
78b. Yamamoto Y, Maruyama K (1980) Tetrahedron Lett 21:4607
79. Mikami K, Matsukawa S, Nagashima M, Funabashi H, Morishima H (1997) Tetrahedron Lett 38:579
80. Mikami K, Matsukawa S, Sawa E, Harada A, Koga N (1997) Tetrahedron Lett 38:1951
81. Mikami K, Matsukawa S (1994) J Am Chem Soc 116:4077
82a. Kolb HC, Bennari YL, Sharpless KB (1993) Tetrahedron Asymmetry 4:133
82b. Larcheveque M, Henrot S (1990) Tetrahedron 46:4277
83. Keck GE, Krishnamurthy D. (1995) J Am Chem Soc 117:2363
84a. Carreira EM, Singer RA, Lee W (1994) J Am Chem Soc 116:8837
84b. Carreira EM, Lee W, Singer RA (1995) J Am Chem Soc 117:3649
85. Keck GE, Li X-Y, Krishnamurthy D (1995) J Org Chem 60:5998

Chapter 33   Cycloaddition Reactions

Chapter 33.1
Diels-Alder Reactions

David A. Evans · Jeffrey S. Johnson

Department of Chemistry and Chemical Biology, Harvard University, Cambridge,
Massachusetts 02138, USA
e-mail: evans@chemistry.harvard.edu

Keywords: Diels-Alder reaction, Cycloaddition, Lewis acids, Enantioselective catalysis

1
Introduction

In the hierarchy of carbon-carbon bond constructions, the Diels-Alder reaction has attained a preeminent position [1]. This cycloaddition process allows for the stereoselective formation of cyclohexene rings possessing as many as four contiguous stereogenic centers while intramolecular [2] and transannular [3] variants facilitate the simultaneous formation of as many as three carbocyclic rings. It has long been recognized that the reaction facilitates the rapid development of molecular complexity and has been duly exploited in organic synthesis [4].

Given the prominent role of the Diels-Alder reaction in organic chemistry, it is not surprising that the search for enantioselective variants of this process has captured the attention of numerous researchers. Although chiral auxiliary-based reactions [5] retain a position of central importance, complementary catalytic variants are developing rapidly. Among these, chiral Lewis acid complexes that selectively activate one component (diene or dienophile) while providing a stereodefined environment are maturing as effective catalysts [6]. Accordingly, the ensuing discussion focuses on advances that have been made in the design and application of chiral Lewis acids for the Diels-Alder reaction.

2
Mechanistic Considerations in Lewis Acid Catalysis of the Diels-Alder Reaction [7]

2.1
General

In 1960, Yates and Eaton reported that an approximate rate acceleration of 10^5 was observed for the Diels-Alder reaction of anthracene and maleic anhydride in the presence of aluminum chloride (Scheme 1) [8]. This finding had important practical ramifications since it demonstrated that the Diels-Alder reaction could be conducted under mild conditions when an electropositive metal was used to lower the energy of activation.

In a subsequent study, Inukai and Kojima determined that the enthalpy of activation in the thermal reaction of butadiene with methyl acrylate was 18.0± 1.0 kcal/mol, while aluminum chloride catalysis reduced the activation energy to 10.4±1.9 kcal/mol; little effect on the entropy of activation was observed [9]. Cal-

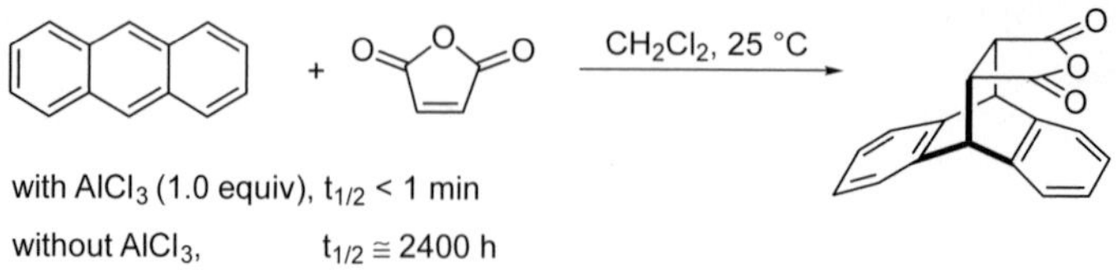

with AlCl$_3$ (1.0 equiv), t$_{1/2}$ < 1 min

without AlCl$_3$,   t$_{1/2}$ ≅ 2400 h

Scheme 1

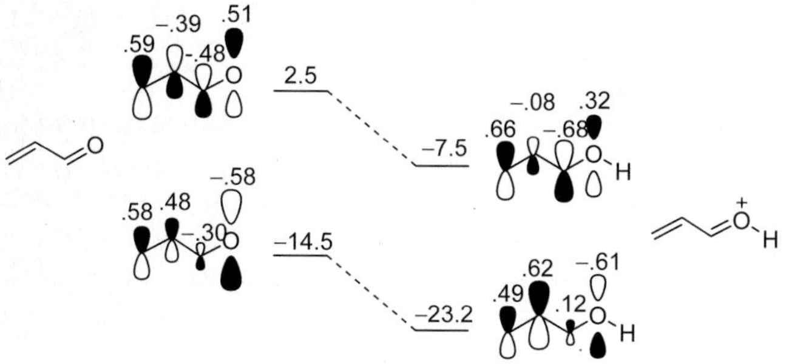

Fig. 1. Frontier orbital energies (eV) and coefficients for acrolein and protonated acrolein

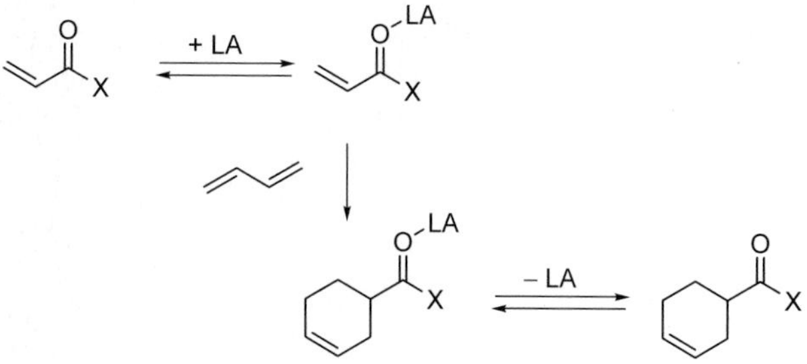

Scheme 2

culations for related cycloadditions suggest that Lewis acid catalysis will usually contribute approximately a 10 kcal/mol drop in the activation energy [10]. This positive attribute of Lewis acid catalysis has been explained on the basis of frontier molecular orbital theory by Houk and Strozier who showed that the coordination of an acid (a proton in this case) to a typical dienophile substantially lowers the LUMO energy (Fig. 1), thereby enhancing interaction with the diene HOMO and lowering the activation energy for the process [11].

In a simplified catalytic cycle, reversible coordination of the dienophile to the Lewis acid (LA) activates the substrate toward diene cycloaddition. In the catalyst turnover event, the Lewis acid-product complex dissociate to reveal the decomplexed cycloadduct and regenerated catalyst (Scheme 2). While this catalytic cycle neglects issues of product inhibition and nonproductive catalyst binding for dienophiles having more than one Lewis basic site, the gross features of this process are less convoluted than many other enantioselective reactions (e.g., olefin dihydroxylation, aldol reactions), a fact which may provide insight as to why this process is frequently used as a test reaction for new Lewis acid catalysts.

2.2
Diastereoselectivity and Transition State Issues

Subsequent studies have demonstrated another attractive feature of Lewis acid activation: enhanced *endo* diastereoselectivity (Fig. 2) [12,13]. Augmented secondary orbital interactions, an extension of Alder's notion of "maximum accumulation of unsaturations" [14], stemming from a "tighter" transition state for the catalyzed process relative to the thermal variant have been postulated as the source of enhanced *endo* diastereoselection [11]. However, this picture has since been refined, in particular for the Lewis acid-mediated cycloaddition. One study contends that stabilizing HOMO (diene)-LUMO (dienophile) interactions lead to destabilizing charge donation. The geometry of the *endo* transition state allows for the minimum induced charge separation and is thus favored [15]. It is generally agreed that Lewis acid dienophile activation results in a more asynchronous transition state: bond formation at the dienophile terminus is more advanced than for the internal carbon [16].

More recent research has uncovered some unusual attributes of the transition states for Lewis acid-catalyzed Diels-Alder reactions. Of note, a [4+3] transition state has been postulated as the low energy pathway for the borane-catalyzed Diels-Alder reaction between acrolein and 1,3-butadiene [17]; that is, a stabilizing interaction between the terminal carbon of the diene and the carbonyl carbon of the dienophile appears to be more important than the classically-invoked interaction between C-2 of the diene and the carbonyl carbon (Fig. 3). While this argument was originally advanced for only the *endo s-trans* transition state, it has been subsequently broadened in scope to include each of the four possible diastereomeric transition states for the boron trifluoride-promoted process [10]; thus, the energy differences for these reaction pathways are determined by the strength and number of the secondary interactions.

An even more unusual mechanistic hypothesis has arisen from calculations conducted at the Hartree-Fock level of theory which concluded that the boron trifluoride-promoted reaction of acrolein and 1,3-butadiene proceeds *via* a

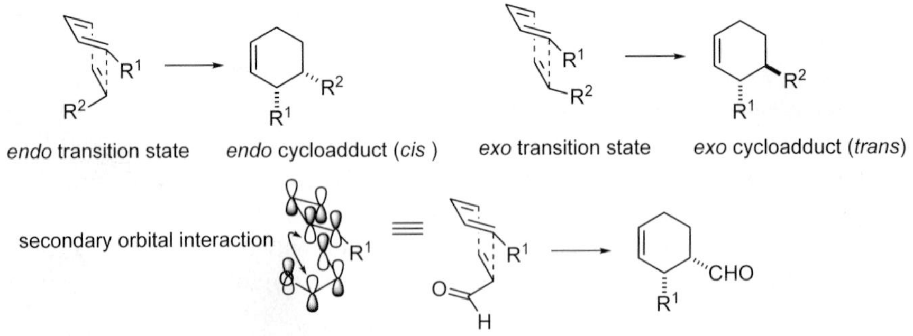

endo transition state   *endo* cycloadduct (*cis*)   *exo* transition state   *exo* cycloadduct (*trans*)

secondary orbital interaction

Fig. 2. Diels-Alder transition states and secondary orbital interaction for the *endo* transition state

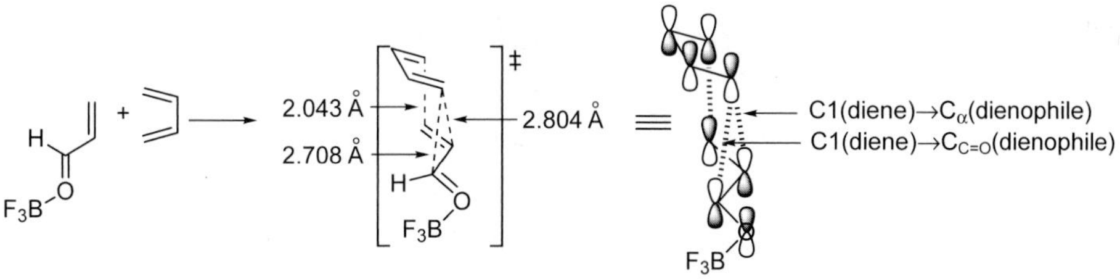

Fig. 3. Proposed [4+3] transition state

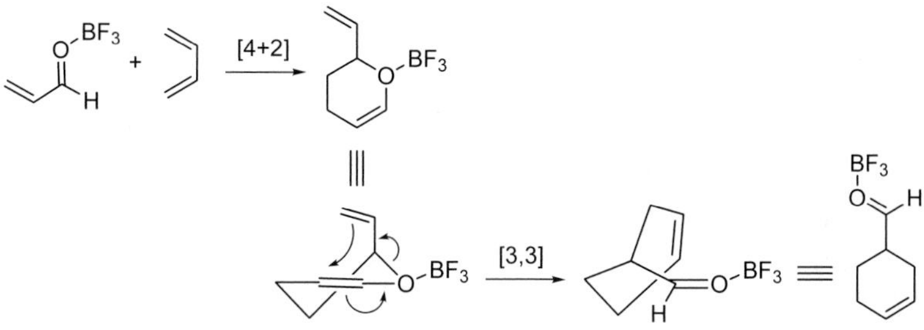

Fig. 4. Proposed hetero [2+4] cycloaddition-Claisen rearrangement mechanism

[2+4] hetero Diels-Alder reaction to afford a high energy boron-coordinated vinyl dihydropyran which undergoes a [3,3] sigmatropic rearrangement to give the observed carbocyclic Diels-Alder adduct (Fig. 4) [18]. When electron correlation effects are included by means of Density Functional Theory (DFT) calculations, however, the expected [4+2] cycloadduct is observed [10].

The preceding discussion is not meant to imply that stereoelectronic effects alone are responsible for determining diastereoselection in the Diels-Alder reaction. Indeed, examples of reactions that do not conform to the *endo* rule abound, and these cases are not easily explained without invoking alternative hypotheses. For instance, it has been demonstrated that 1,1-disubstituted dienophiles can favor formation of the *exo* product with cyclopentadiene, sometimes to the complete exclusion of the electronically favored *endo* isomer [19]. There appears to be subtle interplay between steric and electronic factors, as simply switching the diene to cyclohexadiene or an acyclic diene results in a turnover in selectivity to favor the *endo* isomer. While the exact source of stereocontrol for a given cycloaddition is still a source of debate, this review will emphasize the practical ramifications of diastereoselection, namely, prototypical dienophiles such as α-methacrolein and α-bromoacrolein can be relied on to deliver *exo* cycloadducts preferentially with cyclopentadiene (*endo* otherwise), while acrylate, crotonate, and cinnamate-derived dienophiles will generally favor the *endo* tran-

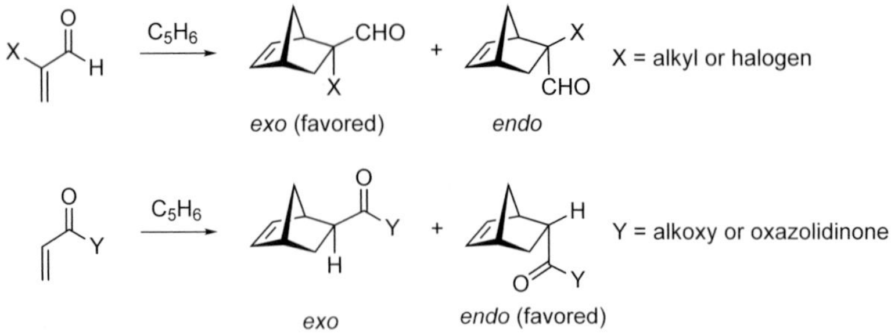

Fig. 5. Generalized stereochemical preferences as a function of dienophile

sition state in catalyzed reactions with dienes (Fig. 5). The reader is referred to reviews of this topic for a more exhaustive discussion [14, 20, 21].

2.3
The Nature of the Lewis Acid-Dienophile Complex

The realization of high enantioselectivity for the catalyzed Diels-Alder reaction (or any enantioselective process) relies on effective funneling of the reactants through a transition state that is substantially lower in energy relative to competing diastereomeric transition states. For the process at hand, a high level of transition state organization is required, necessitating control of several factors: 1) mode of binding (η^2 vs. η^1) of the carbonyl group to the Lewis acid; 2) for η^1 complexes, the regiochemistry of complexation to two or more available lone pairs; 3) conformation of the dienophile (s-cis vs. s-trans).

Control of these variables poses a formidable challenge to those engaged in reaction design, since all regiochemical and conformation issues must be addressed, independent of enantiofacial bias (Fig. 6). All three of these topics have been surveyed extensively elsewhere; the salient points will be summarized for the purpose of the ensuing discussion.

2.3.1
Mode of Complexation

Analysis of solid state and solution structures of metal-bound carbonyl complexes reveals two distinct modes of interaction. The carbonyl component may associate with the Lewis acid through its non-bonding electron pairs, or it may complex in a π sense through the C-O π bond. The interaction between an electron deficient Lewis acid and a carbonyl will likely result in a η^1 complex, while metal complexes with greater electron density have a higher propensity to form η^2 complexes with carbonyl compounds that are sufficiently π acidic [22]. For instance, Gladysz has shown that cationic rhenium complex [(η^5-C_5H_5)(PPh_3) (NO)Re]PF_6 binds an aldehyde (high π acidity) η^2, while the same complex

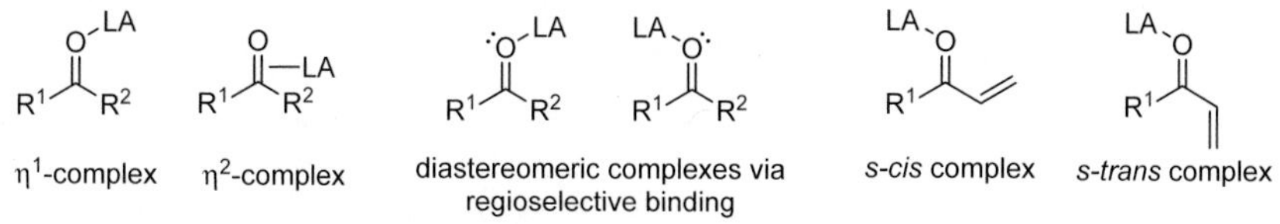

Fig. 6. Key factors in Lewis acid-dienophile complexation

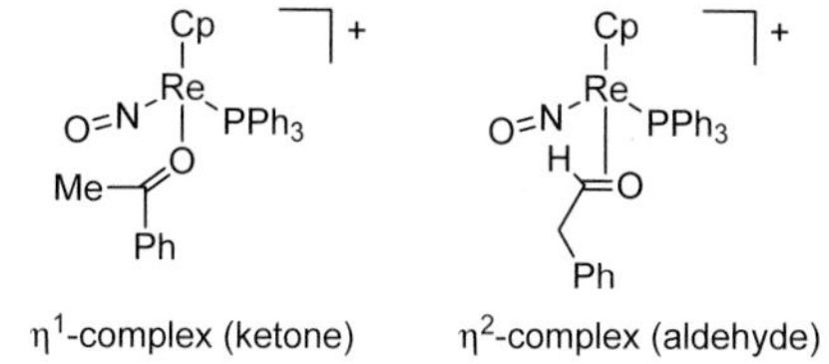

η^1-complex   η^2-complex   diastereomeric complexes via regioselective binding   *s-cis* complex   *s-trans* complex

η^1-complex (ketone)   η^2-complex (aldehyde)

Fig. 7. Turnover in binding mode (η^1 vs η^2) as a function of carbonyl group

binds a ketone (lower π acidity) η^1 (Fig. 7) [23, 24]. Because the former case results in increased electron density on the carbonyl due to a HOMO (metal)-LUMO (carbonyl) interaction, this is less useful with respect to activation of α,β unsaturated carbonyls toward electron-rich dienes (normal electron demand). As a consequence, η^1 complexes are thought to be operative in catalytic enantioselective Diels-Alder reactions. The reader is referred to an excellent review of this topic by Schreiber and co-workers for a thorough treatment of the literature associated with Lewis acid-carbonyl complexation [25].

2.3.2
Regioselection in Lewis-Acid/Carbonyl Complexation

Extensive spectroscopic and theoretical work has laid a solid foundation for predicting how a given carbonyl compound will bind to a Lewis acid. The case of unsaturated aldehydes is the most straightforward, as Lewis acid complexation has only been observed *syn* to the formyl proton, both in the solid state and in solution. Reetz and co-workers have reported that the benzaldehyde-BF$_3$ complex exhibits the expected E geometry both in solution and the solid state by means of heteronuclear Overhauser effect (HOE) experiments and X-ray crystallography (Fig. 8A) [26], while Corey and co-workers showed crystallographically that BF$_3$ likewise coordinates methacrolein *syn* to the formyl proton (Fig. 8B) [27]. By the observation of an HOE between the metal center and formyl proton, Denmark and Almstead deduced that a number of aldehydes coordinate to SnCl$_4$ in the E geometry (Fig. 8C) [28]. They further observed that α,β-unsaturated aldehydes were significantly more Lewis basic that saturated or alkynyl aldehydes.

While this geometrical preference likely results from the impact of steric effects, hypotheses which suggest electronic effects as biasing elements have been

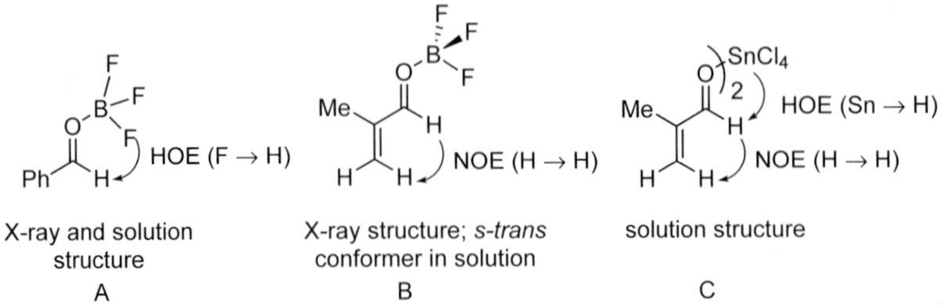

Fig. 8. *Anti* aldehyde-Lewis acid complexes

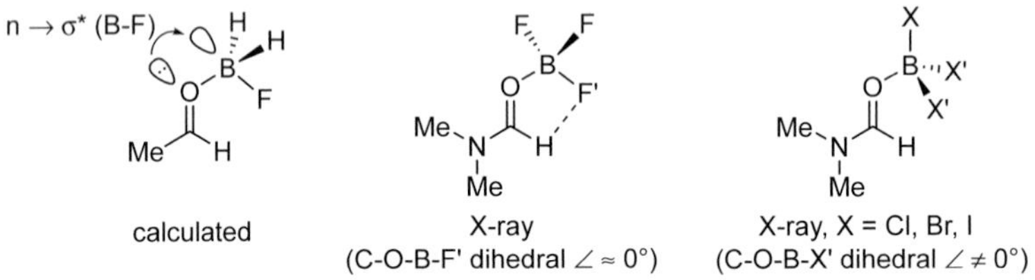

Fig. 9. Conformation of *anti* aldehyde-Lewis acid complexes; proposed anomeric effect and formyl hydrogen bond

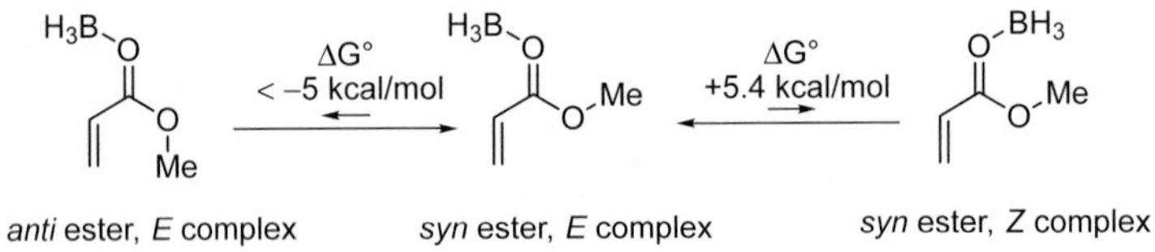

Fig. 10. Borane-methyl acrylate complexes (*ab initio*)

proposed. Goodman has suggested that the computationally-indicated conformational preference (Fig. 9) for boron-bound aldehydes is a consequence of an anomeric effect between the uncomplexed oxygen lone pair and the B-F antibonding orbital ($n \rightarrow \sigma^*$ (B-F)) [29]. Corey and co-workers have argued that the issue is not one of stereoelectronics, but rather a previously unappreciated hydrogen bond between the boron-bound fluoride and the formyl hydrogen [30]. This argument is derived from the fact that only the B-F bond eclipses the C-O bond in the X-ray structures of dimethylformamide-BX_3 complexes (Fig. 9, X= F, Cl, Br, I). The heteroatom-formyl hydrogen bond concept has been extended by analogy to explain other enantioselective processes [31].

The lack of spectroscopic evidence for Lewis acid complexation *anti* to the formyl hydrogen in aldehyde-derived complexes does not imply that such complexes do not exist. Indeed, *ab initio* molecular orbital calculations suggest that the energy difference for E and Z BF_3·aldehyde complexes can be as small as 1.2 kcal/mol, indicating that the Z conformer is present at equilibrium [32].

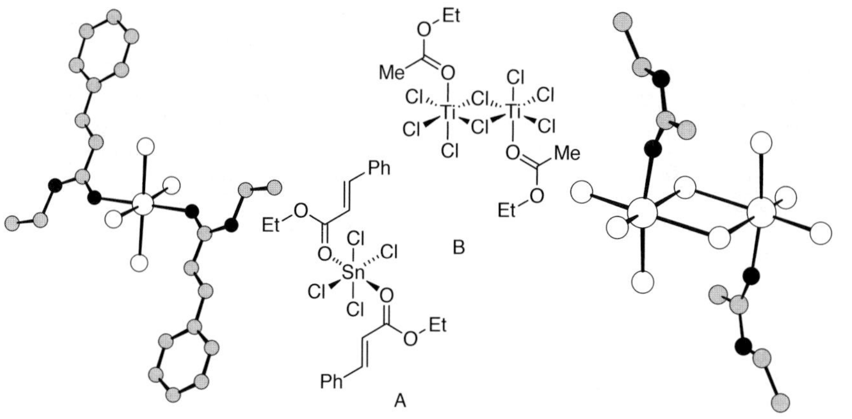

Fig. 11. X-ray structures of metal-bound esters

Ab initio calculations for the borane-methyl acrylate complex indicate that complexation of the lone pair *anti* the OMe group (*E* complex) is favored by 5.4 kcal/mol; the *syn* conformation of the ester is strongly favored over the *anti* (Fig. 10, for a discussion of the *s-cis/s-trans* issue, see Sect. 2.3.3) [33].

For the most part, the results of these calculations are reinforced by solid state structures. A structure of (ethyl cinnamate)$_2$·SnCl$_4$ indicates that both esters are disposed *syn* and favor the *E* complex (Fig. 11A) [34]. Similarly, ethyl acetate complexes with TiCl$_4$ to afford a dimeric structure with bridging chlorides; the esters are *syn* and complexation occurs *anti* to the ethoxy group (Fig. 11B) [35].

Those carbonyl compounds discussed above are prototypical "one-point binding" substrates. That is, coordination to the Lewis acid occurs in a monodentate fashion. In enantioselective catalysis of the Diels-Alder reaction, frequent use is made of bidentate dienophiles, substrates containing two Lewis basic sites capable of forming a chelate to the metal center (this is an extension of concepts which originated in the study of auxiliary-based Diels-Alder reactions; for leading references, see [36]). Such chelating interactions contribute an important organizational constraint to the transition state, and some effort has been made to understand the interaction of such substrates with Lewis acids. Shown in Fig. 12A is the representation of an X-ray structure of a complex between TiCl$_4$ and an acryloyl lactate dienophile known to afford cycloadducts in high diastereomeric excess [37]. An interesting feature of this structure is the somewhat unusual partial π coordination of the acryloyl carbonyl moiety, although the source of this out-of-plane bonding is not completely clear. While both ester groups are disposed *syn*, coordination of the acryloyl moiety is *anti* to the alkene, indicating that formation of a chelate is sufficiently favored to override the preference for the normal coordination mode *syn* to the alkene. Oppolzer's titanium-bound crotonyl sultam (Fig. 12B) also exhibits chelation in the solid state; the most convincing corroborating evidence for the existence of the chelate in solution was a

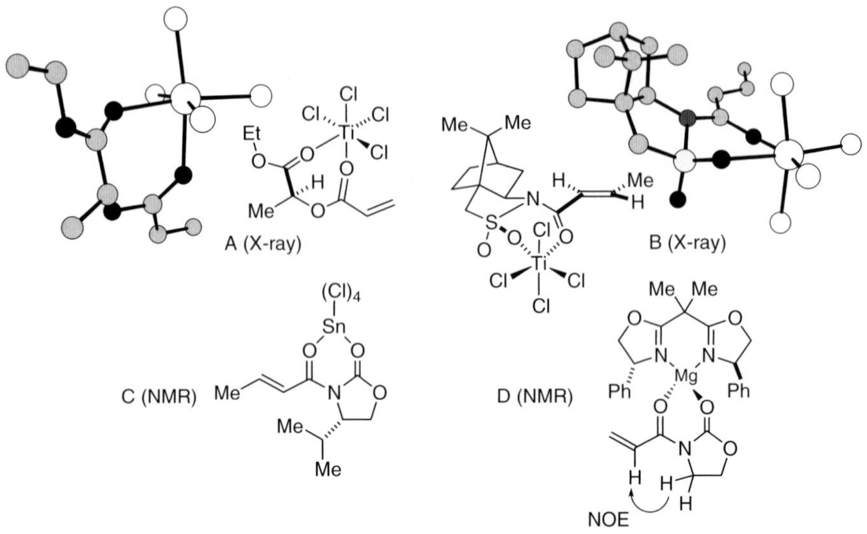

Fig. 12. Solution and solid state structures of metal-bound chelating dienophiles

decrease in IR stretching frequency for both the carbonyl and sulfonyl vibrations relative to the free sultam [38]. Castellino has performed ^{1}H-, ^{13}C-, and ^{119}Sn-NMR spectroscopic studies on $SnCl_4$-bound crotonyl oxazolidinone and found that, to the limits of detection, only the chelated complex is formed (Fig. 12C) [39]. Achiral acyl oxazolidinones are among the most commonly employed chelating dienophiles in catalytic enantioselective Diels-Alder reactions, although in cases where the metal center cannot accommodate two additional ligands, one point-binding has been invoked in discussions of asymmetric induction.

2.3.3
Dienophile Conformation

The *s-cis/s-trans* dienophile conformational issue is critical to the analysis of any given enantioselective process since the interconversion of the two conformers in any well-defined chiral environment results in a reversal in the predicted enantiofacial bias. Consequently, considerable effort has been expended in studying this equilibrium.

   Ab initio calculations and experimental measurements suggest that coordination of an α,β-unsaturated carbonyl to a Lewis acid results in an increase in the barrier to rotation about the C_1-C_2 single bond from 4–9 kcal/mol to 8–12 kcal/mol as a result of augmented C_1-C_2 double bond character (Fig. 13) [33]. This energy barrier is in the same regime as the measured energy of activation for a typical catalyzed Diels-Alder reaction.

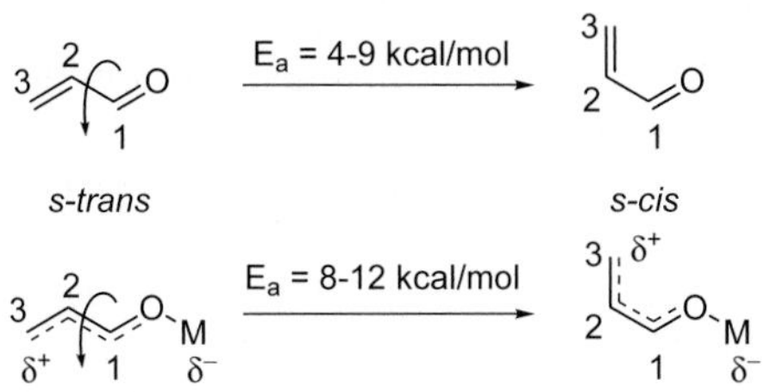

Fig. 13. Barrier to rotation of free and coordinated dienophiles

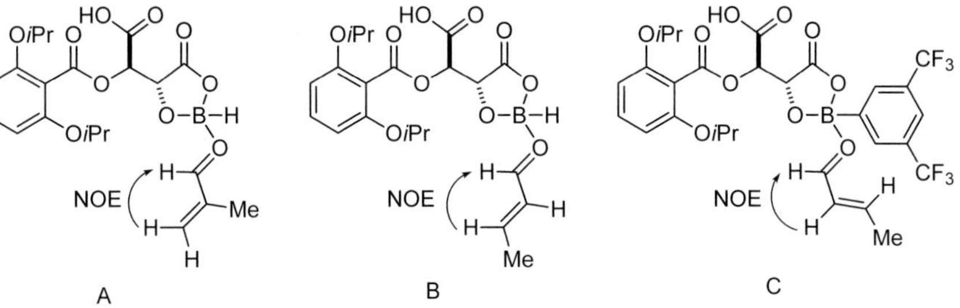

Fig. 14. Solution structures of CAB-aldehyde complexes

Uncomplexed acrolein, methacrolein, and crotonaldehyde all favor the *s-trans* conformer, and this preference is enhanced upon complexation to a Lewis acid. For example, Corey showed that the BF_3-methacrolein complex adopts the *s-trans* conformation in the solid state as well as in solution by crystallographic and NMR spectroscopic methods (Fig. 8B) [27], while Denmark and Almstead found that methacrolein adopts the *s-trans* geometry upon complexation with $SnCl_4$ (Fig. 8C) [28]. Yamamoto demonstrated that methacrolein is also observed in the *s-trans* conformation upon complexation to his chiral acyloxyborane (CAB) catalyst (Fig. 14A and Sect. 3.1.2) [40]. Interestingly, with the same CAB system, crotonaldehyde exhibited varying preferences for the two possible conformers depending on the exact substituents on the boron. On the basis of NOE enhancements, the *s-trans* conformer was observed exclusively with a hydrogen substituent on boron (Fig. 14B); the *s-cis* conformer was the only one detected in the case of the aryl-substituted acyloxyborane (Fig. 14C).

The general preference for the *s-trans* conformer carries over to some extent for carboxylic esters. *Ab initio* calculations for the borane-methyl acrylate complex show a 1.4 kcal/mol preference for the *s-trans* conformer (Fig. 10), presumably due to reduced steric interactions (B-H for the *s-trans* vs. B-CH$_2$ for the *s-cis*) [33]. Solid state structures, however, show that both conformers can be observed for esters (Figs. 11 and 12).

The *s-cis* conformation observed for Oppolzer's chelating sultam (Fig. 12B) reflects a general dispositional preference for amides, consistent with lanthanide metal-induced shift NMR studies by Montaudo *et al.*, who provided an empirical equation for assessing the conformational distribution for a given α,β-unsaturated carbonyl compound [41]. In the context of studies on chiral magnesium catalysts, Desimoni and coworkers disclosed that acryloyl oxazolidinones exhibit NOE enhancements between the nitrogen-bearing methylene group of the heterocycle and the α-vinyl proton (Fig. 12D) [42]. Largely on the basis of these NMR experiments and those of Collins (Sect. 3.2.4), and crystallographic data provided by Jørgensen (Sect. 3.2.4), the *s-cis* conformation of chelated acyl oxazolidinone dienophiles has been inferred.

However, as a harbinger of the danger in predicting transition state structures on the basis of preferred ground state conformations, Houk has found in *ab initio* calculations that borane-bound acrolein preferentially adopts the *s-trans* configuration, but the activation energy for reaction (with 1,3-butadiene) from the *s-cis* configuration is decidedly lower [15]. This finding has been reinforced with the DFT calculations of Garcia *et al.* [10] and foreshadows the ubiquity of the Curtin-Hammett principle in catalytic Diels-Alder reactions: numerous proposed transition structures that appear in this review are derived from higher energy dienophile-catalyst complexes.

3
Chiral Lewis Acid Catalysis of the Diels-Alder Reaction

Promotion of the Diels-Alder reaction by a substoichiometric amount of chiral Lewis acid has developed to a relatively high level of sophistication as a result of the extensive research in this field. In the interest of providing mechanistic insight into highly efficient systems, the discussion will be limited to systems which provide synthetically useful levels of enantioselection (typically greater than 90%) [43]. Even with this restriction, the reader will note remarkable breadth in the chiral complexes that have been studied. As a result of the unique characteristics different metals confer to Lewis acidic complexes, it is advantageous to discuss each metal in turn.

3.1
Main Group Lewis Acids

3.1.1
Aluminum

In 1979, Koga and coworkers disclosed the first practical example of a catalytic enantioselective Diels-Alder reaction [44] promoted by a Lewis acidic complex, presumed to be "menthoxyaluminum dichloride" (1), derived from menthol and ethylaluminum dichloride, whose structure remains undefined [45]. This complex catalyzed the cycloaddition of cyclopentadiene with acrolein, methyl acrylate, and methacrolein with enantioselectivities as high as 72% ee. Oxidation of 2 (predominantly *exo*) followed by recrystallization actually lowered the ee;

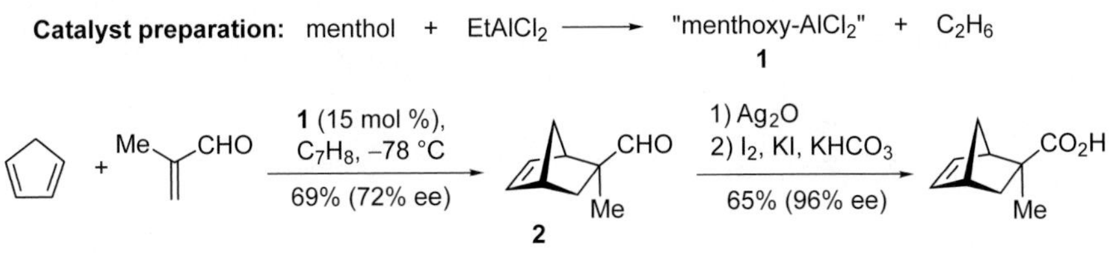

Catalyst preparation: menthol + EtAlCl$_2$ $\longrightarrow$ "menthoxy-AlCl$_2$" + C$_2$H$_6$

1

Scheme 3

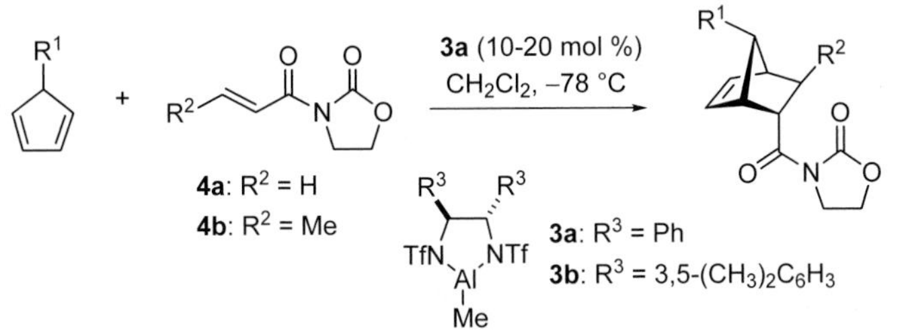

4a: R^2 = H
4b: R^2 = Me

3a: R^3 = Ph
3b: R^3 = 3,5-(CH$_3$)$_2$C$_6$H$_3$

R^1	R^2	Yield [%]	endo/exo	ee [%]
H	H	92	>50:1	91
H	Me	88	96:4	94
CH$_2$OBn	H	94	–	95

Scheme 4

however, isolation of the mother liquors gave product of 96% ee (Scheme 3). On the basis of a proposed transition state, Koga and coworkers made systematic changes to the cyclohexanol moiety and the aluminum substituents, but the highest ee was realized for the original system [46, 47].

A decade later, Corey introduced an effective aluminum-diamine controller for Diels-Alder and aldol additions. The C_2-symmetric stilbenediamine (stien) ligands are available in good yield from substituted benzils, which are in turn derived from benzoic acids, aryl aldehydes, or aryl bromides [48]. Formation of the active catalyst **3** is achieved by treatment of the bis(sulfonamide) with trimethylaluminum; recovery of the ligand was essentially quantitative. Acryloyl and crotonyl imides **4** are particularly effective dienophiles for this system, as shown in Scheme 4.

The transition structure depicted in Fig. 15 was suggested by the authors based on a dimeric X-ray structure of the catalyst and NMR spectroscopic data showing an NOE enhancement between the α-vinyl proton of the dienophile and the benzylic proton of the catalyst. While imide **4** is typically viewed as a chelating Lewis base, the presumed tetracoordinate aluminum would prevent this mode of activation. As noted previously, imide **4** is generally assumed to pre-

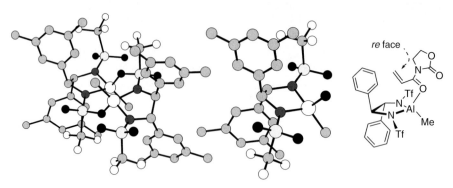

Fig. 15. X-ray structure of catalyst **3b** (dimeric); simplified view of the X-ray structure with one-half of the dimer excised; proposed transition structure for aluminum-stien catalyzed Diels-Alder reactions

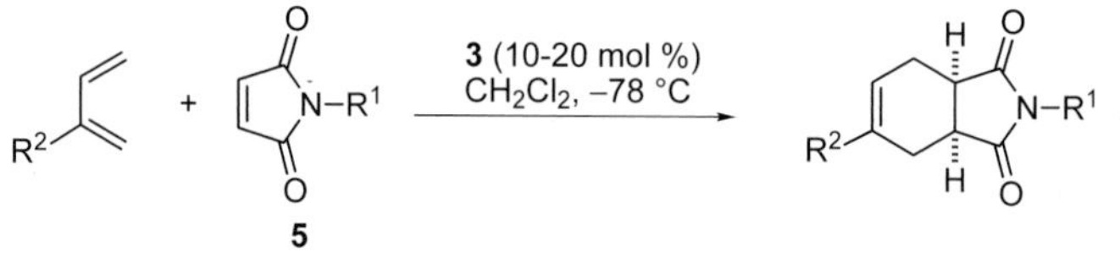

R^1	R^2	Catalyst	ee [%]
2-CH$_3$C$_6$H$_4$	OMe	3b	93
2-CH$_3$C$_6$H$_4$	OMe	3a	58
Ph	OMe	3b	62
2-CMe$_3$C$_6$H$_4$	OMe	3b	95
2-I C$_6$H$_4$	OMe	3b	93
2-Me-4-BrC$_6$H$_4$	OMe	3b	>97
2-CMe$_3$C$_6$H$_4$	CH$_2$SiMe$_3$	3b	95

Scheme 5

fer the *s-cis* conformer in the ground state; the proposed *s-trans* geometry and potential electrostatic repulsion between the two carbonyls should be noted, but the absolute stereochemistry of the adducts is consistent with Fig. 15 and the aforementioned experimental data.

Subsequent studies have expanded the scope of this catalyst to include maleimides **5** [49]. In order to obtain enantiomerically enriched cycloadducts with this symmetrical dienophile an unsymmetrical diene was used (Scheme 5); this constitutes the first example of such a process. *Ortho* substitution on the *N*-aryl group was found to be crucial to the realization of high enantioselectivity, perhaps to discourage catalyst binding the carbonyl lone pair *syn* to the *N*-aryl moiety; a transition state analogous to that depicted for the imides (Fig. 15) was invoked. The fact that the dienophile is locked in the *s-trans* conformer could lend

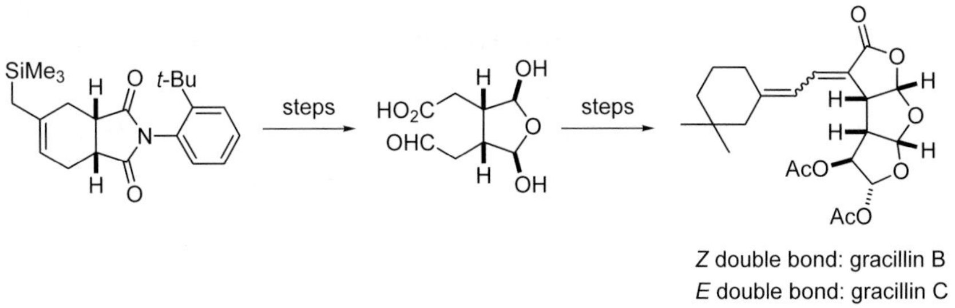

Z double bond: gracillin B
E double bond: gracillin C

Fig. 16. Transformation of maleimide-derived cycloadduct into gracillins B and C

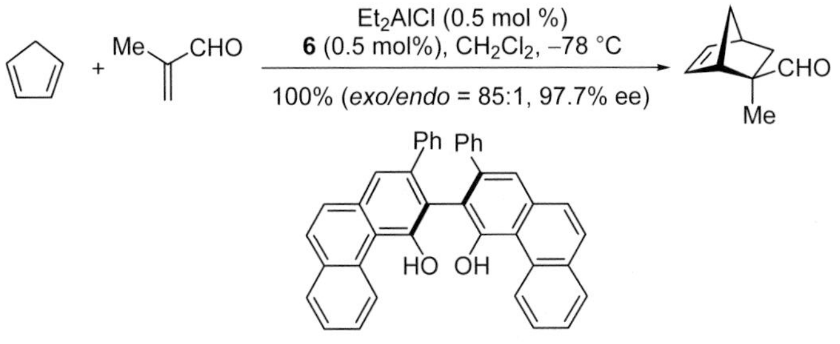

Et_2AlCl (0.5 mol %)
6 (0.5 mol%), CH_2Cl_2, −78 °C

100% (exo/endo = 85:1, 97.7% ee)

6

Scheme 6

support to the Curtin-Hammett scenario necessary for the transition state depicted in Fig. 11 to be operative. This methodology was exploited in elegant syntheses of gracillins B and C (Fig. 16) [50].

In 1993, Wulff and coworkers reported their finding that a complex derived from diethylaluminum chloride and "vaulted" biaryl ligand **6** catalyzed the enantioselective Diels-Alder reaction between cyclopentadiene and methacrolein (Scheme 6) [51, 52]. Although somewhat lengthy, the ligand preparation is amenable to preparative scale synthesis (11–12 mmol). This possible detraction is attenuated by the 0.5 mol % catalyst loading, which is the lowest reported for any enantioselective carbocyclic Diels-Alder reaction. Further, the chiral ligand is recovered quantitatively by silica gel chromatography.

A notable feature of this catalytic system is that asymmetric induction is lower at the early stages of the reaction. A subsequent study revealed that in the reaction of methyl acrylate and cyclopentadiene, the cycloadduct interacts with the catalyst in a fashion such that the enantioselectivity is intimately tied to the percent conversion (Scheme 7).

The result of their exploratory effort was the determination that an enantioselective autoinductive mechanism is operative. Only one other example of such a mechanism exists in the context of the Diels-Alder reaction [53]. In a series of clever experiments, the authors found that achiral additives achieve the same end, facilitating uniformly high asymmetric induction throughout the course of

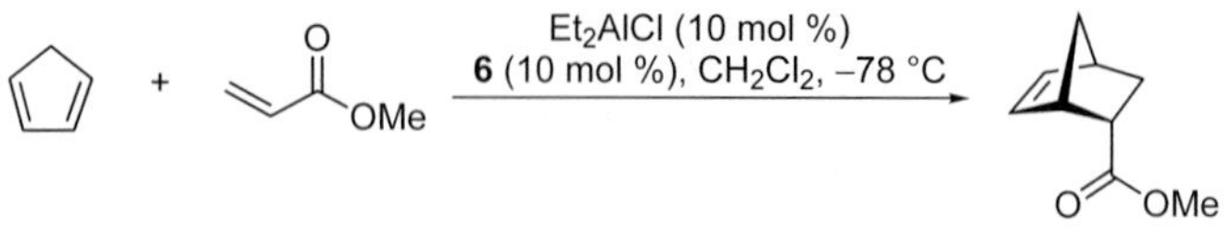

Et$_2$AlCl (10 mol %)
6 (10 mol %), CH$_2$Cl$_2$, −78 °C

Conversion [%]	ee [%]
21	48
43	72
61	81
>95	82

Scheme 7

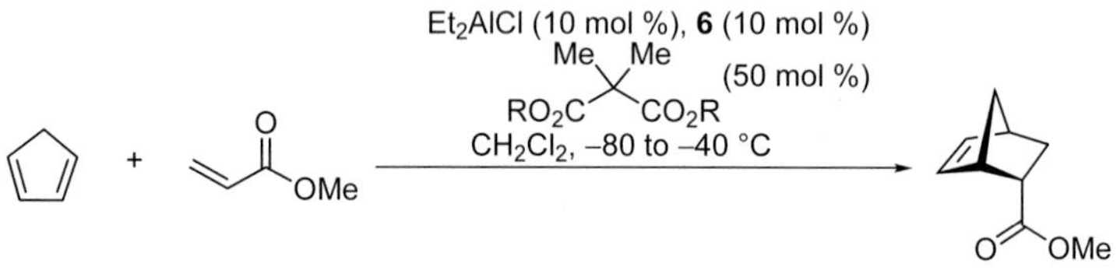

Et$_2$AlCl (10 mol %), 6 (10 mol %)
Me   Me
(50 mol %)
RO$_2$C   CO$_2$R
CH$_2$Cl$_2$, −80 to −40 °C

R	Temp [°C]	Yield [%]	endo/exo	ee [%]
Me	−80	49	99:1	98
i-Pr	−80	70	99:1	97.5
t-Bu	−80	76	99:1	>99
1-adamantyl	−40	100	98.1:1	92.5

Scheme 8

the reaction (Scheme 8). The efficacy of malonate additives suggests that the catalytically active species might be a hexacoordinate aluminum center; future work is aimed at determining whether this is in fact the case. It should be noted that selective cycloadditions of acrylate esters are rarer than for their aldehyde or imide counterparts [54]; therein lies an attractive attribute of the Wulff system.

3.1.2
Boron [55]

Yamamoto and coworkers have developed a practical Diels-Alder catalyst for aldehyde dienophiles. Treatment of a monoacylated tartaric acid with borane released ca. 2.2 equiv of H$_2$ gas, affording a complex that has been assigned structure 7. Circumstantial evidence for structure 7 was found in the comparable enantioselectivity of a catalyst in which the free carboxyl group was esterified (see below). The chiral (acyloxy)borane (CAB) complex is effective in catalyzing a number of aldehyde-based Diels-Alder reactions (Scheme 9) [56]. Reactions with

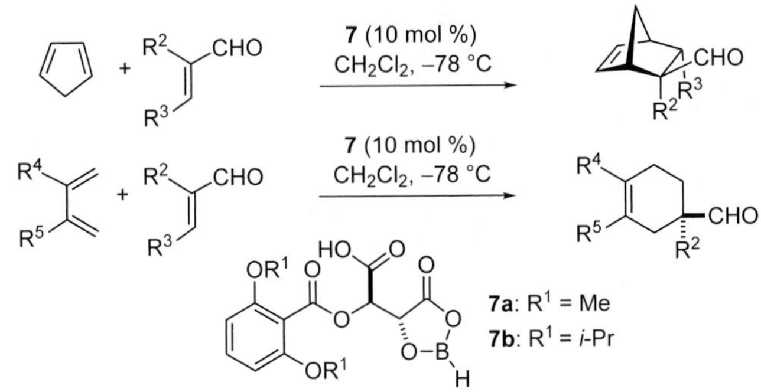

7 (10 mol %)
CH$_2$Cl$_2$, −78 °C

7 (10 mol %)
CH$_2$Cl$_2$, −78 °C

7a: R^1 = Me
7b: R^1 = i-Pr

Catalyst	R^2	R^3	R^4	R^5	Yield [%]	exo/endo	ee [%]
7a	Me	H	–	–	85	89:11	96
7a	H	H	–	–	90	12:88	84
7a	Me	Me	–	–	91	97:3	90
7b	Br	H	–	–	100	94:6	95
7b	Br	Me	–	–	100	>99:1	98
7b	Br	H	Me	Me	80	–	95
7a	Me	H	Me	Me	61	–	97
7a	Me	H	Me	H	65	–	91

Scheme 9

cyclopentadiene are fairly general with respect to the aldehyde, with the exception of crotonaldehyde (2% ee). Less reactive dienes such as isoprene and 2,3-dimethyl-1,3-butadiene may be successfully employed with bromoacrolein and methacrolein dienophiles.

A series of NMR spectroscopic experiments established that the preferred ground state conformation for both crotonaldehyde and methacrolein is *s-trans* when complexed to **7b** (Fig. 14) [40]. Additionally, NOE experiments indicated close proximity of the aldehyde and the aryl ring; π-stacking between the aryl group and aldehyde was suggested as an organizational feature which imparted high enantioselectivity to the cycloaddition event (Fig. 17) [57]. A crystal structure of the uncomplexed monoacylated tartaric acid revealed a folded rather than extended structure, further suggesting the possibility of this arrangement.

As illustrated in Scheme 10, the CAB catalyst also effectively catalyzes the intramolecular Diels-Alder reaction of trienal **8** to afford bicyclic product **9** in high diastereo- and enantioselectivity [58]. In a single step, this *endo*-selective reaction achieves the formation of a tetrahydroindane ring system containing a stereogenic quaternary center.

A tryptophan-derived oxazaborolidine has been shown to be an effective catalyst for aldehyde-based Diels-Alder reactions. Complex **10**, prepared from α-methyl tryptophan and BuB(OH)$_2$ with removal of water, effects the cycloaddi-

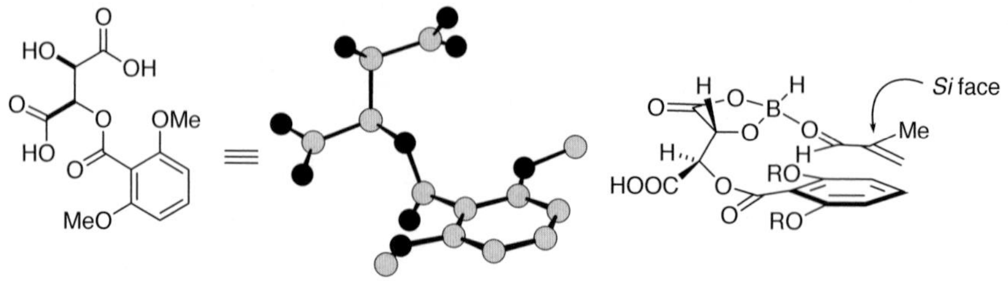

Fig. 17. X-ray structure for monoacylated tartaric acid precursor for complex **7a** and proposed transition state assembly for CAB catalyst **7** [40]

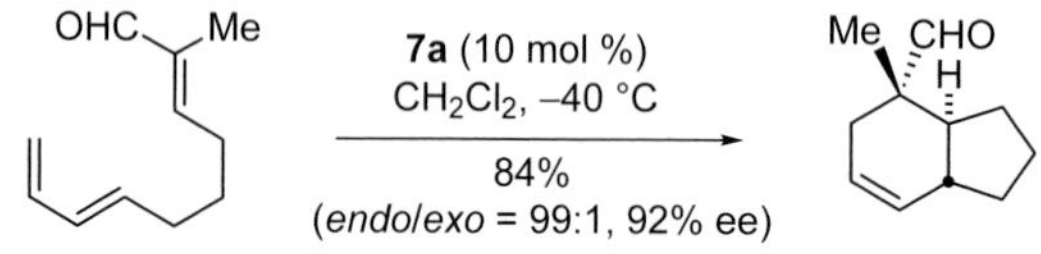

Scheme 10

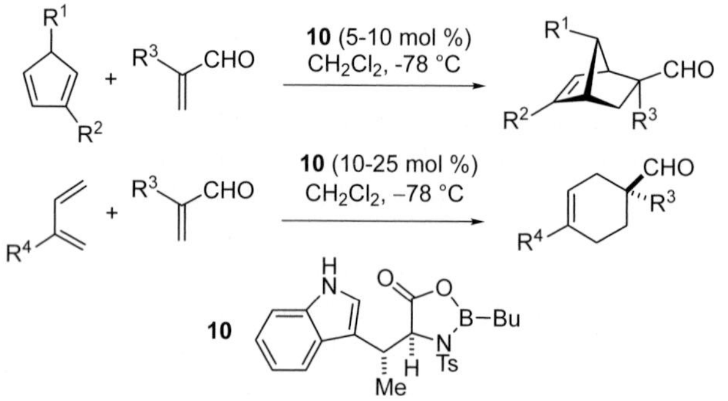

R^1	R^2	R^3	R^4	Yield [%]	ee [%]
H	H	Br	–	95	99
CH_2OBn	H	Br	–	81–83	>92
H	$CH_2C(Br)CH_2$	Br	–	81	99
–	–	Cl	OTIPS	–	94
–	–	Br	Me	76	92

Scheme 11

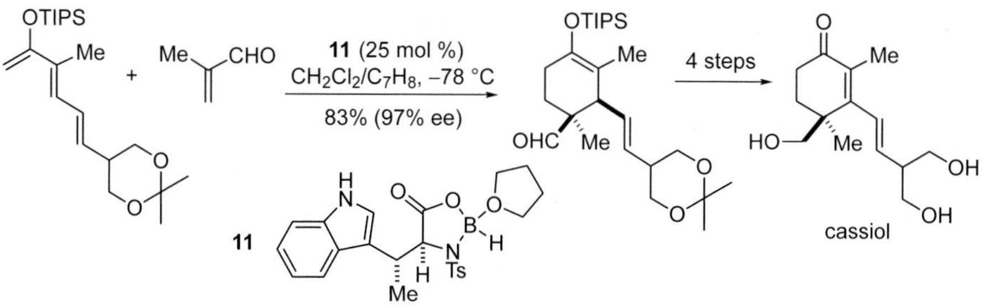

Scheme 12

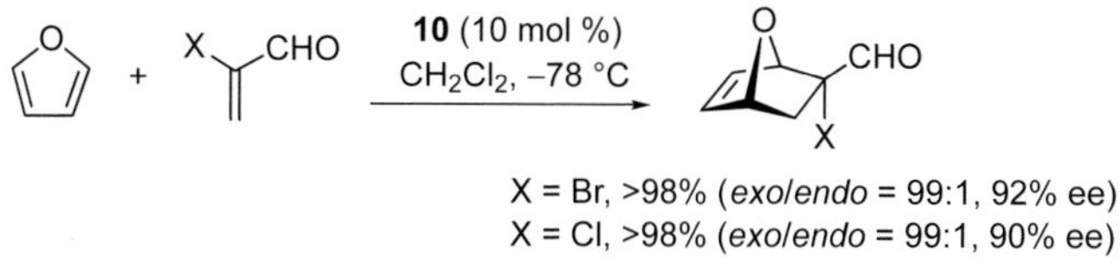

Scheme 13

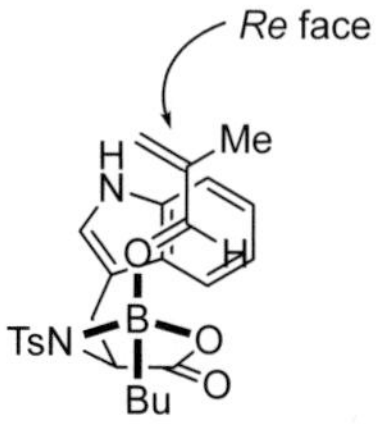

X = Br, >98% (*exo/endo* = 99:1, 92% ee)
X = Cl, >98% (*exo/endo* = 99:1, 90% ee)

Fig. 18. Proposed transition state assembly for oxazaborolidine catalyst **10** (Me group omitted for clarity)

tion of α-halo- and α-alkylacroleins with cyclic and acyclic dienes in high stereoselectivity (Scheme 11) [59].

The utility of such cycloadditions has been demonstrated by the elaboration of the cycloadducts to complex natural products [60]. For example, the adduct derived from a cyclopentadiene having a 2-bromoallyl sidechain has been converted to an intermediate employed in a previous (racemic) synthesis of gibberellic acid. As illustrated in Scheme 12, an exceptionally efficient synthesis of cassiol is realized by the successful execution of a rather difficult *endo*-selective Diels-Alder reaction using a slightly modified oxazaborolidine (**11**). The high catalyst loading is balanced by the fact that all the carbons and the quaternary center of the natural product are introduced in a single step.

It has been further demonstrated that furan may be successfully employed as a diene using catalyst **10** and α-halo acroleins as the 2π component (Scheme 13) [61].

A rationalization for the sense of induction for this system has been advanced and is illustrated in Fig. 18 (methyl group omitted for clarity) [62]. Salient ob-

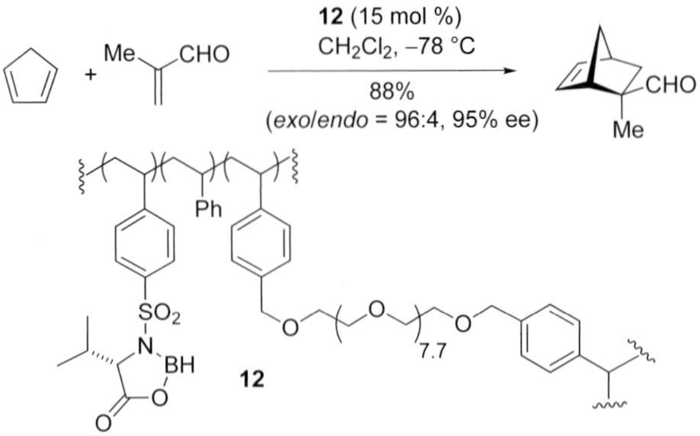

Scheme 14

servations on the ground state complex include: the appearance of a bright orange-red color on addition of methacrolein at 210 K that was attributed to an electron donor-acceptor complex; NOE's that imply a rigidified catalyst structure upon addition of the dienophile; a preferred *s-trans* conformer of the complexed aldehyde based on NMR spectroscopic observations. The Curtin-Hammett principle is invoked, and the *s-cis* conformer is proposed to be the active catalytic species. A subsequent publication has suggested a hydrogen bond between the formyl hydrogen and the carboxylate oxygen as an additional organizational feature [31a].

A related catalyst reported by Itsuno and coworkers offers some exciting practical benefits to the oxazaboroline system. A valine-derived cross-linked copolymer, when treated with borane-methyl sulfide, serves as an effective catalyst for the methacrolein-cyclopentadiene Diels-Alder reaction (Scheme 14) [63]. The polyether in the cross-linking unit is particularly important for realizing maximum selectivity. The advantages of heterogeneous catalysis are realized: the catalyst was easily recovered from reaction mixtures and reused multiple times without deleterious effects to the enantioselectivity or yield. As an added benefit, the polymeric catalyst in some cases conferred higher levels of enantioselectivity than the solution analogs which were reported independently by Yamamoto and Helmchen [64]. The absolute configuration of the product is opposite that obtained from the tryptophan-derived oxazaborolidine catalyst 10, suggesting that the mechanism of asymmetric induction is probably different for the two systems.

It is evident that minimization of the degrees of freedom of the dienophile in the transition state is an important criterion for reaction selectivity. A unique catalyst system designed by Hawkins and coworkers takes advantage of two distinct binding interactions to rigidify the catalyst-substrate complex [65]. The aromatic alkyldichloroborane 13 is an effective cycloaddition catalyst for acrylate dienophiles (Scheme 15) [66]; however, reports utilizing this catalyst are strictly confined to ester substrates with either cyclopentadiene or cyclohexadiene.

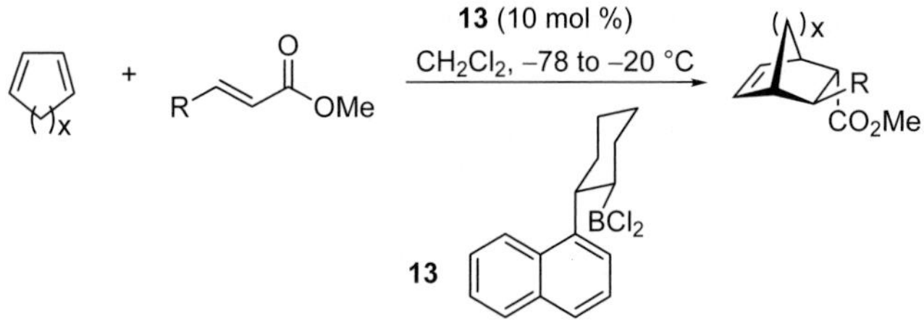

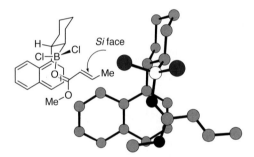

R	x	Yield [%]	ee [%]
H	1	97	99.5
Me	1	91	93
CO$_2$Me	1	92	90
H	2	83	86

Scheme 15

Fig. 19. Proposed transition state assembly for catalyst **13**; catalyst-methyl crotonate complex (X-ray)

The X-ray structure of the indicated borane-methyl acrylate complex (Fig. 19) unequivocally confirms the design concept. The solid state structure shows a close contact (3.40 Å from the center of the substituted phenyl ring to the carbonyl carbon) between the electron-rich arene and boron-bound methoxycarbonyl group, an arrangement which also exists in solution. As the polarizability of the aryl group is increased, the dienophile is drawn closer to the arene, suggestive of a dipole-induced attractive interaction. The air-sensitive catalyst **13** was synthesized by way of a resolution in 5 steps.

Yamamoto and co-workers have introduced a conceptually interesting series of catalysts that incorporate an acidic proton into the active catalyst. Termed Brønsted acid-assisted chiral Lewis acid (BLA), catalyst **14** selectively catalyzes a number of diene-aldehyde cycloadditions reactions (Scheme 16) [67]. While extremely selective for the substrates shown, no aldehydes lacking an α-substituent were reported to be effective in this reaction. This feature was addressed in

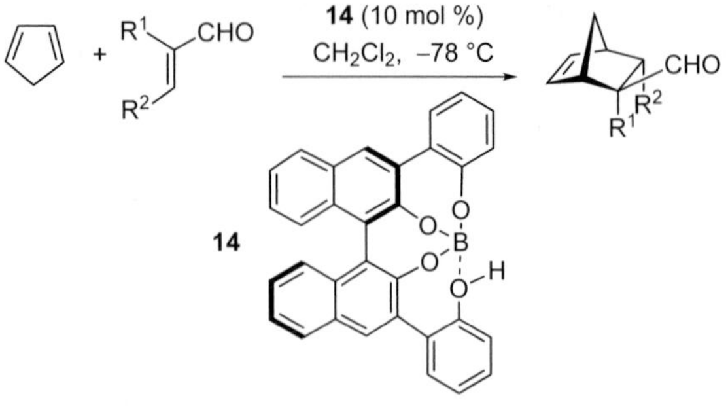

R¹	R²	Yield [%]	exo/endo	ee [%]
Br	H	>99	>99:1	99
Me	H	>99	>99:1	99
Me	Me	>99	>99:1	98
(CH$_2$)$_3$	–	>99	98:2	93

Scheme 16

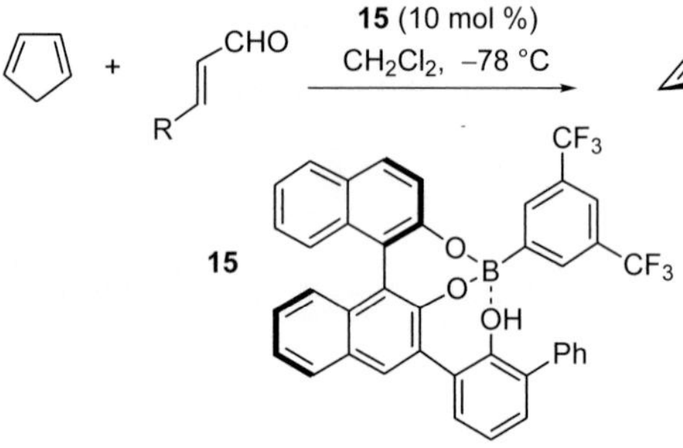

R	Yield [%]	endo/exo	ee [%]
H	84	97:3	95
Me	94	90:10	95
CO$_2$Et	91	98:2	95

Scheme 17

a second-generation BLA (**15**), which was general with respect to the aldehyde component (Scheme 17) [68]. Despite this uniformly high selectivity, the lack of spectral or solid state characterization of the active catalyst makes stereochemical models speculative at this point. One particularly relevant observation is that formation of a monoether corresponding to **14** gives a far less selective catalyst, implicating the active proton in the catalytic event. Dienes other than cyclopentadiene may be employed with both catalysts and recovery of the chiral ligand is quantitative.

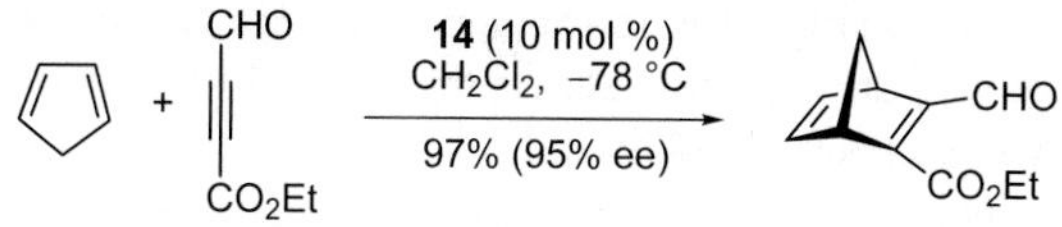

Scheme 18

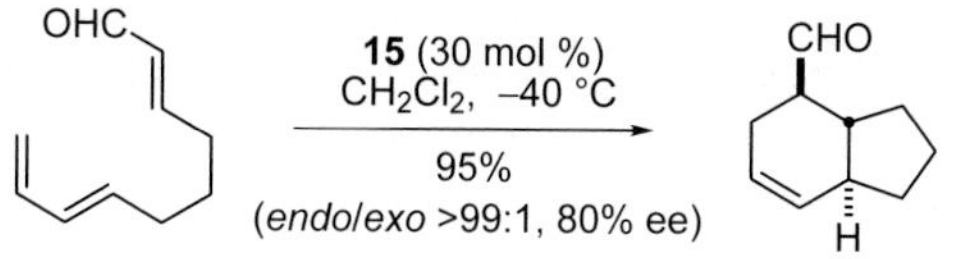

Scheme 19

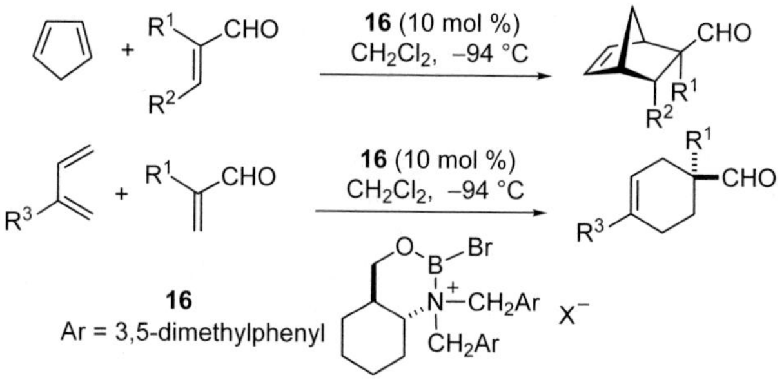

16

Ar = 3,5-dimethylphenyl

R^1	R^2	R^3	X	Yield [%]	*exo:endo*	ee [%]
Br	H	–	B(3,5-(CF$_3$)$_2$Ph)$_4$	99	91:9	98
Me	H	–	Br	99	88:12	90
Br	Me	–	B(3,5-(CF$_3$)$_2$Ph)$_4$	99	>98:2	96
Me	Me	–	B(3,5-(CF$_3$)$_2$Ph)$_4$	97	>98:2	89
-(CH$_2$)$_4$-	–	–	Br	99	>98:2	96
Br	H	H	B(3,5-(CF$_3$)$_2$Ph)$_4$	99	–	94
Br	H	Me	B(3,5-(CF$_3$)$_2$Ph)$_4$	99	–	96

Scheme 20

Cycloadditions between acetylenic aldehydes and dienes are effected by catalyst **14**, the best case being illustrated in Scheme 18 [69]. This is one of only two
reports of highly enantioselective Diels-Alder reactions using alkynes. *Ab initio*
calculations propose that the reaction is proceeding via an *exo* transition state.
On the basis of FMO theory, the authors suggest that a secondary antibonding
interaction between the lobes on C-2 of cyclopentadiene and the carbonyl oxygen accounts for the higher relative energy of the *endo* transition state.

An enantioselective intramolecular Diels-Alder reaction of α-unsubstituted
2,7,9-decatrienal afforded the corresponding bicyclic aldehyde in high yield and

good enantioselection using BLA **15** (Scheme 19). Alternatively, when CAB catalyst **7a** was employed in the same reaction, the adduct was obtained in lower yield and selectivity (74% yield, 46% ee).

Promising results have been reported by Corey using cationic oxazaborinane complex **16** as an aldehyde-diene cycloaddition catalyst (Scheme 20) [70]. α-Substituted aldehydes and four dienes are reported to undergo low-temperature (–94 °C) Diels-Alder reaction to give adducts in high *exo* selectivity and excellent enantioselection. The catalyst is prepared in seven steps and ligand recovery after the reaction is 85%; catalyst decomposition occurs above –60 °C.

Catalyst **16** has also been reported to effect cycloaddition of propargyl aldehydes with cyclopentadiene (Scheme 21) [71]. While simple β-alkyl substituted alkynyl aldehyde dienophiles proved to be unreactive, the derived silyl- or stannyl-substituted analogues proceeded with good levels of enantioselectivity. It was further demonstrated that the derived cycloadducts are useful chiral building blocks by virtue of their ability to undergo transition metal-catalyzed cross-coupling reactions. Despite the somewhat elevated catalyst loading, this system does not require two activating substituents on the alkyne, in contrast to BLA catalyst **14**. As with that system, an *exo* transition state is proposed as the favored reaction pathway.

An enantioselective Diels-Alder reaction between methacrolein and cyclopentadiene with 3 mol % of borate catalyst **17** (Fig. 20) proceeds with good selectivity (–78 °C, 85% yield, *exo/endo*=97.4:2.6, 90% ee) [72]. The catalyst is available in one step from BINOL and the ligand may be recovered in nearly quantitative yield after the reaction.

Proline-derived boron complex **18** catalyzes the enantioselective cycloaddition of methacrolein and cyclopentadiene (–78 °C, 84% yield, *exo/endo*>99:1,

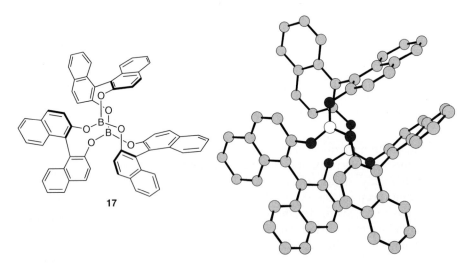

Fig. 20. Borate "propeller" catalyst **17** and X-ray structure

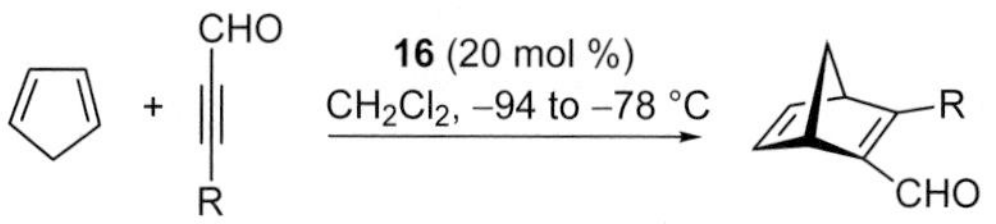

Scheme 21

R	X	Yield [%]	ee [%]
$SiMe_3$	$B(3,5-(CF_3)_2Ph)_4$	68	87
$SiEt_3$	$B(3,5-(CF_3)_2Ph)_4$	37	85
$SiMe_2Ph$	$B(3,5-(CF_3)_2Ph)_4$	50	87
$SnBu_3$	$B(3,5-(CF_3)_2Ph)_4$	83	80

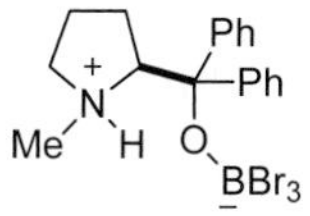

18

Fig. 21. Zwitterionic proline-based Lewis acid

97% ee) [73]. ^{11}B-, ^{1}H-, and ^{13}C-NMR spectroscopy were instructive in assigning the structure of **18** (Fig. 21): the methyl group appeared as a doublet and was shifted downfield from its position in the prolinol ligand. Efforts to study complexation between methacrolein and **18** were not successful due to the unfavorable equilibrium. It is evident that further work will be needed to elucidate the role that **18** plays in this reaction.

3.1.3
Magnesium

Magnesium-derived Lewis acids, while not attracting as much attention as their boron counterparts, have been developed as selective Diels-Alder catalysts. A significant point of divergence between the two metals in their applications should be noted: enantioselective Diels-Alder reactions with boron Lewis acids utilize aldehyde or ester dienophiles without exception, while successful cycloadditions with magnesium complexes always employ a dicarbonyl compound as the activating moiety of the 2π component. The magnesium center is typically viewed as being amenable to chelating substrates, while the boron center is attractive for single-point binding dienophiles.

Bis(oxazoline)-magnesium complex **20** (10 mol %) catalyzes the indicated cycloaddition (Scheme 22) to give **19** (2R) in 82% yield and 91% ee (*endo/exo*= 97:3) [74]. The absolute stereochemistry of the product is consistent with bidentate activation of the substrate through a tetrahedral metal geometry with reaction out of the *s-cis* conformer. Complex **21**, derived from the opposite enantio-

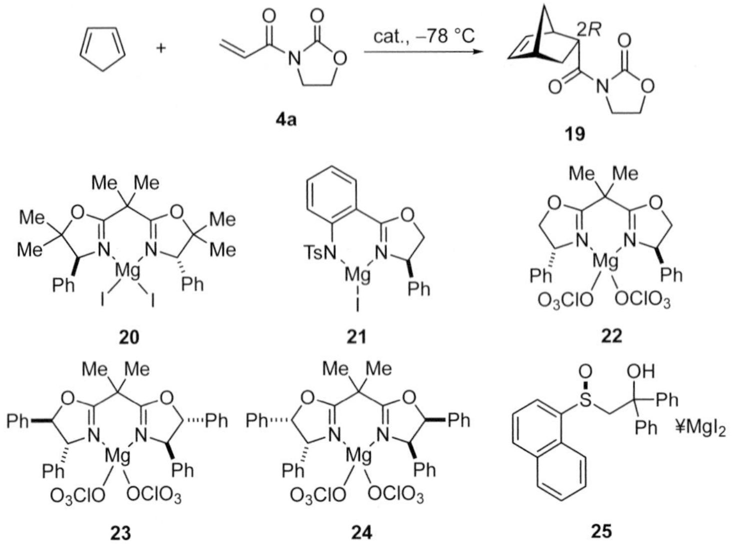

Scheme 22

meric series of chiral amino alcohol provides the same enantiomer (2R) as **20**, proceeding in 81% yield (50 mol % **21**) and 91% ee [75]. Hydroxysulfoxide-derived catalyst **25** mediates the same reaction (10 mol %) and delivers *ent*-**19** (2S) in 88% ee [76]. Structural and mechanistic investigations on these complexes are less developed than for many boron catalysts; accordingly, hypotheses pertaining to selectivity issues are still quite speculative.

The most detailed work for magnesium catalysts has been performed by Desimoni and coworkers with complexes **22**, **23**, and **24** [42, 77]. NMR spectroscopic studies with **22** and imide **4a** suggest that the metal center in the dienophile-catalyst complex adopts the expected tetrahedral coordination geometry, with the imide disposed in an *s-cis* configuration. Upon addition of 2 equivalents of methanol-d_1, the complex is transformed to an octahedral geometry with the endocyclic carbonyl bound out of the plane of the chiral ligand (based on chemical shift data and observed NOE's). The practical consequence of this geometry change is an alteration of the exposed enantioface of the dienophile with the addition of auxiliary ligands (Fig. 22). In support of this hypothesis, complex **23** catalyzes (5 mol %) the reaction of cyclopentadiene and acryloyl imide to afford *ent*-**19** (2S) in 97% ee (*endo/exo*=199:1). The enantiomeric product **19** (2R, 89% ee, *endo/exo*=94:6) was obtained using complex **24** (5 mol %) with water as the auxiliary ligand (10 mol %). In the absence of water, catalyst **24** preferentially delivers the 2S product (22–43% ee), making **24** one of two Diels-Alder catalysts which can deliver either product enantiomer with proper choice of addend (see also Sect. 3.3). Complicating mechanistic analysis somewhat is the fact that no turnover in stereochemistry occurs using complex **23** and auxiliary ligands. The importance of the *cis* or *trans* relationship of the phenyl groups in complex-

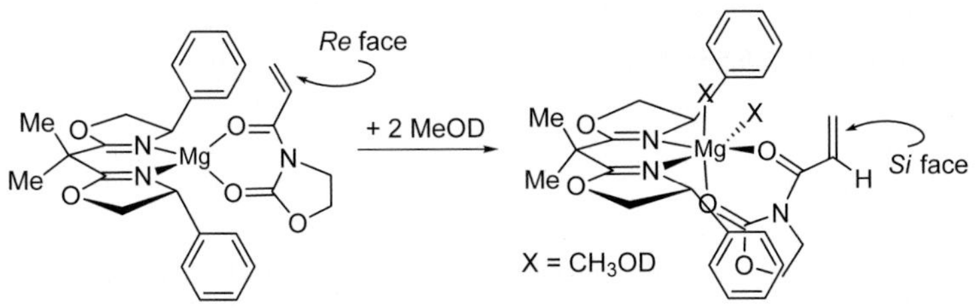

Fig. 22. Turnover in selectivity for Mg(bisoxazoline) catalysts with the addition of coordinating ligands (tetrahedral→octahedral transposition)

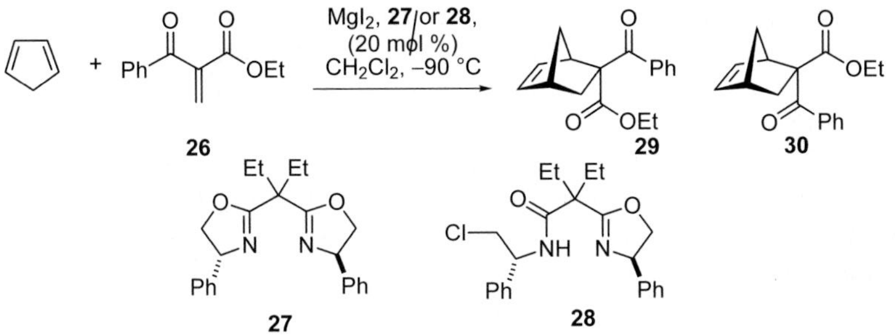

Ligand	Yield [%]	29:30	ee of 29 [%]
26	81	>99:1	85.3
27	88	>99:1	87.0

Scheme 23

es 23 and 24 points to possibility of an electronic effect which could either reinforce or partially cancel the steric bias provided by the proximal phenyl group. While not mentioned explicitly by the authors, the relative stereochemistry of the phenyl groups could also play a significant role in gearing that could significantly affect the chiral environment about the metal center (for general references on gearing effects, see [78]). As a final note on mechanism, no nonlinear effects were observed with catalyst 22, indicative of a putative mononuclear catalyst-dienophile complex.

A magnesium-based catalyst system which employs either a bis(oxazoline) or amido-mono(oxazoline) ligand derived from phenylglycine has been reported to effect an interesting Diels-Alder between unsymmetrical alkylidene 26 and cyclopentadiene (Scheme 23) [79]. The reaction generates a quaternary center and is noteworthy in its ability to preferentially deliver cycloadduct 29, despite a superficial similarity between the two carbonyl substituents (OEt vs. Ph). This selectivity has been rationalized on the basis of an inferred steric preference for the phenyl group to reside perpendicular to the alkylidene, thus creating a marked bias in the diastereomeric transition states [80].

3.2
Transition Metal Lewis Acids [81]

3.2.1
Copper

Evans and coworkers have reported that cationic copper(II)-bis(oxazoline) complexes derived from *tert*-leucine are effective Lewis acids for a wide range of enantioselective Diels-Alder reactions. While initial investigations employed cyclopentadiene as the diene and triflate catalyst **31a** (Scheme 24) as the Lewis acid [82], subsequent studies revealed that the reaction rate is strongly dependent on the counterion X [83]. The hexafluoroantimonate catalyst **31b** is approximately 20 times more reactive than **31a** and is typically more stereoselective. The heightened reactivity and selectivity conferred by catalyst **31b** allows access to more substituted adducts in uniformly high enantioselectivity. The active catalyst is easily prepared and robust: exposure to air is not deleterious and the reactions may be conducted in the presence of free hydroxy groups. However, reduction of the metal center can be problematic with electron-rich dienes; this side reaction may be controlled by a judicious choice of temperature.

A stereoselective Diels-Alder reaction between furan and acrylimide yielded bicyclic adduct **32** that could be recrystallized to isomeric purity (Scheme 25) [84]. The cycloaddition reaction was reversible at higher temperatures and resulted in preferential formation of racemic *exo* isomer. Cycloadduct **32** was elaborated in six steps to *ent*-shikimic acid.

The synthetic utility of this copper(II) system has been subsequently expanded to include a number of less reactive dienes, several of which have not been previously used in enantioselective Diels-Alder reactions [85]. Functionalized buta-

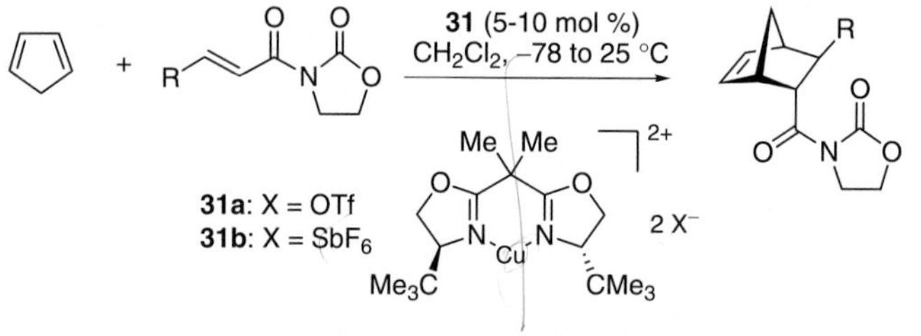

R	Catalyst	Yield [%]	*endo/exo*	ee [%]
H	1a	86	98:2	>98
CO₂Et	1a	92	94:6	95
Me	1a	85	96:4	97
Ph	1b	96	91:9	96
Cl	1b	96	86:14	95

Scheme 24

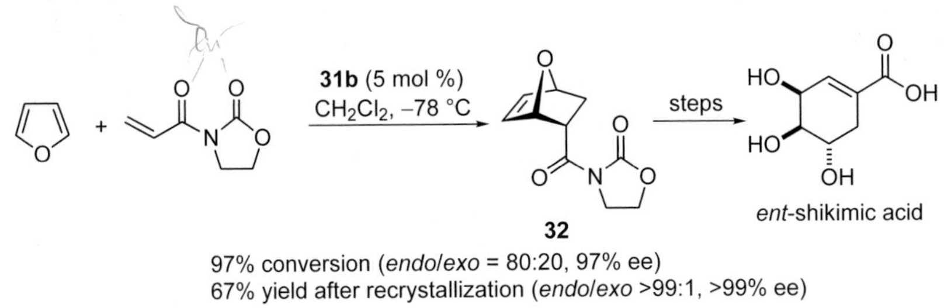

32

97% conversion (*endo*/*exo* = 80:20, 97% ee)
67% yield after recrystallization (*endo*/*exo* >99:1, >99% ee)

Scheme 25

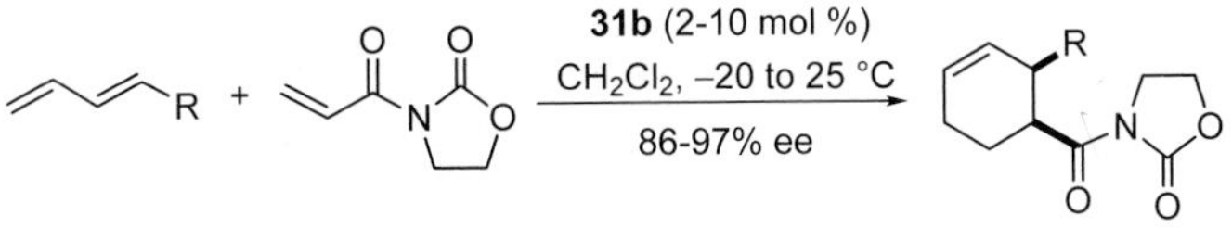

R	Yield [%]	*endo*/*exo*	ee [%]
Me	70	91:9	94
OAc	75	85:15	96
Ph	95	85:15	97
SPh	84	98:2	98
NHCbz	54	72:28	90

Scheme 26

dienes are particularly good substrates: alkyl, aryl, oxygen, nitrogen, and sulfur substitution at the terminal position may be tolerated with no loss in stereoselectivity for the favored *endo* product (Scheme 26). The adducts also exhibit a high incidence of crystallinity which greatly simplifies purification efforts. Catalyst loadings of 2 mol % are generally sufficient to achieve complete reaction and scale-up occurs without incident, making this one of the more efficient Diels-Alder catalysts. A study by Jørgensen disclosed an apparent accelerating effect using nitromethane as a solvent (vs. dichloromethane) for this catalyst system [86].

The reaction between 1-acetoxy-3-methylbutadiene preferentially affords *exo* adduct **33** in high enantioselectivity (Scheme 27); **33** was elaborated in four steps to *ent*-Δ^1-tetrahydrocannabinol [87]. The turnover in diastereoselectivity is thought to be a result of a steric interaction between the 3-methyl group of the diene and the chiral ligand, a repulsion which is not present for the parent 1-acetoxybutadiene (an *endo* selective diene).

Hexafluoroantimonate catalyst **31b** mediates a number of enantioselective intramolecular Diels-Alder reactions as well (Scheme 28) [88]. The marine natural product isopulo'upone was assembled in a straightforward fashion from the bicyclo[6.5.0] skeleton possessing a functionalized side chain. From an acyclic

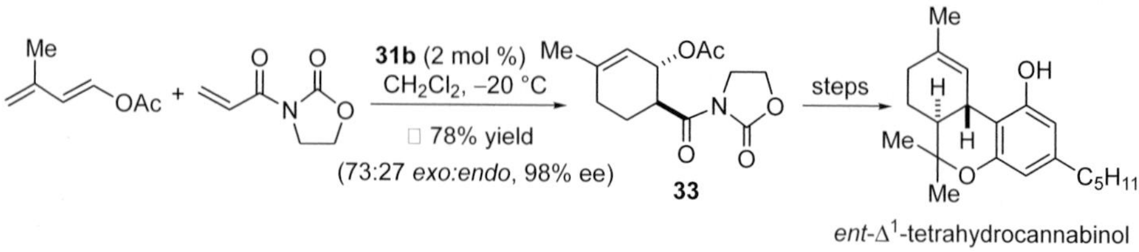

Scheme 27

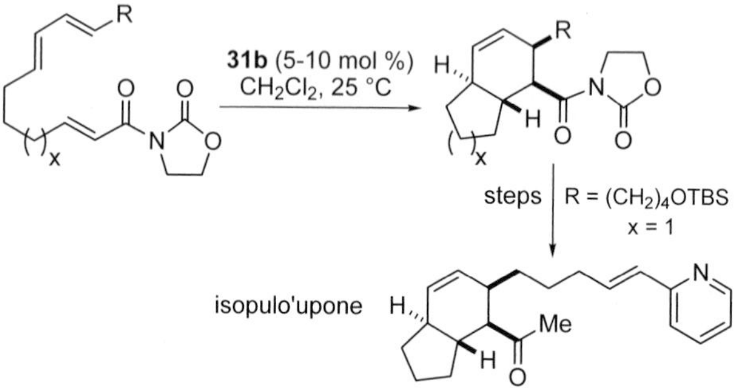

R	X	Yield [%]	*endo/exo*	ee [%]
H	1	89	>99:1	86
Ph	1	86	>95:5	92
Ph	2	97	84:16	97
(CH$_2$)$_4$OTBS	1	81	>99:1	96

Scheme 28

precursor, all four of the natural product's contiguous stereocenters are correctly installed in a single step.

In every case for copper catalyst **31**, the absolute stereochemistry of the cycloadducts is accounted for by the intervention of the substrate-catalyst complex depicted in Fig. 23, in which the *s-cis* configured dienophile is bound to the catalyst in the plane of the ligand in a bidentate fashion. The *tert*-butyl group shields the top face and cycloaddition occurs from the exposed *si* enantioface. Support for this model derives from X-ray structures of aquo complexes of catalysts **31a** and **31b** which show that the complex possesses a distorted square planar geometry; EPR spectroscopy on the binary catalyst-dienophile complex indicates that this geometry carries over from the solid state into solution. Calculations at the PM3 level of theory further favor the indicated reactive assembly [85].

Double stereodifferentiating experiments [89] using chiral dienophiles have effectively ruled out the intervention of a tetrahedral copper center or a reactive *s-trans* conformer (Scheme 29) [82]. It is noteworthy that in the mismatched

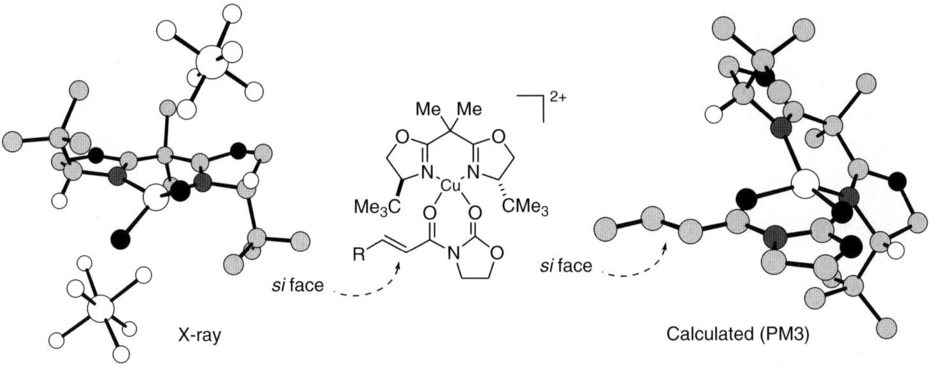
Me Me
2+
O
O
N
Cu
N
Me3C
CMe3
O
O
R
N
O
si face
si face
X-ray
Calculated (PM3)

Fig. 23. X-ray structure of **31b**·$(H_2O)_2$, proposed transition state assembly for bis(oxazoline)Cu catalyst **31**, and PM3 calculated structure of the substrate·catalyst complex

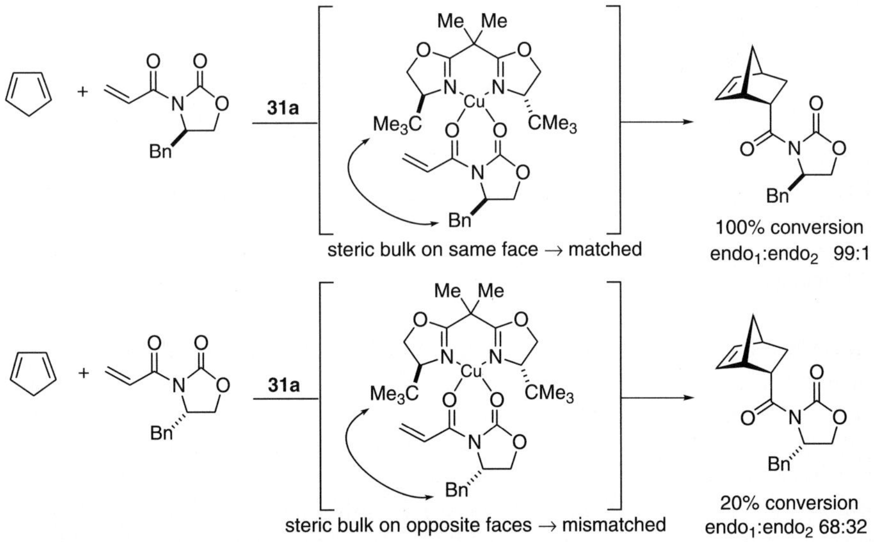
31a
Me Me
Me3C
CMe3
Cu
Bn
steric bulk on same face → matched
100% conversion
endo1:endo2 99:1
31a
steric bulk on opposite faces → mismatched
20% conversion
endo1:endo2 68:32

Scheme 29

case, the stereochemical preference exhibited by the catalyst overrides that of the auxiliary.

Modifications to the parent bis(oxazoline) structure have been subsequently disclosed (Fig. 24). Spirobis(oxazoline) **34** derived from amino indanol catalyzes (10 mol %) the enantioselective cycloaddition between cyclopentadiene and acrylimide (–78 °C) in 96.3% ee (*endo*/*exo*=44:1) [90]. When the size of the spiro ring (*e.g.*, cyclobutyl, -pentyl, -hexyl) is increased, the resulting structural change progressively degrades the reaction enantioselectivity, demonstrating a relation-

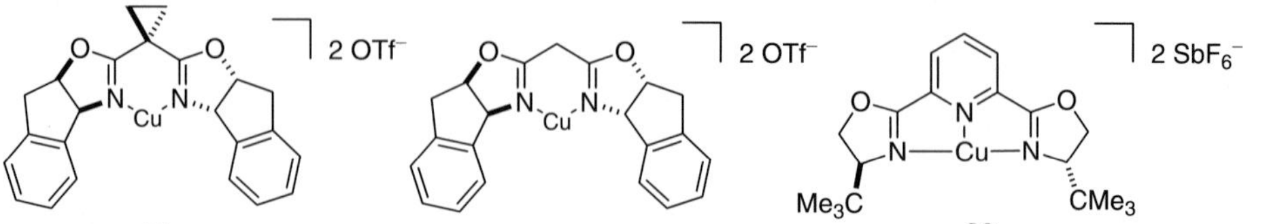

Fig. 24. Bis(oxazoline)Cu complexes for Diels-Alder reactions

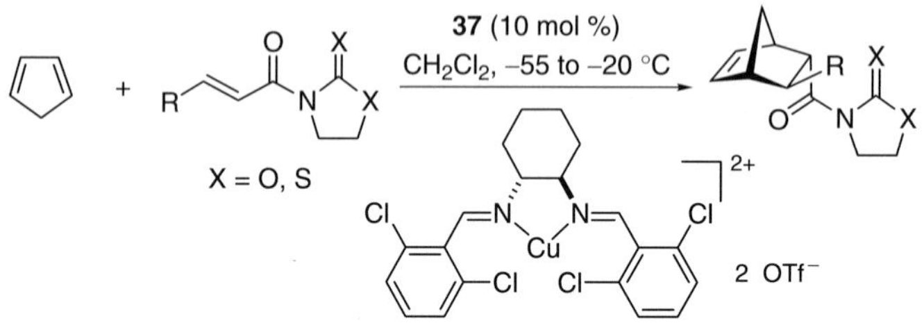

Scheme 30

R	X	Yield [%]	endo/exo	ee [%]
H	O	87	80:20	92
Me	S	86	93:7	91
Ph	S	84	92:8	92
CO_2Et	S	99	90:10	88

ship between ligand bite angle and enantiomeric excess. A simple *gem*-dimethyl bridge (rather than cycloalkane) delivers the adduct in 95% ee [91], while unsubstituted **35** has been shown to catalyze the same reaction in 99% ee (dr>99:1) [92]. In spite of the ligand modifications introduced in complexes **34** and **35**, it is not evident their performance is superior to the parent *tert*-butyl-bis(oxazoline)Cu(SbF$_6$)$_2$ complex **31b** (Scheme 24). Finally, bis(oxazolinyl)pyridine (pybox) complex **36** is a selective catalyst for Diels-Alder reactions between unsaturated aldehydes and cyclopentadiene [83].

Cationic copper(II) complex **37** derived from a chiral bis(imine) ligand has also been shown to be an effective catalyst for reactions between cyclopentadiene and acylated thiazolidine-2-thione dienophiles, albeit with slightly lower selectivities than for the bis(oxazoline) complex **31** (Scheme 30) [93]. The bis(2,6-dichlorophenylimine) was found to be optimal among a number of electron-rich and -poor aryl imines screened. The reaction exhibits a positive non-linear effect which suggests that the minor ligand enantiomer can be sequestered by the formation of a catalytically less active (R,R)/(S,S)Cu(II) dimer.

A recent disclosure by Helmchen has demonstrated that (phosphino-oxazoline)copper(II) complexes are also good chiral templates for asymmetric cataly-

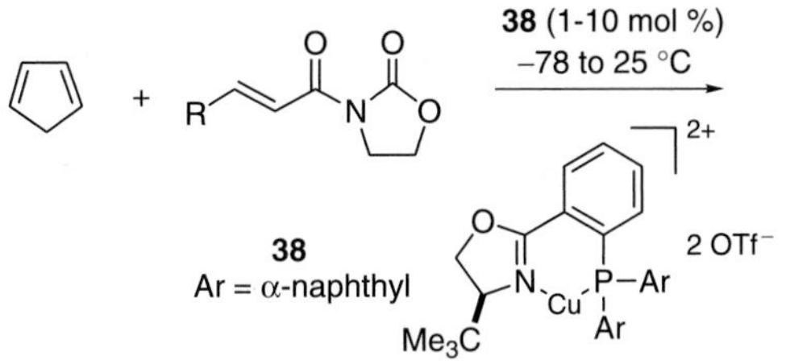

Scheme 31

R	Catalyst (mol%)	Solvent	Yield	endo:exo	ee [%]
H	10	CH_2Cl_2	92	94:6	97
H	1	$EtNO_2$	86	95:5	92
Me	10	CH_2Cl_2	98	88:12	86
Ph	10	$EtNO_2$	74	40:60	85
CO_2Et	10	CH_2Cl_2	95	60:40	75

sis of the Diels-Alder reaction (Scheme 31) [94]. As with the bis(oxazoline)Cu(II) complex 31, the *tert*-leucine-derived variant was found to be optimal, and bulky aryl groups on the phosphorous center were crucial as well. Dichloromethane and nitroethane were found to function well as solvents, allowing access to cycloadducts of good enantiomeric excess with acryloyl, crotonyl, cinnamoyl, and fumaroyl imide dienophiles. A turnover in diastereoselectivity occurs with the cinnamoyl imide dienophile, and the *exo* cycloadduct is formed in moderate excess. Interestingly, for this system the more associating triflate counterion was found to afford a more selective catalyst than the hexafluoroantimonate-derived catalyst, in contrast to 31. Additionally, it appears that more sterically restrictive catalysts are more active catalysts.

3.2.2
Iron

Kündig's cationic iron(II) complex 39a, derived from *trans*-1,2-cyclopentanediol, is a stable, isolable brown solid that possesses sufficient Lewis acidity to catalyze Diels-Alder reactions between unsaturated aldehydes and dienes [95]. The highest selectivities and yields were realized using bromoacrolein as the dienophile (Scheme 32). Further inspection reveals that dienes less reactive than cyclopentadiene give cycloadducts in higher yield and enantioselectivity, a characteristic that is even more impressive when one considers that the *endo* and *exo* transition states produce enantiomeric products for isoprene and 2,3-dimethylbutadiene. Cyclohexadiene may be used in the reaction with bromoacrolein to afford the cycloadduct in 80% de and >99% ee. In the case of cyclopentadiene,

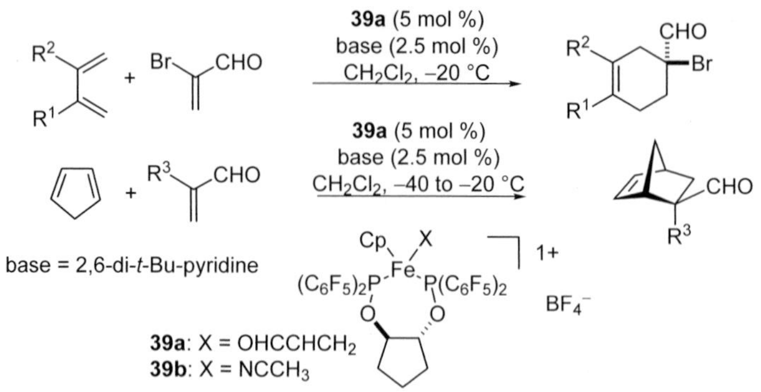

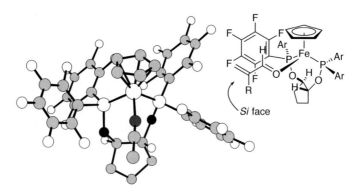

R^1	R^2	R^3	Yield [%]	ee [%]
Me	H	–	99	96
Me	Me	–	92	97
–	–	Me	62	90
–	–	Br	87	95

Scheme 32

Fig. 25. X-ray structure of **39b** and proposed transition state assembly for cationic iron complex **39a**

diastereomeric excesses are greater than 90% for the two cases shown. In all cases, low catalyst loadings are feasible.

An undefined catalyst is recovered after the conclusion of the reaction by precipitation with hexane and filtration, but no mention is made of recycling. The presence of the acid scavenger is important, as irreproducible results (variable reaction rate, diminished selectivity) are obtained in the absence of the pyridine base. While **39a** gradually decomposes in solution above –20 °C, spectroscopic observations (^{1}H-NMR and IR) support the assigned structure and the mode of binding (Fe-O=C η^1 complex). An X-ray structure of **39b** wherein acetonitrile

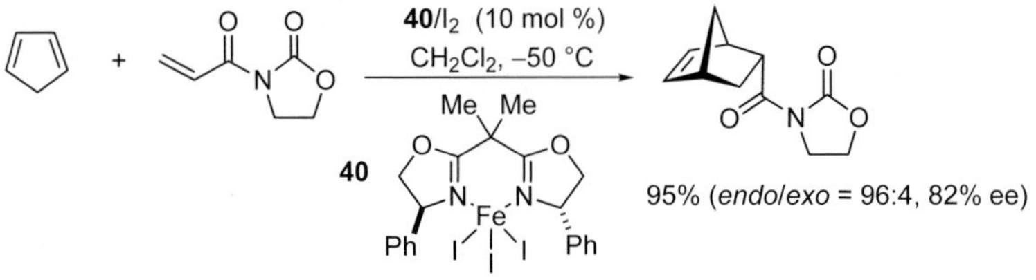

Scheme 33

has replaced acrolein as a ligand has provided the basis for a transition state model (Fig. 25) which suggests that the *Re* face of the aldehyde is blocked by the pentafluorophenyl ring of the ligand. The presence of this electron-poor moiety lies in contrast to oxazaborolidine catalyst **10** and dichloroalkylborane catalyst **13** which employ electron-rich aromatics as key constituents of the complexes. That these three electronically diverse catalysts are all able to deliver cycloadducts in high enantioselectivities highlights that our understanding pertaining to substrate-catalyst interaction is still in its infancy.

A bis(oxazoline)Fe(III) complex has also been shown to function as an effective catalyst for an enantioselective Diels-Alder reaction between cyclopentadiene and acryloyl imide (Scheme 33) [96]. Recovery of the chiral ligand proceeded in >85% yield. The scope of this catalyst has not been evaluated against less reactive dienes and dienophiles that require higher reaction temperatures.

3.2.3
Other Late Transition Metal Catalysts

While copper and iron Lewis acids are the most prominent late transition metal Diels-Alder catalysts, there are reports on the use of other chiral complexes derived from ruthenium [97, 98], rhodium [99], and zinc [100] in enantioselective cycloaddition reactions, with variable levels of success. As a comparison study, the reactions of a zinc(II)-bis(oxazoline) catalyst **41** and zinc(II)-pyridylbis(oxazoline) catalyst **42** were evaluated side-by-side with their copper(II) counterparts (Scheme 34) [101]. The study concluded that zinc(II) Lewis acids catalyzed a few cycloadditions selectively, but, in contrast to the [Cu(*t*-Bubox)](SbF$_6$)$_2$ complex **31b** (Sect. 3.2.1), enantioselectivity was not maintained over a range of temperatures or substitution patterns on the dienophile. An X-ray crystal structure of [Zn(Ph-box)](Cl)$_2$ revealed a tetrahedral metal center; the absolute stereochemistry of the adduct was consistent with the reaction from that geometry and opposite that obtained with Cu(II) complex **31**.

A C_2-symmetrical tridentate ligand that employs a benzofuran backbone, recently reported by Kanemasa, is also an effective chiral controller for asymmetric Diels-Alder reactions [102]. Dubbed DBFOX, the ligand forms catalytically competent complexes with a wide range of transition metal salts. Remarkably, complexes derived from Fe(ClO$_4$)$_2$, Co(ClO$_4$)$_2$·6H$_2$O, Ni(ClO$_4$)$_2$·6H$_2$O, Ni(ClO$_4$)$_2$,

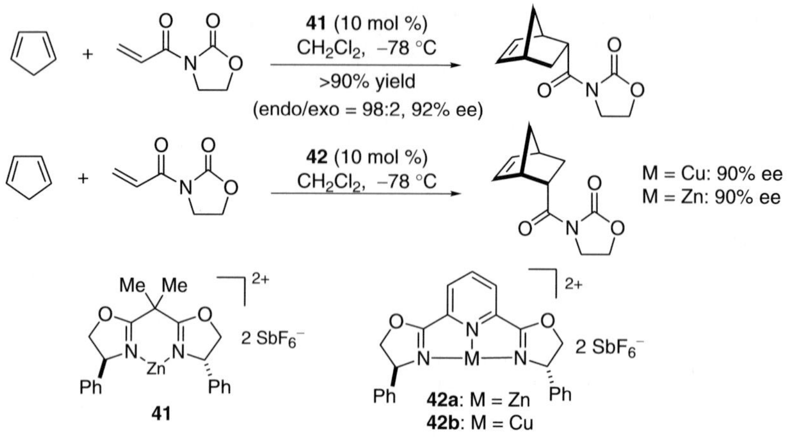

Scheme 34

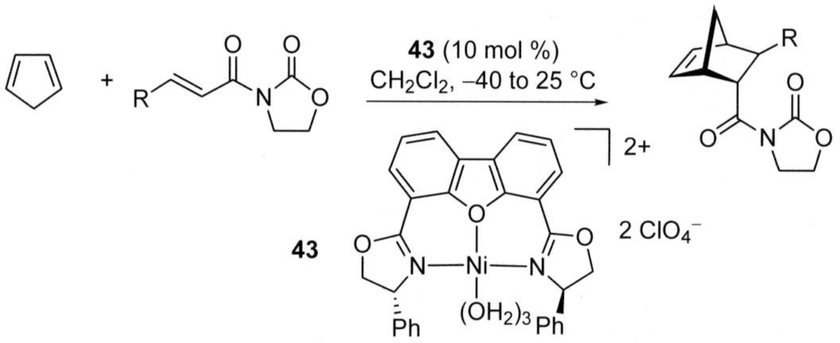

R	Yield [%]	endo/exo	ee [%]
H	95	98:2	96
Me	90	92:8	93
Pr	90	93:7	94
Ph	52	–	74

Scheme 35

$Cu(ClO_4)_2 \cdot 3H_2O$, and $Zn(ClO_4)_2 \cdot 3H_2O$ all catalyze the reaction of acryloyl imide and cyclopentadiene in >96% ee (Scheme 35). This generality with respect to the metal center of the Lewis acid complex is unprecedented and quite extraordinary. The results also point to a typical advantage of transition metal catalysts over boron complexes: insensitivity to moisture. Catalyst **43** can be stored at ambient temperature and atmosphere for weeks with no deleterious effects and may also be used with dienophiles bearing an alkyl group at the β-position.

An X-ray structure of catalyst **43** reveals the nickel disposed in an octahedral geometry. From this solid state structure a transition state model was fashioned

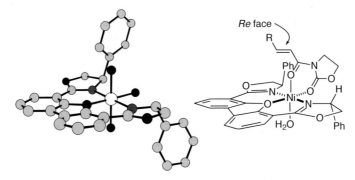

Fig. 26. X-ray structure and proposed transition state assembly for DBFOX-Ni(II) complex **43**

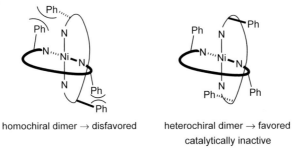

homochiral dimer → disfavored   heterochiral dimer → favored
                                          catalytically inactive

Fig. 27. Dimeric DBFOX complexes; formation of the heterochiral dimer is irreversible and sequesters the minor enantiomer as a catalytically inactive complex

(Fig. 26) in which the exocyclic carbonyl group is bound in the apical position and the dienophile reacts out of the *s-cis* conformation. Steric shielding of the *Si* face by the ligand phenyl group would favor diene attack on the exposed *Re* face. As with Mg(II)bis(oxazoline) complexes (Sect. 3.1.3), the presence of ligands other than the dienophile appear to be important in the creation of a stereodefined environment about the metal center.

Dramatic nonlinear effects are observed for this system, as the employment of ligand at 20% ee affords the cycloadduct in 91% ee. Preferential formation of a heterochiral dimer serves to sequester the minor enantiomer and it has been proposed that this amplification is augmented by aggregation of the heterochiral dimeric complex in solution (Fig. 27).

3.2.4
Early Transition Metal Lewis Acids

Titanium rivals boron for the amount of attention it has received in the development of catalytic enantioselective Diels-Alder reactions (for enantioselective Diels-Alder reactions promoted by stoichiometric amounts of chiral titanium

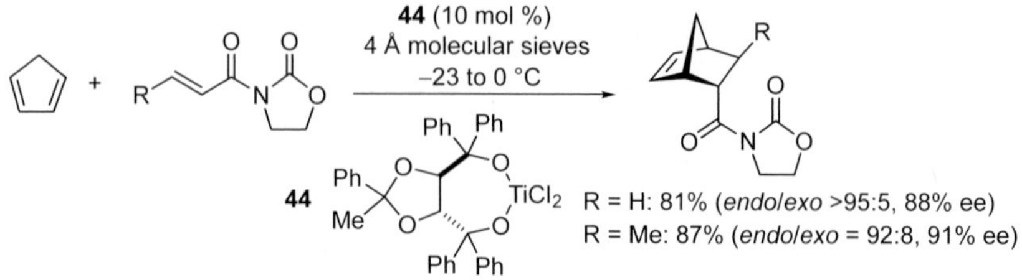

44 (10 mol %)
4 Å molecular sieves
−23 to 0 °C

R = H: 81% (*endo/exo* >95:5, 88% ee)
R = Me: 87% (*endo/exo* = 92:8, 91% ee)

Scheme 36

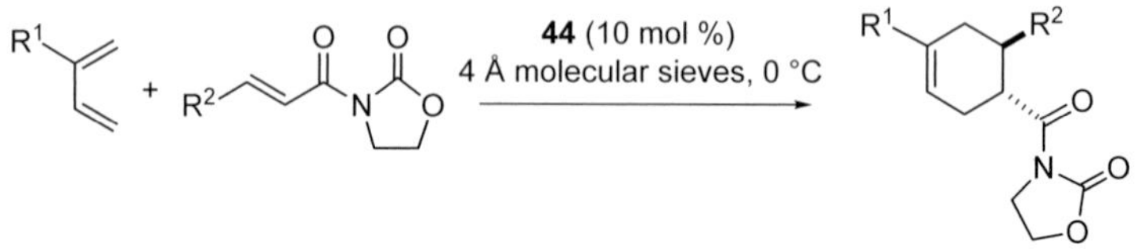

44 (10 mol %)
4 Å molecular sieves, 0 °C

R^1	R^2	Yield [%]	ee [%]
H	CO_2Me	84	91
Me	CO_2Me	94	94
H	H	81	93
Me	H	93	96
SEt	H	72	91

Scheme 37

complexes, see [103]); however, the similarities between the two metals cease at that point. While many boron complexes exhibit tetracoordinacy and have been shown to be monomeric in solution, titanium(IV) accommodates up to six ligands, and the derived complexes frequently feature bridging ligands and attendant aggregation. As will be noted, such behavior has frustrated efforts to probe catalyst structure and address issues of stereoinduction. It has been demonstrated that a wealth of ligands create an effective chiral environment around boron to induce asymmetry; in contrast, primarily one ligand has proven successful for titanium. As a consequence, this particular system has been studied quite extensively.

Tetraaryl-1,3-dioxolane-4,5-dimethanol (TADDOL) ligands synthesized from tartaric acid have been extensively employed by Narasaka as the chiral control element in selective Diels-Alder reactions. Initial experiments were conducted with simple dienes and α,β-unsaturated imides using complex **44** (Scheme 36) [104, 105]. Several rather subtle features have contributed to the success of these endeavors: 1) the use of the acetophenone-derived dioxolane rather than the acetonide resulted in an increase of 20% ee; 2) the use of alkyl-substituted benzenes as solvent augmented enantioselectivities relative to more common organic solvents (*e.g.*, CH_2Cl_2, THF) [106]; 3) use of 4 Å molecular sieves was typically required to achieve maximum enantioselectivity.

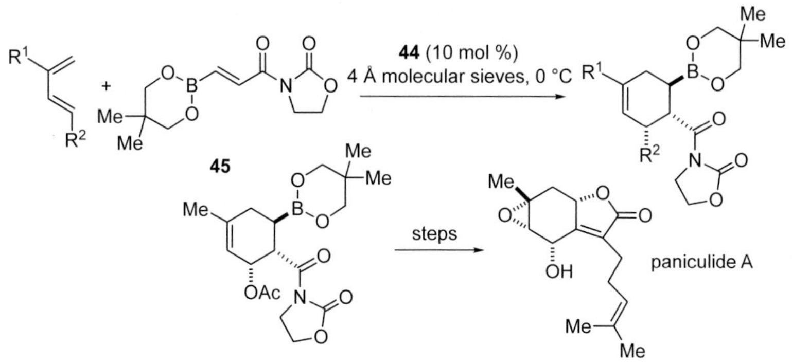

paniculide A

R[1]	R[2]	Yield [%]	ee [%]
H	H	76	>98
Me	H	74	>98
Me	Me	92	94
Me	OAc	71	95

Scheme 38

The titanium-TADDOL system is notable for its breadth of reacting partners. Fumaroyl [104b] and acryloyl [107] imide dienophiles may be employed with substituted and unsubstituted butadienes to afford cyclohexenes in high enantiomeric excess (Scheme 37). In the case of 2-thioethylbutadiene, the lower yield is accounted for by the intervention of a competing [2+2] cycloaddition pathway.

As noted in Scheme 38, 3-borylpropenoic acid derivative **45** functions as an effective dienophile with several butadienes [108]. The impetus for the development of this particular dienophile was the low reactivity observed for the corresponding 3-acetoxypropenoic acid derivative. Subsequent to cycloaddition, the boryl moiety may be stereospecifically oxidized to the corresponding alcohol and, as such, dienophile **45** effectively functions as a β-hydroxyacrylic acid surrogate. An asymmetric synthesis of (+)-paniculide A relied on this strategy as the key transformation [109].

A substituted furan has been demonstrated to afford oxabicyclo[2.2.1]heptene cycloadducts in high enantioselectivity under the influence of the Ti-TADDOL catalyst (Scheme 39) [110]. Reversibility at elevated temperatures was apparently not a problem in this case, in contrast to the reaction mediated by complex **31** (Sect. 3.2.1).

Titanium(IV)-TADDOL complexes are competent catalysts for intramolecular Diels-Alder reactions as well (Scheme 40) [111]. While a highly functionalized product is obtained, reaction times are on the order of days (68–257 h). The presence of the dithiane in the alkyl tether appears to be necessary not only for reasonable reactivity but also for high diastereoselectivity; the latter apparently results from unfavorable interactions between the dithiane and the diene in the

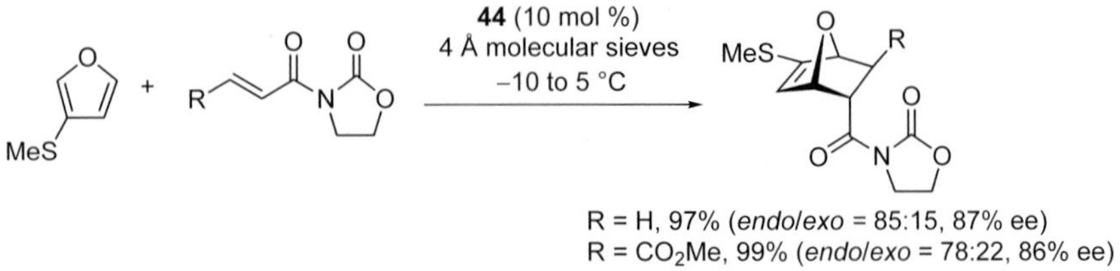

R = H, 97% (*endo/exo* = 85:15, 87% ee)
R = CO$_2$Me, 99% (*endo/exo* = 78:22, 86% ee)

Scheme 39

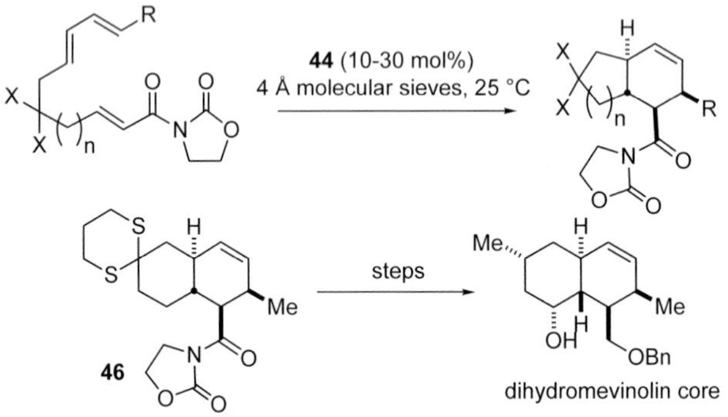

dihydromevinolin core

R	X	n	Yield [%]	ee [%]
H	H	1	87	87
H	S(CH$_2$)$_3$S	1	62	95
H	S(CH$_2$)$_3$S	2	64	86
Me	S(CH$_2$)$_3$S	2	70	87

Scheme 40

exo transition state, while the former is thought to be a manifestation of the Thorpe-Ingold effect [112]. As a demonstration of the synthetic utility of the process, cycloadduct **46** (R=Me, n=2, X=S(CH$_2$)$_3$S) was elaborated to the hydronaphthalene core of the mevinic acids.

The only highly enantioselective (>90% ee) Diels-Alder reaction using a ketone as a dienophile has been reported by Wada using the modified Ti(IV)-TADDOL catalyst **47** (Scheme 41) [113]. The important design feature is the use of a β-sulfonyl ketone, which presumably provides a chelating substrate to enhance catalyst-dienophile organization. The only diene used in this study was cyclopentadiene, and a limited number of dienophiles were employed, but the selectivities observed are noteworthy. As an added bonus, the phenylsulfonyl group may be excised to afford the corresponding methyl ketone in good yield.

As a result of the high level of success enjoyed by this family of catalysts, substantial effort has been invested in the study of the mechanism of asymmetric

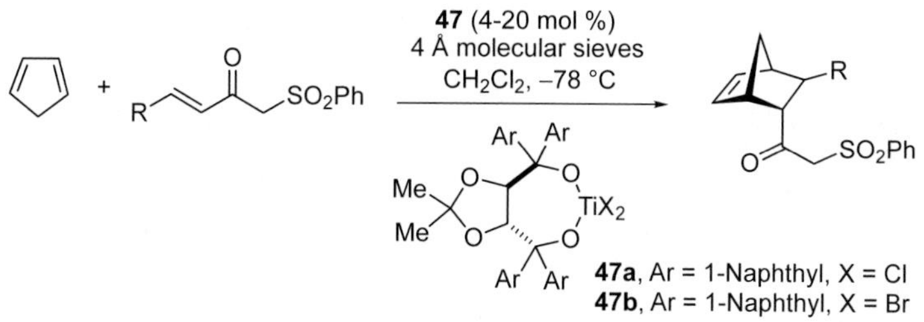

Catalyst	R	Yield [%]	endo/exo	ee [%]
47a	Me	80	>99:1	>99
47b	Pr	90	>99:1	94
47b	Ph	65	83:17	78

Scheme 41

induction. ^{1}H-NMR spectral studies have shown that catalysts such as **44** (typically formed from $TiCl_2(OiPr)_2$ and the chiral diol) are in fact in equilibrium with the starting materials [114]. Not unexpectedly, $TiCl_2(OiPr)_2$ promotes the reaction between a fumaroyl imide and isoprene at a substantially higher rate than complex **44**. In toluene-d_8 at 25 °C, the ratio of complex **44** to free diol is 87:13; the addition of 4 Å molecular sieves changes this ratio to 94:6, perhaps pointing to the role this addend is playing.

A 1:1 complex of a Ti-TADDOL catalyst and imide dienophile has been crystallographically characterized and implicated as the reactive species in enantioselective Diels-Alder reactions [115]. The imide is chelated to the metal center in the same plane as the chiral ligand (**48**, Fig. 28); this arrangement places the prochiral alkene in a position remote from the resident chirality and would presumably result in little stereochemical communication between the ligand and approaching diene. From the X-ray structure, Jørgensen has proposed that the pseudoequatorial phenyl group shields the *Si* face of the olefin in the *endo* transition state (*i.e.*, **49**) [116]; however, others have argued (*vide infra*) that binary complex **48** may simply be the most thermodynamically stable species (and most likely to crystallize from solution), not the dominant reactive species.

^{1}H-NMR spectral studies of a 1:1 mixture of imide dienophile and [Ti(TADDOL)]Cl_2 have revealed the presence of three species in solution, the geometry of the major complex being **49**, the same as in the solid state structure [117]. Complex **50** is proposed as one of the minor components, owing to shielding effects observed for some of the oxazolidinone protons and hindered rotation for the pseudoaxial aryl group of the ligand; such a complex is further postulated to be the reactive species. DiMare has convincingly argued that intermediate **50** in which the activated carbonyl is *trans* to a chloride ligand should experience a higher level of Lewis acid activation than in **49** where the *trans* substituent is an alkoxy group.

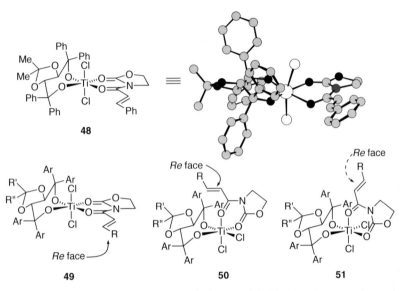

Fig. 28. X-ray structure of Ti-TADDOL-bound cinnamoyl imide **48** and proposed transition state assemblies for Ti-TADDOL mediated Diels-Alder reactions

Corey has proposed that the dienophile is activated in the apical position, but reacts via an *s-trans* configuration as illustrated in **51**. A donor-acceptor interaction between the pseudoequatorial aryl group and the bound dienophile was proposed as an organizational element due to the correlation between enantiomeric excess and aryl substituents [118]. It becomes necessary to invoke the Curtin-Hammett principle twice to validate this transition state: reaction occurs from the less favored metal geometry and the higher energy dienophile conformation. Incorporation of a *gem*-dimethyl group on the nitrogen-bearing carbon of the oxazolidinone led to nearly racemic product and was interpreted as evidence for reaction out of the *s-trans* conformer. While it appears that double stereodifferentiating experiments of the type carried out with catalyst **31** (Sect. 3.2.1) would be informative in differentiating transition structure **51** from **49** and **50**, no such studies have been disclosed.

As a final cautionary note regarding mechanistic interpretation of this system, Seebach has noted positive non-linear effects for the Diels-Alder reaction using Ti(IV)-TADDOL, indicating the possibility of either an aggregated transition state or the formation of catalytically inactive 1:1 (*R,R*)/(*S,S*)-titanium complexes [119].

Another tartaric acid-derived complex catalyzes the Diels-Alder reaction of *tert*-butyl acrylate and cyclopentadiene with good levels of enantiomeric excess (Scheme 42) [120]. The use of a smaller ester substituent resulted in lower enantioselectivity for the derived cycloadduct.

Despite their high reactivity as dienophiles and the potential utility of the derived cycloadducts, quinones have rarely been utilized in catalytic enantioselec-

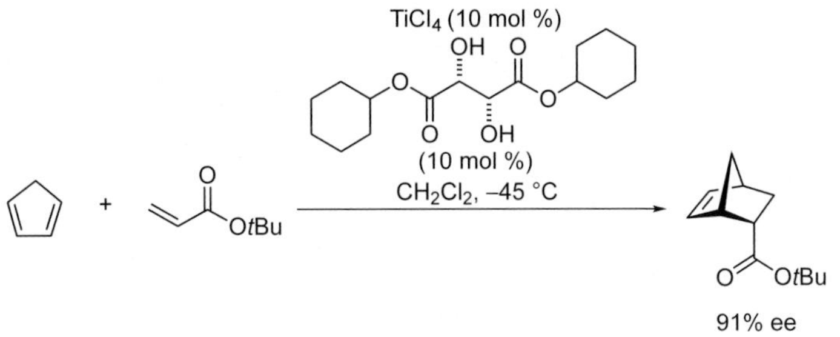

Scheme 42

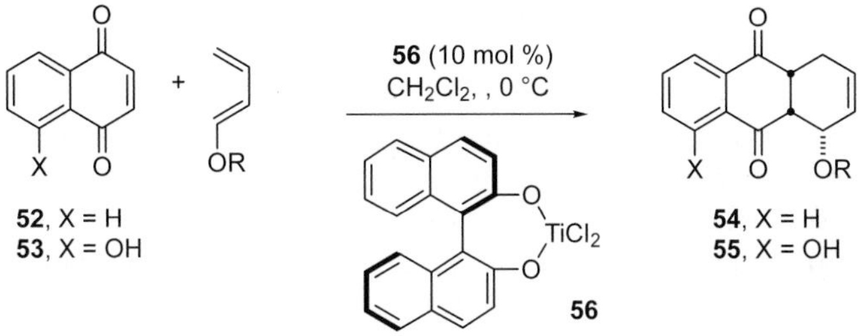

X	R	ee [%]	Comment
H	Me	86	Molecular sieves added
OH	COMe	76–96	Molecular sieve-free catalyst

Scheme 43

tive Diels-Alder reactions (for enantioselective quinone-diene Diels-Alder reactions with stoichiometric amounts of a chiral Lewis acid, see [121]). This is interesting in light of the fact that one of the variables which could contribute to low selectivity has been effectively deleted: quinones must react out of an *s-trans* configuration since the dienophile is locked in a ring. One of the few successful examples of an enantioselective quinone-Diels-Alder reaction was realized by Mikami using naphthoquinone (**52**, X=H) and a Ti(IV)-binaphthol complex in the presence of 4 Å molecular sieves (Scheme 43) [122, 123]. Tricyclic product **54** was formed with complete *endo* selectivity in 85% ee using catalyst **56**. When a similar reaction was attempted with juglone (**53**, X=OH), the cycloadduct **55** was obtained in only 9% ee. It was speculated that the molecular sieves were aiding in a deleterious phenol/chloride exchange; NMR experiments did seem to suggest that the phenol was bound to the catalyst. Accordingly, an alternate catalyst was prepared in which the molecular sieves were removed by centrifugation prior to the start of the reaction. With this modification, the desired cycloaddition could be executed between juglone and 1-acetoxybutadiene with high levels of selectivity, although the authors report that the enantioselectivity is vari-

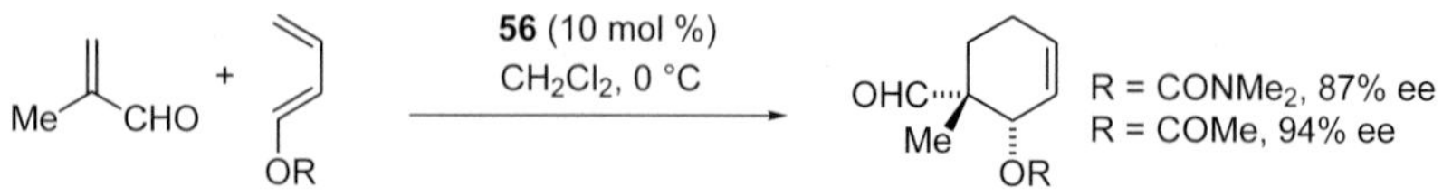

Scheme 44

able and catalyst batch-dependent. The derived cycloadducts are noteworthy as they provide a potential entry into the asymmetric syntheses anthracycline and tetracycline families of antibiotics.

Further work with the molecular sieve-free Ti(IV)-binaphthol catalyst **56** showed that 1-alkoxydienes react with methacrolein to afford cyclohexene products possessing a quaternary center adjacent to a stereochemically defined secondary urethane in near diastereomeric purity and high enantiomeric excess (Scheme 44).

Mikami and coworkers conducted the Diels-Alder reaction with a catalyst prepared by mixing enantiomerically pure (R)-**56** and racemic **56** and observed a positive nonlinear effect; however, they found no asymmetric amplification when they prepared the catalyst by mixing enantiomerically pure (R)-**56** and enantiomerically pure (S)-**56** (*i.e.*, linear correlation between catalyst and product ee). Introduction of molecular sieves restores the asymmetric amplification in the latter case, apparently by equilibration of (R)(R) and (S)(S) dimers into catalytically less active (R)(S) dimers. As expected, the reaction rate was faster for (R)-**56** than for (±)-**56** derived from racemic binaphthol ligand (*ca.* 5-fold faster).

Yamamoto has disclosed that another binaphthol-derived complex is an effective catalyst for enantioselective Diels-Alder reactions of aldehydes and cyclopentadiene (Scheme 45). Azeotropic removal of 2-propanol from a mixture of ligand **57** and Ti(O*i*Pr)$_4$ affords a Lewis acid capable of catalyzing Diels-Alder reactions between cyclopentadiene and acrolein, methacrolein, and crotonaldehyde, delivering cycloadducts with enantioselectivities in excess of 94%; however, diastereoselectivity is moderate in two cases [124].

The authors contend that the Lewis acid complex is helical, but characterization of the catalyst is limited to cryoscopic molecular weight measurements of a related complex in benzene. Two attributes of this system deserve attention: 1) the tetraalkoxytitanium species still possesses sufficient Lewis acidity to catalyze the reactions of interest at low temperatures; 2) the catalyst exhibits a fairly flat enantioselectivity-temperature profile (88% ee at 0 °C for the acrolein-cyclopentadiene reaction). The ligand was synthesized in five steps from (R)-(+)-3,3'-dibromobinaphthol dimethyl ether, and while other groups may be used in lieu of the tri-o-tolylsilyl group, the highest levels of enantioselectivity were realized with ligand **57**.

A rather different titanium(IV) Diels-Alder catalyst employed a *cis*-amino indanol, prepared in five steps from indene, as the chiral control element [125]. The amino indanol is regioisomeric to the one incorporated into a bisoxazolinyl

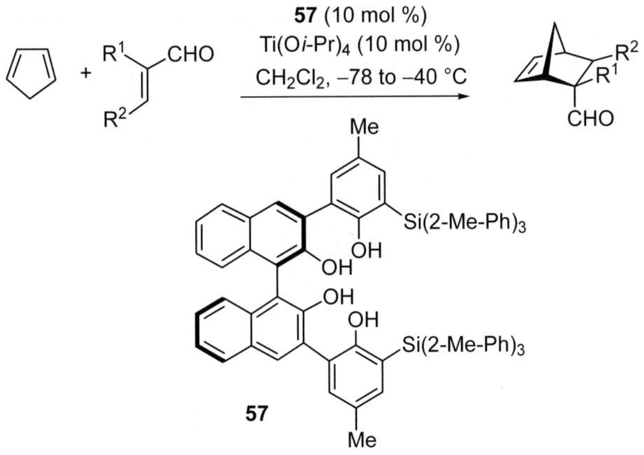

Scheme 45

R¹	R²	Yield [%]	*endo/exo*	ee [%]
H	H	70	85:15	96
Me	H	75	1:99	94 (2*S*)
H	Me	76	70:30	95

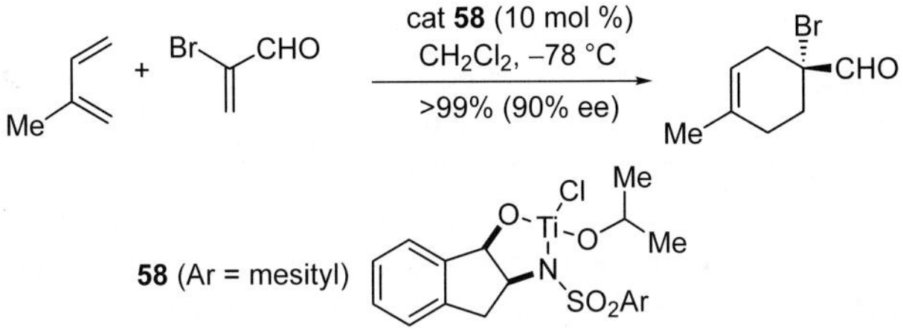

Scheme 46

ligand for copper(II) Lewis acids (**34, 35**). Treatment of the ligand with Ti(O*i*Pr)$_4$ in toluene at elevated temperature, followed by azeotropic removal of 2-propanol and subsequent treatment with one equivalent of SiCl$_4$ yielded a metal complex, tentatively formulated as **58**, as an amorphous yellow solid. By spectroscopic inspection the complex was not monomeric in solution, but aggregated. Nonetheless, **58** functioned as a stereoselective Lewis acid, catalyzing the cycloaddition of bromoacrolein with cyclopentadiene (93% ee) or isoprene (Scheme 46).

Keck and Krishnamurthy have shown that the Diels-Alder reaction of cyclopentadiene and bromoacrolein is facilitated by a Lewis acid derived from titanium tetraisopropoxide and *S*-BINOL (**59**) (Scheme 47) [126]. The cycloaddition may be conducted with isoprene at slightly lower levels of enantioselectivity; methacrolein-cyclopentadiene Diels-Alder reactions are only moderately selective.

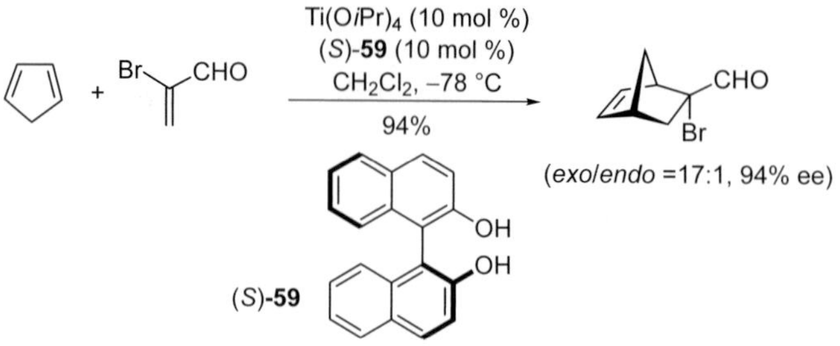

Scheme 47

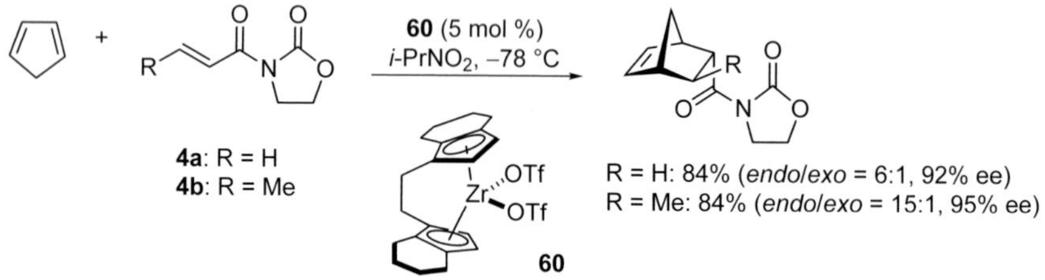

Scheme 48

While metallocenes are ubiquitous in organometallic and polymer chemistry, few such complexes have been reported to catalyze the Diels-Alder process in high enantioselectivity [127, 128, 129]. The bis(tetrahydroindenyl)zirconium triflate **60** and the corresponding titanocene are electrophilic to the extent that they catalyze the low-temperature cycloadditions of acrylate and crotonate imides with cyclopentadiene with good diastereoselectivity and excellent enantioselection (Scheme 48). The reactivity of **60** is noteworthy since the corresponding reaction using the crotonyl imide with highly reactive catalysts **31a** or **44** requires temperatures of −15 and 25 °C, respectively.

Collins and coworkers uncovered a truly dramatic solvent effect during these investigations. In the most telling example, the reaction of **4a** with cyclopentadiene proceeded in CH_2Cl_2 to afford racemic material, while the same reaction, conducted in 2-nitropropane, allows the adduct to be prepared in 92% ee. Only slightly lower enantiomeric excess was observed in nitromethane. ^{1}H- and ^{19}F-NMR spectroscopy of an equimolar mixture of **60** and acryloylimide **4a** in CD_3NO_2 established that very little unbound **4a** was present in solution; rather, at −30 °C, two complexes in a 2:1 ratio were observed. On the basis of the ^{13}C-NMR spectrum, it was concluded that the carbonyls in both complexes were bound to the metal center. In CD_2Cl_2, the ratio of complexes was altered (6:1), leading the authors to surmise that the dramatic shift in selectivity resulted from a change in stability of the catalyst-dienophile complexes, which could be assigned on the basis of NOE enhancements (Fig. 29). Further NOE studies showed that the unsaturated imide resided in the *s-cis* conformation for both bound

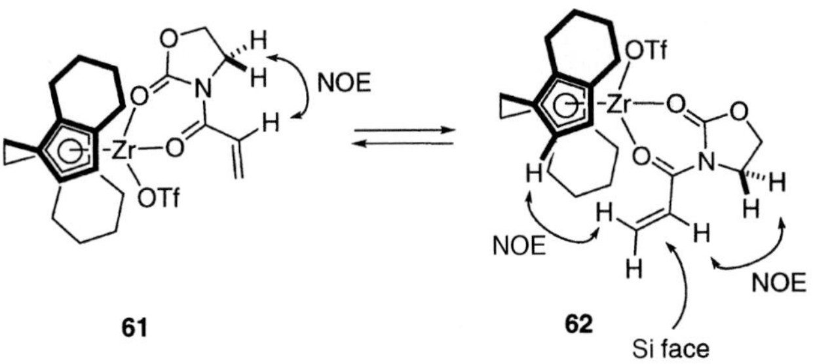

Fig. 29. Diastereomeric complexes formed between zirconocene **60** and imide **4a**

complexes. Since the unsaturated imide lies in a more defined chiral environment in **62**, it may be reasonably assumed that this is the species which leads to the enantioenriched product; the absolute configuration of the adduct is consistent with shielding of the *Re* face by the cyclohexyl ring and reaction from the exposed *Si* face.

3.3
Lanthanide Lewis Acids

Many researchers have refrained from using lanthanide complexes in stereoselective Diels-Alder reactions, perhaps due to large coordination spheres which can accommodate up to a dozen ligands. The rather daunting task of interpreting the identity of active catalysts and substrate-catalyst complexes among the myriad possible options has not hampered the development of some quite useful chiral lanthanide catalysts.

Kobayashi and coworkers have reported that a chiral complex derived from scandium(III)triflate, *R*-(+)-BINOL ((*R*)-**59**), and 1,2,6-trimethylpiperidine in the presence of 4 Å molecular sieves catalyzes the reaction of unsaturated imides with cyclopentadiene in 96–97% ee (Scheme 49) [130].

The particular trialkylamine additive was uniquely effective in securing maximum enantioselectivity, as both more and less steric demand on the amine afforded products having lower ee. ^{13}C-NMR and IR spectral studies have shed some light on the role of the amine additive. Rather than acting as a ligand, it has been suggested that the basic amines are interacting weakly with the acidic hydrogens of the phenols to form a hydrogen bond. The working hypothesis is that this hydrogen bond extends the axial chirality of the binaphthol (Fig. 30). In principle, this organizational motif provides an attractive alternative to the covalent modification the binaphthol ligand, since the amine additive can be easily varied, but the singular effectiveness of 1,2,6-trimethylpiperidine is not apparent from the working model. A second experimental nuance that merits mention is that the enantiomeric excess of the product is eroded as the catalyst ages [131]. While the cause is not known, the diminution in selectivity can be arrested with the addition of certain dicarbonyl additives. As with some other catalyst systems

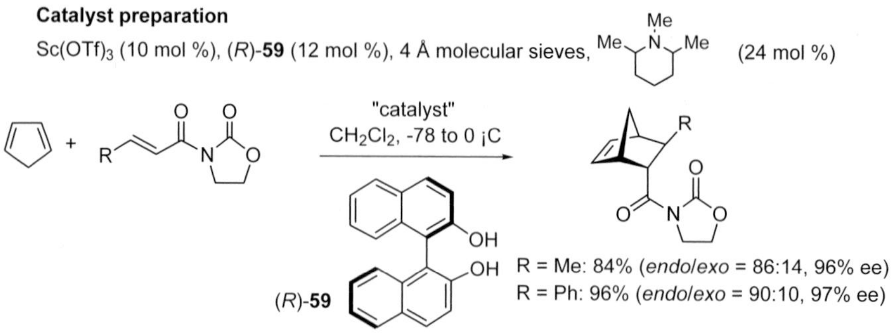
Catalyst preparation
Sc(OTf)3 (10 mol %), (R)-59 (12 mol %), 4 Å molecular sieves,
Me
Me N Me
(24 mol %)
"catalyst"
CH2Cl2, -78 to 0 ¡C
OH
OH
(R)-59
R = Me: 84% (endo/exo = 86:14, 96% ee)
R = Ph: 96% (endo/exo = 90:10, 97% ee)

Scheme 49

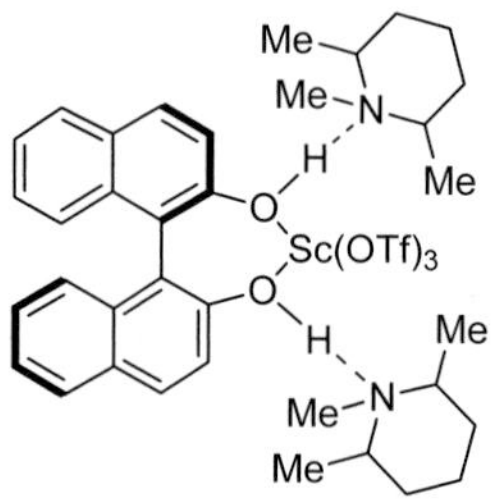
Me
Me N
H
Me
O
Sc(OTf)3
O
H
Me
Me N
Me

Fig. 30. Proposed extended axial chirality by interaction of amine base with binaphthol-$Sc(OTf)_3$ catalyst

presented previously, asymmetric amplification was observed with catalysts prepared from optically impure binaphthol ligand, suggesting aggregative catalyst behavior.

An exciting result with a related system ($Yb(OTf)_3$ [132] vs. $Sc(OTf)_3$) is illustrated in Scheme 50. If, in addition to the amine additive, one equivalent (relative to metal salt) of a dicarbonyl compound was included in the reaction, a turnover in enantioselectivity was observed [133]. In the case of the crotonyl imide, without added ligand the 2*S*,3*R* enantiomer **63** was obtained selectively (95% ee); however, with the addition of 3-phenylacetylacetone (**66**), the 2*R*,3*S* isomer **64** was formed in 81% ee. As was noted previously, the addition of dicarbonyl compounds to the scandium(III) triflate catalyst deters catalyst aging, but no turnover in enantioselectivity was observed. The difference between the two systems is thought to lie in a change in coordination number. To rationalize the reversal in facial bias, it has been postulated that the imide and the acetylacetone possess differential affinity for diastereotopic binding sites. It is conjectured that the stronger binding acetylacetone ligand forces the dienophile into a site in which the opposite enantioface is exposed, but no spectroscopic evidence supporting this turnover in binding has been disclosed. In contrast to the Sc(III)-binaphthol catalyst, the Yb(III)-binaphthol catalyst with added 3-phenylacetylacetone exhibits a negative nonlinear effect, while in the absence of any added ligand, no

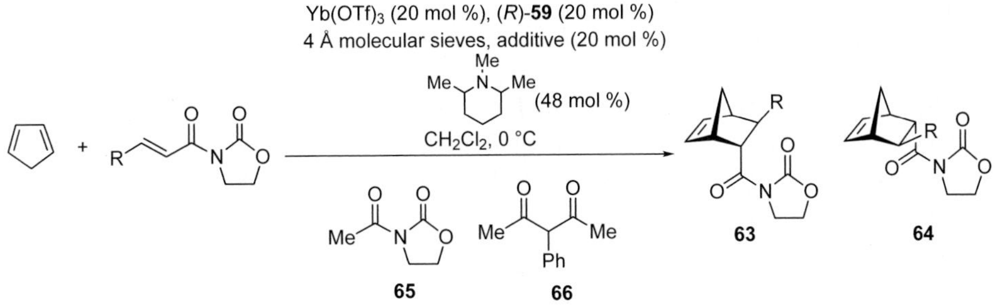

Yb(OTf)$_3$ (20 mol %), (R)-**59** (20 mol %)
4 Å molecular sieves, additive (20 mol %)

(48 mol %)

CH$_2$Cl$_2$, 0 °C

65   **66**

63   **64**

R	Comment	Yield [%]	endo/exo	er (**63**:**64**)
Me	no additive	77	89:11	97.5:2.5
Me	Added **66**	83	93:7	9.5:90.5
n-Pr	no additive	81	80:20	91.5:8.5
n-Pr	added **66**	81	91:9	10:90

Scheme 50

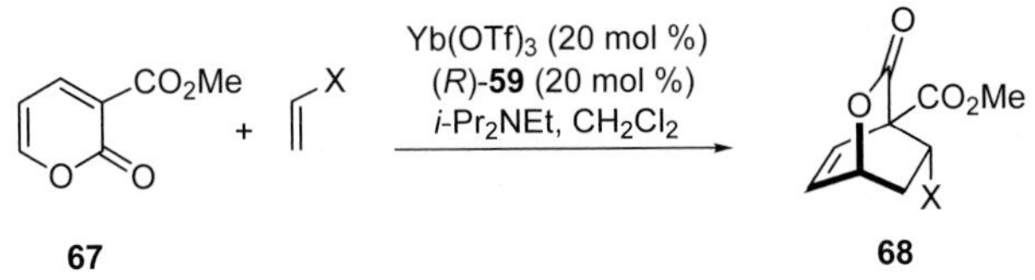

Yb(OTf)$_3$ (20 mol %)
(R)-**59** (20 mol %)
i-Pr$_2$NEt, CH$_2$Cl$_2$

67   **68**

X	Yield [%]	ee [%]
Oet	90	27
OC$_6$H$_{11}$	90	92
Oadamantyl	91	93
SPh	91	>95

Scheme 51

nonlinear effects are observed (at >60% ee) [134]. Thus, it appears that the active catalyst possesses a different degree of aggregation for each case.

A conceptually different [4+2] cycloaddition catalyzed by a chiral lanthanide complex has been disclosed. The inverse electron demand Diels Alder reaction of 3-methoxycarbonyl-2-pyrone (**67**) and enol ethers or sulfides [135] was catalyzed by a chiral ytterbium(III) triflate-binaphthol complex in the presence of diisopropylethylamine (Scheme 51) [136]. Thermal decarboxylations of bicyclic lactones such as **68** are known to yield dienes which may undergo subsequent pericyclic reactions [137]; thus, the adducts of this process are potentially useful chiral building blocks. The nature of the substituent on the 2π component was found to be crucial for the realization of high enantioselectivity.

4
Alternative Methods

The vast majority of strategies aimed at effecting enantioselective Diels-Alder reactions rely on complexation of an unsaturated carbonyl compound to a chiral Lewis acid, but this is not the only catalytic method for achieving enantiofacial bias. A unique approach outlined in Scheme 52 takes advantage of a diene (or precursor) possessing an acidic proton [138]; treatment with a catalytic amount of a chiral base results in transient formation of an oxidodiene which undergoes oxyanion-accelerated cycloaddition with a maleimide [139].

Asymmetric induction is thought to arise from an organized transition state in which the chiral amine base is associated with the oxidodiene (ion pairing) and the dienophile (hydrogen bonding, Fig. 31).

Mechanistic studies have discounted the possibility that cycloadduct 70 arises from a tandem Michael-aldolization pathway [140]. Stereospecificity is observed using fumaronitrile (*trans* double bond) or maleonitrile (*cis* double bond) as dienophiles, indicating either a concerted reaction or a rapid second step (aldol) relative to internal bond rotation. Upon treatment with triethylamine in methanol, ring opening to the formal Michael adduct, a thermodynamic sink, is observed; this Michael adduct was not formed in the enantioselective catalytic reaction. Further, the Michael adduct was not converted to cycloadduct 70 upon treatment with quinidine in chloroform; in fact, access to 70 from the Michael product could be achieved only under fairly special conditions.

The only other example of an enantioselective base-catalyzed Diels-Alder reaction is illustrated in Scheme 53. A hydroxypyrone (71) is the substrate which undergoes activation by a catalytic amount of cinchonine (69a, R=H), subsequently reacting with N-methylmaleimide to form the derived tricyclic adduct with good selectivity [141].

Use of a *Cinchona* alkaloid in which the hydroxy group had been acylated resulted in formation of cycloadducts of low enantiomeric excess, leading the authors to conclude that bidentate activation (Fig. 32) was important in providing

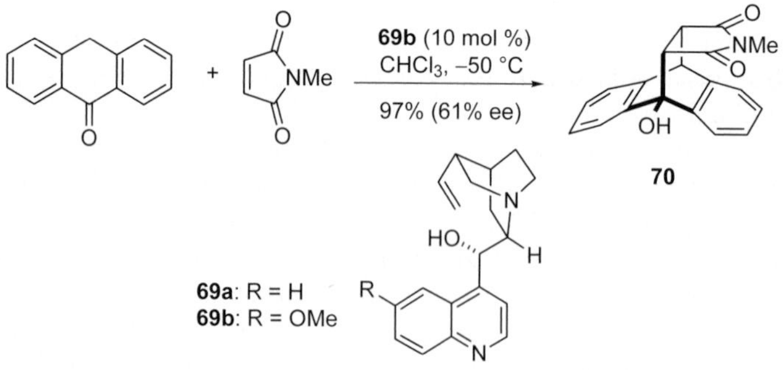

Scheme 52

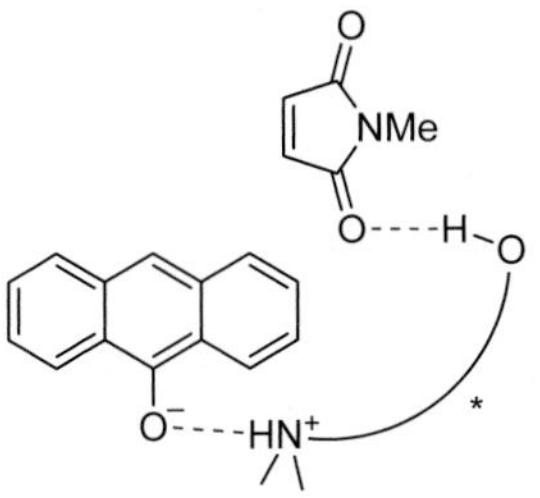

Fig. 31. Postulated ion-pairing in chiral base-catalyzed Diels-Alder reactions

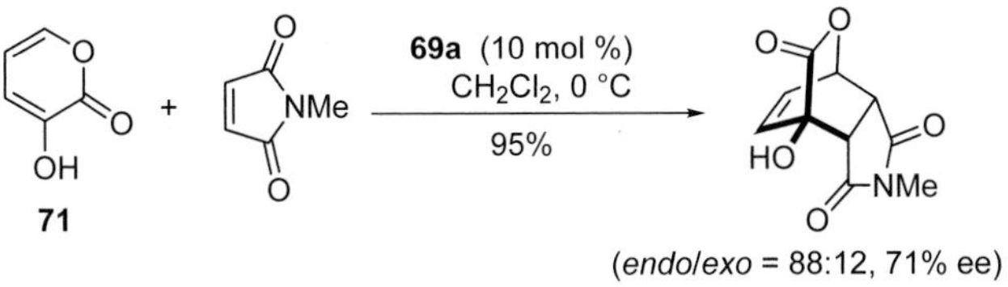

(*endo/exo* = 88:12, 71% ee)

Scheme 53

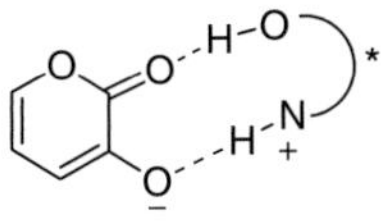

Fig. 32. Postulated ion pairing of hydroxy pyrone **71** with *Cinchona* alkaloid base

a high level of transition state organization. It is important to note that the control experiments which were performed in the anthrone/maleimide system (Scheme 52) were not performed for the reaction in Scheme 53; thus, a tandem Michael-aldolization pathway, while unlikely, has not been strictly excluded.

Enantiocontrol in the base-catalyzed Diels-Alder reaction has not yet reached the level of its Lewis acid-catalyzed counterpart; time will tell if the method can be generalized to a wider scope of substrates.

A fundamentally different approach to asymmetric induction in the Diels-Alder process entails the use of catalytic antibodies generated from transition state analogs [142]. Outlined conceptually in Fig. 33, haptens mimicking the *endo* and *exo* transition states were separately utilized to elicit catalytic antibodies which were used as Diels-Alder catalysts [143]. Bicyclo[2.2.2]hexanes modeling the boat-like Diels-Alder transition state were designed to minimize product inhibition, since the low energy product conformation is the twist chair, which presumably will not bind competitively to the catalytic site.

The catalytic antibodies were found to be exceptionally selective for the indicated cycloaddition reactions (Scheme 54). It is remarkable that either diastereomeric product may be obtained enantiomerically pure simply by selecting the

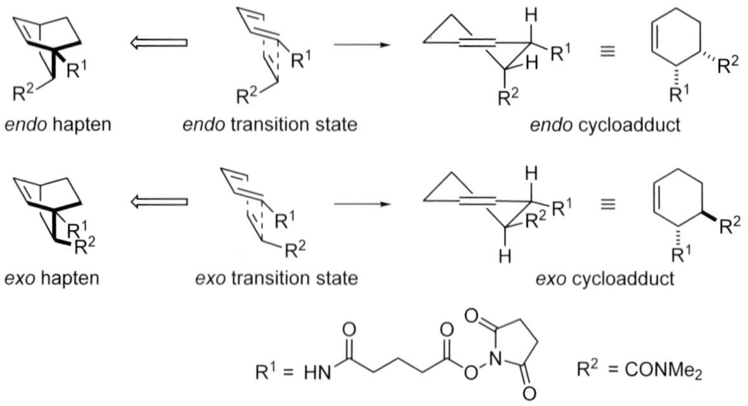

Fig. 33. Generation of catalytic antibodies by haptens designed to mimic the Diels-Alder transition state geometry, but not the product conformation

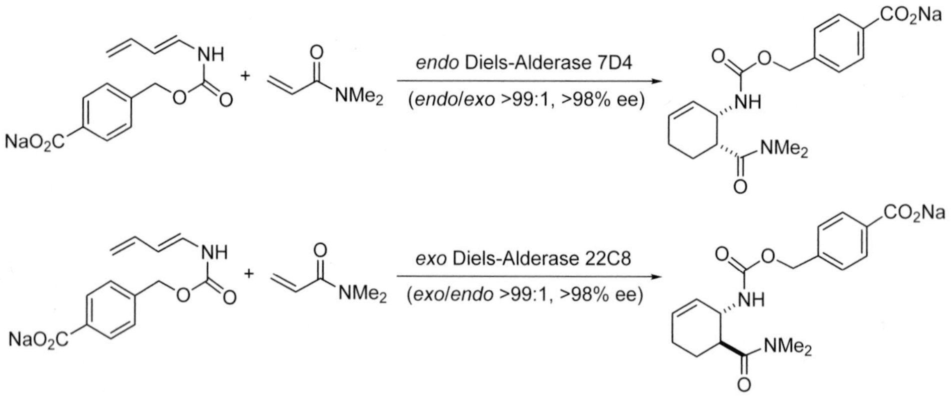

Scheme 54

correct catalytic antibody (for structural studies of Diels-Alderase antibodies, see [144]). Especially interesting is the fact that the energetically disfavored *exo* product may be obtained preferentially. This type of selectivity is rare in enantioselective Lewis acid-catalyzed processes [145, 146]. Based on the reported reaction conditions, it appears that the catalytic antibody turns over roughly four times for the *endo* adduct and five times for the *exo* [147]. With regard to practical considerations, separate antibodies must be prepared for each desired cycloadduct and it is not clear whether this process is amenable to large scale (20 μM of antibody reported); nonetheless, the levels of stereoselectivity render this concept immediately useful.

5
Conclusions

From the preceding discussion, some generalizations may be drawn with respect to the development of catalytic enantioselective Diels-Alder reactions and some conservative predictions pertaining to the future of the field may be proffered.

While astonishing diversity has been realized in the development of chiral complexes which will catalyze the Diels-Alder reaction with high attendant enantioselection (Fig. 34), the pool of reactions which has been sampled is relatively small. A plethora of catalysts has been reported to catalyze the cycloaddition of cyclopentadiene with acrolein or acrylate derivatives, but realization of generality with respect to reacting partners has been more difficult. The goal of this chapter is to provide a comprehensive review of advances in the field and in so doing suggest to the reader the untapped potential which remains. Those interested in executing these enantioselective processes must take into account factors of ligand synthesis, scalability, selectivity, and generality. For unsaturated aldehydes, oxazaborolidine (10), CAB (7) and BLA (14, 15) catalysts distinguish themselves in their general applicability. A similar demarcation is possible for two-point binding substrates, as both Ti-TADDOL (44) and Cu(II)bis(oxazoline) (31) systems are both effective for a wide range of dienes and dienophiles and are amenable to preparative scale processes. A potential marker for synthetic utility of catalytic systems is actual application in multistep natural product synthesis. The aforementioned oxazaborolidine 10, Cu(II)-bis(oxazoline) 31, and Ti-TADDOL 44 have been successfully used in that context, as well as the aluminum-stien complex 3.

Despite the development of a multitude of efficient, selective catalysts, the field is still at a fledgling stage. What does the future hold? In answering this query, it is important to consider two critical issues which are inexorably intertwined: mechanism and synthetic utility.

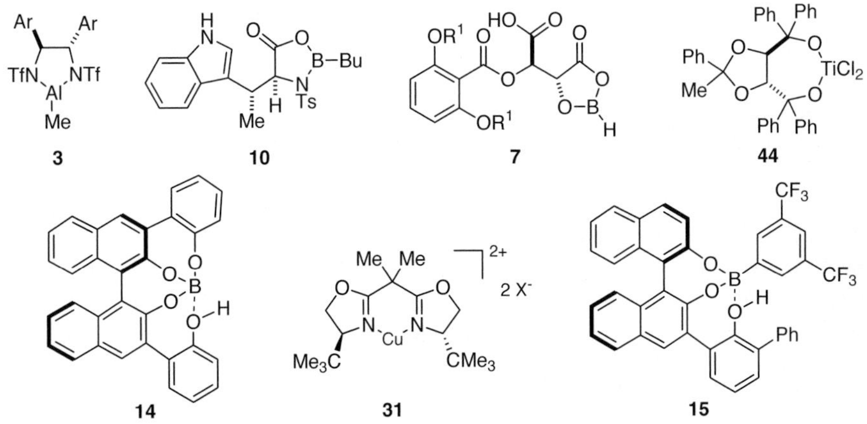

Fig. 34. Diels-Alder catalysts applicable to a breadth of diene/dienophile combinations

Mechanism. A strong argument can be made that the rigor that characterized earlier physical organic studies of the Diels-Alder reaction (and others) has been at least partially supplanted by more qualitative approaches. While exceptions certainly exist, kinetic analyses and isotopic labeling are no longer *de rigueur* and a brief survey of the literature reveals that structural characterization of catalysts and catalyst-substrate complexes does not appear to be a prerequisite for the formulation of transition structures. It is not coincidence that the mostly broadly useful Diels-Alder catalysts are those which are best understood mechanistically. As the field continues to expand, it will become even more critical to obtain a complete understanding of the minute details and nuances of each new catalyst system. Spectroscopic and solid state characterization of catalysts and activated complexes, as well as solution behavior from reaction kinetics will be indispensable in this regard. Without this rigor, mechanistic understanding and advances that extend from such insight will be slow coming, but with innovative approaches to the study of the intimate details of these processes, the field should continue to flourish.

Synthetic Utility. The enantioselective catalytic Diels-Alder reaction will continue to grow in usefulness to the synthetic organic chemist. Broadly speaking, current endeavors seek to expand the scope of this reaction through the development of complexes that effectively catalyze the cycloaddition of an entire spectrum of reacting partners. In simplest terms, more reactive catalysts and dienophiles will provide access to more highly functionalized products and facilitate the rapid assembly of molecular complexity. Even as this development occurs, questions will arise which will demand creative solutions. For example, can catalysts be designed to differentiate between the lone pairs of a simple α,β-unsaturated ketone and deliver cycloadducts in high enantioselectivity? Can alkenes lacking a carbonyl substituent be activated by chiral complexes in a face-selective fashion toward cycloaddition with dienes? Mechanistic studies will be crucial in assessing the feasibility of these and other processes. Practical considerations will lead to the development of more robust catalysts which can operate over a broad temperature range without special experimental precautions (inert atmosphere and the like). The Diels-Alder reaction will likely benefit from general advances being made in the field of asymmetric synthesis: generation of new catalyst leads will be facilitated by the continuing evolution of combinatorial chemistry, while catalyst immobilization in the solid and liquid phase can serve to greatly simplify product isolation and catalyst recycling. Additionally, the aforementioned simplicity of the catalytic cycle assure the continued prominence of the Diels-Alder process as an attractive test reaction for newly developed chiral catalysts. In this context, successful application of chiral Lewis acids to the Diels-Alder reaction is frequently a reliable indicator of potential utility to other classes of reactions. Notably, facile extension to aldol, ene, Michael, dipolar cycloaddition, and hetero-Diels-Alder reactions are a common outgrowth of studies in the enantioselective catalysis of the carbocyclic Diels-Alder reaction. The reader need only briefly scan other chapters of this monograph to find corroborating evidence for this point.

Undoubtedly, the axiom that *the constraints of the multistep synthesis experience provide the impetus for reaction development* will continue to be pertinent to the Diels-Alder reaction. The realization of more reactive and more general catalysts will continue to be a goal for the field and will yield an ever-growing arsenal of tools for use in the synthetic endeavors which require highly functionalized, enantioenriched carbocyclic building blocks.

References

1a. Oppolzer W (1991) Intermolecular Diels-Alder Reactions. In: Trost BM, Fleming I (eds) Comprehensive Organic Synthesis. Pergamon Press, Oxford, vol 5 chap 4.1
1b. Carruthers W (1990) Cycloaddition Reactions in Organic Synthesis. Pergamon Press, New York
2a. Roush WR (1991) Intramolecular Diels-Alder Reactions. In: Trost BM, Fleming I (eds) Comprehensive Organic Synthesis. Pergamon Press, Oxford, vol 5 chap 4.4
2b. Fallis AG (1984) Can J Chem 62:183
2c. Craig D (1987) Chem Soc Rev 16:187
2d. Brieger G, Bennett JN (1980) Chem Rev 80:63; (e) Ciganek E (1984) In: Dauben WG (ed) Org React 32:1 (f) Oppolzer W (1977) Angew Chem Int Ed Eng 16:10
3. Deslongchamps P (1991) Aldrichimica Acta 24:43
4. Desimoni G, Tacconi G, Barco A, Pollini GP (1983) Natural Products Synthesis Through Pericyclic Reactions. ACS Monograph 180, Washington DC
5a. Oppolzer, W (1984) Angew Chem Int Ed Eng 23:876
5b. Helmchem G, Karge R, Weetman J (1986) Asymmetric Diels-Alder Reactions with Chiral Enoates as Dienophiles. In: Scheffold R (ed) Modern Synthetic Methods. Springer-Verlag, Berlin Heidelberg, p 262
6a. Dias LC (1997) J Braz Chem Soc 8:289
6b. Oh T, Reilly M (1994) Org Prep Proc Int 26: 129
6c. Kagan HB, Riant O (1992) Chem Rev 92:1007
7. Santelli M, Pons J-M (1996) Lewis Acids and Selectivity in Organic Synthesis. CRC Press, New York
8. Yates P, Eaton P (1960) J Am Chem Soc 82:4436
9. Inukai T, Kojima T (1967) J Org Chem 32:872
10. Garcia JI, Martinez-Merino V, Mayoral JA, Salvatella L (1998) J Am Chem Soc 120:2415
11. Houk KN, Strozier RW (1973) J Am Chem Soc 95:4094
12. Inukai T, Kojima T (1966) J Org Chem 31:2032
13a. Houk KN (1973) J Am Chem Soc 95:4092
13b. Eisenstein O, Lefour J-M, Ahn NT (1971) J Chem Soc Chem Commun 969
14. Alder K, Stein G (1937) Angew Chem 50:510
15. Birney DM, Houk KN (1990) J Am Chem Soc 112:4127
16. Garcia JI, Mayoral JA, Salvatella L (1997) Tetrahedron 53:6057
17. Singleton DA (1992) J Am Chem Soc 114:6563
18. Yamabe S, Dai T, Minato T (1995) J Am Chem Soc 117:10994
19a. Martin JG, Hill RK (1961) Chem Rev 61:537
19b. Kobuke Y, Fueno T, Furukawa J (1970) J Am Chem Soc 92:6548
20. Gleiter R, Böhm MC (1983) Pure Appl Chem 55:237
21. For an excellent summary and discussion of this issue, as well as other mechanistic aspects of the Diels-Alder reaction, see: Sauer J, Sustmann R (1980) Angew Chem Int Ed Eng 19:779
22. Gladysz JA, Boone BJ (1997) Angew Chem Int Ed Eng 36:550
23a. Fernández JM, Emerson K, Larsen RD, Gladysz JA (1986) J Am Chem Soc 108:8268

23b. Fernández JM, Emerson K, Larsen RD, Gladysz JA (1988) J Chem Soc Chem Commun 37
24. There is also a steric component to this turnover in binding mode
25. Shambayati S, Crowe WE, Schreiber SL (1990) Angew Chem Int Ed Eng 29:256
26. Reetz MT, Hüllmann M, Massa W, Berger S, Rademacher P, Heymanns P (1986) J Am Chem Soc 108:2405
27. Corey EJ, Loh T-P, Sarshar S, Azimioara M (1992) Tetrahedron Lett 33:6945
28. Denmark SE, Almstead NG (1993) J Am Chem Soc 115:3133
29. Goodman JM (1992) Tetrahedron Lett 33:7219
30. Corey EJ, Rohde JJ, Fischer A, Azimioara MD (1997) Tetrahedron Lett 38:33
31a. Corey EJ, Rohde JJ (1997) Tetrahedron Lett 38:37
31b. Corey EJ, Barnes-Seeman D, Lee TW (1997) Tetrahedron Lett 38:1699
32. Gung BW, Wolf MA (1992) J Org Chem 57:1370
33. Loncharich RJ, Schwartz TR, Houk KN (1987) J Am Chem Soc 109:14
34. Lewis FD, Oxman JD, Huffman JC (1984) J Am Chem Soc 106:466
35. Brun L (1966) Acta Crystallogr 20:739
36a. Evans DA, Chapman KT, Bisaha J (1984) J Am Chem Soc 106:4261
36b. Oppolzer W, Chapuis C, Bernardinelli G (1984) Helv Chim Acta 67:1397
36c. Evans DA, Chapman KT, Bisaha J (1988) J Am Chem Soc 110:1238
37. Poll T, Metter JO, Helmchen G (1985) Angew Chem Int Ed Eng 24:112
38. Oppolzer W, Rodriguez I, Blagg J, Bernardinelli G (1989) Helv Chim Acta 72:123
39a. Castellino S (1990) J Org Chem 55:5197
39b. For related NMR reactions with an aluminum Lewis acid, see: Castellino S, Dwight WJ (1993) J Am Chem Soc 115:2986
40. Ishihara K, Gao Q, Yamamoto H (1993) J Am Chem Soc 115:10412
41. Montaudo G, Librando V, Caccamese S, Maravigna P (1973) J Am Chem Soc 95:6365
42. Desimoni G, Faita G, Invernizzi AG, Righetti P (1997) Tetrahedron 53:7671
43. This is not meant to imply that processes which proceed in somewhat lower enantiose-lectivity cannot find practical applications in organic synthesis by virtue of enantiomer-ic enrichment via recrystallization or alternative methods
44. Reaction of cyclopentadiene and methyl acrylate mediated by BF_3·menthyl ethyl ether-ate gave cycloadduct in 70% yield (dr 95:5, 3% ee): Guseinov MM, Akhmedov IM, Mamedov EG (1976) Azerb Khim Zh 1:46
45. Hashimoto S-I, Komeshima N, Koga K (1979) J Chem Soc Chem Commun 437
46. Takemura H, Komeshima N, Takahashi I, Hashimoto S-I, Ikota N, Tomioka K, Koga K (1987) Tetrahedron Lett 28:5687
47. Stoichiometric amounts of aluminum Lewis acids induce highly enantioselective Diels-Alder reactions between a crotonyl imide dienophile and cyclopentadiene in the pres-ence of 0.5–1.0 equivalent of a number of chiral ligands: Chapuis C, Jurczak J (1987) Helv Chim Acta 70:436
48a. Corey EJ, Imwinkelried R, Pikul S, Xiang YB (1989) J Am Chem Soc 111:5493
48b. Corey EJ, Imai N, Pikul S (1991) Tetrahedron Lett 32:7517
48c. Corey EJ, Sarshar S, Bordner J (1992) J Am Chem Soc 114:7938
48d. Corey EJ, Lee D-H, Sarshar S (1995) Tetrahedron: Asymmetry 6:3
49. Corey EJ, Sarshar S, Lee D-H (1994) J Am Chem Soc 116:12089
50. Corey EJ, Letavic MA (1995) J Am Chem Soc 117:9616
51a. Bao J, Wulff WD, Rheingold AL (1993) J Am Chem Soc 115:3814
51b. Bao J, Wulff WD, Dominy JB, Fumo MJ, Grant EB, Rob AC, Whitcomb MC, Yeung S-M, Ostrander RL, Rheingold AL (1996) J Am Chem Soc 118:3392
51c. Heller DP, Goldberg DR, Wulff WD (1997) J Am Chem Soc 119:10551
52. Complexes of **6** and bromoborane are less stereoselective catalysts: Bao J, Wulff WD (1995) Tetrahedron Lett 36:3321
53. Rebiere F, Riant O, Kagan HB (1990) Tetrahedron: Asymmetry 1:199
54. An aluminum-binaphthol complex catalyzes the Diels-Alder reaction of cyclopentadi-ene and methyl acrylate in 82% yield and 67% ee: Maruoka K, Concepcion AB, Yamamo-to H (1992) Bull Chem Soc Jpn 65:3501

55. For a review of asymmetric boron-catalyzed reactions, see: Deloux L, Srebnik M (1993) Chem Rev 93:763
56a. Furuta K, Shimizu S, Miwa Y, Yamamoto H (1989) J Org Chem 54:1481
56b. Ishihara K, Gao Q, Yamamoto H (1993) J Org Chem 58:6917
57. The topic of π-stacking in asymmetric synthesis has been reviewed: Jones GB, Chapman BJ (1995) Synthesis 475
58. Furuta K, Kanematsu A, Yamamoto H (1989) Tetrahedron Lett 30:7231
59a. Corey EJ, Loh T-P (1991) J Am Chem Soc 113:8966
59b. For a tartrate-derived dioxaborolidine Diels-Alder catalyst, see: Loh T-P, Wang R-B, Sim K-Y (1996) Tetrahedron Lett 37:2989
60. Corey EJ, Guzman-Perez A, Loh T-P (1994) J Am Chem Soc 116:3611
61. Corey EJ, Loh T-P (1993) Tetrahedron Lett 34:3979
62. Corey EJ, Loh T-P, Roper TD, Azimioara MD, Noe MC (1992) J Am Chem Soc 114:8290
63a. Itsuno S, Kamahori K, Watanabe K, Koizumi T, Ito K (1994) Tetrahedron: Asymmetry 5:523
63b. Kamahori K, Tada S, Ito K, Itsuno S (1995) Tetrahedron: Asymmetry 6:2547
63c. Kamahori K, Ito K, Itsuno S (1996) J Org Chem 61:8321
64a. Takasu M, Yamamoto H (1990) Synlett 194
64b. Sartor D, Saffrich J, Helmchen G (1990) Synlett 197
65a. Hawkins JM, Loren S (1991) J Am Chem Soc 113:7794
65b. Hawkins JM, Loren S, Nambu M (1994) J Am Chem Soc 116:1657
66a. A modified isopinocampheyldibromoborane catalyzes the reaction of cyclopentadiene with methyl acrylate in 48% ee: Bir G, Kaufmann D (1990) J Organomet Chem 390:1
66b. A Lewis acid derived from N-tosyl tryptophan and 1,8-naphthalenediylbis(dichloroborane) is reported to catalyze the Diels-Adler reaction of methacrolein and cyclopentadiene in 100% ee for the *endo* isomer (*endo/exo*=37:63): Reilly M, Oh T (1994) Tetrahedron Lett 35:7209
66c. Reilly M, Oh T (1995) Tetrahedron Lett 36:221
67. Ishihara K, Yamamoto H (1994) J Am Chem Soc 116:1561
68. Ishihara K, Kurihara H, Yamamoto H (1996) J Am Chem Soc 118:3049
69a. Ishihara K, Kondo S, Kurihara H, Yamamoto H (1997) J Org Chem 62:3026
69b. For a related reaction with an aluminum catalyst, see ref. [54]
70. Hayashi Y, Rohde JJ, Corey EJ (1996) J Am Chem Soc 118:5502
71. Corey EJ, Lee TW (1997) Tetrahedron Lett 38:5755
72. Kaufmann D, Boese R (1990) Angew Chem Int Ed Eng 29:545
73a. Kobayashi S, Murakami M, Harada T, Mukaiyama T (1991) Chem Lett: 1341
73b. Aggarwal VK, Anderson E, Giles R, Zaparucha A (1995) Tetrahedron: Asymmetry 6:1301
74. Corey EJ, Ishihara K (1992) Tetrahedron Lett 33:6807
75a. Fujisawa T, Ichiyanagi T, Shimizu M (1995) Tetrahedron Lett 36:5031
75b. Ichiyanagi T, Shimizu M, Fujisawa T (1997) J Org Chem 62:7937
76. Ordoñez M, Guerrero-de la Rosa V, Labastida V, Llera JM (1996) Tetrahedron: Asymmetry 7:2675
77a. Desimoni G, Faita G, Righetti PP (1996) Tetrahedron Lett 37:3027
77b. Carbone P, Desimoni G, Faita G, Filippone S, Righetti P (1998) Tetrahedron 54:6099
78a. Chandrasekhar K, Bürgi H-B (1983) J Am Chem Soc 105:7081;
78b. Roussel C, Lidén A, Chanon M, Metzger J, Sandström J (1976) J Am Chem Soc 98:2847
79. Honda Y, Date T, Hiramatsu H, Yamauchi M (1997) Chem Commun 1411
80. For a more in-depth discussion of this observation, see: Yamauchi M, Honda Y, Matsuki N, Watanabe T, Date K, Hiramatsu H (1996) J Org Chem 61:2719
81. For a general reference on catalysis of the Diels-Alder reaction by achiral transition metal Lewis acids, see: Bonnesen PV, Puckett CL, Honeychuck RV, Hersh WH (1989) J Am Chem Soc 111:6070
82 Evans DA, Miller SJ, Lectka T (1993) J Am Chem Soc 115:6460
83. Evans DA, Murry JA, von Matt P, Norcross RD, Miller SJ (1995) Angew Chem Int Ed Eng 34:798

84. Evans DA, Barnes DM (1997) Tetrahedron Lett 38:57
85. Barnes DM (1997) PhD thesis, Harvard University
86. Johannsen M, Jørgensen KA (1997) J Chem Soc Perkin Trans 2 1183
87. Evans DA, Shaughnessy EA, Barnes DM (1997) Tetrahedron Lett 38:3193
88. Evans DA, Johnson JS (1997) J Org Chem 62:786
89. For a review of this topic: Masamune S, Choy W, Peterson JS, Sita LR (1985) Angew Chem Int Ed Eng 24:1
90. Davies IW, Gerena L, Castonguay L, Senanayake CH, Larsen RD, Verhoeven TR, Reider PJ (1996) J Chem Soc Chem Commun 1753
91a. Davies IW, Gerena L, Cai D, Larsen RD, Verhoeven TR, Reider PJ (1997) Tetrahedron Lett 38:1145
91b. Davies IW, Senanayake CH, Larsen RD, Verhoeven TR, Reider PJ (1996) Tetrahedron Lett 37:1725
92. Ghosh AK, Mathivanan P, Cappiello J (1996) Tetrahedron Lett 37:3815
93. Evans DA, Lectka T, Miller SJ (1993) Tetrahedron Lett 34:7027
94. Sagasser I, Helmchen G (1998) Tetrahedron Lett 39:261
95. Kündig EP, Bourdin B, Bernardinelli G (1994) Angew Chem Int Ed Eng 33:1856
96. Corey EJ, Imai N, Zhang H-Y (1991) J Am Chem Soc 113:728
97. Davies DL, Fawcett J, Garratt SA, Russell DR (1997) Chem Commun 1351
98. For a crystal structure of methacrolein bound to a chiral cationic ruthenium complex, see: Carmona D, Cativiela C, Elipe S, Lahoz FJ, Lamata MP, López-Ram de Víu MP, Oro LA, Vega C, Viguri F (1997) Chem Commun 2351
99. Davenport AJ, Davies DL, Fawcett J, Garratt SA, Lad L, Russell DR (1997) Chem Commun 2347
100a. Takacs JM, Lawson EC, Reno MJ, Youngman MA, Quincy DA (1997) Tetrahedron: Asymmetry 8:3073
100b. Takacs JM, Quincy DA, Shay W, Jones BE, Ross CR (1997) Tetrahedron: Asymmetry 8:3079
101. Evans DA, Kozlowski MC, Tedrow JS (1996) Tetrahedron Lett 37:7481
102a. Kanemasa S, Oderaotoshi Y, Yamamoto H, Tanaka J, Wada E, Curran DP (1997) J Org Chem 62:6454
102b. Kanemasa S, Oderaotoshi Y, Sakaguchi S-i, Yamamoto H, Tanaka J, Wada E, Curran DP (1998) J Am Chem Soc 120:3074
103a. Bienaymé H (1997) Angew Chem Int Ed Eng 36:2670
103b. Devine PN, Oh T (1992) J Org Chem 57:396
104a. Narasaka K, Inoue M, Yamada T (1986) Chem Lett 1967
104b. Narasaka K, Iwasawa N, Inoue M, Yamada T, Nakashima M, Sugimori J (1989) J Am Chem Soc 111:5340
104c. Seebach D, Beck AK, Imwinkelried R, Roggo S, Wonnacott (1987) Helv Chim Acta 70:954
105. For use of dendrimer- and polymer-bound TiTADDOLates in these cycloadditions, see: Seebach D, Marti RE, Hintermann T (1996) Helv Chim Acta 79:1710
106. Narasaka K, Inoue M, Yamada T, Sugimori J, Iwasawa N (1987) Chem Lett 2409
107. Narasaka K, Tanaka H, Kanai F (1991) Bull Chem Soc Jpn 64:387
108. Narasaka K, Yamamoto I (1992) Tetrahedron 48:5743
109. Yamamoto I, Narasaka K (1994) Bull Chem Soc Jpn 67:3327
110. Yamamoto I, Narasaka K (1995) Chem Lett: 1129
111a. Iwasawa N, Sugimori J, Kawase Y, Narasaka K (1989) Chem Lett 1947
111b. Narasaka K, Saitou M, Iwasawa N (1991) Tetrahedron: Asymmetry 2:1305
112. Jung ME, Gervay J (1991) J Am Chem Soc 113:224 and references therein
113. Wada E, Pei W, Kanemasa S (1994) Chem Lett 2345
114a. Iwasawa N, Hayashi Y, Sakurai H, Narasaka K (1989) Chem Lett 1581
114b. Tietze LF, Ott C, Frey U (1996) Liebigs Ann 63
115. Gothelf KV, Hazell RG, Jørgensen KA (1995) J Am Chem Soc 117:4435
116. Gothelf KV, Jørgensen KA (1995) J Org Chem 60:6847

117a. Haase C, Sarko CR, DiMare M (1995) J Org Chem 60:1777

117b. Quantum chemical studies support the contention that complexes **50** and **51** are activated to a significantly greater exent than **49**: García JI, Martínez-Merino V, Mayoral JA (1998) J Org Chem 63:2321

118. Corey EJ, Matsumura Y (1991) Tetrahedron Lett 32:6289

119. Seebach D, Dahinden R, Marti RE, Beck AK, Plattner DA, Kühnle FNM (1995) J Org Chem 60:1788

120. Ketter A, Glahsi G, Herrmann R (1990) J Chem Res (S) 278; (M) 2118

121a. Kelly TR, Whiting A, Chandrakumar NS (1986) J Am Chem Soc 108:3510

121b. Maruoka K, Sakurai M, Fujiwara J, Yamamoto H (1986) Tetrahedron Lett 27:4895

122a. Mikami K, Terada M, Motoyama Y, Nakai T (1991) Tetrahedron: Asymmetry 2:643

122b. Mikami K, Motoyama Y, Terada M (1994) J Am Chem Soc 116:2812

123. The effect of the biaryl torsional angle on enantioselection has been studied: Harada T, Takeuchi M, Hatsuda M, Ueda S, Oku A (1996) Tetrahedron: Asymmetry 7:2479

124. Maruoka K, Murase N, Yamamoto H (1993) J Org Chem 58:2938

125. Corey EJ, Roper TD, Ishihara K, Sarakinos G (1993) Tetrahedron Lett 34:8399

126. Keck GE, Krishnamurthy D (1996) Synth Commun 26:367

127a. Hong Y, Kuntz BA, Collins S (1993) Organomet 12:964

127b. Jaquith JB, Guan J, Wang S, Collins S (1995) Organomet 14:1079

128. A stable diaquo titanocene catalyzes aldehyde-cyclopentadiene Diels-Alder reactions in moderate to good enantioselectivities: Odenkirk W, Bosnich B (1995) J Chem Soc Chem Commun 1181

129. A C_2-symmetric bridged ferrocene catalyzes the methacrolein-cyclopentadiene Diels-Alder reaction in 10% ee: Gibis K-L, Helmchen G, Huttner, G, Zsolnai L (1993) J Organomet Chem 445:181

130. Kobayashi S, Araki M, Hachiya I (1994) J Org Chem 59:3758

131. Kobayashi S, Ishitani H, Araki M, Hachiya I (1994) Tetrahedron Lett 35:6325

132. Kobayashi S, Hachiya I, Ishitani H, Araki M (1993) Tetrahedron Lett 34:4535

133. Kobayashi S, Ishitani H (1994) J Am Chem Soc 116:4083

134. Kobayashi S, Ishitani H, Hachiya I, Araki M (1994) Tetrahedron 50:11623

135. For an example of this reaction with a Ti(IV) complex, see: Posner GH, Carry J-C, Lee JK, Bull DS, Dai H (1994) Tetrahedron Lett 35:1321

136a. Markó IE, Evans GR (1994) Tetrahedron Lett 35:2771

136b. Markó IE, Evans GR, Declercq J-P, Tinant B, Feneau-Dupont J (1995) Acros Org Acta 1:63

136c. Markó IE, Evans GR, Seres P, Chellé I, Janousek Z (1996) Pure Appl Chem 68:113

137. Swarbrick TM, Markó IE, Kennard L (1991) Tetrahedron Lett 32:2549

138. Koerner M, Rickborn B (1990) J Org Chem 55:2662

139. Riant O, Kagan HB (1989) Tetrahedron Lett 30:7403

140. Riant O, Kagan HB (1994) Tetrahedron 50:4543

141. Okamura H, Nakamura Y, Iwagawa T, Nakatani M (1996) Chem Lett 193

142a. Hilvert D, Hill KW, Nared KD, Auditor M-TM (1989) J Am Chem Soc 111:9261

142b. Braisted AC, Schultz PG (1990) J Am Chem Soc 112:7430

143. Gouverneur VE, Houk KN, de Pascual-Teresa B, Beno B, Janda KD, Lerner RA (1993) Science 262:204

144a. Romesberg FE, Spiller B, Schultz PG, Stevens RC (1998) Science 279:1929

144b. Heine A, Stura EA, Yli-Kauhaluoma JT, Gao C, Deng Q, Beno BR, Houk KN, Janda KD, Wilson IA (1998) Science 279:1934

145. For an example of an achiral complex capable of delivering an *exo* cycloadduct preferentially, see: Maruoka K, Imoto H, Yamamoto H (1994) J Am Chem Soc 116:12115

146. Chiral imidazolidinone carbene complexes undergo *exo* selective Diels-Alder reactions: Powers TS, Jiang W, Su J, Wulff WD (1997) J Am Chem Soc 119:6438

147. An enzyme-promoted intramolecular Diels-Alder reaction proceeds in high enantioselectivity but indeterminate turnover: Oikawa H, Katayama K, Suzuki Y, Ichihara A (1995) J Chem Soc Chem Commun 1321

Chapter 33.2
Hetero-Diels-Alder and Related Reactions

Takashi Ooi · Keiji Maruoka

Department of Chemistry, Graduate School of Science, Hokkaido University, Sapporo, 060–0810 Japan
e-mail: maruoka@sci.hokudai.ac.jp

Keywords: Cyclocondensation, Danishefsky's diene, Aldehydes, Pyrone derivatives, Chiral Lewis acids, Enantioselectivity

1
Introduction

Since Danishefsky demonstrated that activated dienes, such as siloxydiene (commonly referred to as Danishefsky's diene) react with a wide spectrum of aldehydes to afford 5,6-dihydro-γ-pyrones in 1982 [1], the hetero-Diels-Alder reaction has attracted a great deal of attention over the last two decades [2, 3]. The use of asymmetric catalysis in these reactions is overwhelmingly associated with heterodienophiles. Especially, the cyclocondensations of activated dienes with aldehydes or their derivatives are of particular importance, providing a multitude of opportunities for the highly efficient regio- and stereoselective construction of various heterocycles in enantiomerically pure form. In spite of the great potential of this synthetic methodology, its development was for a long time restricted to only a few efforts. However, during the last decade this field has been the subject of intense and successful research [4, 5, 6].

2
The Mechanistic Aspect of the Hetero-Diels-Alder Reactions

The first set of rigorous mechanistic investigations on the hetero-Diels-Alder reaction was performed by Danishefsky and co-workers. When the cyclocondensation between siloxydiene 1 and benzaldehyde was catalyzed by ZnCl$_2$ in THF and quenched with NaHCO$_3$ without aqueous workup conditions, enol ether 2 was isolated in addition to pyrone 3, indicating a pericyclic mechanistic pathway. The enol ether 2 could then be transformed into 3 by acidic treatment. Thus, the reaction catalyzed by ZnCl$_2$ exhibits strong preference for cis (endo) stereochemistry. With BF$_3$·OEt$_2$, the reaction proceeds through Mukaiyama aldol like process, giving a diastereomeric mixture of pyrones 3 and 4 (Scheme 1) [7, 8]. Moreover, Danishefsky demonstrated that lanthanide(III) complexes catalyzed the cycloaddition of activated dienes with aldehydes [9]. The stereoselectivity increases dramatically using these catalysts. When diene 5, for example, reacts with a variety of aldehydes with Eu(fod)$_3$ as catalyst, virtually complete cis (endo) selectivity is observed. The aldehydes that function as dienophiles can be aliphatic or aromatic, see Table 1.

The Lewis acid complexes to the lone electron pair of the aldehyde anti to the R group, forcing the R group into an endo position in the transition state. In the case of lanthanide catalysts, the large steric bulk of the lanthanide metal-ligand complex causes the increase in cis selectivity (Scheme 2).

With the achievement of high stereochemical control resulting from nearly exclusive endo topography in the Lewis acid catalyzed reactions, the effect of chiral Lewis acid complexes on the enantiofacial selectivity of the cyclocondensation reaction should be documented. As an example, one can consider the cycloaddi-

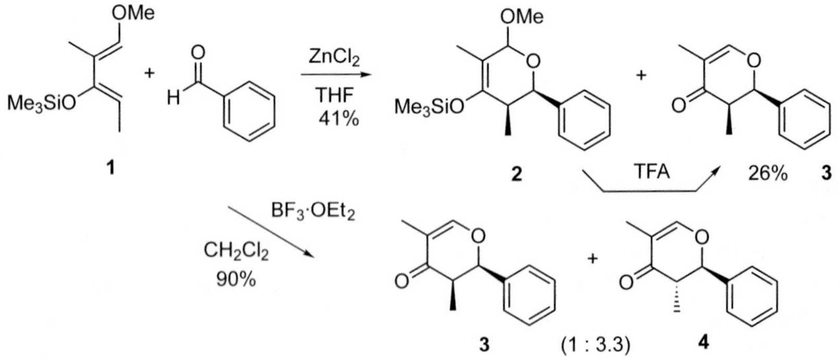

Scheme 1

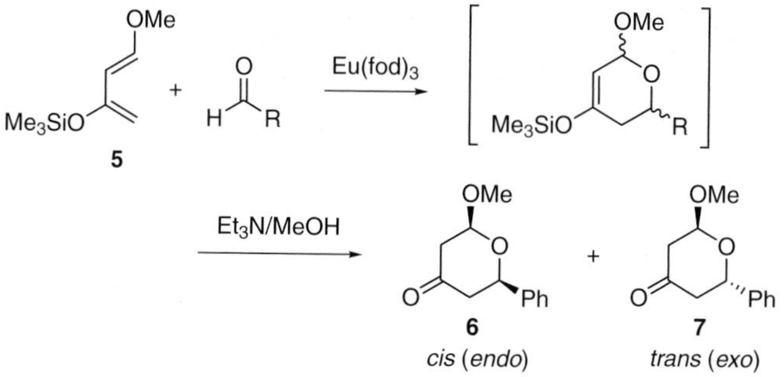

Table 1. Eu(fod)$_3$-catalyzed hetero-Diels-Alder reaction

Entry	Aldehyde (R)	6	7
1	Ph	12	1
2	2-Furyl	6	1
3	trans-Styryl	8	1
4	Me	2.8	1
5	n-C$_5$H$_{11}$	1.2	1
6	i-Pr	1.5	1

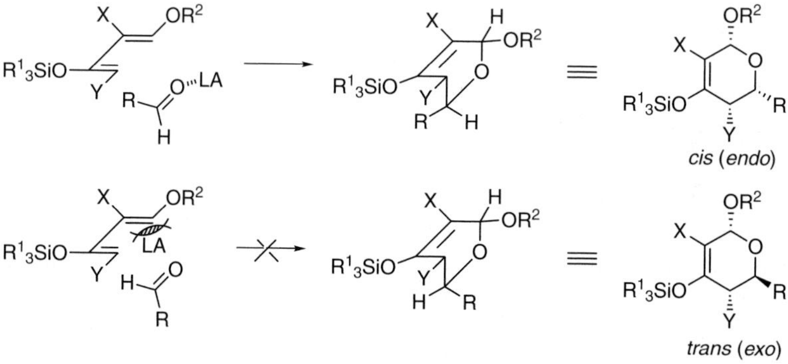

Scheme 2

tion of benzaldehyde to diene **5** under the influence of a chiral catalyst. Assuming a pericyclic reaction mode and high *endo* selectivity, a mixture of two possible products can result. Compound **8** is designated as a D-pyranose and compound **9** as an L-pyranose, based on carbohydrate nomenclature (**8** and **9** are enantiomers). A simple and practical method to synthesize compounds such as **8** and **9** in either enantiomeric form and in high optical purity is of great synthetic value (Scheme 3).

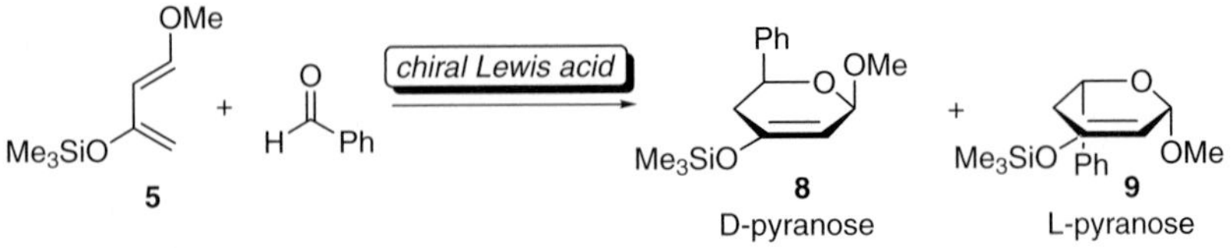

Scheme 3

3
Chiral Eu(hfc)₃ Catalysis

The first asymmetric catalyst to be evaluated was the commercially available chiral lanthanide β-diketonate complex Eu(hfc)₃ [10]. The initial result was obtained from the reaction of benzaldehyde with diene **5**, which showed 18% enantiomeric excess [11]. Attempts to improve the asymmetric induction by varying the substituents on the diene were undertaken (Scheme 4). Substitution at either C2 or C4 of the diene seemed to increase the asymmetric induction. By using the 2,4-dimethyldiene **10a**, for example, the ee is increased to 36% [11]. Interestingly, both the 2-methyldiene **10b** and 2-acetoxydiene **10c** show no significant improvement in facial selectivity over the parent diene.

Since the dienes with different C1 functionalities were obtained by enol silylation of the corresponding alkoxyenones [12] which, in turn, are readily prepared by acid catalyzed exchange with the commercially available *trans*–1-methoxy-1-buten-3-one, investigations were focused on the effects of the C1 alkoxy variation on the asymmetric induction. The presence of a large achiral alkoxy group at C1 of the diene results in a significant increase in facial selectivity [11]. When using the 2-acetoxydiene and the corresponding 2-siloxydiene, substantial increases in ee are observed on replacing the methoxy group with the *t*-butoxy group (Scheme 5).

Finally, with two of the dienes the induction was maximized by modifying the experimental conditions. By conducting the reaction of dienes **11b** and **11d** in the absence of solvent and at reduced temperature up to55% and 58% ee's, respectively, could be obtained [11].

4
Chiral Aluminum Catalysis

The asymmetric hetero-Diels-Alder reaction, which is apparently quite powerful in natural product syntheses, had not been developed to a useful level due to the lack of the well-designed asymmetric catalysts until Yamamoto and Maruoka reported a first solution to the problem by using chiral organoaluminum catalysts of type (*R*)-**12** and (*S*)-**12** (see Structure 1), which were newly devised on the basis of studies on the exceptionally bulky Lewis acid, MAD [13, 14].

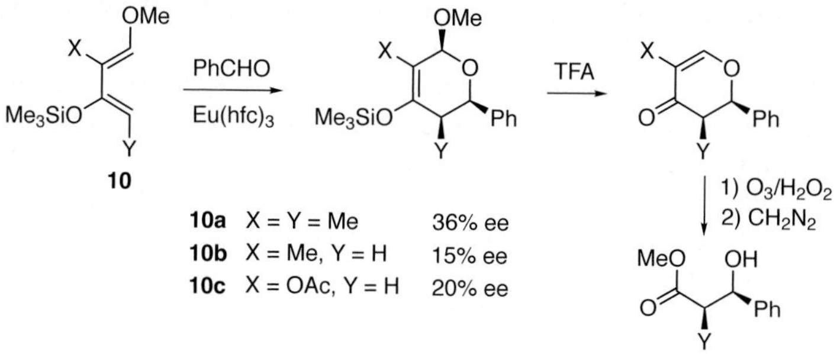
OMe
X
Me3SiO
Y
10
PhCHO
Eu(hfc)3
TFA
O
Ph
1) O3/H2O2
2) CH2N2
MeO
OH
10a X = Y = Me 36% ee
10b X = Me, Y = H 15% ee
10c X = OAc, Y = H 20% ee

Scheme 4

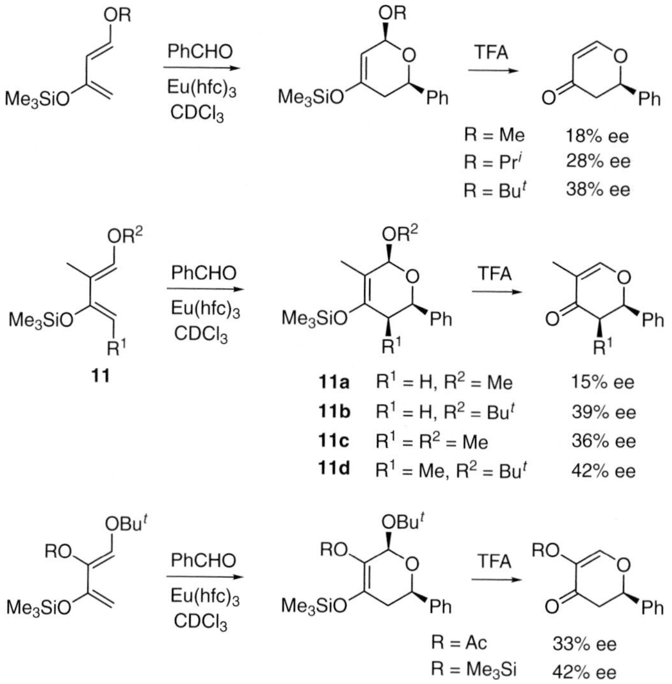
OR
Me3SiO
PhCHO
Eu(hfc)3
CDCl3
TFA
Ph
R = Me 18% ee
R = Pr^i 28% ee
R = Bu^t 38% ee
OR2
R1
11
11a R1 = H, R2 = Me 15% ee
11b R1 = H, R2 = Bu^t 39% ee
11c R1 = R2 = Me 36% ee
11d R1 = Me, R2 = Bu^t 42% ee
OBu^t
RO
R = Ac 33% ee
R = Me3Si 42% ee

Scheme 5

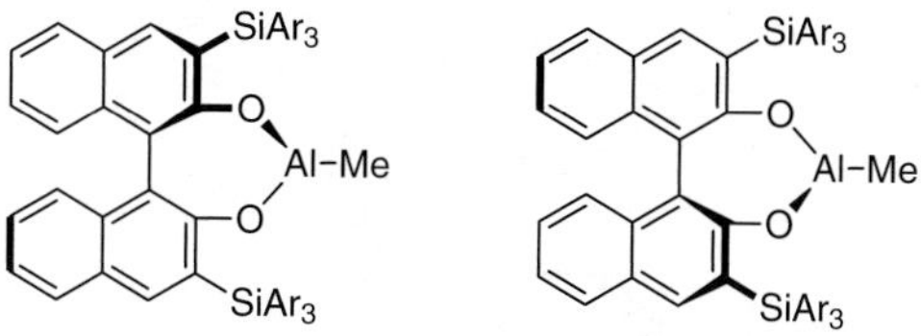
SiAr3
O
Al–Me

(*R*)-**12** (*S*)-**12**

Structure 1

The optically pure (R)-(+)-3,3'-bis(triarylsilyl)binaphthol, (R)-13, requisite for preparation of (R)-12 can be synthesized in two steps from (R)-(+)-3,3'-dibromobinaphthol [15]. Reaction of (R)-13 in toluene with Me$_3$Al produced the chiral organoaluminum reagent (R)-12 as a pink to wine-red solution. Treatment of a mixture of benzaldehyde and siloxydiene 10a in toluene under the influence of catalytic (R)-12 (Ar=Ph: 10 mol %) at −20 °C for 2 h furnished, after exposure of the resulting hetero-Diels-Alder adducts to trifluoroacetic acid in CH$_2$Cl$_2$, cis-dihydropyrone 14 (77%) and its trans isomer 15 (7%) (Scheme 6). The major cis adduct 14 was shown to exist in 95% ee. Furthermore, use of the sterically more hindered aluminum reagent (R)-12 (Ar=3,5-xylyl) has proved to exhibit excellent cis- and enantioselectivitires (93% yield; cis/trans=30:1; 97% ee in 14), indicating the importance of the choice of the bulky triarylsilyl moiety in 12 for obtaining the high enantioface differentiation of prochiral aldehydes.

Use of nonpolar solvents such as toluene produced higher enantioslectivities than polar solvents such as CH$_2$Cl$_2$ and ether solvents, and lowering the temperature gradually increased the optical yield.

The chiral oxygenophilic organoaluminum catalyst 12 bearing such a sterically hindered chiral auxiliary unit may form a stable 1:1 complex with benzaldehyde, allowing enantioselective activation of the carbonyl moiety as illustrated in Fig. 1. Then the diene 10a would approach benzaldehyde with an endo alignment of the aldehyde phenyl residue and 10a in order to minimize the steric

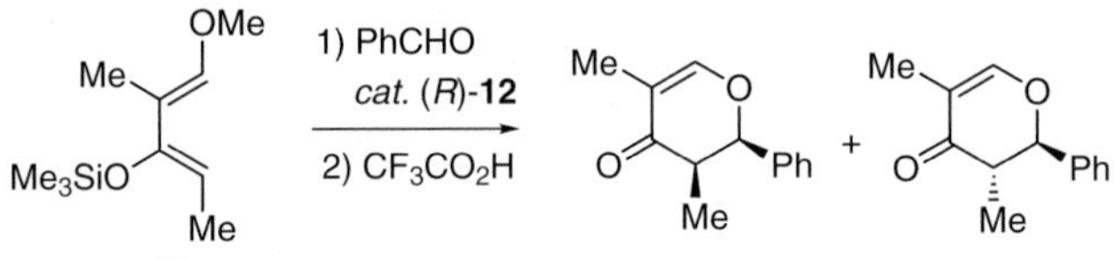

Scheme 6

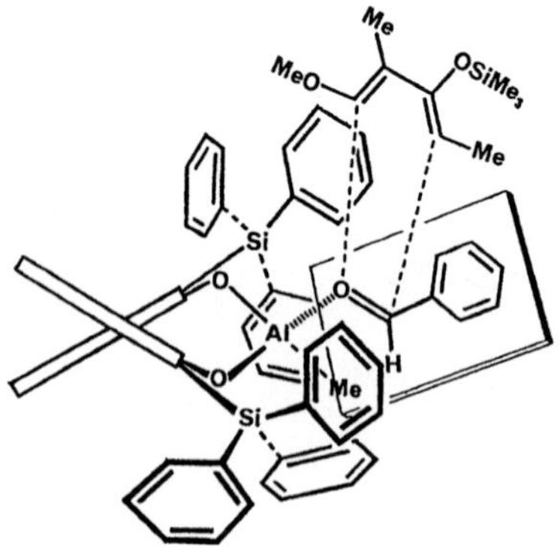

Fig. 1

repulsion between the incoming diene and the front triarylsilyl moiety, thereby yielding the *cis* adduct **14** predominantly in accord with the experimental findings. It was emphasized that the hetero-Diels-Alder adduct, once formed, readily split off from the aluminum center in view of the steric release between the adduct and the aluminum reagent, resulting in regeneration of the catalyst **12** for further use in the catalytic cycle of the reaction [13].

Yamamoto and Maruoka also reported a conceptually new method of in situ generating the chiral catalyst **12** for asymmetric hetero-Diels-Alder reactions by discrimination of the racemic **12** with a chiral ketone [16]. As shown in Scheme 7, sequential treatment of (±)-**12** (0.1 equivalent) with *d*-3-bromocamphor (0.1 equivalent), the diene **10a** (1.05 equivalent), and benzaldehyde (1 equivalent) at –78 °C in CH$_2$Cl$_2$ and stirring of the mixture at this temperature for 3 h afforded the hetero-Diels-Alder adducts **14** and **15**, after acidic workup, in 78% and 19% yields, respectively. The optical yield of the major *cis* isomer **14** was 82%. Although the extent of asymmetric induction is not as satisfactory as that with optically pure **12**, one recrystallization of the *cis* adduct **14** of 82% ee from hexane gave the essentially pure **14** (>98% ee with ~60% recovery), thereby enhancing the practicability of this method.

Jorgensen and co-workers achieved the first Lewis acid-catalyzed chemoselective reaction of conjugated dienes having allylic C-H bonds with glyoxylate esters, which mainly leads to formation of the hetero-Diels-Alder product with

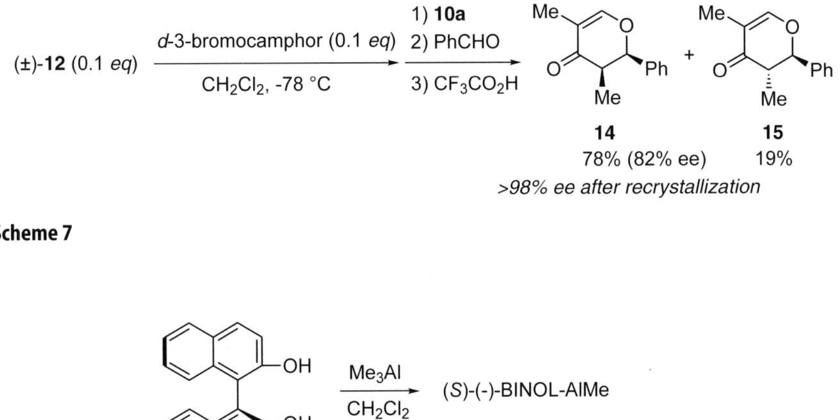

Scheme 7

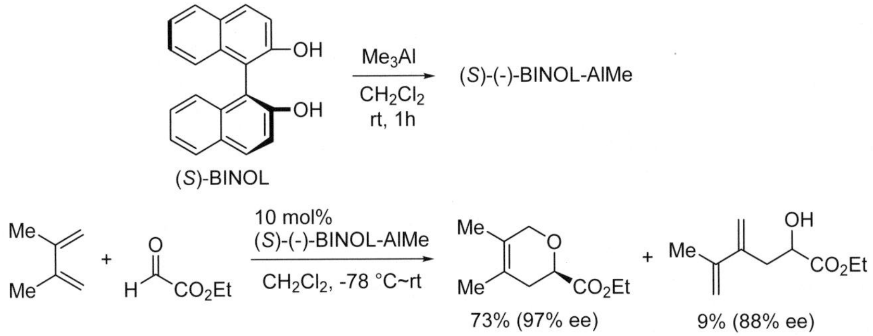

Scheme 8

very high enantioselectivity [17]. The choice of catalyst is crucial for the hetero-Diels-Alder selectivity and it was found that a combination of Me₃Al and BINOL gives a very high chemo- and enantioselective complex [(S)-(−)-BINOL-AlMe] (Scheme 8).

5
Chiral Boron Catalysis

In1989, Yamamoto introduced the chiral (acyloxy)borane (CAB) complex for catalytic asymmetric Diels-Alder reactions [18], which has been utilized as a magic hand catalysis for the aldol synthesis and for the Sakurai-Hosomi reaction so far [19, 20]. In contrast to R=H of **17**, which is both air and moisture sensitive, the *B*-alkylated catalyst, R=Ph or alkyl, is stable and can be stored in closed containers at room temperature. This catalyst is easily prepared from phenyl- or alkylboric acid and **16**: simple mixing of a 1:1 molar ratio of the ester **16** and phe-

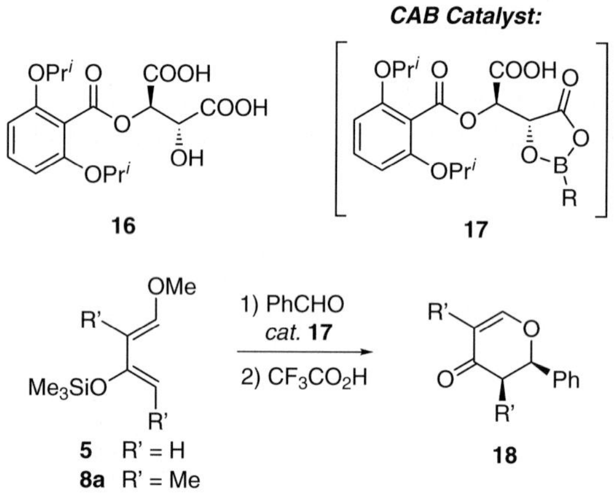

Table 2. CAB-mediated asymmetric hetero-Diels-Alder reaction

Entry	Diene	R of boric acid	Product yield [%]	% ee (confign)
1	5	Bu	67	73 (R)
2		Ph	63	75 (R)
3		2,4,6-Me₃Ph	47	95 (R)
4		2,4,6-i-Pr₃Ph	55	95 (R)
5		o-MeOPh	80	79 (R)
6		o-i-PrOPh	63	84 (R)
7	8a	Bu	56 (12)	93 (2R,3R)
8		Ph	65 (29)	87 (2R,3R)
9		2,4,6-Me₃Ph	<5	
10		o-MeOPh	95 (5)	97 (2R,3R)

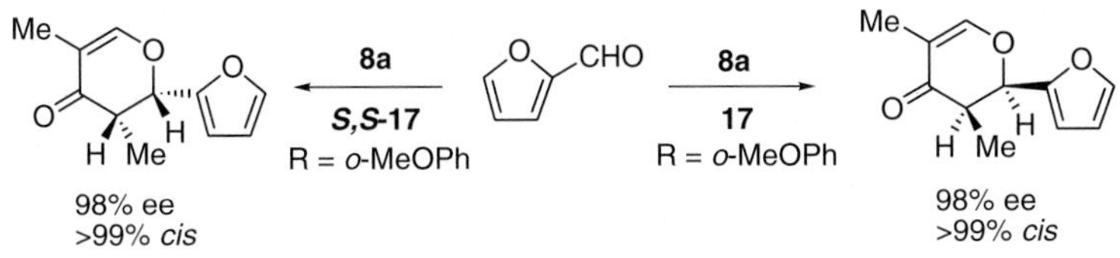

Scheme 9

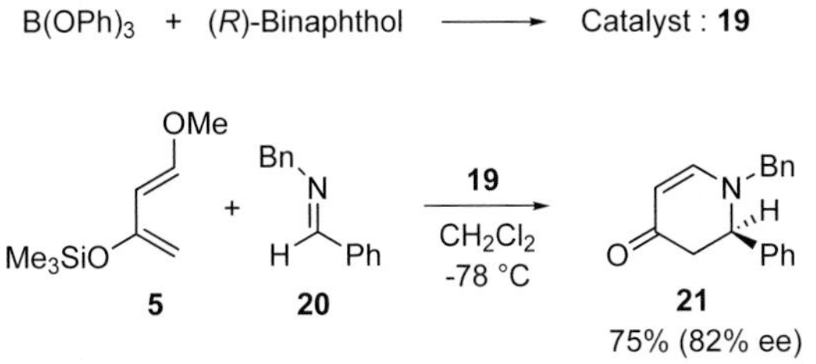

Scheme 10

nylboric acid in freshly distilled propionitrile at room temperature for 0.5 h smoothly produced the reactive catalyst. This catalyst solution is sufficiently reactive and catalyzes hetero-Diels-Alder reaction of aldehydes with Danishefsky diene to yield dihydropyrones of excellent optical purity (Table 2) [21].

The power of the CAB catalytic reaction for the enantioselective route to carbon-branched pyranose derivatives is also seen from the following example (Scheme 9).

Yamamoto also reported an asymmetric aza Diels-Alder reaction of an imine mediated by a stoichiometric amount of the chiral boron complex 19 which is conveniently prepared in situ simply by mixing a 1:1 molar ratio of optically active binaphthol and triphenyl borate in CH$_2$Cl$_2$ at room temperature [22]

For example, the reaction of aldimine 20 with Danishefsky's diene 5 is promoted by 19 in the presence of 4 Å molecular sieves at –78 °C, producing dehydropyridone 21 in 75% yield and 82% ee (Scheme 10). This method is successful with several aldimines and affords products of up to 90% ee.

6
Chiral Titanium Catalysis

Based on their early study in the enantioselective carbonyl-ene reaction [23], Nakai, Mikami, and Terada have found that the asymmetric hetero-Diels-Alder reaction of prochiral glyoxylate with methoxydiene can be catalyzed by the chiral titanium complex, producing the *cis*-dihydropyran carboxylate as a major product in high enantiomeric purity (Scheme 11) [24].

The observed *cis*-selectivity provides a mechanistic insight into the state of complexation between glyoxylate and the chiral titanium catalyst. Of the two

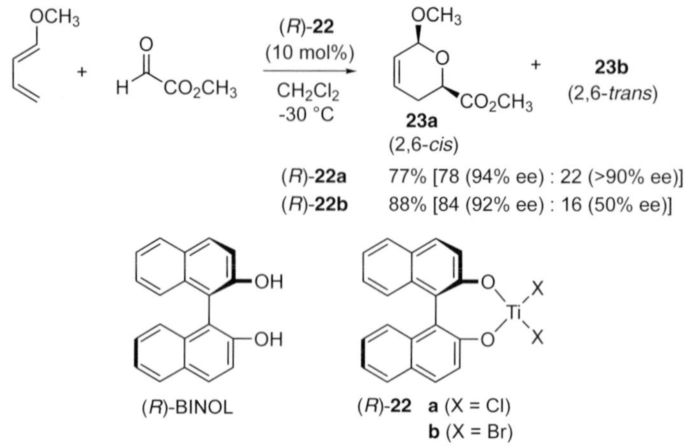

Scheme 11

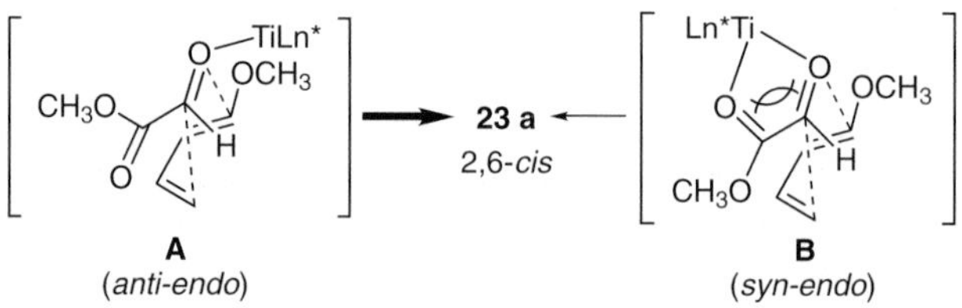

Structure 2

transition states leading to the *cis*-product, the *syn-endo* transition state **B** should be less favorable because of the steric repulsion in the sterically demanding titanium complex. Thus, the titanium catalyst should be complexed in an *anti* fashion and then the reaction proceeds through an *endo*-orientation (Structure 2).

The hetero-Diels-Alder adduct thus obtained by the use of (*S*)-**22a** can readily be converted not only to monosaccharides but also to the lactone portion **24** of mevinolin or compactin in a short step as shown in Scheme 12.

Hetero-Diels-Alder reactions of 1-oxa-1,3-butadienes with vinyl ethers, which lead to 3,4-dihydro-2*H*-pyran derivatives, are synthetically equivalent to Michael type conjugate additions. Wada and coworkers presented the first examples of a catalytic asymmetric intermolecular hetero-Diels-Alder reaction by the use of (*E*)-2-oxo-1-phenylsulfonyl-3-alkenes **25** and vinyl ethers **26** (Table 3) [25].

The reaction of enone **25** with a large excess of ethyl vinyl ether **26a** was performed in the presence of titanium catalyst **27** (10 mol %) at −78 °C for 20 h to give *cis*-isomer **28a** as a single isomer in 59% yield (59% ee). The enantioselectivity was effectively enhanced by increasing the bulkiness of the alkoxy substituent R of the dienophiles **26a-c** (Table 3). This methodology offers a very effective synthetic route for the enantiomers of 4-substituted 2,4-*cis*-2-alkoxy-3,4-dihydro-2*H*-pyrans.

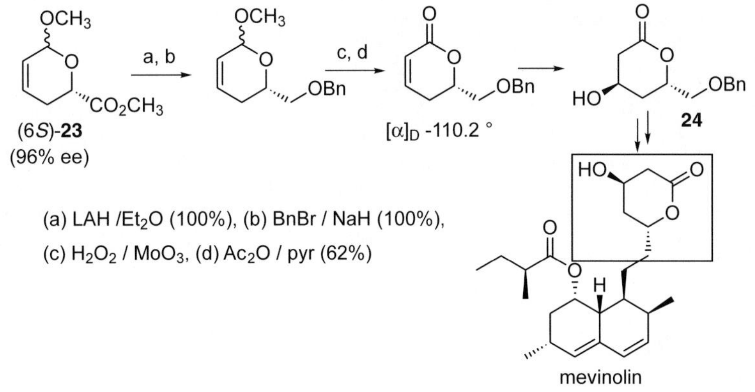

(a) LAH /Et$_2$O (100%), (b) BnBr / NaH (100%),
(c) H$_2$O$_2$ / MoO$_3$, (d) Ac$_2$O / pyr (62%)

mevinolin

Scheme 12

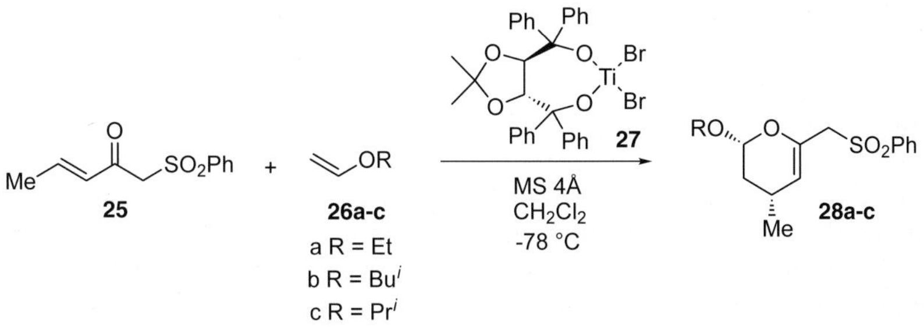

Table 3. Chiral Lewis acid-catalyzed asymmetric hetero-Diels-Alder reactions of enones with vinyl ethers **26a-c**

Entry	Vinyl ether	Yield [%]	% ee (config)
1	26a	**28a** (91)	59 (2R,4R)
2	26b	**28b** (92)	88 (2R,4R)
3	26c	**28c** (90)	97 (2R,4R)

Although it is a stoichiometric procedure, the intramolecular hetero-Diels-Alder reaction of 1-oxa-1,3-butadienes, obtained in situ by a Knoevenagel condensation of aromatic aldehydes and N,N'-dimethylbarbituric acid, is mediated by a chiral titanium Lewis acid **29** which has 1,2:5,6-di-O-isopropylidene-α-D-glucofuranose as a ligand. The highest ee-value was obtained in the reaction of **30** with **31** in isodurene as illustrated in Scheme 13[26].

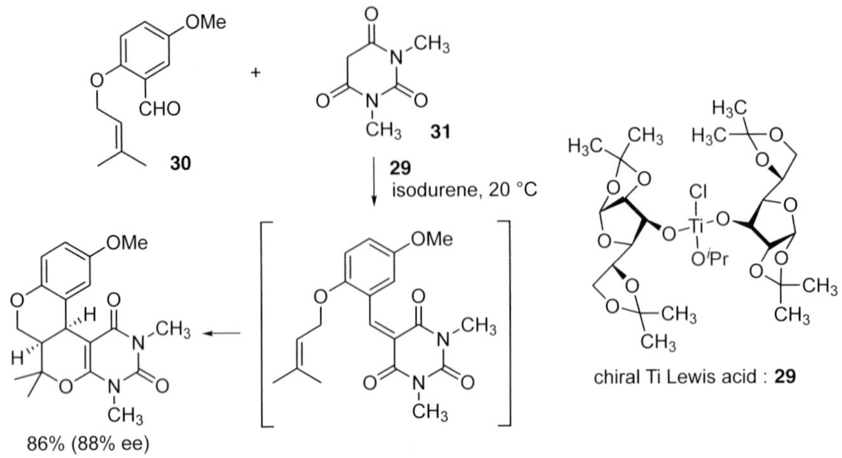

86% (88% ee)

chiral Ti Lewis acid : 29

Scheme 13

7
Chiral Lanthanide Catalysis

Although the utility of rare earth metal complexes as Lewis acid catalysts in organic synthesis has received much attention, only a limited number of investigations has been reported on isolable chiral rare earth metal complex-catalyzed asymmetric reactions [11, 27, 28, 29, 30].

Inanaga and coworkers prepared a series of tris[(R)-(-)-1,1'-binaphthyl-2,2'-diylphosphato]lanthanides(III) Ln[(-)BNP]₃} as new chiral and stable Lewis acids by the simple treatment of lanthanide(III) chlorides with three equivalent of the optically active sodium phosphate at room temperature, and reported the observed catalytic activity [31, 32]. The asymmetric hetero-Diels-Alder reactions of the Danishefsky diene 5 with benzaldehyde or with 2-naphthaldehyde were successfully performed at 0 °C in the presence of 10 mol % of Sc[(-)BNP]₃ to give the corresponding adducts in 77 and 69% chemical yields with 68 and 74% enantiomeric excesses of (R)-(-)-isomers, respectively (Scheme 14).

When the reaction was conducted at room temperature under the catalysis of Yb[(-)BNP]₃, the asymmetric induction was improved to 73% ee. The effect of the central metal ion of the chiral catalysts on the optical yield of the product, 2-phenyl-2,3-dihydro-4H-pyran-4-one, is shown in Fig. 2. The degree of enantioselection is highly sensitive to and dependent on the ionic radius of lanthanide ions [31].

Since the reaction proceeded under heterogeneous conditions, further elaboration was made to make a clear solution by adding a variety of ligands and the efficacy of them was tested for the above reaction by analyzing the enantiomeric excess of the final product. The best result was obtained when the reaction was carried out at room temperature using 10 mol % each of the Yb-catalyst and 2,6-

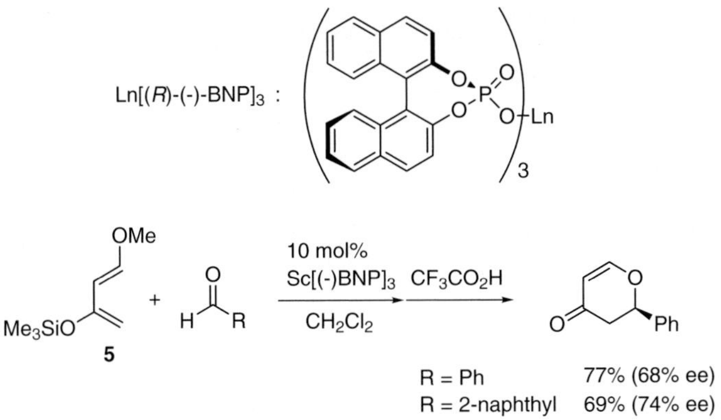

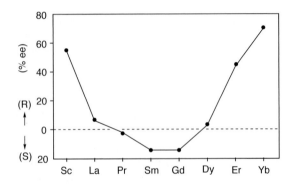

Scheme 14

$Ln[(R)-(-)-BNP]_3$:

OMe

Me₃SiO

5

+

O‖H–R

10 mol%
Sc[(-)BNP]₃ CF₃CO₂H
————————————
CH₂Cl₂

O ... Ph

R = Ph   77% (68% ee)
R = 2-naphthyl 69% (74% ee)

Fig. 2

(% ee)

80
60
40
(R) 20
0
(S) 20

Sc La Pr Sm Gd Dy Er Yb

OMe

Me₃SiO

5

+

O‖H–Ph

10 mol%
Yb[(-)BNP]₃
2,6-lutidine CF₃CO₂H
————————————
CH₂Cl₂
rt

O ... Ph

94% (89% ee)

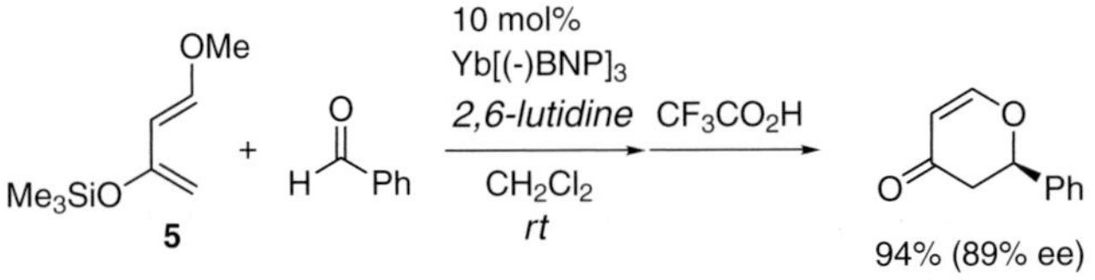

Scheme 15

lutidine as an additive (Scheme 15). It should be noted that the observed enantiomeric excess of 89% is rather high for the reaction temperature (23 to 25 °C), since most catalytic asymmetric hetero-Diels-Alder reactions required a rather low temperature (usually –78 °C) to attain this level of enantioselection [32].

The precise structure of the catalyst is not clear. Since the addition of two equivalents of 2,6-lutidine to the catalyst slightly diminished the enantiomeric

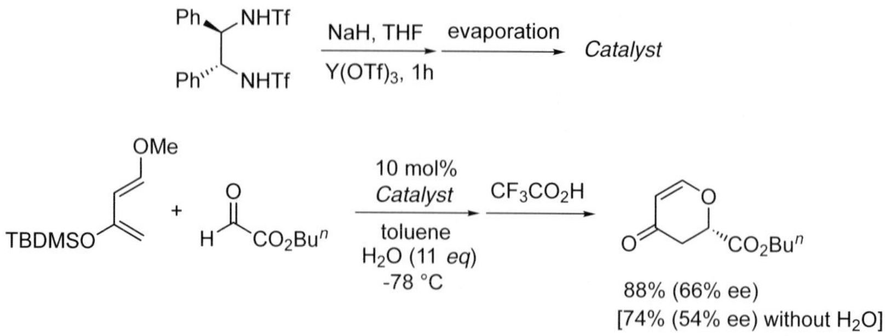

Scheme 16

excess, the active catalyst which leads to high enantioselection is thought to be the 1:1 lutidine-ytterbium complex rather than the 2:1 complex at the stage where enantioselection is made.

Mikami and coworkers also reported the development of lanthanide bis(trifluoromethanesulfonyl)amides (bistrifylamides) as a new type of asymmetric catalysts for the hetero-Diels-Alder reaction of Danishefsky's diene, wherein the significant effect of water as an additive is observed in increasing not only the enantioselectivity but also the chemical yield. Bistrifylamides can be used as effective bidentate ligands to increase the Lewis acidity of their chiral metal complexes on account of the higher acidity of the conjugated acids than those of aliphatic and aromatic diols, which are commonly used as chiral bidentate ligands [33].

Chiral lanthanide bistrifylamides were prepared through the reaction of lanthanide(III) triflates and chiral bistrifylamides, which are deprotonated with 2 equivalents of sodium hydride in THF for 1 h. Dichloromethane or toluene was introduced, after evaporation of THF, to the residual complex. The resultant suspension of the lanthanide complex was then used for the hetero-Diels-Alder reaction (Scheme 16).

8
Chiral Transition Metal Catalysis

The reaction between a diene, such as 2,3-dimethylbuta-1,3-diene, and a carbonyl compound can lead to the formation of both the hetero-Diels-Alder product and the ene product. Copper(II) bisoxazoline-catalyzed reactions of glyoxylate esters with dienes leading to the hetero-Diels-Alder product and the ene product in high yield and with a high enantiomeric excess, have been developed by Johannsen and Jorgensen [34]. The hetero-Diels-Alder product:ene product ratio is in the range 1:0.6 to 1:1.8 and is dependent on both the chiral ligand attached to the metal, the glyoxylate ester, and the reaction temperature. Notably, the use of a polar solvent such as nitromethane leads to a significant improvement of the catalytic properties of a cationic copper-Lewis acid in the hetero-Diels-Alder reaction of alkyl glyoxylates with dienes.

For instance, the reaction of cyclohexa-1,3-diene and ethyl glyoxylate in the presence of (S)-**31** proceeds smoothly to give the cycloadduct in 66% yield with 97% ee. The synthetic application of this process was demonstrated by the preparation of a highly interesting synthon for sesquiterpene lactones in high yield and diastereoselectivity, and with a very high ee as illustrated in Scheme 17 [35].

Ghosh and coworkers also reported that the reaction of Danishefsky's diene **5** and the glyoxylate esters catalyzed by a (1R,2S)-bis(oxazoline)-metal complex afforded the corresponding aldol adduct which, on treatment with TFA, furnished the enantiomerically enriched hetero-Diels-Alder product in good yield [36]. Among various ligand-metal complexes examined, conformationally constrained bis(oxazoline)-Cu(II)-triflates of type **33** afforded 72% ee and 70% isolated yield. Such constrained ligands like **32** are particularly attractive because of their ready availability in both enantiomeric forms from the corresponding commercially available optically active cis-1-amino-2-indanols (Scheme 18).

This methodology was also found to be very effective for the hetero-Diels-Alder reaction of benzyloxyacetaldehyde with Danishefsky's diene **5** producing dihydropyran derivatives appropriately functionalized for the synthesis of the C_3-C_{14} segment of the novel antitumor agent laulimalide (Scheme 19) [37].

Novel optically active oxovanadium(IV) complexes bearing camphor-derived 1,3-diketonato ligands have been prepared by Togni [38]. The complex bis(3-heptafluorobutyryl)camphorato)oxovanadium (**34**) was found to be a very efficient catalyst for the cycloaddition of aldehydes to activated dienes to give pyrone derivatives. Thus, the reaction of benzaldehyde with 1-methoxy-2,4-dimethyl-3-(triethylsiloxy)butadiene in the presence of 5 mol % of (+)-**34** at –78 °C

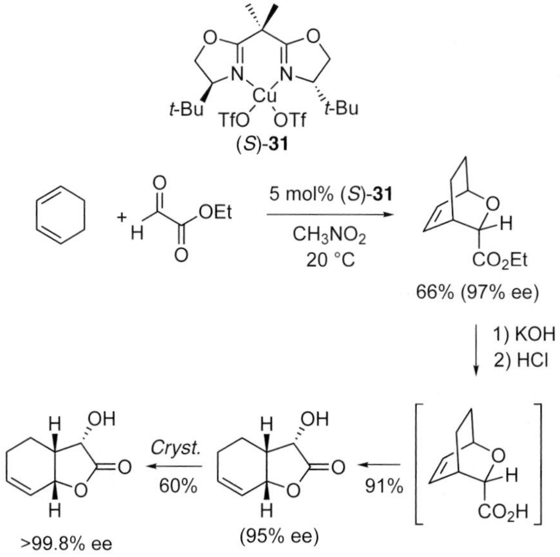

Scheme 17

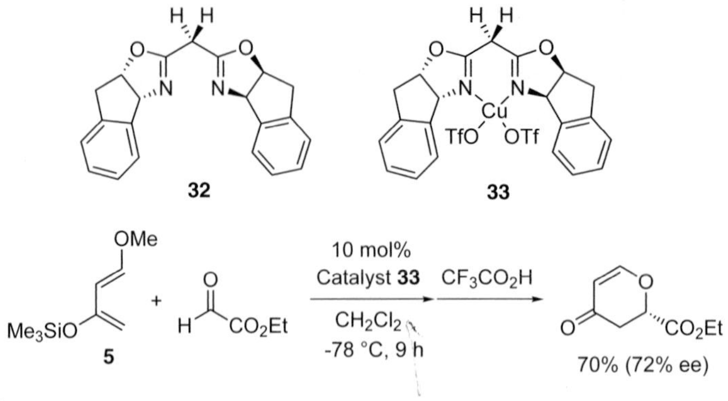

Scheme 18

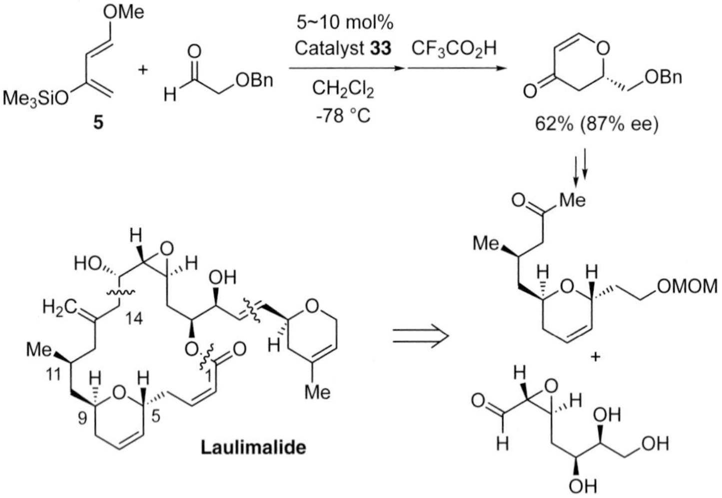

Laulimalide

Scheme 19

gave, after protolytic workup, *cis*-3,5-dimethyl-6-phenyl-5,6-dihydro-4*H*-pyran-4-one with 99% diastereoselectivity and 85% ee (Scheme 20).

Interestingly, the reactions of (*R*)-2,3-*O*-isopropylidene-D-glyceraldehyde (**35**) with 1-methoxy-2,4-dimethyl-3-(trimethylsiloxy)butadiene, catalyzed by (+)-**34** and (−)-**34**, respectively, involved a high degree of double stereodifferentiation. The matched combination of (−)-**34** with (*R*)-**35** gave one of the four possible diastereomeric pyrone products in 93.1% selectivity. On the other hand, the mismatched pair showed almost no selectivity (Scheme 21) [38].

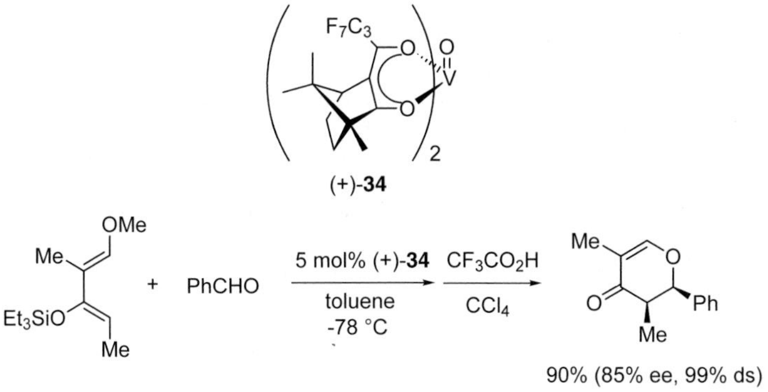

F_7C_3
O
V
2
(+)-34
OMe
Me
Et_3SiO
Me
+
PhCHO
5 mol% (+)-34
toluene
-78 °C
CF_3CO_2H
CCl_4
Me
O
O
Ph
Me
90% (85% ee, 99% ds)

Scheme 20

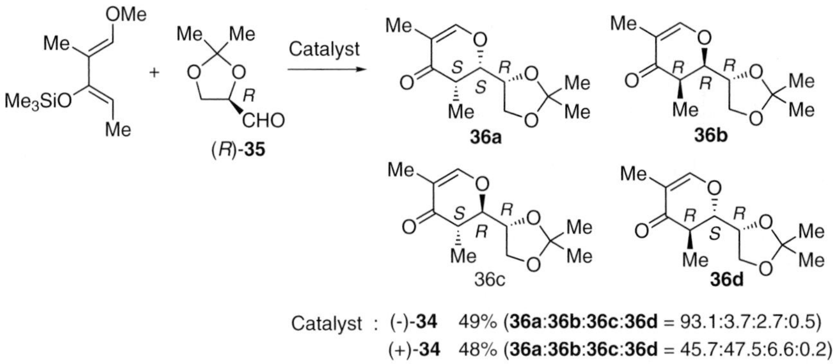

OMe
Me
Me_3SiO
Me
+
Me
Me
O
O
R
CHO
(R)-35
Catalyst
36a
36b
36c
36d
Catalyst : (-)-34 49% (36a:36b:36c:36d = 93.1:3.7:2.7:0.5)
(+)-34 48% (36a:36b:36c:36d = 45.7:47.5:6.6:0.2)

Scheme 21

9 Alternatives

Asymmetric hetero-Diels-Alder reactions utilizing chiral substrates or stoichiometric amounts of chiral auxiliaries constitute a indispensable part of this field and also are very useful methods with potential synthetic importance. Selected recent advances are included in the following references [6, 39, 40, 41, 42, 43, 44, 45, 46, 47, 48].

References

1. Danishefsky SJ, Kerwin JF, Kobayashi S (1982) J Am Chem Soc 104:358
2. Schmidt RR (1986) Acc Chem Res 19:250
3. Bednarski MD, Lyssikatos JP In: Comprehensive Organic Synthesis, Trost BM, Fleming I, Paquette LA Eds, Pergamon Press: New York (1991) vol 2, Chapter 2.5:661

4. Danishefski SJ (1986) Aldrich Acta 19:59
5. Danishefski SJ, DeNinno MP (1987) Angew Chem Int Ed Engl 26:15
6. Waldmann H (1994) Synthesis 535
7. Danishefsky SJ, Larson E, Askin D, Kato N (1985) J Am Chem Soc 107:1246
8. Larson ER, Danishefsky SJ (1982) J Am Chem Soc 104:6458
9. Bednarski MD, Danishefsky SJ (1983) J Am Chem Soc 105:3716
10. Bednarski MD, Danishefsky SJ (1983) J Am Chem Soc 105:6968
11. Bednarski MD, Maring CJ, Danishefsky SJ (1983) Tetrahedron Lett 24:3451
12. Emde H, Domsch P, Feger H, Gotz H, Hofmann K, Kober W, Krageloh H, Oesterle T, Steppan W, West W, Simchen G (1982) Synthesis 1
13. Maruoka K, Itoh T, Shirasaka T, Yamamoto H (1988) J Am Chem Soc 110:310
14. Maruoka K, Itoh T, Sakurai M, Nonoshita K, Yamamoto H (1988) J Am Chem Soc 110:3588
15. Maruoka K, Itoh T, Araki Y, Shirasaka T, Yamamoto H (1988) Bull Chem Soc Jpn 61:2975
16. Maruoka K, Yamamoto H (1989) J Am Chem Soc 111:789
17. Graven A, Johannsen M, Jorgensen KA (1996) J Chem Soc Chem Commun 2373
18. Furuta K, Shimizu S, Miwa Y, Yamamoto H (1989) J Org Chem 54:1481
19. Ishihara K, Maruyama T, Mouri M, Gao Q, Furuta K, Yamamoto H (1993) Bull Chem Soc Jpn 66:3483
20. Ishihara K, Mouri M, Gao Q, Maruyama T, Furuta K, Yamamoto H (1993) J Am Chem Soc 115:11490
21. Gao Q, Maruyama T, Mouri M, Yamamoto H (1992) J Org Chem 57:1951
22. Hattori K, Yamamoto H (1992) J Org Chem 57:3264
23. Mikami K, Terada M, Nakai T (1990) J Am Chem Soc 112:3949
24. Terada M, Mikami K, Nakai T (1991) Tetrahedron Lett 32:935
25. Wada E, Yasuoka H, Kanemasa S (1994) Chem Lett 1637
26. Tietze L, Saling P (1992) Synlett 281
27. Sasai H, Arai S, Sato Y, Houk K N, Shibasaki M (1995) J Am Chem Soc 117:6194
28. Yokomatsu Y, Yamagishi T, Shibuya S (1993) Tetrahedron Asymmetry 4:1783
29. Kobayashi S, Ishitani H (1994) J Am Chem Soc 116:4083
30. Kobayashi S, Ishitani H, Araki M, Hachiya I (1994) Tetrahedron Lett 35:6325
31. Inanaga J, Sugimoto Y, Hanamoto T (1995) New J Chem 19:707
32. Hanamoto T, Furuno H, Sugimoto Y, Inanaga J (1997) Synlett 79
33. Mikami K, Kotera O, Motoyama Y, Sakaguchi H (1995) Synlett 975
34. Johannsen M, Jogensen KA (1995) J Org Chem 60:5757
35. Johannsen M, Jogensen KA (1996) Tetrahedron 52:7321
36. Ghosh A K, Mathivanan P, Cappiello J, Krishnan K (1996) Tetrahedron Asymmetry 7:2165
37. Ghosh A K, Mathivanan P, Cappiello J (1997) Tetrahedron Lett 38:2427
38. Togni A (1990) Organometallics 9:3106
39. Midland MM, Koops RW (1990) J Org Chem 55:4647
40. Arnold T, Orschel B, Reissig H-U (1992) Angew Chem Int Ed Engl 31:1033
41. Ritter AR, Miller MJ (1994) J Org Chem 59:4602
42. Saito T, Karakasa T, Fujii H, Furuno E, Suda H, Kobayashi K (1994) J Chem Soc Perkin Trans 1 1359
43. Marchand A, Mauger D, Guingant A, Pradere J-P (1995) Tetrahedron Asymmetry 6:853
44. Defoin A, Sarazin H, Streith J (1996) Helv Chim Acta 79:560
45. Baldoli C, Buttero PD, Ciolo D, Maiorana S, Papagni A (1996) Synlett 258
46. Lehmler H-J, Nieger M, Breitmaier E (1996) Synthesis 105
47. Badorrey R, Cativiela C, Diaz-de-Villegas MD, Galvez JA (1997) Tetrahedron Lett 38:2547
48. Vogt PF, Hansel J-G, Miller MJ (1997) Tetrahedron Lett 38:2803

Chapter 33.3
[2+2] Cycloaddition Reactions

Yujiro Hayashi · Koichi Narasaka

Department of Chemistry, School of Science, The University of Tokyo, Hongo, Bunkyo-ku, Tokyo 113-0033, Japan
e-mail: narasaka@chem.s.u-tokyo.ac.jp

Keywords: [2+2] Cycloaddition, Cyclobutane, Chiral titanium catalyst, Propionolactone, Cinchona alkaloid

1
General Introduction

[2+2] Cycloaddition reaction is one of the powerful synthetic methods for the construction of 4-membered carbo- and heterocyclic rings [1, 2, 3, 4, 5, 6, 7]. In spite of the potential utility for the 4-membered ring compounds as synthetic intermediates, there have been only a few reports on catalytic asymmetric [2+2] cycloaddition reaction, which can be categorized into two reaction types. One is the Lewis acid-catalyzed [2+2] cycloaddition for the preparation of cyclobutanes, and the other is the [2+2] cycloaddition of ketenes and aldehydes catalyzed by chiral tertiary amines or chiral Lewis acid for the preparation of β-propionolactone derivatives.

2
[2+2] Cycloadditions of Alkenyl Sulfides and Electron-Deficient Alkenes

2.1
Introduction

Photochemical [2+2] cycloaddition reactions and [2+2] cycloaddition reactions of ketenes have been widely used for the preparation of cyclobutane derivatives. The thermal [2+2] cycloaddition reaction is known to proceed between highly electrophilic and nucleophilic alkenes; alkenes having cyano, fluoro, and trifluoromethyl groups react with electron-rich alkenes such as alkenyl ethers and sulfides [8]. As for the catalyst-mediated [2+2] cycloaddition reactions, Lewis acids are known to promote [2+2] cycloadditions [9, 10, 11, 12, 13, 14, 15, 16, 17, 18, 19, 20, 21, 22, 23, 24]. However, the applicability of this cycloaddition is rather limited because of the side reactions such as ene reactions, conjugate addition reactions, and ring opening reactions of the produced cyclobutane derivatives. There was no general Lewis acid-catalyzed [2+2] cycloaddition reaction, until Takeda [25] and Narasaka et al. [26, 27, 28, 29, 30, 31, 32, 33] found that alkenyl sulfides react with a wide variety of electron-deficient olefins in the presence of a Lewis acid.

   The first catalytic asymmetric [2+2] cycloaddition reaction was reported in 1989 by the use of the chiral titanium reagent prepared from the tartrate-derived chiral 1,4-diol 1 and TiCl$_2$(O-i-Pr)$_2$ [26]. Treatment of methyl (E)-4-oxo-4-(2-oxo-1,3-oxazolidin-3-yl)-2-butenoate (2a) and 1,1-bis(methylthio)ethylene (3a) with a 10 mol % amount of the chiral titanium reagent in a mixed-solvent of toluene and petroleum ether (P.E.) at 0 °C afforded the cyclobutane derivative 4a in 96% yield in nearly optically pure form (98% ee) (Scheme 1). In this section we

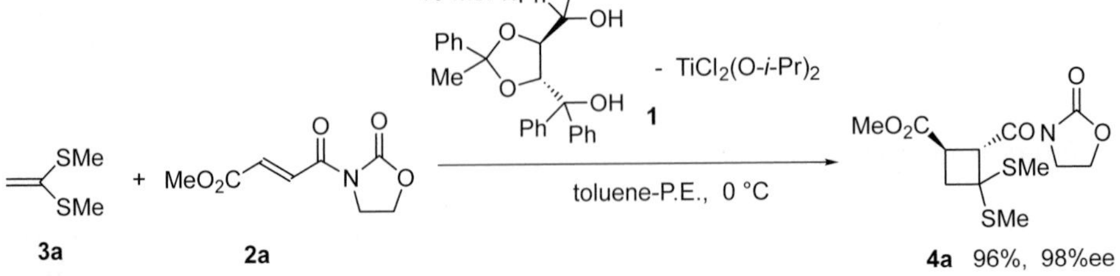

Scheme 1

describe the chiral titanium reagent-promoted asymmetric [2+2] cycloaddition reaction between electron-deficient alkenes and alkenyl sulfides.

2.2
Mechanism of the Catalytic Reactions and Basis of Stereoinduction

The titanium catalyst is prepared in situ by mixing the chiral 1,4-diol 1 and $TiCl_2(O\text{-}i\text{-}Pr)_2$ in toluene in the presence of 4 Å molecular sieves (MS) (Scheme 2). Because of the high asymmetric induction observed in the asymmetric [2+2] cycloaddition reaction described above, $TiCl_2(O\text{-}i\text{-}Pr)_2$ is supposed to be converted completely to the chiral titanium species. An NMR study of a mixture of equimolar amounts of $TiCl_2(O\text{-}i\text{-}Pr)_2$ and the 1,4-diol 1 in toluene-d_8, however, reveals that the mixture consists of a chiral titanium complex and achiral $TiCl_2(O\text{-}i\text{-}Pr)_2$ and that the ratio of the complexed 1 to free 1 is about 84:16 at a concentration of 0.17 mol/l.

This observation does not coincide with the high enantioselectvity in the asymmetric reactions, since the remaining achiral $TiCl_2(O\text{-}i\text{-}Pr)_2$ is considered to be a more effective Lewis acid than the chiral titanium complex. When $TiCl_2(O\text{-}i\text{-}Pr)_2$ and the diol 1 are mixed together, alkoxy exchange takes place and a chiral cyclic titanium alkoxide is generated with elimination of isopropyl alcohol. In the NMR spectrum of the mixture in toluene-d_8, the methine proton of isopropyl alcohol appears at lower field as compared with that of isopropyl alcohol itself. This lower field shift suggests the complexation of isopropyl alcohol with $TiCl_2(O\text{-}i\text{-}Pr)_2$. It is considered that the coordination of isopropyl alcohol causes aggregation of the achiral titanium species $[TiCl_2(O\text{-}i\text{-}Pr)_2]$ which decreases its activity as a Lewis acid [34].

A chiral titanium complex with 3-cinnamoyl-1,3-oxazolidin-2-one (2d) was isolated by Jørgensen et al. from a mixture of $TiCl_2(O\text{-}i\text{-}Pr)_2$ with (2R,3R)-2,3-O-isopropylidene-1,1,4,4-tetraphenyl-1,2,3,4-butanetetrol (1-Me) which is an isopropylidene acetal analogue of 1 [35]. The structure of the complex was determined by X-ray methods. The complex consists of the isopropylidene diol 1-Me and the cinnamoyloxazolidinone 2d in the equatorial plane and the two chloride ligands in the apical (*trans*) position as depicted in structure A (Scheme 3). It appears from this structure that the pseudo-axial phenyl group of the chiral ligand seems to block one face of the coordinated 2d. In contrarst, from an NMR study of the complex in the solution, Di Mare et al. reported that the above *trans*

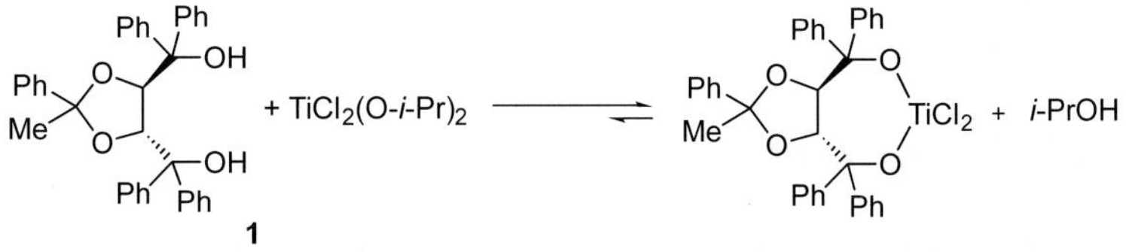

Scheme 2

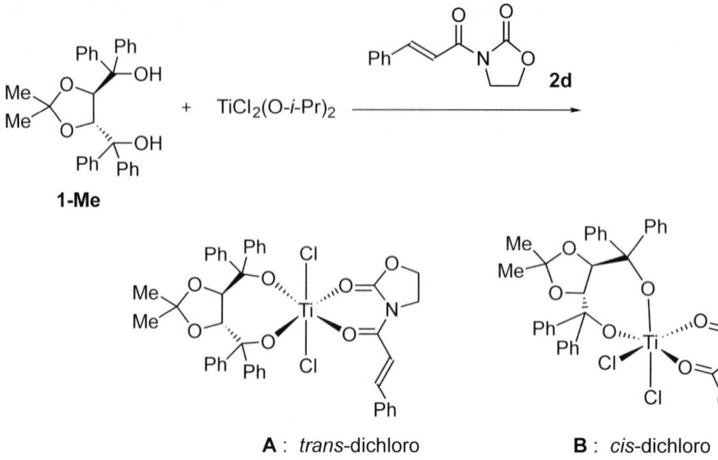

Scheme 3

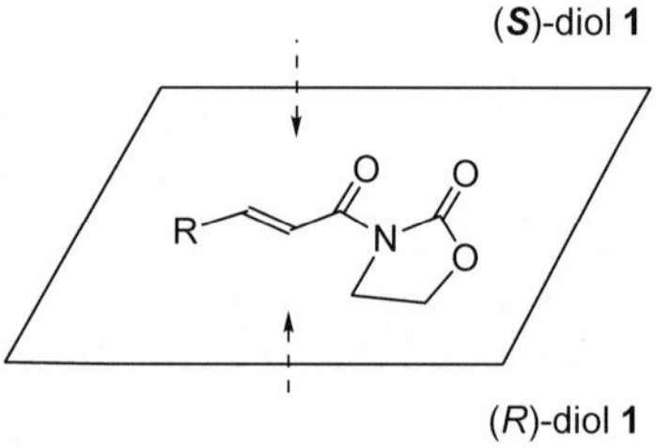

A : *trans*-dichloro   B : *cis*-dichloro

Scheme 4

dichloro complex **A** is a major complex in the solution. He proposes another minor complex **B** (Scheme 3) with the two chlorides facing *cis* to each other to be the most reactive intermediate in this chiral titanium-catalyzed reaction [36].

At the present stage, it is not clearly confirmed whether the *trans* and/or the *cis* complex are the real reactive intermediate. The absolute configuration of the cycloadducts is, however, predicted by both models. When the (*R*)-1,4-diol **1** is employed as a chiral auxiliary, the *Re* face of the α carbon of the alkenoyl moiety of **2** is attacked so far without exception (Scheme 4).

2.3
Practical Aspects

2.3.1
Reaction of Ketene Dithioacetal [26, 29]

As mentioned in the introduction, the chiral titanium-catalyzed asymmetric [2+2] cycloaddition reaction proceeds between methyl (*E*)-4-oxo-4-(2-oxo-1,3-

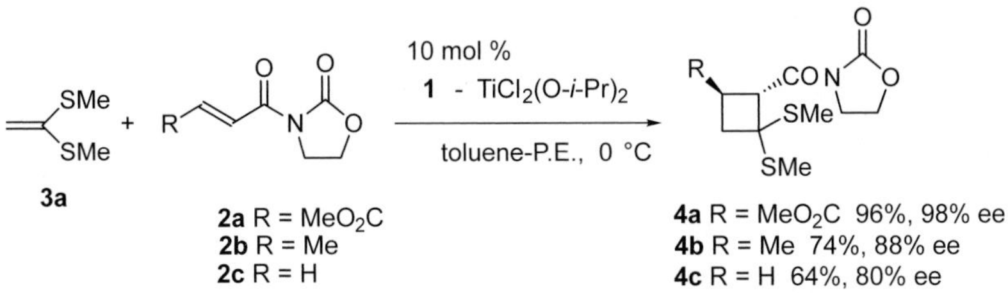

Scheme 5

oxazolidin-3-yl)-2-butenoate (**2a**) and ketene dimethyldithioacetal to afford the cyclobutane derivative **4a** in high optical purity. Ketene dimethyldithioacetal (**3a**) also reacts with 3-acryloyl- or 3-crotonoyloxazolidinone **2b** or **2c**, giving the corresponding cyclobutanone dimethylthioacetals **4b** and **4c** in 88% ee and 80% ee, respectively (Scheme 5). Optically pure cyclobutanes **4a** and **4c** are easily prepared by recrystallization. The chiral auxiliary **1** is completely recovered without loss of the optical purity. Since a 5% molar amount of the catalyst is enough for completion of the reaction, the total turnover number of the catalyst in this reaction is ranging from 10 to 20.

2.3.2
Reaction of Alkynyl Sulfides [27, 29]

As is shown in Scheme 6 and Table 1, alkynyl sulfides can be employed in the asymmetric [2+2] cycloaddition reaction; however, the reactivity of alkynyl sulfides is largely dependent on the substituent at sulfur. A phenyl sulfide, 1-phenylthio-1-hexyne (**5e**), does not react with **2a**, while the alkynyl methyl sulfides **5a-d** react smoothly with fumaric and acrylic acid derivatives **2a,c**, yielding cyclobutenes **6**. Trisubstituted cyclobutenes are prepared in good yield and in almost enantiomerically pure forms with only a catalytic amount of the chiral titanium reagent. For the preparation of tetrasubstituted cyclobutenes, however, an equimolar amount of the chiral titanium is required for the reaction to go to completion. Compared with the ketene dimethyldithioacetal **3a**, alkynyl methyl sulfides **5** are less reactive and the reaction between the crotonoyloxazolidinone **2b** and **5** fails even in the presence of an equimolar amount of the catalyst.

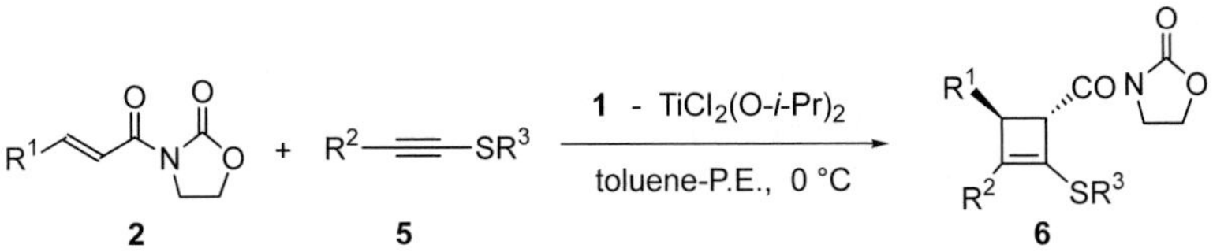

Scheme 6

Table 1. Asymmetric cycloaddition of alkynyl sulfides **5**

2 R^1	5 R^2	R^3		Amount of Ti [mol. amount]	Yield [%]	ee of **6** [%]
CO$_2$Me (2a)	n-Bu	Me	(5a)	1.1	92	>98
	Me	Me	(5b)	1.1	90	>98
	cyclohexyl	Me	(5c)	1.1	84	>98
	H	Me	(5d)	0.1	83	>98
	n-Bu	Ph	(5e)	1.1	0	
Me (2b)	n-Bu	Me	(5a)	1.1	0	
H (2c)	n-Bu	Me	(5a)	0.1	80	>98
	cyclohexyl	Me	(5c)	0.3	65	>98

2.3.3
Reaction of Alkenyl Sulfides [27, 29]

Alkenyl sulfides are known to react with some labile electron-deficient olefins such as methyl vinyl ketone in the presence of AlCl3 to form cyclobutanes [25]. In the present chiral titanium-promoted asymmetric reaction, alkenyl sulfides can also be employed as electron-rich components. 2-Ethylthio-1-propene (**7a**) reacts with **2a** in the presence of a catalytic amount of the chiral titanium reagent, giving the diastereomeric [2+2] cycloaddition products **8a** and **9a** in 51% (>98% ee) and 19% (79% ee) yields, respectively (Scheme 7 and Table 2). Although 2-ethylthio-1-propene (**7a**) is known as a good ene component in the reaction with carbonyl compounds, 3-(3-(methoxycarbonyl)-5-ethylthio-5-hex-enoyl)-1,3-oxazolidin-2-one, an ene product, is obtained only in 16% yield as a side product.

The reaction of **2a** with an allylsilane-type sulfide, 3-trimethylsilyl-2-methyl-thio-1-propene (**7b**), also afforded the cyclobutanes **8b** and **9b** in 54% (>98% ee) and 17% yields, respectively, without any formation of allylation and ene reaction products. The diastereoselectivity of these two reactions using 2-alkylthio-propene derivatives **7a,b** is not high, but the major isomers **8a,b** are obtained in nearly enantiomerically pure forms.

Cycloalkenyl sulfides **10** are converted into bicyclo[n.2.0]alkane compounds with almost complete enantioselectivity by the reaction with the fumaric or acrylic acid derivatives **2a,c** as listed in Scheme 8 and Table 3. As the reactivity of **10** is not as high as that of acyclic alkenyl sulfides, the use of an equimolar amount of the chiral titanium reagent is required to attain a good chemical yield in some cases. Diastereoselectivity is generally excellent and no ene product is detected.

Table 2. Asymmetric [2+2] cycloaddition reaction of alkenyl sulfides **7**

R^1	R^2	Sulfide	Yield [%]		ee of [%]	
			8	**9**	**8**	**9**
Et	Me	(**7a**)	51 (**8a**)	19 (**9a**)	>98	79
Me	CH_2SiMe_3	(**7b**)	54 (**8b**)	17 (**9b**)	>98	–

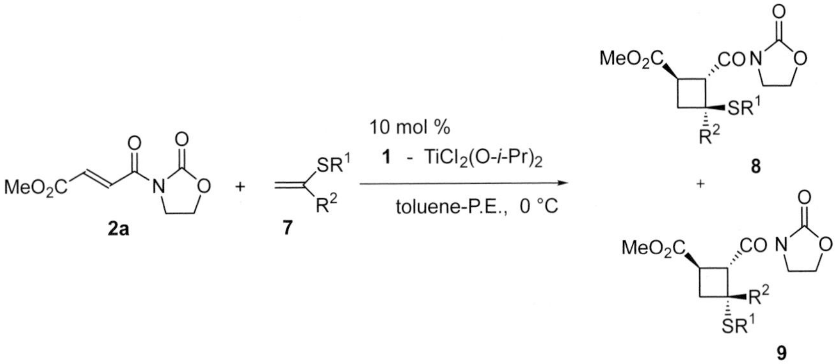

Scheme 7

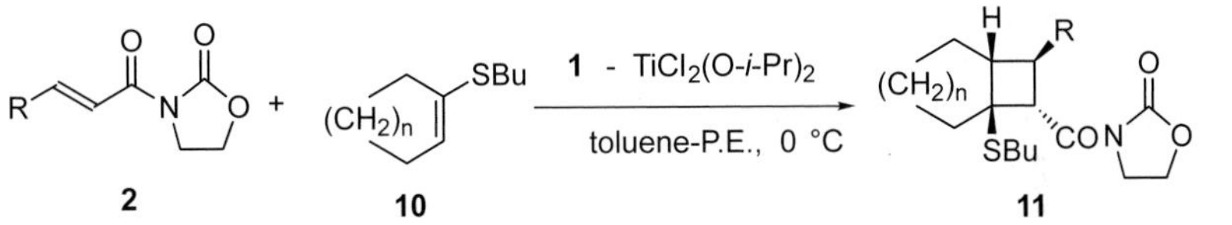

+ the other isomer **12**

Scheme 8

Table 3. Reaction of cyclic alkenyl sulfides **10**

R		**10** n	Amount of Ti [mol. amount]	Yield (**11+12**) [%]	**11:12**	ee of **11** [%]
CO_2Me	(**2a**)	2 (**10a**)	1.1	96	>99:1	>98
			0.15	92	>99:1	>98
		3 (**10b**)	1.1	97	92:8	>98
		4 (**10c**)	1.1	89	91:9	>98
H	(**2c**)	2 (**10a**)	0.25	74	82:18	>98

2.3.4
Reaction of 1,2-Propadienyl Sulfides [28, 29, 32]

1,2-Propadienyl sulfides having an α-trimethylsilyl, trimethylstannyl, or benzyl substituent, **13a,b,c**, react with **2a,c** to give methylenecyclobutane derivatives **14** and **15** in good chemical yields in nearly enantiomerically pure forms (Scheme 9 and Table 4). The diastereoselectivity is high in the reactions of **2a** with the α-trimethylsilyl- and α-trimethylstannylallenes **13a,b**, in which these bulky substituents and the oxazolidinylcarbonyl group have a *trans* geometry in the products. Although the trimethylstannylallene **13b** reacts in high yield, the corresponding tributylstannyl analogue does not react with **2a**, presumably because of the steric hindrance.

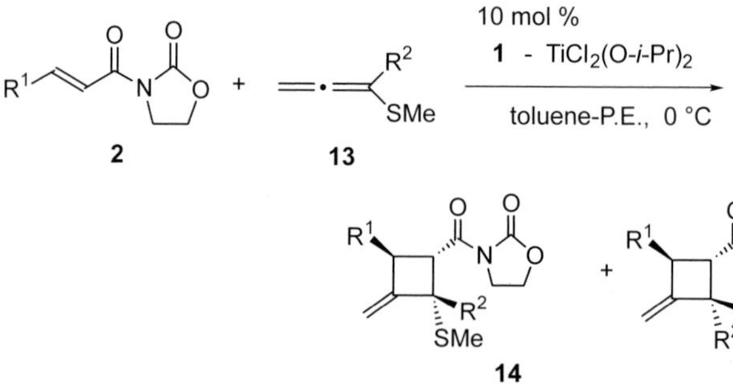

Scheme 9

Table 4. Asymmetric [2+2] cycloaddition reaction of propadienyl sulfides **13**

R¹		R²	Yields [%]		ee's [%]	
			14	15	14	15
CO₂Me	(2a)	SiMe₃ (13a)	quant.[a]	–	>98	–
		SnMe₃ (13b)	93[a]	–	96	–
		SnBu₃	0	0		
		CH₂Ph (13c)	30	57	94	>98
H	(2c)	SiMe₃ (13a)	41	21	>98	–

[a]**15** was not detected by ¹H-NMR spectroscopy

2.3.5
Reaction of Styrenes and 1,4-Benzoquinones

The chiral titanium catalyst prepared by mixing the diol **1**, $TiCl_4$, and $Ti(O\text{-}i\text{-}Pr)_4$ in a 1:1:1 ratio promotes the asymmetric [2+2] cycloaddition reaction between styrenes **16** and 1,4-benzoquinones **17** to afford cyclobutane derivatives in good optical purity (Scheme 10 and Table 5) [37]. This reaction is not a truly catalytic reaction because excess amounts of the catalyst (5 molar amounts) have to be employed in order to obtain a high enantiomeric excess. The catalyst in this reaction is thought to be different from the one prepared from the diol **1** and $TiCl_2(O\text{-}i\text{-}Pr)_2$ in a 1:1 ratio as described previously.

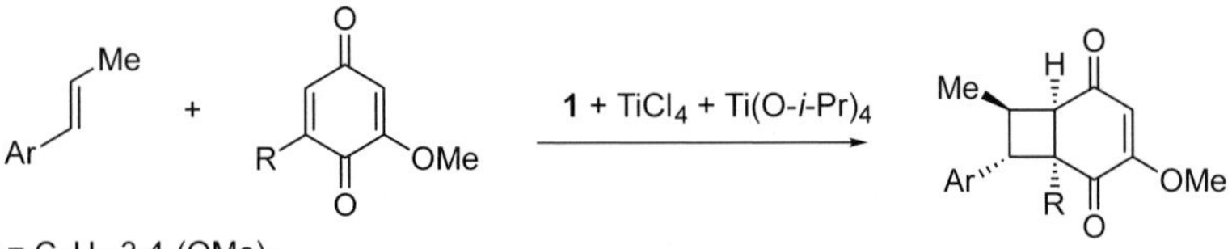

16a Ar = C_6H_3-3,4-$(OMe)_2$ **17a** R = H
16b Ar = C_6H_4-4-OMe   **17b** R = Me
16c Ar = C_6H_4-4-Me

Scheme 10

Table 5. Asymmetric [2+2] cycloaddition of styrenes and quinones

Styrene	Quinone	Yield [%]	ee [%]
16a	17a	88	92
16b	17a	86	90
16c	17a	71	86
16a	17b	43	88
16b	17b	72	90

2.4
Principle Alternatives

Highly enantiomerically enriched cyclobutane derivatives are prepared via diastereoselective [2+2] cycloaddition reactions of chiral ethylene derivatives; for instance, by photochemical (Scheme 11 and Scheme 12) [38, 39] and thermal [2+2] cycloaddition (Scheme 13) [40, 41] reactions of chiral alkenes and by the [2+2] cycloaddition of chiral keteneiminium ions (Scheme 14) [42].

Chiral cyclobutane derivatives are also synthesized by the enantioselctive alkylation [43], by the chemical [44] and enzymatic [45] resolution of racemic precursors (Scheme 15), and by the ring enlargement of the corresponding chiral cyclopropylidene oxide (Scheme 16)[46].

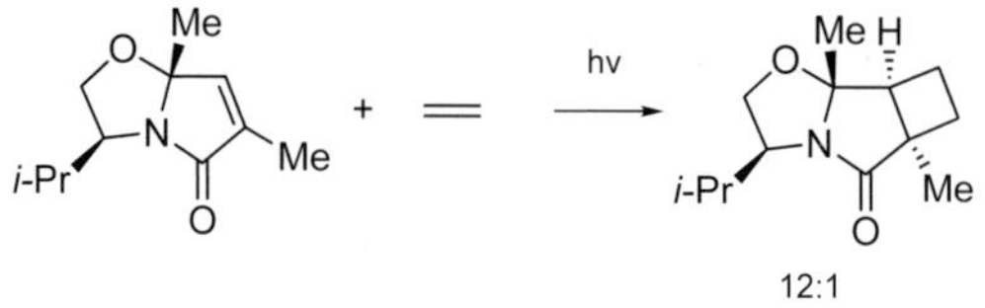

Scheme 11

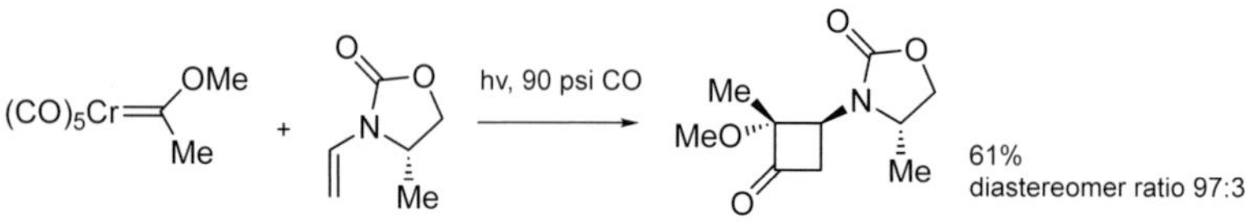

Scheme 12

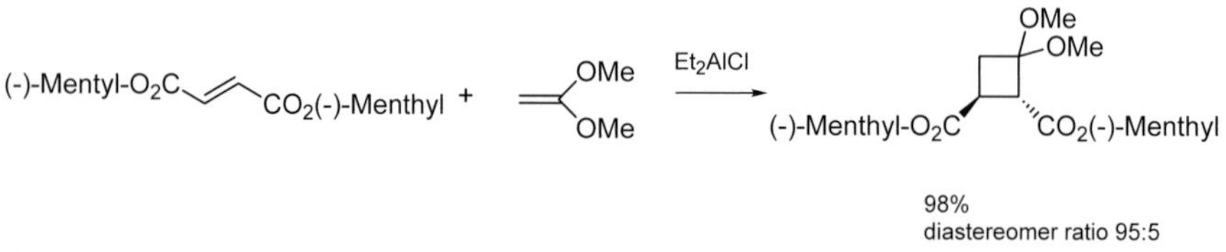

Scheme 13

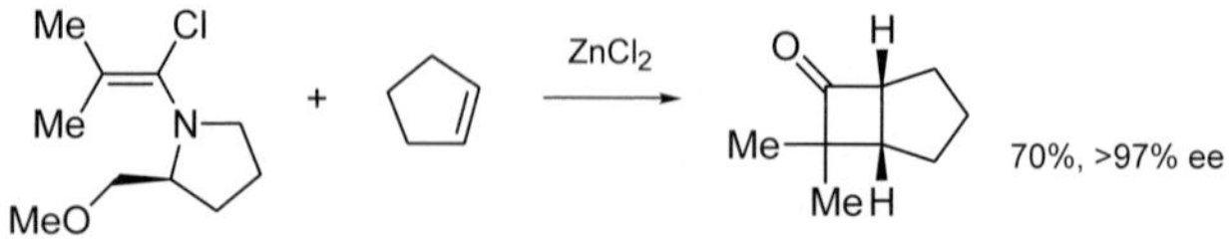

Scheme 14

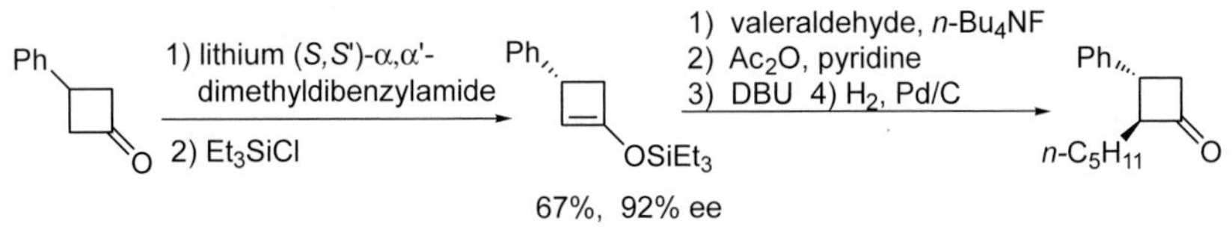

Scheme 15

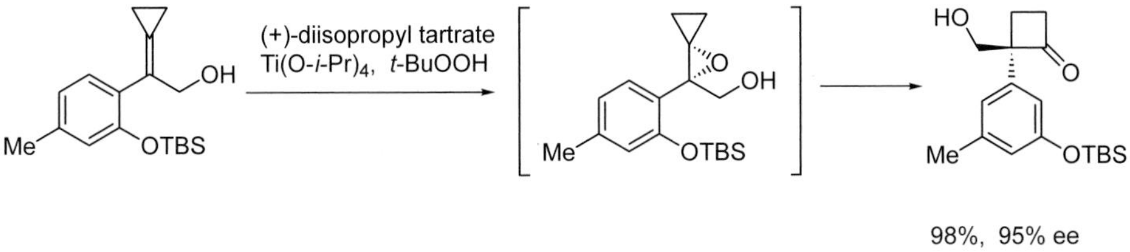

98%, 95% ee

Scheme 16

3
[2+2] Cycloadditions of Ketenes and Aldehydes

3.1
Introduction

The hetero [2+2] cycloaddition reaction is a synthetically important reaction for the construction of 4-membered heterocyclic compounds. As far as the catalytic asymmetric reaction is concerned, however, only the cycloaddition between ketenes and aldehydes has been reported. The thus synthesized chiral oxetan-2-ones are employed as monomer precursors for the biologically degradable co-polyesters and also as chiral building blocks for natural product synthesis. Two types of catalysts, *Cinchona* alkaloids and a chiral Lewis acid, are known to promote this reaction.

3.2
Cinchona-Alkaloids-Catalyzed Reaction

3.2.1
Mechanism of the Catalytic Reaction

Cinchona alkaloids are the effective asymmetric catalysts in the [2+2] cycloaddition reaction of ketenes and aldehydes [47]. That is, the reaction between ketenes and chloral proceeds with a catalytic amount (2.5 mol %) of quinidine in toluene at -50 °C to provide the β-propionolactone $(-)$-**18** in quantitative yield with 98% ee (Scheme 17).

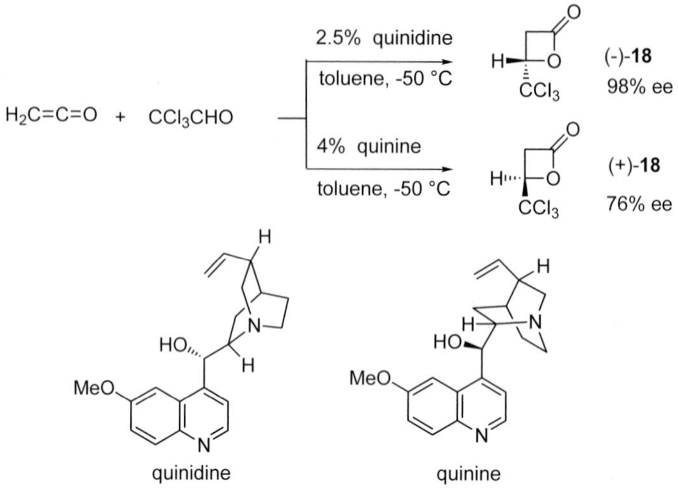

Scheme 17

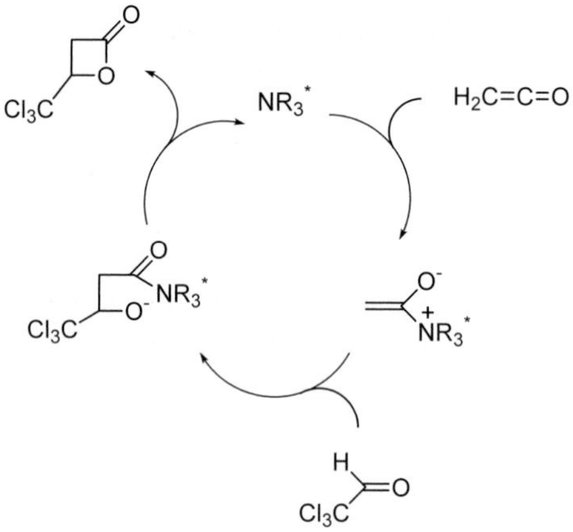

Scheme 18

By the use of quinine instead of quinidine, the opposite enantiomer (+)-**18** can be synthesized with good optical purity (76% ee) (Scheme 17). It is also note-worthy that exceedingly simple catalysts such as 1,2-dimethylpyrrolidine and N,N-dimethyl-α-phenylethylamine gave good enantiomeric excess (60 and 77% ee's, respectively). Alhough the reaction mechanism and the basis of the stereoinduction are not clear, the reaction is thought to proceed in a stepwise manner via a tertiary amine-ketene complex as initial intermediate and not via a tertiary amine-chloral complex (Scheme 18).

3.2.2
Generality of the Reaction

Not only chloral but also other polyfunctionalized aldehydes can be employed in this reaction to afford chiral 4-substituted-2-oxetanones **19** (Scheme 19) and the results are listed in Table 6 [48]. Highly electron-deficient ketones react with ketene, for example, trichloroacetophenone does not react with ketene but the reaction does proceed when an electron-withdrawing substituent such as a chloro or nitro group is introduced into the phenyl group.

3.2.3
Principle Alternatives

The alternative procedure for chiral 4-substituted-2-oxetanones is the ring closure of optically active β-hydroxycarboxylic acid derivatives [49].

3.3
Chiral Lewis Acid-Catalyzed Reaction

A C_2-symmetric *N,N'*-di-3,5-bis(trifluoromethyl)benzenesulfonyl-(1*R*,2*R*)-1,2-diphenylethylenediamine **20**-Et$_3$Al complex promotes the [2+2] cycloaddition reaction between ketene and aldehydes to afford optically active 4-substituted oxetan-2-ones **21** (Scheme 20) [50]. The catalyst is prepared by mixing the bissulfonamide **20** and Et$_3$Al, and the reaction proceeds by the coordination of the aldehyde to the chiral Lewis acid.

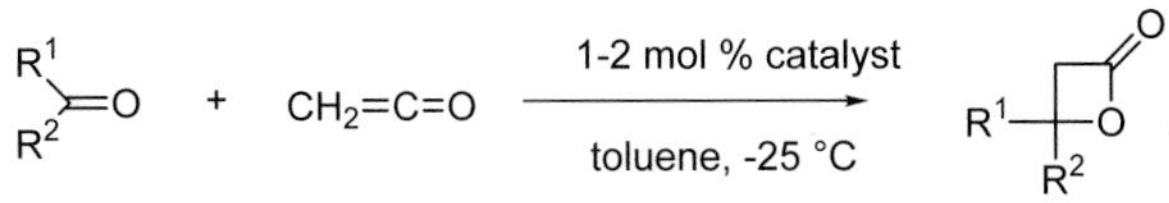

Scheme 19

Table 6. Reaction catalyzed by *Cinchona* alkaloids

R^1	R^2	Yield [%]	ee [%][a]	ee [%][b]
CCl$_3$	H	89	98 (R)	76 (S)
CCl$_2$CH$_3$	H	95	91 (R)	76 (S)
CCl$_2$C$_6$H$_5$	H	89	90	68
CCl$_3$	CH$_3$	72	94 (R)	85 (S)
CCl$_3$	C$_6$H$_4$Cl-*p*	68	90	65
CCl$_3$	C$_6$H$_4$NO$_2$-*p*	95	89	65

[a]Catalyst is quinidine
[b]Catalyst is quinine

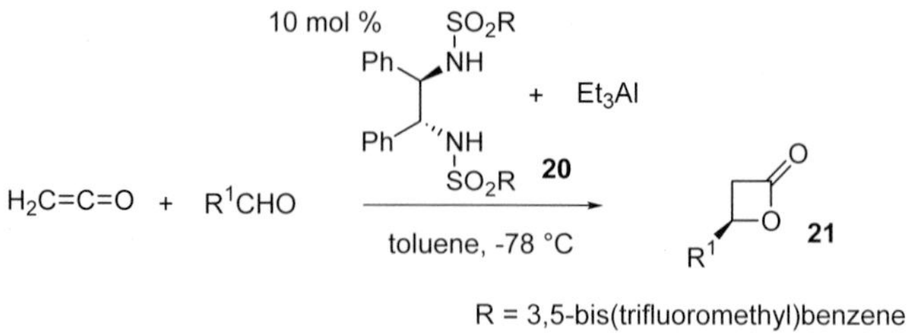

$$R = 3,5\text{-bis(trifluoromethyl)benzene}$$

Scheme 20

Table 7. Reaction catalyzed by chiral aluminum reagent

R^1	Yield [%]	ee [%]
Me	59	30
Et	77	33
n-Bu	82	41
i-Pr	76	56
cyclohexyl	75	74
t-Bu	77	65
Ph	11	14

In the presence of 10 mol % of the chiral catalyst, aldehydes react with ketene at –78 °C in toluene for 1 h to afford oxetan-2-one derivatives **21** in 14 to 74% ee as listed in Table 7. This suggests that the reactions of bulkier aldehydes proceed in better optical purity. Although the obtained enantioselectivity is not high, there is a room for the future improvement.

References

1. Baldwin JE (1991) Thermal Cyclobutane Ring Formation. In: Trost BM (ed) Comprehensive Organic Synthesis. Pergamon Press, Oxford, chap 2.1
2. Ghosez L, Marchand-Brynaert J (1991) Formation of Four-membered Heterocycles. In: Trost BM (ed) Comprehensive Organic Synthesis. Pergamon Press, Oxford, chap 2.2
3. Crimmins MT (1991) Ptotochemical Cycloadditions. In: Trost BM (ed) Comprehensive Organic Synthesis. Pergamon Press, Oxford, chap 2.3
4. Porco JA, Schreiber SL (1991) The Paterno-Buchi Reaction. In: Trost BM (ed) Comprehensive Organic Synthesis. Pergamon Press, Oxford, chap 2.4
5. Delucchi O (1991) Di-π-methane Photoisomerizations. In: Trost BM (ed) Comprehensive Organic Synthesis. Pergamon Press, Oxford, chap 2.5
6. Demuth M (1991) Oxa-di-π-methane Photoisomerizations. In: Trost BM (ed) Comprehensive Organic Synthesis. Pergamon Press, Oxford, chap 2.6
7. Bellus D, Ernst B (1988) Angew Chem Int Ed Engl 27:797
8. Huisgen R (1977) Acc Chem Res 10:117, 199
9. Snider BB (1976) J Org Chem 41:3061
10. Snider BB, Rodini DJ, Conn RSE, Sealfon S (1979) J Am Chem Soc 101:5283
11. Snider BB, Roush DM, Rodini DJ, Gonzalez D, Spindell D (1980) J Org Chem 45:2773

12. Fienemann H, Hoffmann HMR (1979) J Org Chem 44:2802
13. Snider BB, Hrib NJ (1977) Tetrahedron Lett 20:1725
14. Clark RD, Untch K G (1979) J Org Chem 44:248, 253
15. Quendo A, Rousseau G (1988) Tetrahedron Lett 29:6443
16. Snider BB, Spindell D K (1980) J Org Chem 45:5017
17. Hoffmann H M R, Ismail Z M, Weber A (1981) Tetrahedron Lett 22:1953
18. Padwa A, Filipkowski MA, Meske M. Watterson S H, Ni Z (1993) J Am Chem Soc 115:3776
19. Snider BB, Duncia J V (1980) J Am Chem Soc 102:5926
20. Snider BB, Rodini DJ, Straten van J (1980) J Am Chem Soc 102:5872
21. Engler TA, Ali M H, Velde DV (1989) Tetrahedron Lett 30:1761
22. Engler TA, Combrink KD, Ray JE (1988) J Am Chem Soc 110:7931
23. Scheeren HW, Frissen AE (1983) Synthesis 794
24. Quendo A, Rousseau G (1989) Synth Commun 19:1551
25. Takeda T, Fujii T, Morita K, Fujiwara T (1986) Chem Lett 1311
26. Hayashi Y, Narasaka K (1989) Chem Lett 793
27. Hayashi Y, Narasaka K (1990) Chem Lett 1295
28. Hayashi Y, Niihata S, Narasaka K (1990) Chem Lett 2091
29. Narasaka K, Hayashi Y, Shimadzu H, Niihata S (1992) J Am Chem Soc 114:8869
30. Narasaka K, Kusama H, Hayashi Y (1991) Bull Chem Soc Jpn 64:1471
31. Ichikawa Y, Narita A, Shiozawa A, Hayashi Y, Narasaka K (1989) J Chem Soc Chem Commun 1919
32. Narasaka K, Hayashi K, Hayashi Y (1994) Tetrahedron 50:4529
33. Narasaka K, Hayashi Y (1997) Lewis Acid Catalyzed [2+2] Cycloaddition Reactions of Vinyl Sulfides and Their Analogues: Catalytic Asymmetric [2+2] Cycloaddition Reactions, In: Lautens M (ed) Advances in Cycloaddition. JAI Press Inc, Greenwich, p 87
34. Iwasawa N, Hayashi Y, Sakurai H, Narasaka K (1989) Chem Lett 1581
35. Gothelf K V, Hazell RG, Jørgensen KA (1995) J Am Chem Soc 117:4435
36. Haase C, Sarko C R, DiMare M (1995) J Org Chem 60:1777
37. Engler TA, Letavic MA, Reddy JP (1991) J Am Chem Soc 113:5068
38. Meyers AI, Fleming SA (1986) J Am Chem Soc 108:306
39. Hegedus LS, Bates RW, Soderberg B C (1991) J Am Chem Soc 113:923
40. Ahmad S (1991) Tetrahedron Lett 32:6997
41. Greene AE, Charbonnier F, Luche M-J, Moyano A (1987) J Am Chem Soc 109:4752
42. Houge C, Frisque-Hesbain AM, Mockel A, Ghosez L, Declercq JP, Germain G, Meerssche MV (1982) J Am Chem Soc 104:2920
43. Honda T, Kimura N (1994) J Chem Soc Chem Commun 77
44. For instance, Brunet J-J, Herbowski A, Neibecker D (1996) Synth Commun 26:483
45. For instance, Jung M, Sledeski AW (1993) J Chem Soc Chem Commun 589
46. Nemoto H, Fukumoto K (1997) Synlett 863
47. Wynberg H, Staring EGJ (1982) J Am Chem Soc 104:166
48. Wynberg H, Staring EGJ (1985) J Org Chem 50:1977
49. Adam W, Baeza J, Liu J-C (1972) J Am Chem Soc 94:2000
50. Tamai Y, Yoshiwara H, Someya M, Fukumoto J, Miyano S (1994) J Chem Soc Chem Commun 2281

Chapter 34   **Additions to Enolates**

Chapter 34.1
Alkylation of Enolates

David L. Hughes

Merck and Co., Inc., Mail Drop R80Y-250, Rahway, NJ 07065, USA
e-mail: Dave_Hughes@Merck.com

Keywords: Phase transfer, Chiral ligand, Chiral base, Allylic alkylation, Cinchonine, Enolate alkylation, Asymmetric deprotonation, Asymmetric arylation, Radical alkylation

1
Introduction

Reaction of enolates with alkylating agents is one of the oldest methods of forming carbon-carbon bonds in synthetic organic chemistry. Progress in selective alkylations of enolates has been steadily advancing over the past four decades. Interest in selective alkylations arose in the 1960s with studies directed toward understanding axial vs. equatorial diastereoselective alkylations of cyclohexanones [1], and *erythro* vs. *threo* alkylations of acyclic enolates [2]. The concept of diastereoselective alkylations was taken a step further in the 1970s to provide the first indirect methods for asymmetric alkylations of enolates, with the pio-

neering studies of Meyers [3], Koga [4], Whitesell [5], and Enders [6]. These
methods involved a three-step sequence which included reaction with a stoichiometric covalent chiral auxiliary, deprotonation and alkylation to form product
enriched in one diastereomer, and cleavage to produce the enantioenriched
product with release of the chiral auxiliary, Eq. (1). Over the past two decades a
wide variety of chiral auxiliaries have been introduced that are applicable to substrates of diverse structures [7].

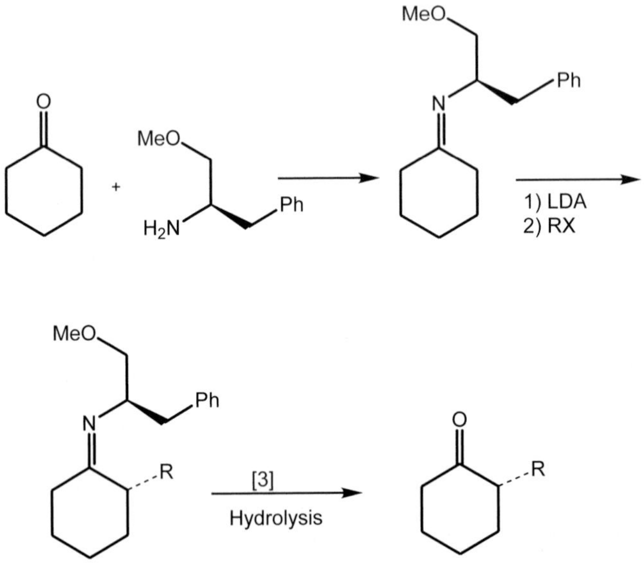

(1)

On the other hand, asymmetric enolate alkylation via catalytic methods is rare,
and the technology for doing so can still be considered in its infancy. While
many other catalytic C-C bond forming methods for enolates have been achieved,
such as aldol and Michael reactions, progress has been slow for alkylation reactions. Part of the reason for this is that, for typical alkylations, no organized
transition state exists which involves all three components of the reaction: the
enolate, the chiral catalyst, and the alkylating agent. Instead, for enantioselective
alkylation to occur, tight binding between the catalyst and enolate must occur,
and the alkylating agent will simply attack the least hindered face. Since achieving a single tightly bound complex requires a perfect match between substrate
and enolate (3-point binding essential for chiral recognition), catalytic enantioselective alkylations are rare. On the other hand, to use an aldol reaction as an
example, all components of the reaction, including the aldehyde, the organometal species, and chiral catalyst, can be organized in a 6-membered transition state
held in place by metal coordination, which facilitates an enantiotopic reaction.
In this case, organization of the TS is the key, not organization of the ground
state molecules, as in an alkylation reaction.

The following sections review the work on catalytic asymmetric alkylations
of enolates that has primarily been carried out in the past decade. Each approach

has seen its successes, although none has found widespread use due to drawbacks which range from narrow substrate specificity to difficult catalyst preparation. The primary alternatives to catalytic routes involve diasteroselective
alkylations using a chiral auxiliary, as described briefly above, and enantioselective protonations, described in the following chapter.

2
Phase Transfer Catalysis

In the late 1970s Wynberg and coworkers pioneered the use of *Cinchona* alkaloids as catalysts for asymmetric reactions, demonstrating 76% ee's for both an
epoxidation of a naphthoquinone [8] and the Michael reaction of an indanone
derivative with methyl vinyl ketone [9]. On the other hand, asymmetric enolate
alkylations using chiral catalysts had limited success during this period. An
asymmetric alkylation of an enolate using chiral phase transfer catalysts was
first reported in 1975 by Fiaud [10] for the reaction of 2-acetylcyclohexanone
with allyl bromide using 10% aq. NaOH and a catalyst derived from ephedrine.
An optical yield of 6% was reported, based on the magnitude of the optical rotation in comparison with a pure sample of the enantiomer. Likewise, ee's of 5 to
7% were reported in 1979 and 1980 for the alkylation of ethyl 2-oxocyclohexane-
carboxylate using chiral phase transfer catalysts [11, 12]. However, even these
modest enantioselectivities were challenged by Dehmlov and coworkers [13],
who determined that, in some cases, most or all of the optical rotations were due
to decomposition products of the catalyst, and that no optical rotation was observed in products that had been purified by chromatography.

The first breakthrough in asymmetric alkylation came in 1984 when Dolling
and coworkers [14] reported a 94% ee in the phase-transfer alkylation of indanone derivatives using *Cinchona* alkaloids as catalysts, Eq. (2).

$$\tag{2}$$

The key to the excellent enantioselectivity was finding conditions to maxi-
mize tight ion pairing between the enolate and the catalyst such that only one
face was available for alkylation. The authors proposed that three-point binding
was necessary for high ee. A hydrogen bond between the alkaloid hydroxy group
and the enolate oxygen provided the directional handle, with additional stabili-
zation and directionality being furnished by π-bonding interactions between
the aromatic rings (complex **1**). Several experiments supported the ion-pair hypothesis. First of all, the selectivity was highest in nonpolar solvents such as toluene, which should enhance tight ion pair formation. Secondly, the selectivity
increased as the concentration of NaOH was increased from 30% to 50%. With
30% NaOH the water content in the organic layer is much higher compared to

1

50% NaOH, and the higher level of water would be expected to disrupt the hydrogen bond of the ion pair. Finally, the electronics of the catalyst had a significant impact on the selectivity. A Hammett plot of log ee/ee$_o$ vs. the substituent constant of the N-benzyl group of the catalyst gave a reaction constant ϱ of 0.21 with a range of ee's from 60% for p-MeO to 92% for p-CF$_3$. This suggests that electron-withdrawing substituents enhance the binding of the enolate and catalyst, as might be expected for a charge-transfer complex between the electron-rich phenyl group of the enolate and the increasingly electron-poor benzyl portion of the catalyst.

A mechanistic study of this reaction produced a number of unanticipated findings [15]. While N-benzylcinchoninium bromide is virtually insoluble in toluene (<10^{-5} M), millimolar quantities were detected by HPLC analysis of the organic layer of aliquots removed from the reaction. Based on titration experiments (acid-base and bromide) and methylation experiments, it was proposed that the species in the organic layer was a dimer comprised of the catalyst and its deprotonated zwitterion. Subsequently, solution NMR experiments [16] corroborated this proposal. The NMR spectrum of a dimer solution in benzene compared to that of the octanesulfonate salt of the monomeric catalyst revealed that the protons on the quinoline rings were shifted downfield while those on the phenyl ring were shifted upfield. This is consistent with the dimer having a structure wherein the quinoline is opposite the phenyl ring, and the NMR shifts indicate the quinoline donates electrons to the phenyl ring. The protons on the face of the quinuclidine ring also shift, while those in the back remain unchanged. Thus, the structure of the catalyst dimer appears to be quite similar to that proposed for the indanone carbanion-catalyst ion pair. Subsequently, an X-ray structure of the catalyst dimer was obtained which was consistent with the proposed solution structure [16].

Due to the low solubility of the catalyst itself, formation of the dimer appears to be a critical part of the chiral alkylation process since it greatly enhances solubility of the catalyst. Kinetic measurements of the alkylation step revealed an order of 0.5 in catalyst, indicating the dimer dissociates to the monomer before complexation with the indanone enolate.

Another curious aspect of these reactions was the differences observed depending on the concentration of the aq. NaOH. Using 50% NaOH, the ee of product was measurably the same regardless of the amount of catalyst used. In contrast, for 30% NaOH, the ee of the product decreased as the catalyst concentration increased. Calculation of the rates of formation of the racemic and chiral product revealed that the order in catalyst was 0.5 for the asymmetric process, analogous to that found with 50% NaOH, but the order for the racemic process was 1.0. Thus, in this case, the racemic and asymmetric reactions must proceed through different mechanisms.

The asymmetric phase transfer reaction was extended by the Merck group to a substrate which lacked the phenyl group, Eq. (3) [17].

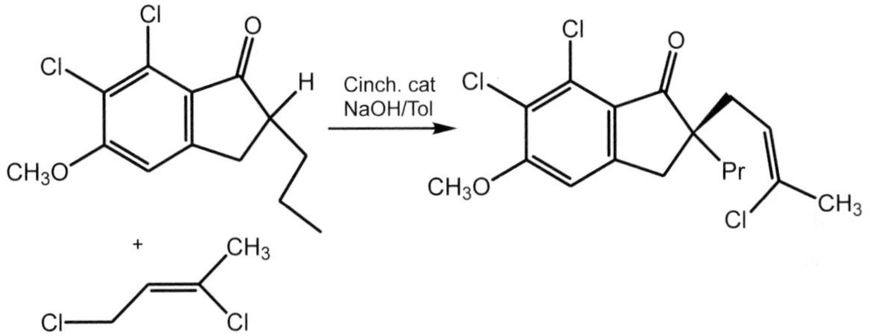

(3)

Using the same catalyst (p-CF$_3$ benzylcinchoninium) a 92% ee was obtained with yield >95%, while use of the diasteromeric catalyst, p-CF-benzylcinchonidinium, provided an ee of 76%, but 30 mol% catalyst was required for the reaction to reach completion. While having opposite configurations at the critical C8 and C9 positions, cinchonidine and cinchonine are diastereomers (Fig. 1, since they have the same configuration at the two other chiral centers, C3 and C4. Examination of space-filling models suggested the preferred conformation of cinchonine had the vinyl group behind the hydroxy group, leaving only one face available for hydrogen bonding. On the other hand, cinchonidine appears to be more flexible, and the vinyl group does not block one side of the hydroxy group, which may account for the lower enantioselectivity observed with this catalyst.

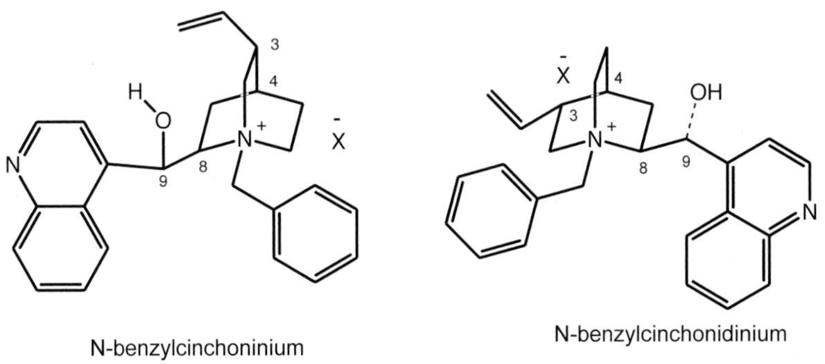

N-benzylcinchoninium   N-benzylcinchonidinium

Fig. 1

The equivalent enantioselectivity for the phenylindanone and propylindanone is somewhat surprising since the propyl group can only provide van der Waals interaction with the benzyl group of the catalyst and this might not be expected to provide adequate interaction to secure the two substrates in an ion-pair complex of the type shown with the phenyl enolate. However, extensive computational studies on enolate:cinchoninium complexes have led to the conclusion that dispersion forces play an important role in determining the shape of the intermolecular complexes, and that Coulombic attractions simply augment the strength of the intermolecular attraction [18]. Thus, these calculations suggest that replacement of phenyl with propyl might not cause major differences in binding, which concurs with the experimental results. In addition, the absence of the phenyl group means that the negative charge is more localized, leading to enhanced Coulombic attraction.

A novel dual catalytic process was also discovered for the reaction with the propylindanone substrate [16]. Since the cinchonidinium catalyst is unstable, 30 mol % was required for the reaction to reach completion using the standard phase-transfer conditions. However, employing a nonionic poly(ethylene oxide) surfactant (Triton X-405) and solid KOH along with the chiral catalyst, the level of the cinchona catalyst could be reduced to 7 mol % with no loss in enantioselectivity. In this process, the Triton X-405 plays two roles. Firstly, it extracts KOH into the organic layer (as measured by titration), where the indanone is deprotonated, as detected by formation of the yellow anion. In addition, the surfactant draws the monomeric catalyst into the organic phase. The apparent strong binding of the enolate and catalyst then causes a tight ion-pair to form such that no racemic alkylation of the Triton-K^+-indanone anion occurs.

The asymmetric phase transfer alkylation has been applied to a few other cyclic substrates, all of which can form planar enolate anions conducive to ion pairing with the cationic catalysts. Nerinckx and Vandewalle reported that tetralones could be alkylated with 1,5-dibromopentane with a yield of 74% and an ee of >70% in benzene/50% NaOH using the p-CF_3-benzylcinchonidinium catalyst, Eq. (4) [19]. The configuration of the

$$(4)$$

resulting alkylated product is consistent with an ion-pair in which the benzyl group of the catalyst interacts with the aromatic portion of the enolate; with the hydrogen bond between the catalyst hydroxy group and the enolate oxygen, two-point binding is invoked to rationalize the product configuration. With the non-fused cyclic substrates 2-phenylcyclopentanone and 2-phenylcyclohexanone, ee's for methylation at ambient temperature with MeBr were only 13% and 36%, respectively.

These results are consonant with the ion-pair proposal of Dolling, since the more flexible non-fused substrates would have more degrees of freedom and hence would not be expected to provide a selective ion pair from one conformer. On the other hand, Michael reactions with methyl vinyl ketone of these substrates at –20 °C provided the adducts in >80% ee. Although the authors provide no rationale, the TS for the Michael reaction may be organized by the cationic alkaloid binding to both the anionic enolate and the developing negative charge in the Michael acceptor.

A further example that demonstrates the usefulness of the phase transfer chemistry for fused aromatic subtrates is the cyanomethylation of a physostigmine precursor, Eq. (5) [20]. High selectivity (78%) was observed only for cinchoninium catalysts with electron-withdrawing groups in the benzyl group, as the unsubstituted benzyl catalyst gave only 10% ee and *ortho* substituents in the benzyl group also gave minimal enantioselectivity.

$$(5)$$

Based on the above results and rationalizations of enantioselectivities, expectations would be low for success in the enantioselective alkylation of flexible, acyclic enolates via chiral phase transfer catalysis. However, in 1989 O'Donnell and coworkers reported the asymmetric alkylation of acyclic Schiff bases with ee's up to 66%, Eq. (6) [21]. Critical to obtaining high enantioselectivity was use of the *t*-Bu ester, 50% NaOH as base, methylene chloride as solvent, and bromide as the leaving group. Further optimization, including 70% toluene/30% dichloromethane as solvent, 4:1 ratio of organic:aqueous, high agitation, and a temperature of 5 °C resulted in an improved enantioselectivity of 81% [22]. In contrast to Dolling's work, either enantiomer could be obtained in nearly equal selectivity by simply changing the catalyst from cinchoninium to cinchonidinium, and no effect was observed by addition of electron-withdrawing groups into the benzyl group of the catalyst.

$$(6)$$

Besides being acyclic, another confounding factor with this substrate is that it is secondary, so the product still contains an active proton such that racemization and dialkylation are concerns. However, neither of these potential obstacles were problems in this chemistry since the monoalkylated product is 4 pKa units less acidic than the starting active methylene compound [23]. A further complication of the acyclic enolate is that both Z- and E-enolates can be formed, and these are likely to complex differently with the catalyst. A computational study suggested that the Z-enolate complexes to the cinchoninium catalyst more strongly than the E-enolate, and this enolate fits into a groove of the catalyst rather than directly in a face-to-face orientation. However, calculations could not determine any trends or patterns to explain the enantioselectivity [18].

The phase transfer alkylation of Schiff bases has been extended to several other alkyl bromides as a route to new amino acids [24], and the enantioselectivities in these cases are comparable (50 to 70% ee) to those reported by O'Donnell. In addition, the methodology has been used for the synthesis of α-methylamino acids, with ee's of about 50% [25]. Futher optimization using an O-alkylated catalyst and a no solvent process with solid KOH/K_2CO_3 led to improved ee's of 70% [26].

Further work by the O'Donnell group led to an unexpected finding that the free OH of the cinchona catalyst is not required for high enantioselectivity [27a]. Under phase transfer conditions using N-benzylcinchonidinium bromide, the hydroxy group of the catalyst reacts with the alkylating agent, Eq. (7). When this O-alkylated catalyst was prepared independently and used as a catalyst, the enantioselectivity in the asymmetric alkylation of the benzophenone Schiff base was the same (60% ee) as in the reaction using the free OH catalyst. Since all hypotheses for the mechanism of enantioselectivity involve hydrogen-bonding from the catalyst OH group to the enolate, this result indicates that other modes of chiral recognition must be occurring, at least in some cases.

N-Benzylcinchonidinium

While O'Donnell and co-workers offered no explanation for these unusual observations, recent work by Corey [27b] and Lygo [27c] has provided a rationale along with the most impressive results in this field to date. As with the O'Donnell work, the catalyst used by the Corey group is O-alkylated (allyl), while the new twist is introduction of an N-anthracenylmethyl group on the quinucline nitrogen of cinchonidine. The Lygo group also used the anthracenylmethyl catalyst, and showed no difference whether the hydroxy group was alkylated or not, the assumption being that alkylation occurred during the reaction. Using solid cesium hydroxide as base at –60 °C to –78 °C and the same substrate as shown in Eq. 6, enantioselectivities in the 95–99.5% range were achieved by the Corey group. The Lygo group used liquid-liquid phase transfer conditions with KOH as base and achieved ee's up to 94%. With the hydroxy group allylated, no possibility of hydogen bonding exists, so an explanation different from that of Dolling must be invoked. That provided by Corey [27b] involves a tight ion-pair held in a rigid 3-dimensional geometry by van der Waals and coulombic attractions, with the large anthacenyl group providing steric sceening and rigidity. The most stable geometry of the ion pair allows for approach of the electrophile from only one face of the enolate.

In the work described thus far, the chiral catalysts that have been effective have all been derived from *Cinchona* alkaloids, most specifically, cinchonine and cinchonidine. Efforts to use other catalysts have been described, but most have met with very limited success [28, 29]. One extraordinary exception has been reported by Eddine and Cherqaoui, who found that the pyrrolidinehydrazonium salt (**2**) catalyzed the phase transfer alkylation of the imine shown in Eq. (8). with enantioselectivities up to 94% [30]. Due to the low acidity of the imine substrate, the solid base medium of KOH/K_2CO_3 was used which allowed reaction within 24 hours with 2–5% catalyst. Notably, no reaction occurred with standard ammonium phase transfer catalysts. When the hydroxy group of the catalyst was alkylated, the ee in the phase transfer reaction was reduced from 91% to 58%, demonstrating the need for hydrogen bonding for tight enolate-catalyst binding.

Summary. Asymmetric catalytic phase transfer alkylations are effective within a limited pool of substrates. No generalized catalyst is effective with a wide range of substrates; instead, catalyst and conditions must be tuned for each reaction. The rationale for enantioselectivity has been probed by theory and experiment, but much work remains to unravel the details of the chemistry.

3
Transition Metal-Catalyzed Asymmetric Allylic Alkylations of Enolates

Using palladium metal complexed with chiral ligands, allylic alkylations of stabilized enolates have been developed in recent years into highly enantioselective processes [31]. The majority of the work in this area has focused on selectivity at the electrophilic allylic center, not at the prochiral enolate center, Eq. (9).

$$(9)$$

As outlined in the Trost and van Vranken review [31], selectivity can result from a number of different processes, depending on the substrate. In general, a π-allyl-metal-ligand complex is formed, with the chiral ligand binding such that it can influence enantioselectivity at the allylic center. Three X-ray crystal structures have been published of π-allyl-palladium complexes coordinated with chiral ligands which shed light on the enantioselective process [32, 33]. The structure of a Chiraphos-palladium complex with a triarylallyl substrate indicates that the phosphane ligand has steric interactions with the phenyl groups of the allyl moiety, creating a chiral environment around the allyl group such that enantioselectivity can be induced at the allylic center [32]. In the other examples, sparteine was complexed with either cyclohexenyl or 1,1,3-triphenylallyl groups. In both cases palladium complexes with one face of the allyl group, although the orientation of the cyclohexenyl and triarylallyl groups was different for the two substrates. In all three cases the palladium atom and the chiral ligand block one face of the allyl group, leaving the other face open for nucleophilic attack. From these examples, one can see that the incoming nucleophile would have little or no interaction with the chiral ligand, since they are on opposite sides of the allyl group. Thus, expectations are low that conventional chiral ligands could induce asymmetry at the nucleophilic center.

Kagan was the first to study reactions in which enantioselectivity at a prochiral nucleophile was examined. In the reaction of 2-acetyltetralone with allylic ethers in the presence of a chiral DIOP-Pd catalyst, Eq. (10), the allylated products were obtained with ee's of only 10% [34].

$$(10)$$

This reinforced the expectation that simple chiral ligands would not be able to produce useful levels of enantioselectivity at the nucleophilic center. Since then a number of workers have devised chiral ligands specifically designed for inducing enantioselectvivity at the nucleophilic center of prochiral nucleophiles in allylic alkylations [35]. In 1982 Kumada and coworkers constructed a series of ligands with a chiral group remote from the phosphane ligand such that the chiral group could possibly interact with nucleophilic enolate [36]. The best ligand for the allylation of the sodium salt of 2-acetylcyclohexanone was 3, which afforded a 52% ee for a reaction at –50 °C. For comparison the DIOP ligand only gave a 2% ee. The enantioselectivity was rationalized assuming the amide and methoxycarbonyl groups chelate to the sodium ion, which essentially connects the nucleophile to the asymmetric Pd complex and provides an increased ability to differentiate between the enantiotopic faces of the nucleophile.

(+)-DIOP   3

This concept was further expanded to the design of ferrocenylphosphane ligands which possessed functional groups designed to interact with the approaching nucleophile. The design of these ligands differed from the previous example since the phosphane groups are at the center of asymmetry in the catalyst while the pendant group which is to interact with the counterion is achiral. With catalyst 4, an ee of 81% [37] (later revised to 70% [38]) was reported for allylation of 2-acetylcyclohexane. Interestingly, 2-acetylcyclopentanone provided an ee of <5%. The acyclic substrates, phenylacetaldehyde (53% ee) and benzoylacetone (60% ee) gave reasonable enantioselectivities, indicating that a cyclic substrate is not required. The fused-ring substrate, 2-acetyltetralone, gave the highest ee in the series of 82%. Hydrogen-bonding between the hydroxy group on the catalyst ligand and the enolate oxygen was proposed to explain the high selectivity. A X-ray structure of the π-allylpalladium complex with the hydroxylated ligand 4 provided support for this hypothesis [39]. In this structure (5) the hydroxy group is located over the allyl group and is close to one of the allyl carbons; thus, coordination to the metal cation would bring the nucleophile into close proximity to one of the allyl carbon centers.

The idea of using a secondary interaction with the counterion to connect the nucleophile and electophile was taken a step further in a catalyst design that included an aza-crown ether appended to the chiral ferrocenylphosphane ligand, 6. As in the previous example, the chirality of the ligand is in close proximity to the palladium binding site, while the pendant group is achiral. Introduction of the crown ether was designed to enhance coordination to the counterion. A ternary complex including the crown ether, potassium cation, and the enolate anion was proposed in which the bulky crown ether blocks approach of the enolate to C1 and provides a chiral pocket around carbon C3 [40]. Enantioselectivities up to 75% were reported [40], which were later revised to 65% [38].

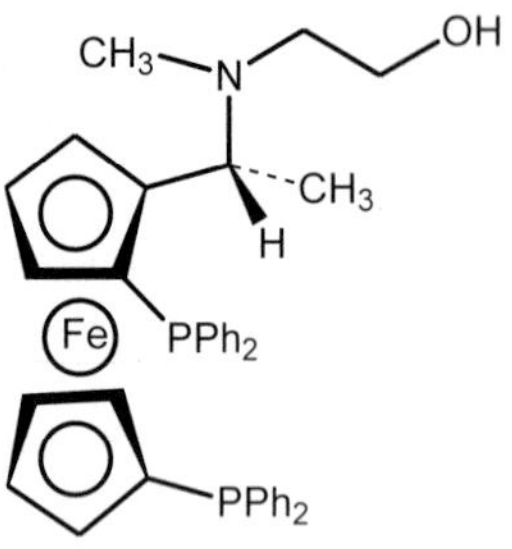

4

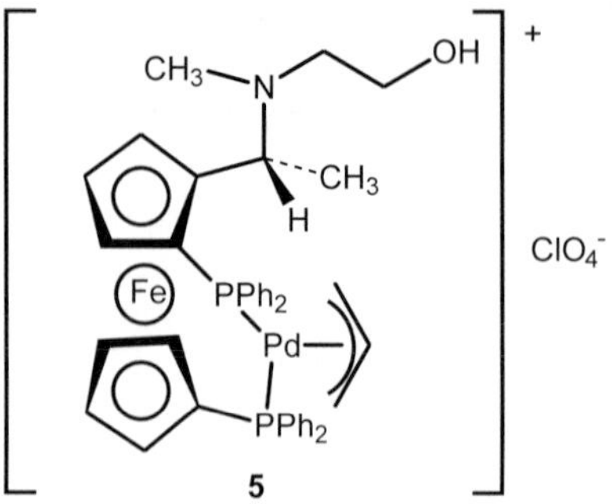

5

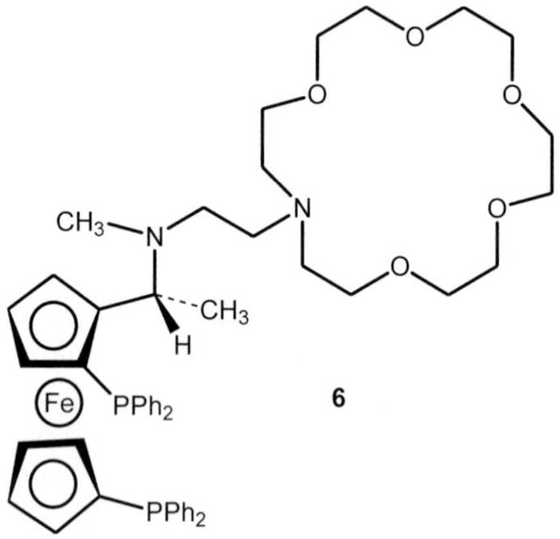

6

Allylations of α-nitroketones and α-nitroesters were also carried out with the crown ether ligands using fluoride salts as base. Moderate ee's in the 40–50% range were obtained using rubidium fluoride and ferrocenylphosphane ligands bearing monoaza-15-crown-5 or 18-crown-6 pendants. With the nitroesters, enantioselectivity increased with increased steric hindrance of the ester alkyl group, with an 80% ee realized in the best case using rubidium perchlorate as co-catalyst, which was proposed to increase the ratio of cation bound ligand to free ligand [41].

Genet has studied the enantioselective allylation of acyclic Shiff bases. Given the discussion and results above, one would not expect significant selectivity to result from use of simple chiral ligands. However, a respectable ee of 57% was de-termined for the reaction at −60 °C using chiral DIOP as ligand with the lithium enolate of the imine, Eq. (11) [42]. Further work indicated that use of two moles of DIOP vs. Pd, use of palladium acetate as catalyst precursor, and lithium hex-amethyldisilazide instead of LDA as base provided an increase in ee to 68% [43, 44]. When the amount of DIOP ligand was reduced to 1 mol equiv. vs. Pd, the op-posite enantiomer was formed, but with lower enantioselectivity [45]. No expla-nation of this unusual finding was given.

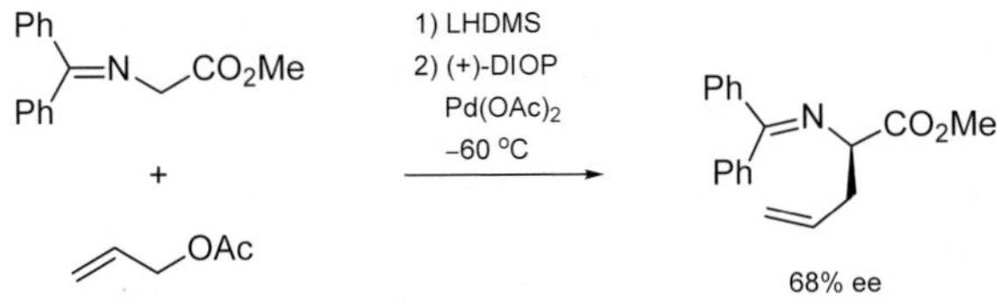

$$(11)$$

68% ee

Enantioselective allylic alkylation of α-cyanoesters was accomplished using a two-component system of Pd and Rh and the chiral ferrocenyl ligand, PhTRAP (7) [46]. The cyanopropanoate is coordinated to the rhodium atom, which is also coordinated to the chiral ligand, via the cyano nitrogen atom, while the allyl group is activated by forming the π-allyl complex with palladium, as depicted in the transition state structure 8. Nucleophilic attack of the enolate on the allyl complex then occurs enantioselectively to produce the allylated product. Ee's up to 99% were obtained using allyl hexafluoro-2-propyl carbonate and the electon-rich Anis-TRAP ligand at –40 °C, Eq. (12).

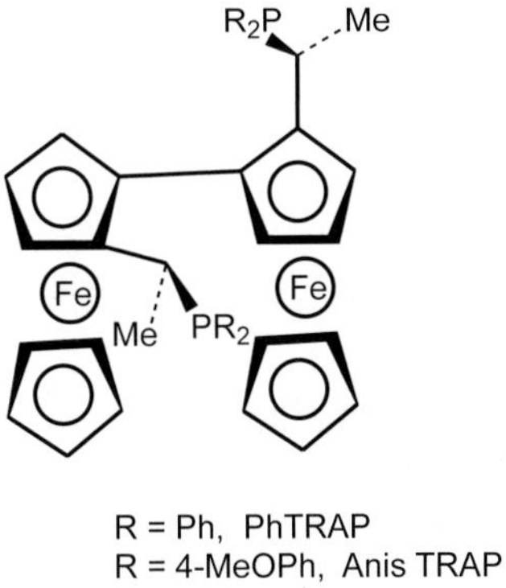

R = Ph, PhTRAP
R = 4-MeOPh, Anis TRAP

7

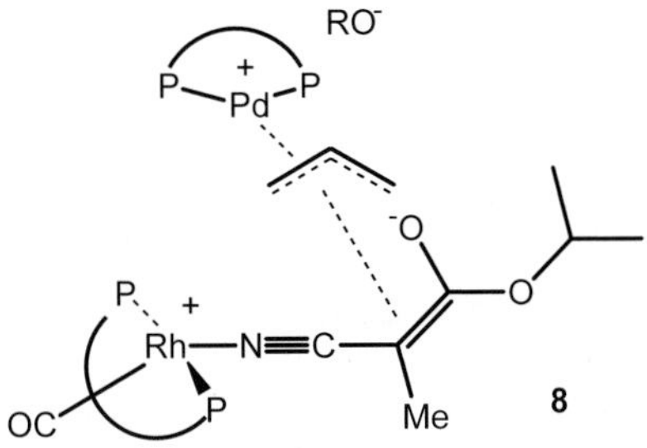

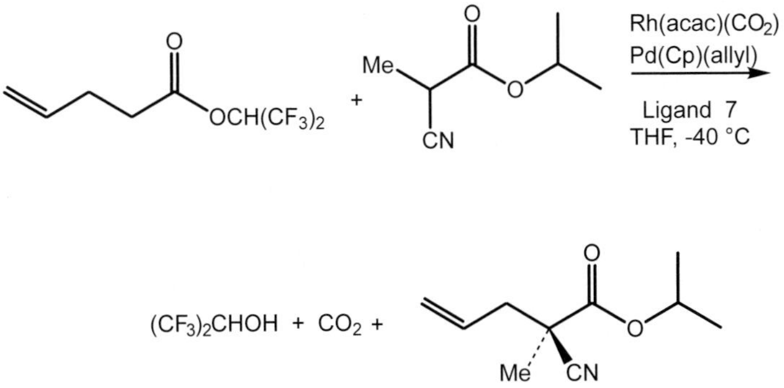

$$(12)$$

A different approach has been recently reported by the Trost group, which involves investigating whether a chiral pocket can transfer its chirality to the nucleophile just through geometric constraints [47]. Success was achieved using the diphosphane catalyst 9 with a series of β-ketoesters. Enantioselectivity of 86% ee was obtained for 2-alkoxycarbonylcyclohexanone, Eq. (13), the first time a high ee has been observed for a ketoester substrate. The examples shown previously in this section were all diketones or imine derivatives. Higher ee's (90–95%) were obtained for the reactions involving the more rigid tetralone substrates.

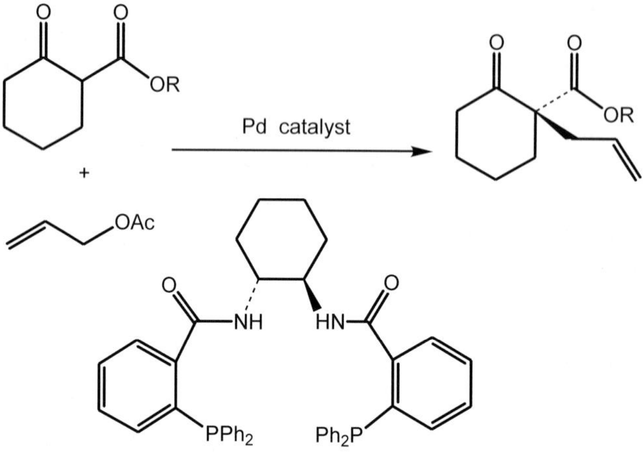

(13)

4
Transition Metal-Catalyzed Asymmetric Arylation of Enolates

Palladium-catalyzed regioselective arylation of enolates was disclosed by Hartwig [48] and Buchwald [49] in 1997, closely followed by a report by Buchwald that high levels of enantioselectivity could be attained in this catalytic process [50]. The enolate is generated using NaOt-Bu in toluene, while the arylation is accomplished using an aryl bromide, chiral BINAP, and either Pd₂(dba)₃ or Pd(OAc)₂ as catalyst. Using 2-methytetralone as substrate, ee's ranged from 61–88% with variously substituted aryl bromides, Eq. (14).

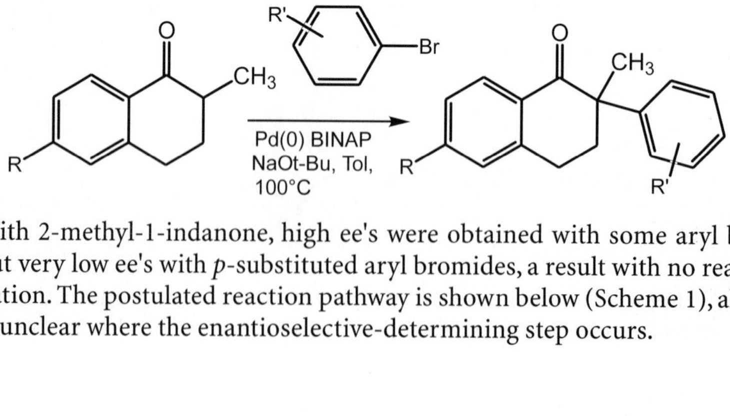

(14)

With 2-methyl-1-indanone, high ee's were obtained with some aryl bromides, but very low ee's with p-substituted aryl bromides, a result with no ready explanation. The postulated reaction pathway is shown below (Scheme 1), although it is unclear where the enantioselective-determining step occurs.

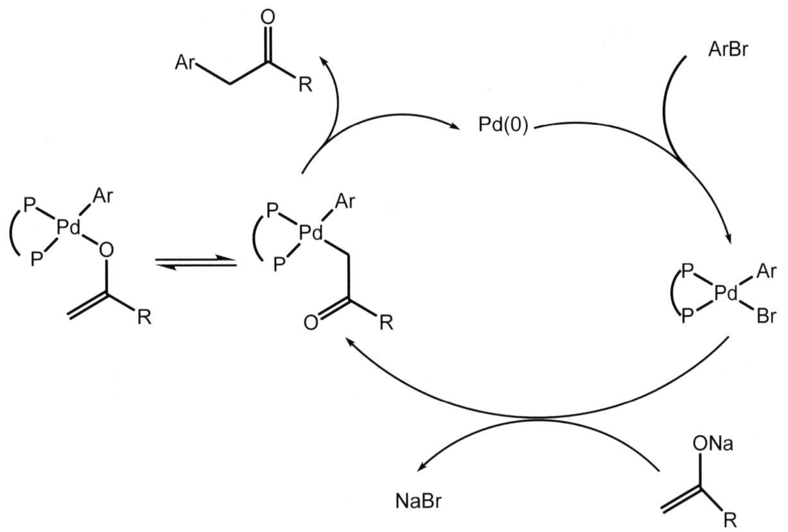

Scheme 1. Catalytic cycle for enolate arylation

5
Asymmetric Alkylation of Enolates Using Chiral Ligands

Koga has pioneered the use of chiral amine ligands in the enantioselective reactions of lithium enolates with achiral electrophiles [51]. In the initial work, a full equivalent of the chiral ligand **10** was used for the benzylation of cyclohexanone and 1-tetralone, Eq. (15) [52]. Rationale for the enantioselectivity was coordination of the lithium ion by the tridentate ligand which provides a chiral environment around the enolate, leading to an enantioselective reaction.

The lithium enolate was generated from the corresponding silyl enol ether using MeLi, then the ligand was added to complex the lithium ion, followed by reaction with the alkyl halide. Since the ee increased as the reaction proceeded, LiBr, which is formed during the reaction, was hypothesized to be important in the enantioselection. This was proven by addition of LiBr initially, which gave an overall higher ee and resulted in a constant ee over the course of reaction. When the chiral amine was added at 20 mol %, less than 1% reaction occurred. It was proposed that the chiral ligand was inactivated by complexation with LiBr. Thus, LiBr was shown to be necessary for high enantioselectivity, yet prevented use of the ligand catalytically. This constraint was overcome by addition of achiral diamines, such as N,N,N',N'-tetramethylethylenediamine or N,N,N',N'-tetramethylpropylenediamine, which were added as traps for LiBr via 5- or 6-membered chelates of Li+ [53]. An optimized ee of 96% was realized using just 5 mol % of catalyst **11** and 2 equiv. of tetramethylpropylenediamine in toluene. That high ee's can be obtained in the presence of a large amount of an achiral amine indicates that the chiral amine-lithium enolate complex must be significantly more

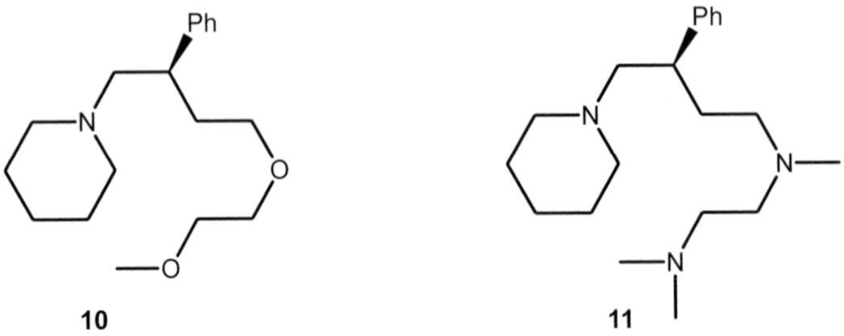

10   **11**

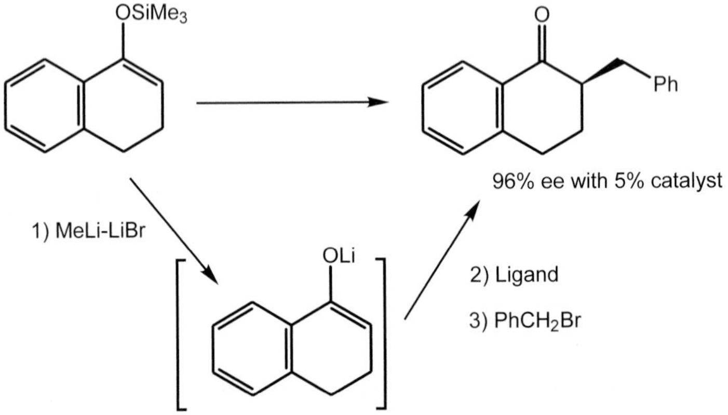

(15)

96% ee with 5% catalyst

1) MeLi-LiBr

OLi

2) Ligand

3) PhCH₂Br

reactive than the bidentate achiral ligand-lithium enolate complex. The catalytic reaction was successfully extended to cyclohexanone enolate, with a yield of 52% and an ee of 90%.

6
Deprotonation by Chiral Lithium Amide Bases

Deprotonation of carbonyl compounds by chiral lithium bases to generate chiral lithium enolates is another method toward asymmetric alkylation of enolates. This method is similar to that discussed in the previous section, in that both generate lithium enolates with the lithium ion complexed to a chiral amine base. In the method described in the previous section, the lithium enolate is generated from the corresponding silyl enol ether using MeLi, then a neutral tertiary amine ligand is added. This differs from chiral deprotonation in that a chiral amide base derived from a secondary amine is used to generate the lithium enolate, and the resulting protonated secondary amine then becomes the chiral ligand associated with the lithium enolate.

The first successful chiral deprotonation was reported in 1980 by Whitesell, who found that epoxides could be deprotonated with chiral amide bases to generate optically active allylic alcohols with ee's up to 31% [54]. In 1986 Simpkins and Koga independently reported stoichiometric asymmetric deprotonations of ketones. Koga studied the deprotonation of prochiral 4-alkyl-cyclohexanones using a series of chelating amide bases, Eq. (16), with the best ee of 97% at −105 °C using 4-*t*-butylcyclohexanone [55]. The use of HMPA increased enantioselectivity, which was shown by NMR to be due to the ability of HMPA to convert the dimeric form to the more reactive monomeric form [56].

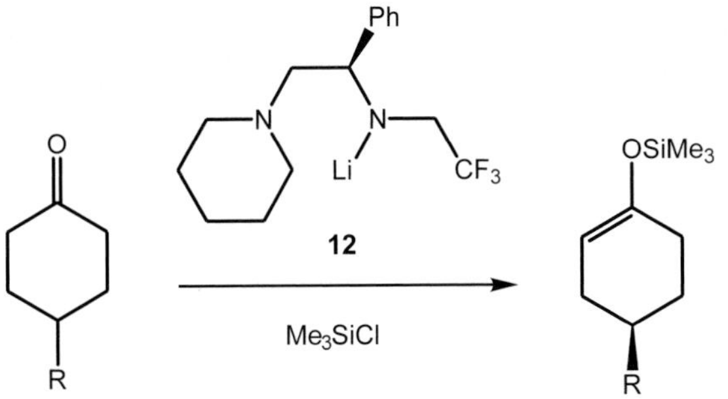

(16)

Simpkins examined the deprotonation of 2,6-dimethylcyclohexanone using a series of chiral amide bases. Enantioselectivity up to 74% was achieved using the bicyclic base **13**, Eq. (17) [57].

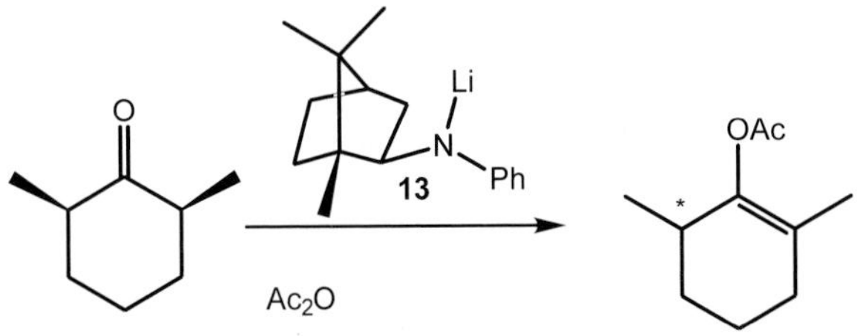

(17)

Koga has recently devised a catalytic asymmetric deprotonation of 4-substituted cyclohexanones using the same substrates and amide bases shown in Eq. (15) [58]. The strategy conceived by Koga is as follows. Since deprotonation of carbonyl compounds is thought to involve coordination of the carbonyl oxygen to lithium, a tridentate base should be inferior to a bidentate base, since the former has an additional coordination ligand that will prevent coordination of lithium to the carbonyl group. The idea was then to use a catalytic amount of chiral bidentate ligand along with a large excess of an achiral tridentate ligand. The chiral base will serve to asymmetrically deprotonate the carbonyl group, and the achiral base will react with the resulting protonated chiral amine to regenerate the chiral base due to the greater acidity of the chiral base induced by the electron-withdrawing trifluoromethyl group (Scheme 2). With 30 mol % chiral base, chemical yields of 83% with 79% ee were achieved.

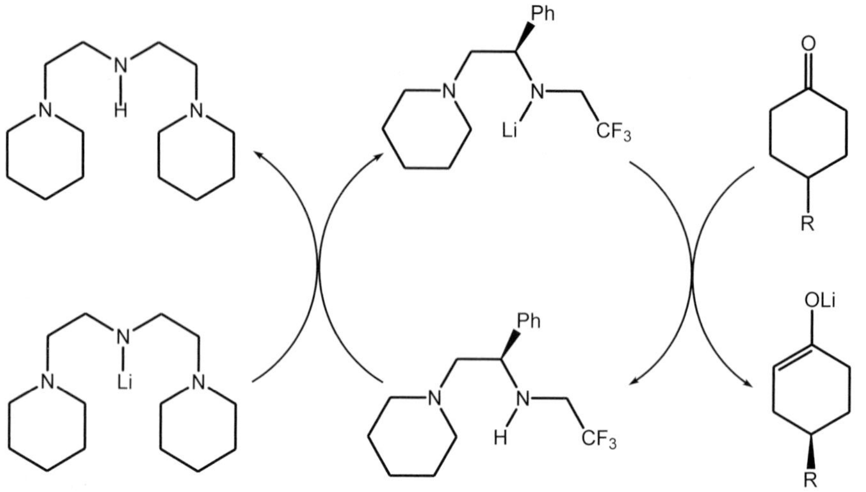

Scheme 2

7
Enantioselective α-Alkylation of Carbonyl Groups via Free Radicals

Within the past decade, diastereosolective radical reactions have become feasible and the factors contolling selectivity defined. Chiral auxiliaries for radical reactions have been recently developed in analogy to those developed for carbanion chemistry in the 1970s and 1980s. The first example of stoichiometric use of a chiral ligand for enantioselective radical additions was recently reported by Porter and coworkers [59, 60, 61]. Reaction of the amide **14** with allyltrimethylsilane at –78 °C, initiated by triethylborane, in the presence of 1 equiv. each of zinc triflate and the chiral bidentate ligand **15,** provided the allylated product in a yield up to 88% and ee of 90%, Eq. (18). The presumed intermediate is the α-keto radical complexed to the chiral Lewis acid.

The first catalytic asymmetric radical-mediated allylation was reported in late 1997 by Hoshino and coworkers, who studied the allylation of an α-iodolactone substrate, Eq. (19) using trimethylaluminum as Lewis acid and a silylated binaphthol as the chiral catalyst, with triethylborane as radical initiator [62]. Use of one equiv. of diethyl ether was crucial for high enantioselectivity, providing an ee up to 91% in the presence of one equiv. of catalyst, with only a 27% ee in the absence of ether, and poorer ee's when other ethers were employed. In the catalytic version, the ee's dropped off vs. the stoichiometric reaction, with an ee of 81% with 0.5 equiv., and 80% with 0.2 equiv., and 72% with 0.1% catalyst. As in the above example, the presumed chiral intermediate involves complexation of the lactone radical with the Lewis acid-binaphthol complex, with the diethyl ether perhaps as a ligand on the aluminum.

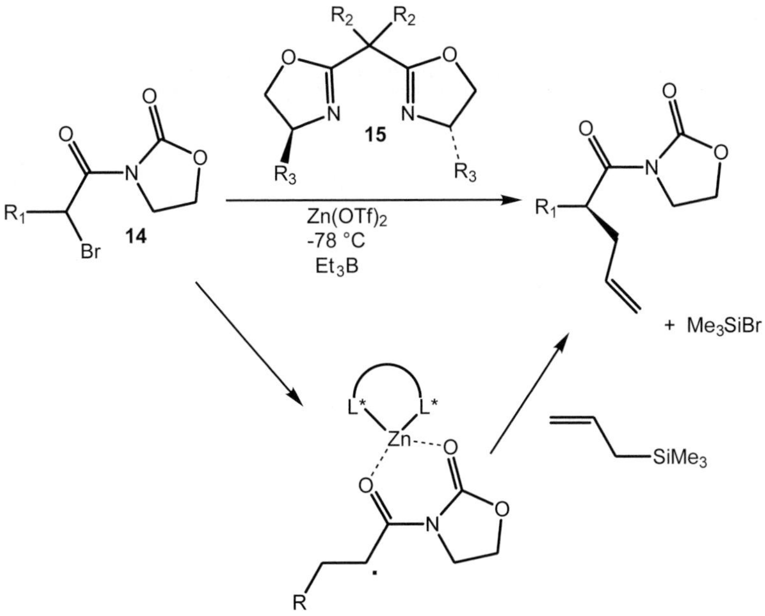

(18)

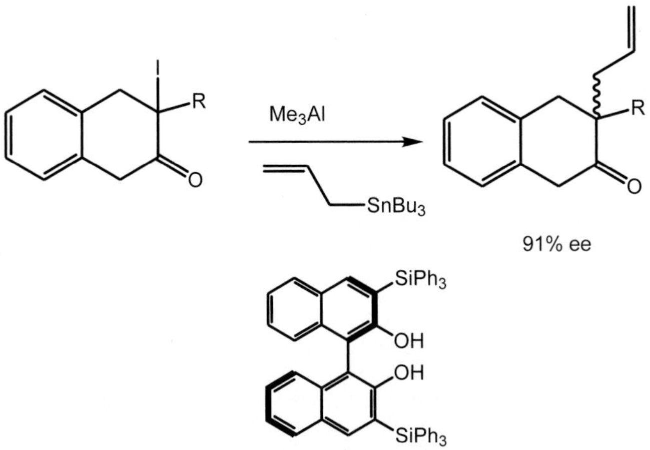

(19)

8
Enantioselective Alkylation via Cyclopropanation of Silyl Enol Ethers

Reissig and coworkers have devised an indirect method of enantioselective alkylation of ketones via cyclopropanation of silyl enol ethers in the presence of the chiral copper catalyst **16**, followed by ring opening to provide the substituted ketones. Overall, the transformation corresponds to alkylation of ketones using methyl diazoacetate as the electrophile. Enantioselectivities up to 88% were realized in the cyclopropanation of aryl substituted olefins, Eq. (20) [63, 64].

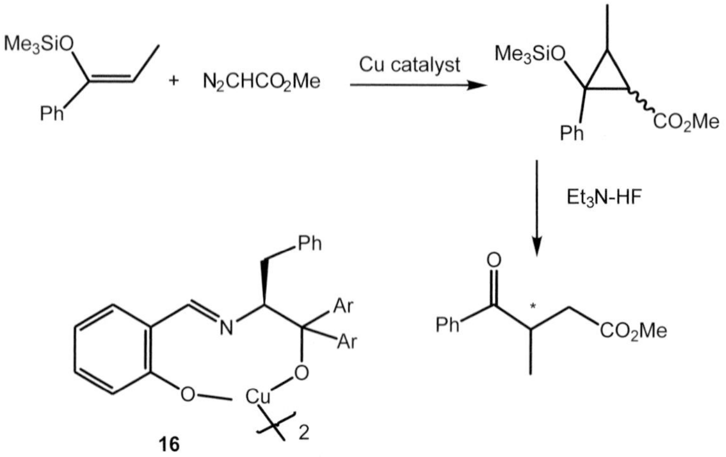

(20)

References

1. House HO. Trost BM (1965) J Org Chem 30:1341, 2502; House HO, Tefertiller BA, Olmstead HD (1968) J Org Chem 33:935; House HO, Bare TM (1968) J Org Chem 33:943
2. Kenyon WG, Meyer RB, Hauser CR (1963) J Org Chem 28: 3108; von Schriltz DM, Hampton KG, Hauser CR (1969) J Org Chem 34:2509
3. Meyers AI, Williams DR, Druelinger M (1976) J Am Chem Soc 98:3032; Meyers AI (1979) Pure Appl Chem 51:1255
4. Hashimoto S, Koga K (1978) Tetrahedron Lett 573; Tomioka K, Ando K, Takemasa Y, Koga K (1984) J Am Chem Soc 106:2718
5. Whitesell JK, Whitesell MA (1977) J Org Chem 42:377
6. Enders D, Eichenauer H (1979) Chem Ber 112:2933; Enders D (1981) Chemtech 504
7. Review: Ager DJ, Prakash I, Schaad DR (1996) Chem Rev 96:835
8. Pluim H, Wynberg H (1980) J Org Chem 45:2498
9. Hermann K, Wynberg H(1979) J Org Chem 44:2238.
10. Fiaud JC (1975) Tetrahedron Lett 3495
11. Saigo K, Koda H, Nohira H(1979) Bull Chem Soc Jpn 52:3119
12. Julia S, Ginebreda A, Guixer J, Tomas A (1980) Tetrahedron Lett. 21:3709
13. Dehmlow EV, Singh PT, Heider J (1981) J Chem Res Syn 292
14. Dolling U-H, Davis P, Grabowski EJJ(1984) J Am Chem Soc 106:446; Dolling U-H, Hughes DL, Bhattacharya A, Ryan KM, Karady S, Weinstock LM, Grabowski EJJ (1987) In: Phase Transfer Catalysis. (ACS Symposium Series: 326), Starks CM (ed) American Chemical Society, Washington, DC, pp 67–81
15. Hughes DL, Dolling U-H, Ryan KM, Schoenewaldt EF, Grabowski EJJ (1987) J Org Chem 52:4745
16. Dolling U-H, Hughes DL, Bhattacharya A, Ryan KM, Karady S, Weinstock LM, Grenda VJ, Grabowski EJJ (1988) In: Catalysis of Organic Reactions, Rylander PN, Greenfield H, Augustine RL (eds), Dekker, New York, pp 65–86
17. Bhattacharya A, Dolling U-H, Grabowski EJJ, Karady S, Ryan KM, Weinstock LM (1986) Angew Chem Int Ed Engl 25:476
18. Lipkowitz KB, Cavanaugh MW, Baker B, O'Donnell MJ (1991) J Org Chem 56:5181
19. Nerinckx W, Vandewalle M(1990) Tetrahedron:Asymmetry 1: 265
20. Lee TBK, Wong GSK (1991) J Org Chem 56:872
21. O'Donnell M J, Bennett WD, Wu S (1989) J Am Chem Soc 111:2353

22. Esikova IA, Nahreini TS, O'Donnell MJ (1997) In: Phase-Transfer Catalysis (ACS Symposium Series); Halpern M (ed), American Chemical Society: Washington, DC, pp 89–96
23. O'Donnell MJ, Bennett WD, Bruder WA, Jacobsen WN, Knuth K, LeClef B, Polt RL, Bordwell FG, Mrozack SR, Cripe TA (1988) J Am Chem Soc 110:8520
24. (a) Tohdo K, Hamada Y, Shiori T (1994) Synlett 247; (b) DeLombaert S, Blanchard L, Tan J, Sakane Y, Berry C, Ghai RD (1995) Biorg Med Chem Lett 5:145; (c) Dehmlow EV, Nachstedt I (1993) J prakt Chem 335:371; (d) Kim MH, Lai JH, Hangauer DG (1994) Int J Peptide Res 44:457; (e) Imperiali B, Fisher SL (1992) J Org Chem 57:757; (f) Imperiali B, Prins TJ, Fisher SL (1993) J Org Chem 58:1613; (g) Imperiali B, Roy RS(1994) J Am Chem Soc 116:12083; (h) Imperiali B, Roy RS (1995) J Org Chem 60:1891; (i) Torrado A, Imperiali B (1996) J Org Chem 61:8940; (j) Pirrung MC, Krishnamurthy N (1993) J Org Chem 58:957
25. O'Donnell MJ, Wu S (1992) Tetrahedron:Asymmetry 3:591
26. O'Donnell MJ, Esikova IA, Mi A, Shullenberger DF, Wu S (1996) In: Phase-Transfer Catalysis (ACS Symposium Series), Halpern M (ed.), American Chemical Society, Washington, DC
27. (a) O'Donnell MJ, Wu S, Huffman JC (1994) Tetrahedron 50:4507; (b) Corey EJ, Xu F, Noe MC (1997) J Am Chem Soc 119:12414; (c) Lygo B, Wainwright PG (1997) Tetrahedron Lett 38:8595
28. Dehmlow EV, Schrader S (1994) Pol J Chem 68:2199; Dehmlow EV, Knufinke V (1992) Liebigs Ann Chem 283
29. Belokon' YN (1992) Pure Appl Chem 64:1917
30. Eddine JJ, Cherqaoui M (1995) Tetrahedron:Asymmetry 6:1225
31. Review: Trost BM, van Vranken DL (1996) Chem Rev 96:395
32. Farrar DH, Payne NC (1985) J Am Chem Soc 107:2054
33. Togni A, Rihs G, Pregosin PS, Ammann C (1990) Helv Chim Acta 73:723
34. Fiaud JC, de Gournay H, Larcheveque M, Kagan HB (1978) J Organomet Chem 154:175
35. Review: Sawamura M, Ito Y (1992) Chem Rev 92:857
36. Hayashi T, Kanehira K, Tsuchiya H, Kumada M (1982) J Chem Soc Chem Commun 1162
37. Hayashi T, Kanehira K, Hagihara T, Kumada M (1988) J Org Chem 53:113
38. Sawamura M, Nakayama Y, Tang W-M, Ito Y (1996) J Org Chem 61:9090
39. Hayashi T, Yamamoto A, Ito Y, Nishioka E, Miura H, Yanagi K (1989) J Am Chem Soc 111:6301
40. Sawamura M, Nagata H, Sakamoto H, Ito Y (1992) J Am Chem Soc 114:2586
41. Sawamura M, Nakayama Y, Tang W-M, Ito Y (1996) J Org Chem 61:9090
42. Genet J-P, Ferroud D, Juge S, Montes JR (1986) Tetrahedron Lett 27:4573
43. Genet J-P, Juge S, Achi S, Mallart S, Montes JR, Levif G (1988) Tetrahedron 44:5263
44. Genet J-P, Juge S, Besnier I, Uziel J, Ferroud D, Kardos N, Achi S, Ruiz-Montes J, Thorimbert S (1990) Bull Soc Chim Fr 127, 781
45. Genet J-P, Juge S, Montes JR, Gaudin J-M(1988) J Chem Soc Chem Commun 718
46. Sawamura M, Sudoh M, Ito Y (1996) J Am Chem Soc 118:3309
47. Trost BM; Radinov R, Grenzer EM(1997) J Am Chem Soc 119:7879; Trost BM (1996) Proc Robert A Welch Found Conf Chem Res 40:13
48. Hamann BC, Hartwig JF (1997) J Am Chem Soc 119:12382
49. Palucki M, Buchwald SL (1997) J Am Chem Soc 119:11108
50. Ahman J, Wolfe JP, Thoutman MV, Palucki M, Buchwald SL (1998) J Am Chem Soc 120:1918
51. Review: Koga K (1994) Pure Appl Chem 66:1487
52. Murakata M, Nakajima M, Koga K (1990) J Chem Soc Chem Commun 1657
53. Imai M, Hagihara A, Kawasaki H, Manabe K, Koga K (1994) J Am Chem Soc 116:8829
54. Whitesell JK, Felman SW (1980) J Org Chem 45:755
55. Shirai R, Tanaka M, Koga K (1986) J Am Chem Soc 108:543; Shirai R, Sato D, Aoki K, Tanaka M, Kawasaki H, Koga K (1997) Tetrahedron 53:5963; Aoki K, Tomioka K, Noguchi H, Koga K (1997) Tetrahedron 53:13641
56. Sato D, Kawasaki H, Shimada I, Arata Y, Okamura K, Date T, Koga K (1992) J Am Chem Soc 114:761

57. Simpkins NS (1986) J Chem Soc Chem Commun 88
58. Yamashita T, Sato D, Kiyoto T, Kumar A, Koga K (1996) Tetrahedron Lett 37:8195
59. Wu JH, Radinov R, Porter NA (1995) J Am Chem Soc 117:11029
60. Wu JH, Zhang G, Porter NA (1997) Tetrahedron Lett 38:2067
61. Porter NA, Wu JH, Zhang G, Reed AD (1997) J Org Chem 62:6702
62. Murakata M, Jono T, Mizuno Y, Hoshino O (1997) J Am Chem Soc 119:11713
63. Kunz T, Reissig H-U (1989) Tetrahedron Lett 30:2079
64. Dammast F, Reissig H-U (1993) Chem Ber 126:2449, 2727

Chapter 34.2
Protonation of Enolates

Akira Yanagisawa · Hisashi Yamamoto

Graduate School of Engineering, Nagoya University, CREST, Japan Science and Technology
Corporation (JST), Chikusa, Nagoya 464–8603, Japan
e-mail: j45988a@nucc.cc.nagoya-u.ac.jp

Keywords: Protonation, Metal enolates, Chiral proton sources, Achiral proton sources

1
Introduction

Asymmetric protonation of enols or enolates is an efficient route as is asymmetric alkylation of enolates to prepare carbonyl compounds which possess a tertiary asymmetric carbon at the α-position (Scheme 1). Numerous successful methods have been developed and applied to organic synthesis. Several reviews of asymmetric protonation have been published [1, 2, 3, 4, 5] and the most recent [4, 5] describe the work in detail up to early in 1995. This chapter is focussed on enantioselective protonation of prochiral metal enolates by a catalytic amount of chiral proton sources. Compounds **1** to **40** [6, 7, 8, 9, 10, 11,12, 13, 14, 15, 16, 17, 18, 19, 20, 21, 22, 23, 24, 25, 26, 27, 28, 29, 30, 31, 32, 33, 34, 35, 36, 37, 38, 39, 40, 41, 42, 43, 44, 45, 46, 47, 48, 49, 50, 51, 52, 53, 54, 55, 56, 57, 58, 59, 60, 61, 62]

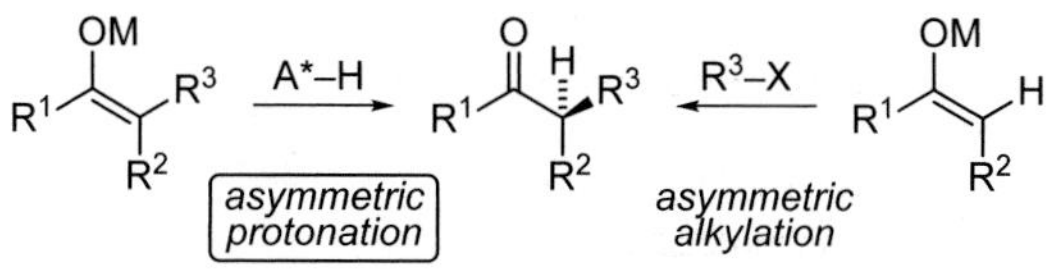

Scheme 1

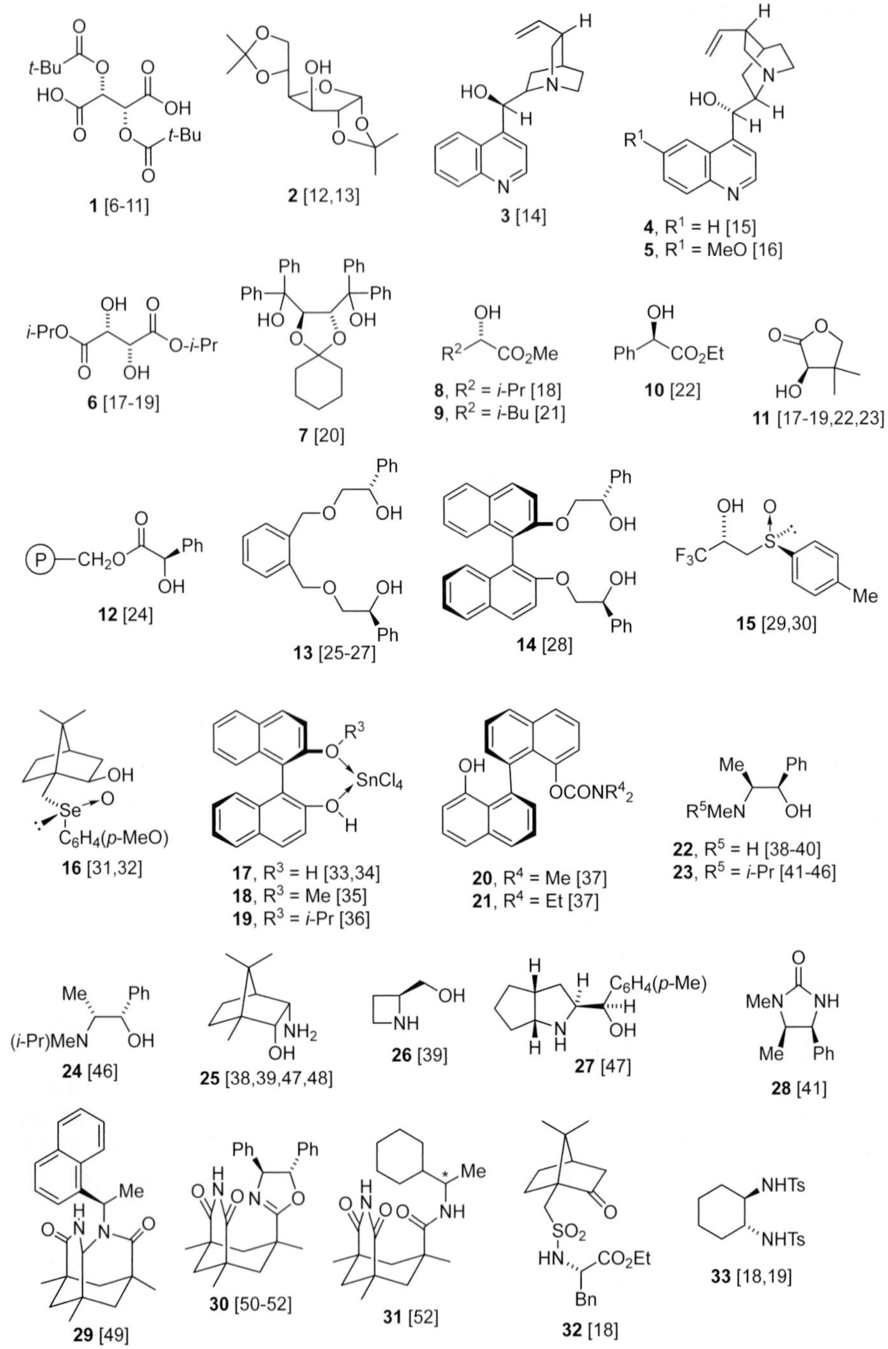

1 [6-11] **2** [12,13] **3** [14] **4**, R^1 = H [15] **5**, R^1 = MeO [16]

6 [17-19] **7** [20] **8**, R^2 = i-Pr [18] **9**, R^2 = i-Bu [21] **10** [22] **11** [17-19,22,23]

12 [24] **13** [25-27] **14** [28] **15** [29,30]

16 [31,32] **17**, R^3 = H [33,34] **18**, R^3 = Me [35] **19**, R^3 = i-Pr [36] **20**, R^4 = Me [37] **21**, R^4 = Et [37] **22**, R^5 = H [38-40] **23**, R^5 = i-Pr [41-46]

24 [46] **25** [38,39,47,48] **26** [39] **27** [47] **28** [41]

29 [49] **30** [50-52] **31** [52] **32** [18] **33** [18,19]

Fig. 1

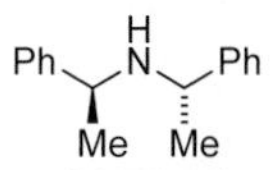

Ph N H Ph
Me Me
34 [53,54]

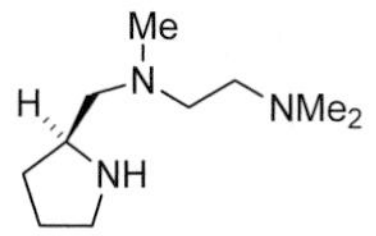

Me
N
H
NH
NMe₂
35 [55,56]

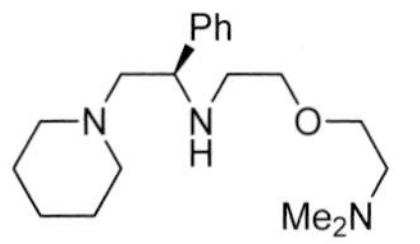

Ph
N
N
H
O
Me₂N
36 [57,58]

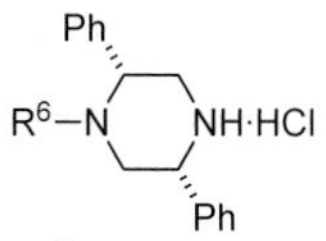

Ph
R⁶—N
NH·HCl
Ph
37, R⁶ = H [59]
38, R⁶ = t-Bu(CH₂)₂ [59]

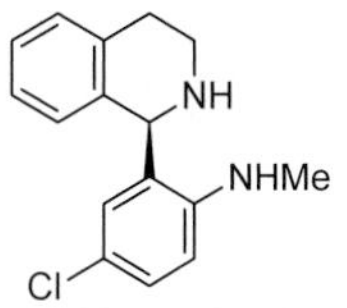

NH
NHMe
Cl
39 [40,56,60]

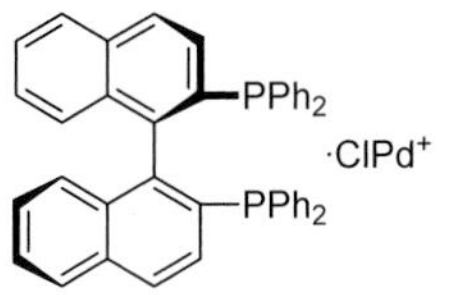

PPh₂
·ClPd⁺
PPh₂
40 [61,62]

Fig. 1 (continued)

shown in Fig. 1 are the chiral proton sources or chiral catalysts reported to date which have been successfully used for the stoichiometric protonation of metal enolates. Some of these have been used to realize the catalytic process in combination with achiral proton sources as described in Section 3. The rest also have great potential for utilization in catalytic protonation.

2
Mechanism of Catalysis

Asymmetric protonation of a metal enolate basically proceeds catalytically if a coexisting achiral acid A-H reacts with the deprotonated chiral acid A*-M faster than with the metal enolate, a concept first described by Fehr et al. [44]. A hypothesis for the catalytic cycle is illustrated in Scheme 2. Reaction of the metal enolate with the chiral acid A*-H produces (R)- or (S)-ketone and the deprotonated chiral acid A*-M. The chiral acid A*-H is then reproduced by proton transfer from the achiral acid A-H to A*-M. Higher reactivity of A*-M toward A-H than that of the metal enolate makes the catalytic cycle possible. When the achiral acid A-H protonates the enolate rapidly at low temperature, selective deprotonation of one enantiomer of the resulting ketone by the metallated chiral acid A*-M is seen as an alternative possible mechanism.

3
Catalytic Protonation of Metal Enolates

Catalytic enantioselective protonations of metal enolates already published can be roughly classified into two methods carried out under basic conditions and acidic conditions. The process under basic conditions is, for example, the protonation of reactive metal enolates such as lithium enolates with a catalytic amount of chiral acid and an excess of achiral acid. The process under acidic conditions employs silyl enol ethers or ketene silyl acetals as substrates. Under the influence

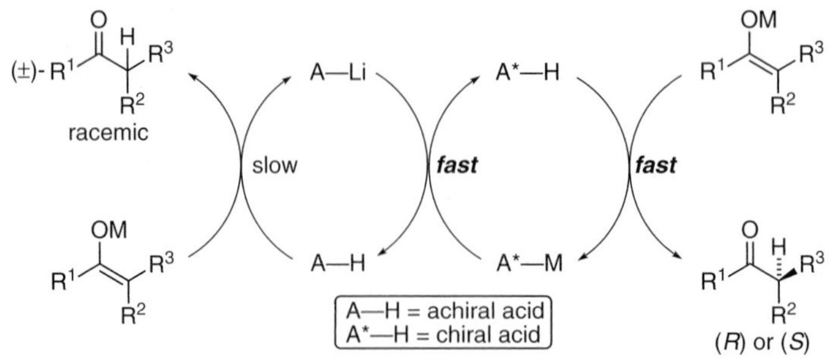

Scheme 2

of a chiral Lewis acid or chiral Brønsted acid catalyst, the silyl ethers are transformed into optically active carbonyl compounds by an achiral acid.

3.1
Protonation under Basic Conditions

The first example of catalytic enantioselective protonation of metal enolates was achieved by Fehr and coworkers (Scheme 3) [44]. They found the enantioselective addition of a lithium thiolate to ketene **41** in the presence of an equimolar amount of (–)-N-isopropylephedrine (**23**) with up to 97% ee. Based on the results, they attempted the catalytic version; for example, slow addition of p-chlorothiophenol to a mixture of ketene **41** (1 equiv) and lithium alkoxide of (–)-N-isopropylephedrine **23-Li** (0.05 equiv) gave thiol ester **43** with 90% ee. First, the thiol is deprotonated by **23-Li** to generate lithium p-chlorothiophenoxide and **23**. The thiophenoxide adds to the ketene **41** leading to Z-thiol ester enolate which is presumed to react with the chiral amino alcohol **23** via a four-membered cyclic transition state **42** to form the product **43** and **23-Li**. The lithium alkoxide **23-Li** is reused in the catalytic cycle. The key to success in the catalytic process is that the rate of introduction of thiophenol to a mixture of the ketene **41** and **23-Li** is kept low, avoiding the reaction of the thiol with the intermediate lithium enolate.

Later, the same group showed that an asymmetric protonation of preformed lithium enolate was possible by a catalytic amount of chiral proton source **23** and stoichiometric amount of an achiral proton source [45]. For instance, when lithium enolate **44**, generated from ketene **41** and n-BuLi, was treated with 0.2 equiv of **23** followed by slow addition of 0.85 equiv of phenylpropanone, (S)-enriched ketone **45** was obtained with 94% ee (Scheme 4). In this reaction, various achiral proton sources including thiophenol, 2,6-di-$tert$-butyl-4-methylphenol, H_2O, and pivalic acid were used to provide enantioselectivity higher than 90% ee. The pK_a value of the achiral acid must be smaller than that of **45** to accomplish a high level of asymmetric induction. The catalytic cycle shown in Scheme 2 is the possible mechanism of this reaction.

Our research group independently found a catalytic enantioselective protonation of preformed enolate **47** with (*S,S*)-imide **30** founded on a similar concept
(Scheme 5) [51]. The chiral imide **30**, which has an asymmetric 2-oxazoline ring
and is easily prepared from Kemp's triacid and optically active amino alcohol, is
an efficient chiral proton source for asymmetric transformation of simple metal
enolates into the corresponding optically active ketones [50]. When the lithium
enolate **47** was treated with a stoichiometric amount of the imide **30**, (*R*)-en-
riched ketone **48** was produced with 87% ee. By a ^{1}H-NMR experiment of a mixture of (*S,S*)-imide **30** and lithium bromide, the chiral imide **30** was found to
form a complex rapidly with the lithium salt. We envisaged that a catalytic asym-

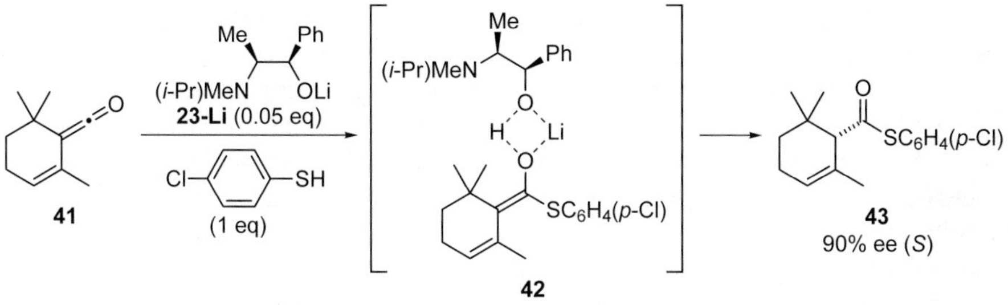

Scheme 3

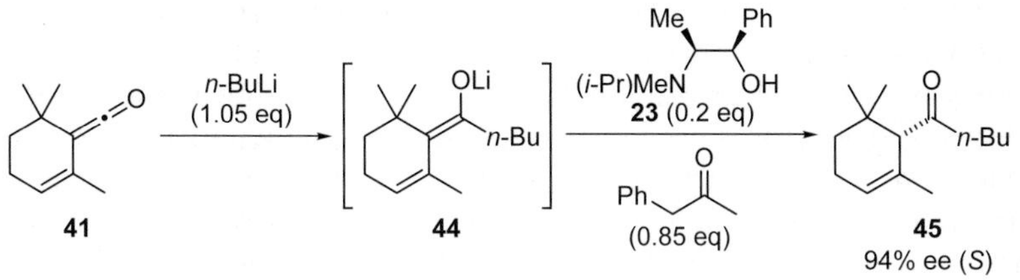

Scheme 4

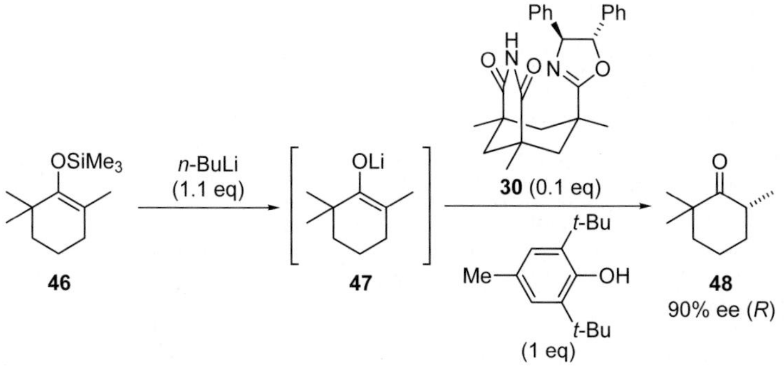

Scheme 5

metric protonation might be possible if the lithium enolate **47** forms a complex with (*S,S*)-imide **30** more rapidly than with the coexisting achiral proton source and the achiral acid is selectively deprotonated by the resulting lithiated (*S,S*)-imide. The catalytic reaction has been realized by addition of 0.1 equiv of (*S,S*)-imide **30** to the lithium enolate **47**, generated from the corresponding silyl enol ether **46** and *n*-BuLi, prior to the addition of a stoichiometric amount of an achiral acid over a period of 2 h. Among the achiral acids examined, 2,6-di-*tert*-butyl-*p*-cresol (BHT) gave the best result (90% ee, Scheme 5).

The aforementioned catalytic process was further applied to diastereoselective protonation of a chiral enolate of (–)-menthone (Scheme 6) [52]. When the lithium enolate **49** was quenched with BHT at –78 °C, an 86:14 mixture of *trans*-product **50** and *cis*-product **51** was obtained. Reaction of the enolate **49** with (*S*)-imide **31** (0.1 equiv), which was derived from (*S*)-1-cyclohexylethylamine, followed by slow addition of BHT (1 equiv) at the same reaction temperature furnished a higher *trans*-selectivity (**50:51**=95:5). The enantiomer of (*S,S*)-imide **30** showed a similar level of *trans*-selectivity, while *cis*-isomer **51** was produced as a major product (**50:51**=31:69) in reaction with (*S,S*)-imide **30**. This is an example of diastereoselective protonation in which a new stereogenic center is formed under the influence of a chiral proton source rather than of the asymmetric carbon of the enolate **49**.

The chiral tetradentate amine **36** was shown to be an efficient chiral source for enantioselective protonation of prochiral lithium enolate **53** by Koga's group [57]. The corresponding silyl enol ether **52** was treated with methyllithium-lithium bromide complex to generate the lithium enolate **53** containing LiBr. Asymmetric synthesis of 2-methyl-1-tetralone (**54**) was achieved with up to 91% ee by protonation of **53** with a stoichiometric amount of the chiral amine **36** and achiral Brønsted acid. LiBr is necessary for attaining high asymmetric induction in the reaction and a ternary complex formed from the enolate **53**, the chiral amine **36**, and LiBr is assumed to be a reactive intermediate. The catalytic version was achieved using 0.2 equiv of the amine **36** and a large excess of powdered

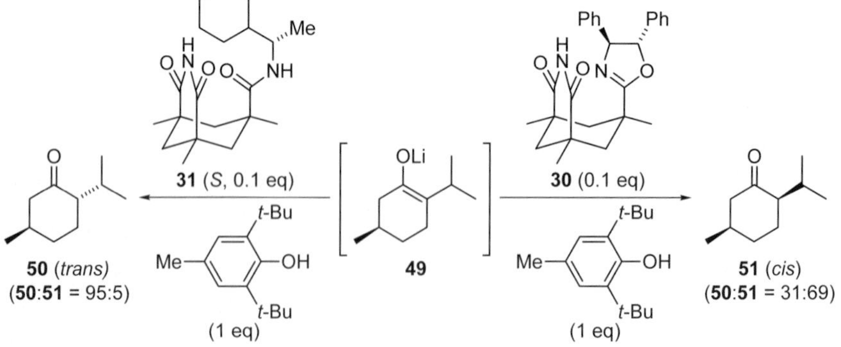

Scheme 6

succinimide, and thus (S)-enriched ketone **54** was obtained with 83% ee (Scheme 7) [58].

A C_2-symmetric homochiral diol **13** (DHPEX) is a chiral proton source developed by Takeuchi et al., for samarium enolates which are readily prepared by SmI_2-mediated allylation of ketenes [25, 26]. In the stoichiometric reaction using DHPEX **13**, they found that –45 °C was the best reaction temperature for the enantioface discrimination, e.g., when methyl (1-methyl-1-phenylethyl)ketene **55** was used as a substrate, the product exhibited 95% ee [27]. The catalytic reaction was carried out using trityl alcohol as an achiral proton source which was added to a mixture of *in situ* generated samarium enolate **56** and DHPEX **13** (0.15 equiv) slowly so as not to exceed the ratio of the achiral proton source to DHPEX **13** of more than 0.7. The highest ee (93% ee) of product **57** was gained when the achiral proton source was added over a period of 26 h (Scheme 8) [27].

Muzart and coworkers have succeeded in a catalytic asymmetric protonation of enol compounds generated by palladium-induced cleavage of β-ketoesters or enol carbonates under nearly neutral conditions [47, 48]. Among the various optically active amino alcohols tested, (+)-*endo*-2-hydroxy-*endo*-3-aminobornane (**25**) was effective as a chiral catalyst for the enantioselective reaction. Treatment of the β-ketoester of 2-methyl-1-indanone **58** with a catalytic amount of the amino alcohol **25** (0.3 equiv) and 5% Pd on charcoal (0.025 equiv) under bubbling of hydrogen at 21 °C gave the (R)-enriched product **59** with 60% ee

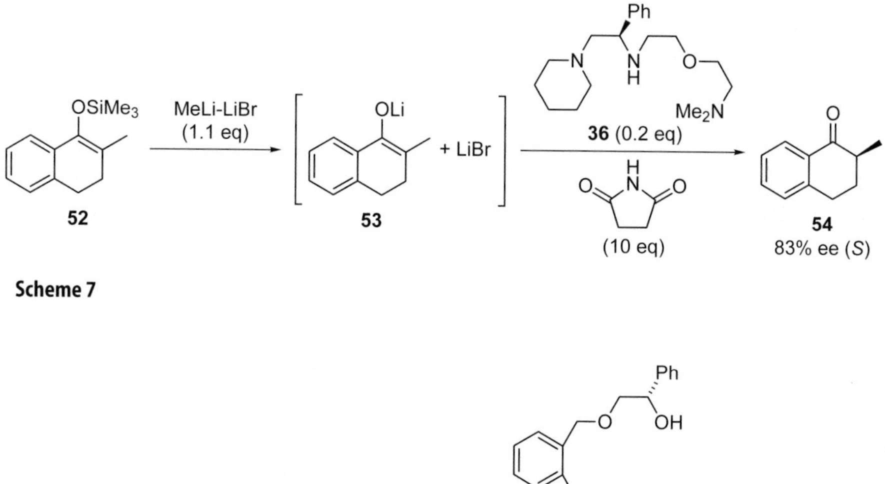

Scheme 7

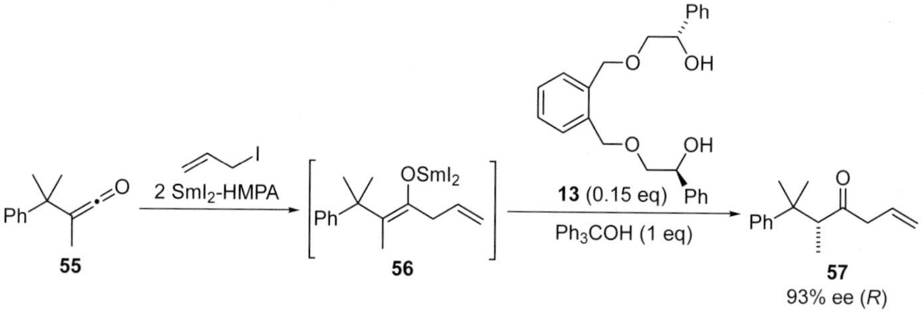

Scheme 8

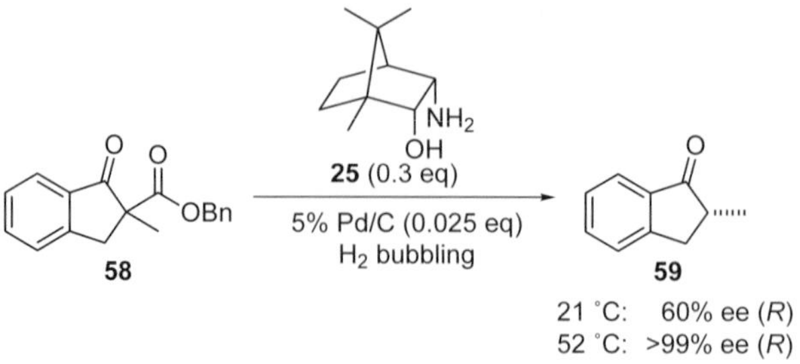

Scheme 9

(Scheme 9) [48]. The enantioselectivity was highly dependent on the reaction temperature and almost enantiopure 2-methyl-1-indanone (**59**) was obtained at 52 °C. The reaction was assumed to proceed via an enol or palladium enolate intermediate which was produced by cleavage of the benzyl-oxygen bond and the subsequent decarboxylation.

3.2
Protonation under Acidic Conditions

Silyl enol ethers, known as chemically stable and easy handled enolates, can be protonated by a strong Brønsted acid. Our group demonstrated that a Lewis acid-assisted Brønsted acid (LBA **17**), generated from optically pure binaphthol and tin tetrachloride, was a chiral proton source of choice for asymmetric protonation of silyl enol ethers possessing an aromatic group at the α-position [33, 34]. Binaphthol itself is not a strong Brønsted acid, however, LBA **17** can protonate less reactive silyl enol ethers since the acidity of the phenolic protons of **17** is enhanced by complexation with tin tetrachloride. The catalytic asymmetric protonation of silyl enol ethers was accomplished for the first time by LBA **18**. Treatment of ketene bis(trimethylsilyl)acetal **60** with 0.08 equiv of LBA **18** and a stoichiometric amount of 2,6-dimethylphenol as an achiral proton source afforded (*S*)-2-phenylpropanoic acid (**61**) with 94% ee (Scheme 10) [35]. LBA **19** derived from binaphthol monoisopropyl ether has been successfully applied to the enantioselective protonation of *meso* 1,2-enediol bis(trimethylsilyl) ethers under stoichiometric conditions [36].

Nakai and a coworker achieved a conceptually different protonation of silyl enol ethers using a chiral cationic palladium complex **40** developed by Shibasaki and his colleagues [61] as a chiral catalyst and water as an achiral proton source [62]. This reaction was hypothesized to progress via a chiral palladium enolate which was diastereoselectively protonated by water to provide the optically active ketone and the chiral Pd catalyst regenerated. A small amount of diisopropylamine was indispensable to accomplish a high level of asymmetric induction and the best enantioselectivity (79% ee) was observed for trimethylsilyl enol ether of 2-methyl-1-tetralone **52** (Scheme 11).

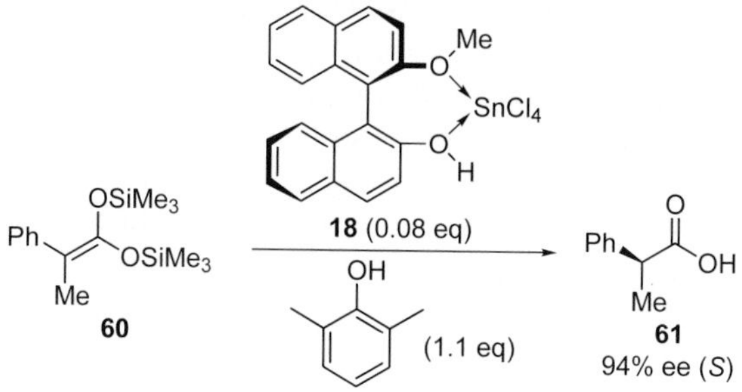

Scheme 10

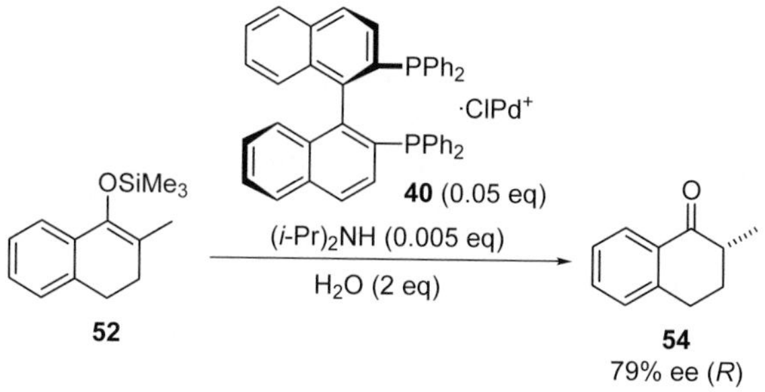

Scheme 11

4
Principal Alternatives

Various methods using a stoichiometric amount of chiral proton sources or chiral ligands are available for enantioselective protonation of metal enolates: e.g., protonation of metal enolates preformed by deprotonation of the corresponding ketones or by allylation of ketenes [6, 7, 8, 9, 10, 11, 13, 17, 18, 19, 21, 22, 25, 26, 29, 30, 31, 32, 37, 40, 41, 42, 43, 49, 50, 53, 54, 55, 56, 57, 59, 60, 63], the Birch reduction of α,β-unsaturated acids in the presence of a sugar-derived alcohol **2** [12], a SmI$_2$-mediated reduction of an α-diketone or 2-aryl-2-methoxyketones with chiral proton sources [16, 28], deracemization of 2-alkylcyclohexanones with chiral diol **7** in alkaline conditions based on host-guest inclusion complexation [20], and decarboxylation of malonic acid derivatives with Cu(I)/alkaloid catalysts of **3** and **4** [14, 15]. A ketene silyl acetal derived from racemic mandelic acid can be enantioselectively protonated by (R)-pantolactone (**11**) in the presence of LiCl [23] or by polymer-supported chiral alcohol **12** [24], which is a substitute for LBA **17–19**. The method employing **12** is temperature dependent and exhibits the highest enantioselectivity (94% ee) at –40 °C [24]. Optically active 2-methyl-1-indanone and 2-methyl-1-tetralone can be synthesized from the corresponding prochiral enol carbonates or racemic β-ketoesters by a multistep reaction: palladium-catalyzed cleavage/decarboxylation/chiral amino alcohol-mediated enantioselective ketonization [38, 39]. Enantioselective photodeconjugation of α,β-unsaturated carbonyl compounds with chiral amino alcohols is a convenient route to the corresponding optically active β,γ-unsaturated carbonyl compounds [64, 65, 66, 67, 68, 69, 70, 71, 72, 73, 74, 75, 76]. Irradiation of α-disubstituted indanones under analogous reaction conditions has led to Norrish type II cleavage followed by asymmetric tautomerization of the resulting enols [77, 78]. Addition of alcohols or amines to ketenes [79, 80, 81, 82, 83, 84, 85, 86] and Michael addition of thiocarboxylic acids or thiols to α,β-unsaturated esters [87, 88] are alternative ways of generating enols. Some other processes of asymmetric protonation of enols have been reported [89, 90]. Racemic ketones can be deracemized via enamines which are converted into optically active enriched ketones by protonation with chiral acids and subsequent hydrolylsis [91, 92, 93, 94, 95, 96]. Enzymatic enantiofacially selective hydrolysis of enol esters is also a promising route to optically active carbonyl compounds [97, 98, 99, 100, 101]. Antibodies are attractive optically active proteins for asymmetric catalysis of stereogenic transformation. Prochiral enol ethers and enol acetates are protonated to produce enantiomerically pure carbonyl compounds by antibody-catalyzed hydrolysis [102, 103, 104, 105, 106]. Multi gram-scale synthesis is possible with catalytic antibodies [107].

References

1. Duhamel L, Duhamel P, Launay J-C, Plaquevent J-C (1984) Bull Soc Chim Fr II 421
2. Fehr C (1991) Chimia 45:253
3. Waldmann H (1991) Nachr Chem Tech Lab 39:413
4. Hünig S (1995) In: Helmchen G, Hoffmann RW, Mulzer J, Schaumann E (eds) Houben-Weyl: Methods of Organic Chemistry, vol E 21. Georg Thieme Verlag, Stuttgart, p 3851
5. Fehr C (1996) Angew Chem Int Ed Engl 35:2566
6. Duhamel L, Plaquevent J-C (1978) J Am Chem Soc 100:7415
7. Duhamel L, Plaquevent J-C (1980) Tetrahedron Lett 21:2521
8. Duhamel L, Plaquevent J-C (1982) Bull Soc Chim Fr II 75
9. Duhamel L, Launay J-C (1983) Tetrahedron Lett 24:4209
10. Duhamel L, Fouquay S, Plaquevent J-C (1986) Tetrahedron Lett 27:4975
11. Duhamel L, Duhamel P, Fouquay S, Eddine JJ, Peschard O, Plaquevent J-C, Ravard A, Solliard R, Valnot J-Y, Vincens H (1988) Tetrahedron 44:5495
12. Kinoshita T, Miwa T (1974) J Chem Soc Chem Commun 181
13. Stoyanovich FM, Zakharov EP, Goldfarb YL, Krayushkin MM (1986) Izv Akad Nauk SSSR Ser Khim 1455
14. Toussaint O, Capdevielle P, Maumy M (1987) Tetrahedron Lett 28:539
15. Brunner H, Kurzwart M (1992) Monats Chem 123:121
16. Takeuchi S, Miyoshi N, Hirata K (1992) Bull Chem Soc Jpn 65:2001
17. Gerlach U, Hünig S (1987) Angew Chem Int Ed Engl 26:1283
18. Gerlach U, Haubenreich T, Hünig S (1994) Chem Ber 127:1969
19. Gerlach U, Haubenreich T, Hünig S (1994) Chem Ber 127:1981
20. Tsunoda T, Kaku H, Nagaku M, Okuyama E (1997) Tetrahedron Lett 38:7759
21. Matsumoto K, Ohta H (1991) Tetrahedron Lett 32:4729
22. Gerlach U, Haubenreich T, Hünig S, Klaunzer N (1994) Chem Ber 127:1989
23. Cavelier F, Gomez S, Jacquier R, Verducci J (1993) Tetrahedron:Asymmetry 4:2501
24. Cavelier F, Gomez S, Jacquier R, Verducci J (1994) Tetrahedron Lett 35:2891

25. Takeuchi S, Miyoshi N, Ohgo Y (1992) Chem Lett 551
26. Takeuchi S, Ohira A, Miyoshi N, Mashio H, Ohgo Y (1994) Tetrahedron:Asymmetry 5:1763
27. Nakamura Y, Takeuchi S, Ohira A, Ohgo Y (1996) Tetrahedron Lett 37:2805
28. Nakamura Y, Takeuchi S, Ohgo Y, Yamaoka M, Yoshida A, Mikami K (1997) Tetrahedron Lett 38:2709
29. Kosugi H, Hoshino K, Usa H (1997) Tetrahedron Lett 38:6861
30. Kosugi H, Abe M, Hatsuda R, Usa H, Kato M (1997) Chem Commun 1857
31. Takahashi T, Nakao N, Koizumi T (1996) Chem Lett 207
32. Takahashi T, Nakao N, Koizumi T (1997) Tetrahedron:Asymmetry 8:3293
33. Ishihara K, Kaneeda M, Yamamoto H (1994) J Am Chem Soc 116:11179
34. Ishihara K, Nakamura S, Yamamoto H (1996) Croat Chem Acta 69:513
35. Ishihara K, Nakamura S, Kaneeda M, Yamamoto H (1996) J Am Chem Soc 118:12854
36. Taniguchi T, Ogasawara K (1997) Tetrahedron Lett 38:6429
37. Fuji K, Kawabata T, Kuroda A, Taga T (1995) J Org Chem 60:1914
38. Hénin F, Muzart J (1992) Tetrahedron:Asymmetry 3:1161
39. Aboulhoda SJ, Hénin F, Muzart J, Thorey C, Behnen W, Martens J, Mehler T (1994) Tetrahedron:Asymmetry 5:1321
40. Martin J, Lasne M-C, Plaquevent J-C, Duhamel L (1997) Tetrahedron Lett 38:7181
41. Fehr C, Galindo J (1988) J Am Chem Soc 110:6909
42. Fehr C, Guntern O (1992) Helv Chim Acta 75:1023
43. Fehr C, Stempf I, Galindo J (1993) Angew Chem Int Ed Engl 32:1042
44. Fehr C, Stempf I, Galindo J (1993) Angew Chem Int Ed Engl 32:1044
45. Fehr C, Galindo J (1994) Angew Chem Int Ed Engl 33:1888
46. Fehr C, Galindo J (1995) Helv Chim Acta 78:539
47. Aboulhoda SJ, Létinois S, Wilken J, Reiners I, Hénin F, Martens J, Muzart J (1995) Tetrahedron:Asymmetry 6:1865
48. Muzart J, Hénin F, Aboulhoda SJ (1997) Tetrahedron:Asymmetry 8:381
49. Potin D, Williams K, Rebek J, Jr (1990) Angew Chem Int Ed Engl 29:1420
50. Yanagisawa A, Kuribayashi T, Kikuchi T, Yamamoto H (1994) Angew Chem Int Ed Engl 33:107
51. Yanagisawa A, Kikuchi T, Watanabe T, Kuribayashi T, Yamamoto H (1995) Synlett 372
52. Yanagisawa A, Watanabe T, Kikuchi T, Kuribayashi T, Yamamoto H (1997) Synlett 956
53. Hogeveen H, Zwart L (1982) Tetrahedron Lett 23:105
54. Eleveld MB, Hogeveen H (1986) Tetrahedron Lett 27:631
55. Vedejs E, Lee N (1991) J Am Chem Soc 113:5483
56. Vedejs E, Lee N (1995) J Am Chem Soc 117: 891
57. Yasukata T, Koga K (1993) Tetrahedron:Asymmetry 4:35
58. Riviere P, Koga K (1997) Tetrahedron Lett 38:7589
59. Fuji K, Tanaka K, Miyamoto H (1993) Tetrahedron:Asymmetry 4:247
60. Vedejs E, Lee N, Sakata ST (1994) J Am Chem Soc 116:2175
61. Sodeoka M, Ohrai K, Shibasaki M (1995) J Org Chem 60:2648
62. Sugiura M, Nakai T (1997) Angew Chem Int Ed Engl 36:2366
63. Haubenreich T, Hünig S, Schulz H-J (1993) Angew Chem Int Ed Engl 32:398
64. Henin F, Mortezaei R, Muzart J, Pete J-P (1985) Tetrahedron Lett 26:4945
65. Mortezaei R, Henin F, Muzart J, Pete J-P (1985) Tetrahedron Lett 26:6079
66. Pete J-P, Henin F, Mortezaei R, Muzart J, Piva O (1986) Pure Appl Chem 58:1257
67. Mortezaei R, Piva O, Henin F, Muzart J, Pete J-P (1986) Tetrahedron Lett 27:2997
68. Piva O, Henin F, Muzart J, Pete J-P (1986) Tetrahedron Lett 27:3001
69. Piva O, Henin F, Muzart J, Pete J-P (1987) Tetrahedron Lett 28:4825
70. Mortezaei R, Awandi D, Henin F, Muzart J, Pete J-P (1988) J Am Chem Soc 110:4824
71. Henin F, Mortezaei R, Muzart J, Pete J-P, Piva O (1989) Tetrahedron 45:6171
72. Piva O, Mortezaei R, Henin F, Muzart J, Pete J-P, (1990) J Am Chem Soc 112:9263
73. Piva O, Pete J-P (1990) Tetrahedron Lett 31:5157
74. Muzart J, Hénin F, Pète J-P, M'Boungou-M'Passi A (1993) Tetrahedron:Asymmetry 4:2531

75. Piva O (1995) J Org Chem 60:7879
76. Henin F, Muzart J, Pete J-P, Piva O (1991) New J Chem 15:611
77. Henin F, Muzart J, Pete J-P, M'Boungou-M'Passi A, Rau H (1991) Angew Chem Int Ed Engl 30:416
78. Henin F, M'Boungou-M'Passi A, Muzart J, Pete J-P (1994) Tetrahedron 50:2849
79. Pracejus H (1960) Liebigs Ann Chem 634:9
80. Pracejus H (1960) Liebigs Ann Chem 634:23
81. Tille A, Pracejus H (1967) Chem Ber 100:196
82. Pracejus H, Kohl G (1969) Liebigs Ann Chem 722:1
83. Anders E, Ruch E, Ugi I (1973) Angew Chem Int Ed Engl 12:25
84. Jähme J, Rüchardt C (1981) Angew Chem Int Ed Engl 20:885
85. Salz U, Rüchardt C (1982) Tetrahedron Lett 23:4017
86. Larsen RD, Corley EG, Davis P, Reider PJ, Grabowski EJJ (1989) J Am Chem Soc 111:7650
87. Gawronski JK, Gawronska K, Kolbon H, Wynberg H (1983) Recl Trav Chim Pays-Bas 102:479
88. Kumar A, Salunkhe RV, Rane RA, Dike SY (1991) J Chem Soc Chem Commun 485
89. Henze R, Duhamel L, Lasne M-C (1997) Tetrahedron:Asymmetry 8:3363
90. Bergens SH, Bosnich B (1991) J Am Chem Soc 113:958
91. Matsushita H, Noguchi M, Saburi M, Yoshikawa S (1975) Bull Chem Soc Jpn 48:3715
92. Matsushita H, Tsujino Y, Noguchi M, Yoshikawa S (1976) Bull Chem Soc Jpn 49:3629
93. Matsushita H, Tsujino Y, Noguchi M, Saburi M, Yoshikawa S (1978) Bull Chem Soc Jpn 51:862
94. Duhamel L (1976) C R Acad Sci Paris 282C:125
95. Duhamel L, Plaquevent J-C (1977) Tetrahedron Lett 2285
96. Duhamel L, Plaquevent J-C (1982) Bull Soc Chim Fr II 69
97. Matsumoto K, Tsutsumi S, Ihori T, Ohta H (1990) J Am Chem Soc 112:9614
98. Kume Y, Ohta H (1992) Tetrahedron Lett 33:6367
99. Katoh O, Sugai T, Ohta H (1994) Tetrahedron:Asymmetry 5:1935
100. Hirata T, Shimoda K, Ohba D, Furuya N, Izumi S (1997) Tetrahedron:Asymmetry 8:2671
101. Ohta H (1997) Bull Chem Soc Jpn 70:2895
102. Fujii I, Lerner RA, Janda KD (1991) J Am Chem Soc 113:8528
103. Reymond J-L, Janda KD, Lerner RA (1992) J Am Chem Soc 114:2257
104. Reymond J-L, Jahangiri GK, Stoudt C, Lerner RA (1993) J Am Chem Soc 115:3909
105. Jahangiri GK, Reymond J-L (1994) J Am Chem Soc 116:11264
106. Sinha SC, Keinan E (1995) J Am Chem Soc 117:3653
107. Reymond J-L, Reber J-L, Lerner RA (1994) Angew Chem Int Ed Engl 33:475

Chapter 35   **Ring Opening of Epoxides and Related Reactions**

Chapter 35
Ring Opening of Epoxides and Related Reactions

Eric N. Jacobsen · Michael H. Wu

Department of Chemistry and Chemical Biology, Harvard University, Cambridge,
MA 02138, USA
e-mail: jacobsen@chemistry.harvard.edu

Keywords: Epoxide, Ring opening, Desymmetrization, Kinetic resolution

1
Introduction

The ready availability of achiral and racemic epoxides from simple alkene precursors renders epoxide ring-opening an appealing approach to asymmetric synthesis. The inherent strain-induced reactivity of epoxides can be enhanced by coordination of the epoxide oxygen to a Lewis acid, thereby creating the possibility for chiral Lewis acids to catalyze enantioselective ring opening events.

The asymmetric ring opening (ARO) of *meso*-epoxides has the potential to generate two contiguous stereogenic centers from an achiral starting material. The wide variety of nitrogen, sulfur, oxygen, carbon, and halogen nucleophiles that have been reported in epoxide desymmetrization reactions underscores the versatility of epoxides and their ring-opened derivatives. Catalytic approaches to effect the enantioselective deprotonation of *meso*-epoxides have also emerged as viable synthetic methods.

The ability to couple ARO with the kinetic resolution of a racemic mixture of epoxides offers another powerful strategy in asymmetric synthesis. Kinetic resolution is an attractive approach when the epoxide is easily accessed in racemic form, and becomes even more valuable when effective enantioselective routes to these epoxides are lacking. The most appealing scenario is one where both the ring-opened product and the unreacted epoxide are valuable chiral products.

The number of catalytic asymmetric transformations involving epoxides has grown considerably, even since this topic was last reviewed in 1996 [1, 2]. This chapter will highlight recent progress in asymmetric catalytic ring-opening methods and their increasing importance in the stereoselective synthesis of enantio-enriched compounds.

2
Enantioselective Ring Opening of *meso*-Epoxides

2.1
Nitrogen Nucleophiles

Highly enantioselective catalytic desymmetrization of *meso*-epoxides through nucleophilic ring opening was first effectively demonstrated by Nugent, who found that a zirconium trialkanolamine complex catalyzed the addition of azidosilanes to *meso*-epoxides (Scheme 1) [3]. Azide has been the most widely explored nitrogen nucleophile [4, 5, 6, 7], in part due to its utility as an amine sur-

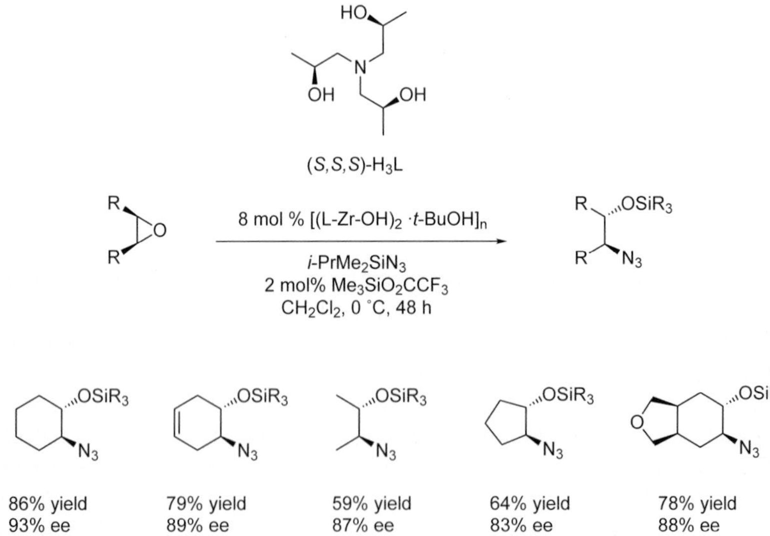

Scheme 1

rogate that requires no protection for subsequent chemical elaboration. The useful levels of optical purity as well as the relatively broad substrate scope validated the ARO of epoxides as a worthy goal in asymmetric catalysis.

A very efficient chiral Cr(salen) catalyst that promoted the enantioselective addition of TMSN$_3$ was reported subsequently by Jacobsen [8]. The Cr complex **1** was conveniently prepared by insertion of CrCl$_2$ or alternatively a CrCl$_3$/Zn mixture into the (salen) ligand and subsequent air oxidation to the stable Cr(III) species. Low catalyst loadings (2 mol %) of this complex effected epoxide ring-opening of a wide variety of *meso*-epoxides containing carbamate, amide, or ester functionality (Scheme 2). In addition to its functional group tolerance, other practical advantages of the Cr(salen) catalyst include its indefinite stability under catalytic conditions which allowed for its repeated recycling. The discovery that the Cr(salen)-catalyzed ARO performed equally well in the absence of solvent rendered the process even more appealing, as simple distillation of the product mixture afforded the azido silyl ether with the highest possible volumetric productivity and no by-products generated (Table 1).

A novel mechanism was elucidated in the Cr(salen)-catalyzed ARO reaction in which the catalyst was discovered to perform dual roles [9]. The synthesis and characterization of the Cr(salen)N$_3$ complex **2** allowed the identification of this species as the active catalyst in the ARO reaction, suggesting that one role of the Cr catalyst was to deliver the azide nucleophile. X-ray crystallographic and IR spectroscopic analysis of **2** revealed a 6-coordinate geometry with the Lewis acid-

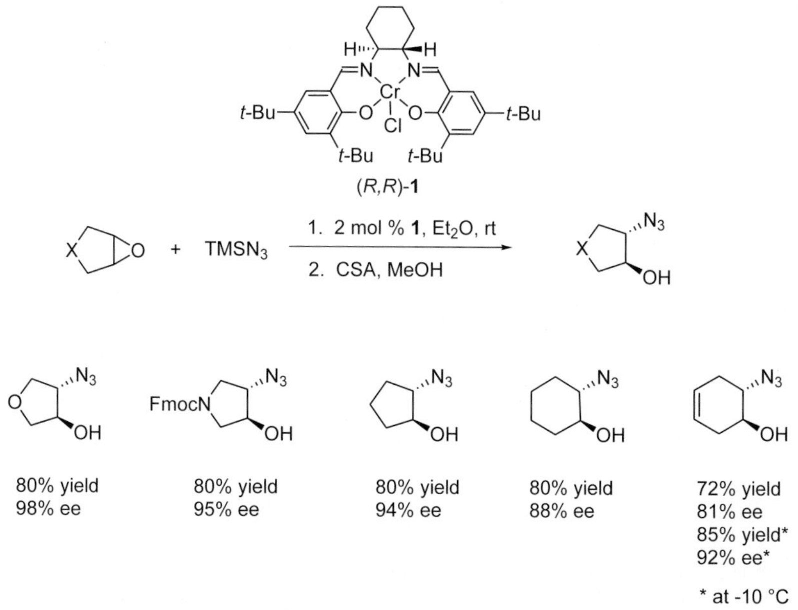

Scheme 2

Table 1. Recycling of Cr(salen) complex 1 in the ARO with TMSN$_3$

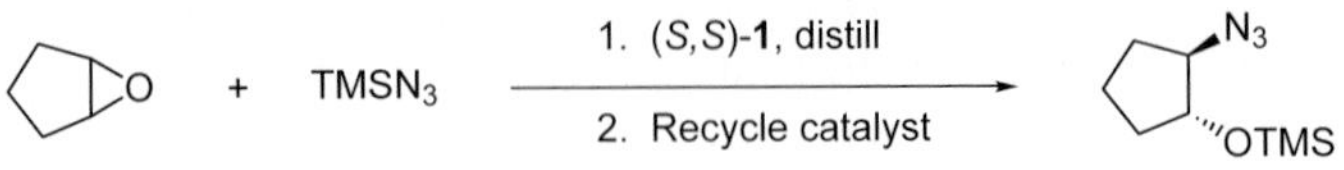

Cycle	Yield (%)	ee (%)
1	84	93
2	92	94
3	93	94
4	95	94
5	91	94
6	95	94
7	95	94
8	94	94
9	95	94
10	95	95
11	92	95

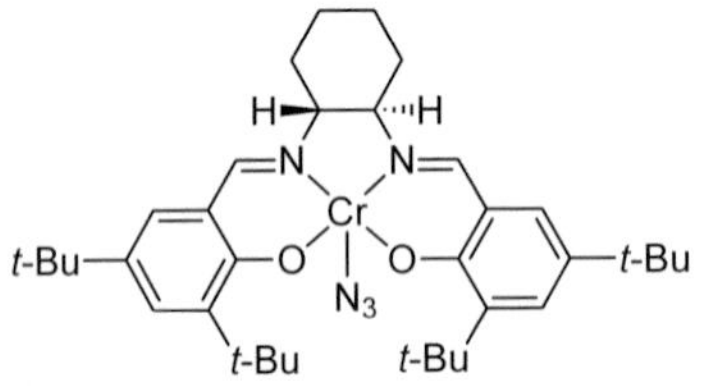

Fig. 1

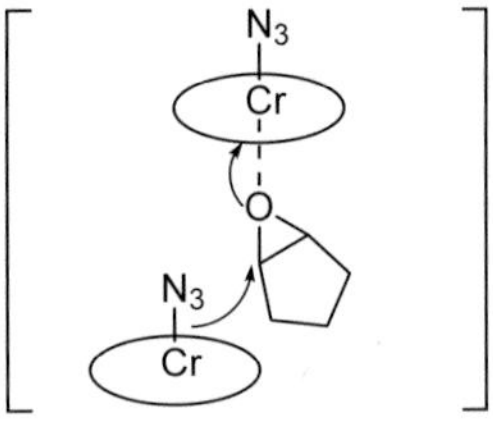

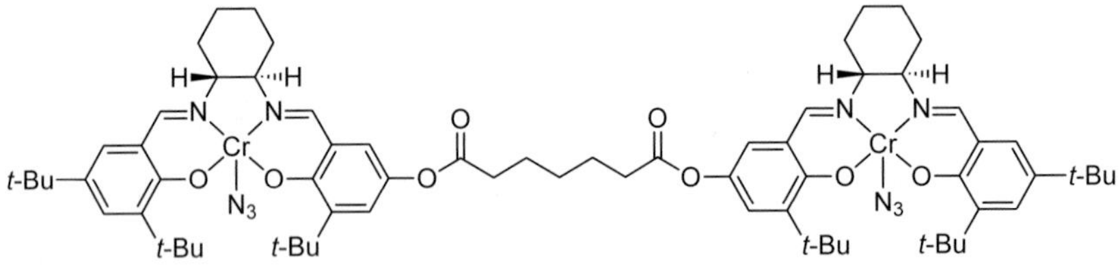

ic Cr center bound to ligands such as THF or cyclopentene oxide. Kinetic analysis of the reaction established a rate law where the reaction rate was proportional to [Cr]^2. This was consistent with epoxide opening occurring through a bimetallic rate- and enantioselectivity-determining step where the Cr-activated azide is delivered to the coordinated epoxide of another Cr(salen) catalyst (Fig. 1). Additional evidence for a cooperative mechanism was provided by constructing dimeric catalysts such as **3** in which two Cr(salen) units were covalently linked. This led to a dramatic rate enhancement in the ARO with TMSN$_3$ with similar levels of enantioselectivity to those obtained with monomeric catalyst **2** [10].

The Cr(salen)-catalyzed ARO could be applied to prepare a range of chiral building blocks useful for the synthesis of biologically important compounds. Practical routes to cyclic *cis*- and *trans*-1,2-amino alcohols have been developed using Cr(salen) catalysis [11]. ARO methodology also enabled the enantioselective synthesis of the core structures of balanol [12], prostaglandin derivatives [13], and a series of carbocyclic nucleoside analogs such as aristeromycin and carbovir (Scheme 3) [14]. The enantioselective addition of TMSN$_3$ to polymer-immobilized epoxides catalyzed by **1** also allowed the facile construction of cyclic RGD peptide derivatives on a solid phase [15].

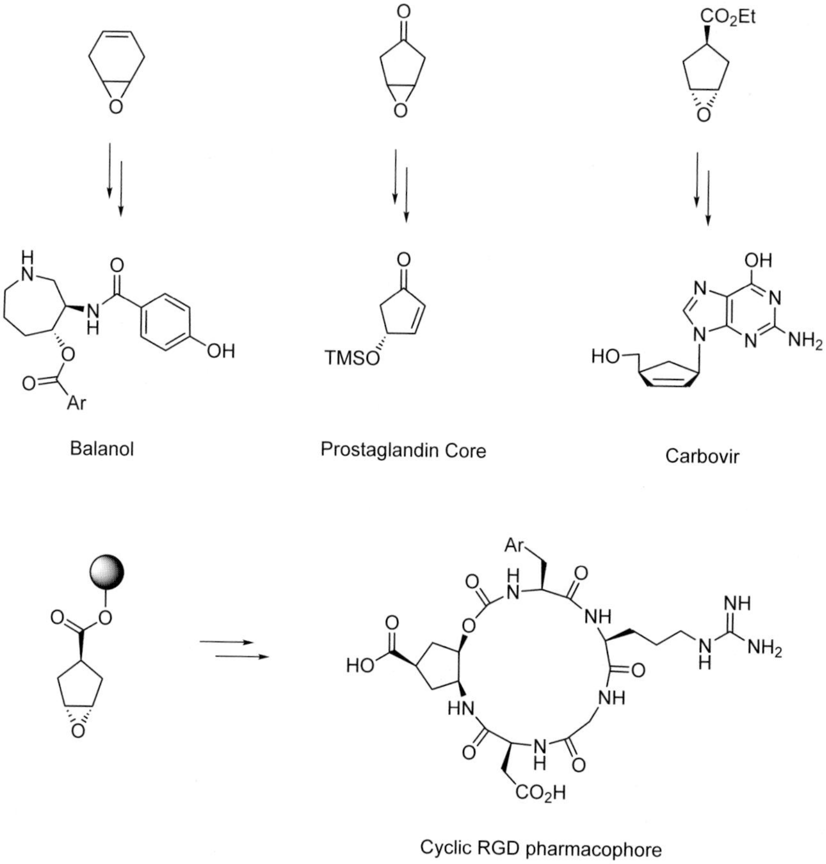

Balanol

Prostaglandin Core

Carbovir

Cyclic RGD pharmacophore

Scheme 3

2.2
Sulfur Nucleophiles

A breakthrough in the ARO with sulfur nucleophiles was made by Shibasaki, who discovered that the Ga-Li-bis(binaphthoxide) complex 4 catalyzed the addition of *tert*-butyl thiol to cyclic and acyclic epoxides in 82–97% ee (Scheme 4) [16]. Ten mol % of the heterobimetallic catalyst promoted the ring opening of a wide range of substrates, and the *tert*-butyl thiol adducts could be converted to enantio-enriched allylic alcohols through an oxidation/elimination sequence.

The addition of other sulfur nucleophiles was reported by Jacobsen to be catalyzed by the same Cr(salen) complex 1 initially reported for the ARO with TMSN$_3$. Benzyl mercaptan afforded the ring-opened hydroxy sulfides in excellent yield and 59–70% ee [17]. The moderate levels of enantioselectivity were improved by use of the dithiol 5, which afforded mixtures of bishydroxy sulfides in which the ee of the chiral product 6 was substantially enriched (Scheme 5). The sulfide products could be easily elaborated into the free thiols by reductive debenzylation, providing access to the β-silyloxy thiol 8 in optically pure form.

A titanium (IV) complex with the identical (salen) ligand was reported recently to effect the addition of thiophenol with moderate enantioselectivity [18]. The complex was formed *in situ* from 5 mol % Ti(O-*i*-Pr)$_4$ and 5.5 mol % of the chiral ligand 9, and this catalyst promoted the ring opening of cyclohexene oxide in 93% yield and 63% ee at −40 to −25 °C (Scheme 6).

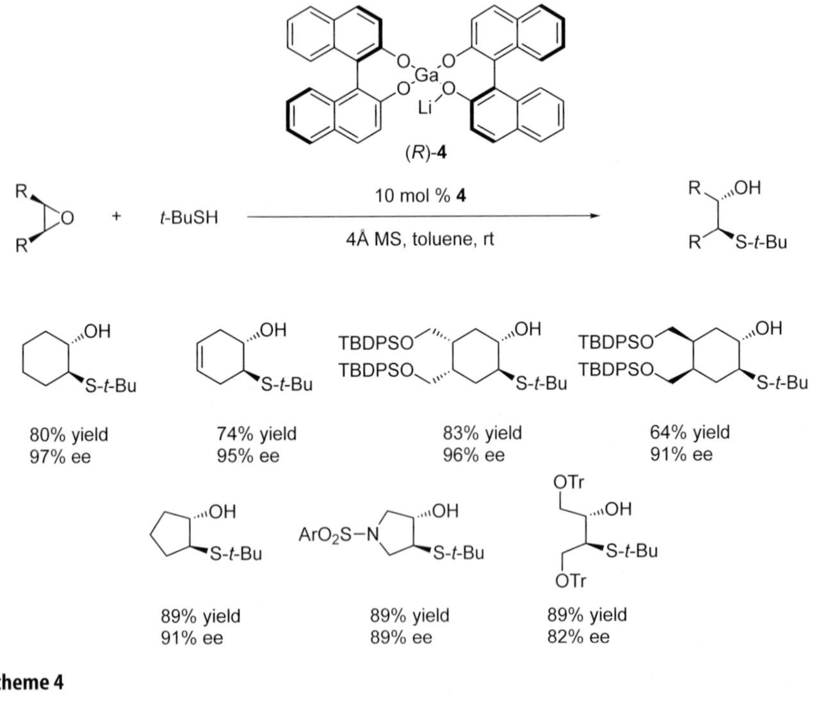

Scheme 4

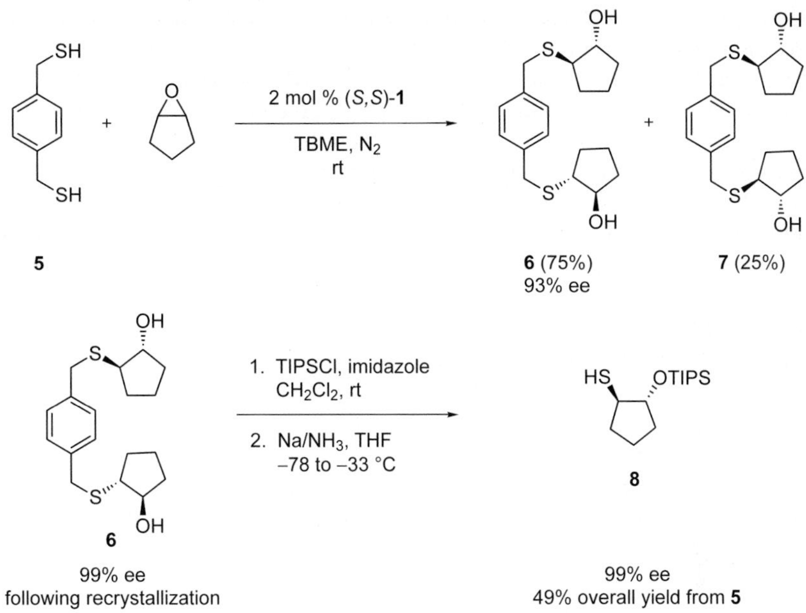

5

6 (75%)
93% ee

7 (25%)

2 mol % (S,S)-**1**

TBME, N$_2$
rt

6
99% ee
following recrystallization

1. TIPSCl, imidazole
CH$_2$Cl$_2$, rt

2. Na/NH$_3$, THF
−78 to −33 °C

HS⟍⟍OTIPS

8

99% ee
49% overall yield from **5**

Scheme 5

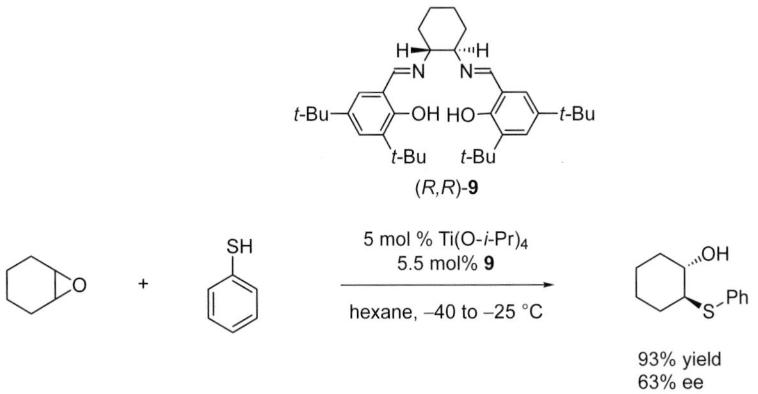

(R,R)-**9**

5 mol % Ti(O-i-Pr)$_4$
5.5 mol% **9**

hexane, −40 to −25 °C

93% yield
63% ee

Scheme 6

2.3
Oxygen Nucleophiles

The reaction of oxygen-containing nucleophiles including alcohols, phenols, and carboxylic acids allows the generation of a 1,2-diol equivalent in which the oxygen atoms are differentially protected. Given the synthetic utility of 1,2-diols

[19], the ability to selectively functionalize either oxygen atom would render these ARO products even more versatile.

Jacobsen reported that the Co(salen) complex **10** catalyzed the addition of carboxylic acids to *meso*-epoxides [20]. An initial screen revealed that benzoic acid and its derivatives were the most useful nucleophiles from the perspective of reactivity and selectivity. Although optical purities exceeding 90% ee were observed only with selected substrates, the crystallinity of the benzoate esters in some cases allowed enhancement of their enantiopurity by recrystallization. The ring opening of cyclohexene oxide, for instance, proceeded on a multigram scale in quantitative yield and 77% ee; subsequent recrystallizations of the monobenzoate ester **11** then afforded 98% ee material isolated in 75% yield (Scheme 7).

The Ga-Li-BINOL complex **4** was discovered by Shibasaki to catalyze the ring opening of epoxides with 4-methoxyphenol. Elevated temperatures and high catalyst loadings render this catalyst system less practical than the *tert*-butyl thiol counterpart, but the hydroxy aryl monoethers produced in this reaction do offer access to valuable monoprotected 1,2-*trans*-diols (Scheme 8) [21].

An intramolecular ring-opening reaction with oxygen nucleophiles was discovered by Jacobsen to be catalyzed by the Co(salen)OAc catalyst **12** [22]. The

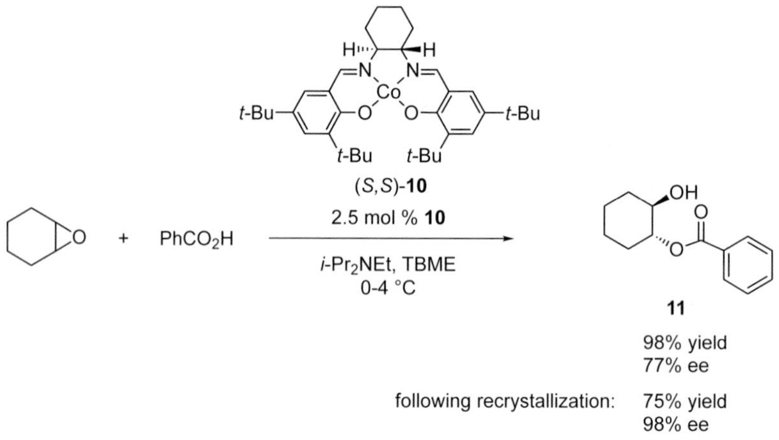

Scheme 7

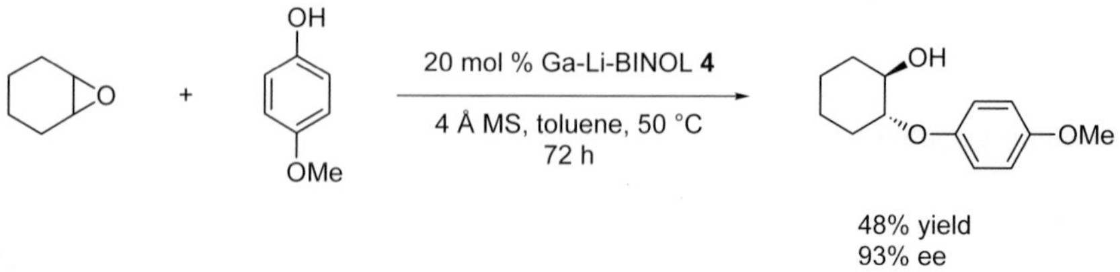

Scheme 8

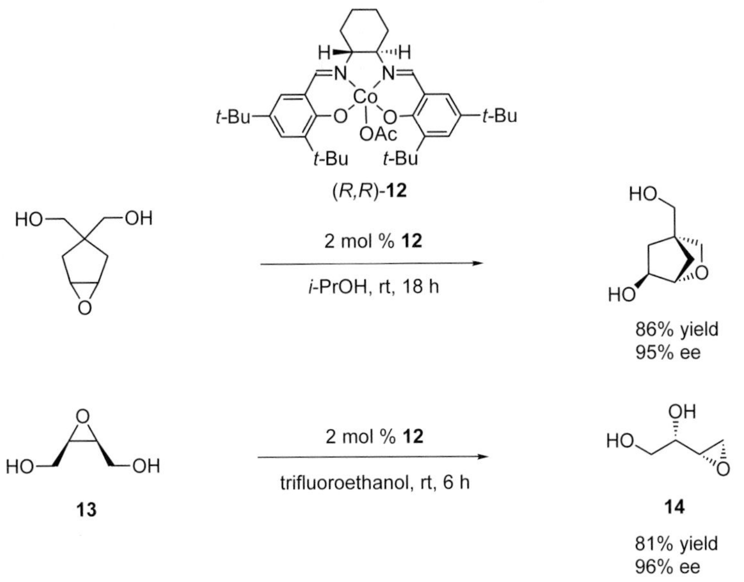

Scheme 9

cyclization of *meso*-epoxy diols produced novel cyclic and bicyclic products in good yields and >95% ee. Complex **12** also catalyzed an asymmetric Payne rearrangement of the *meso*-epoxy diol **13** to afford the enantio-enriched C$_4$ building block **14** in 81% yield and 96% ee (Scheme 9).

2.4
Carbon Nucleophiles

A difficult challenge in developing ARO reactions with carbon nucleophiles is identifying a reagent that is sufficiently reactive to open epoxides but at the same time innocuous to chiral metal catalysts. A recent contribution by Crotti clearly illustrates this delicate reactivity balance. The lithium enolate of acetophenone added in the presence of 20 mol % of the chiral Cr(salen) complex **1** to cyclohexene oxide in very low yield but in 84% ee (Scheme 10) [23]. That less than one turnover of the catalyst was observed strongly suggests that the lithium enolate and the Schiff base catalyst are not compatible under the reaction conditions.

Oguni discovered that phenyllithium in the presence of 5 mol % of chiral Schiff base ligands created a stable and efficient catalyst system. The addition to cyclohexene oxide occurred in quantitative yield to form the phenylcyclohexanol in 90% ee (Scheme 11) [24]. Oguni proposed that deprotonation of the phenol and/or 1,2-addition to the imine ligand **15** formed the catalytically-active species.

Cyanide stands as an appealing candidate for ARO reactions, given its stability toward a variety of metal-ligand complexes, its reactivity toward epoxides

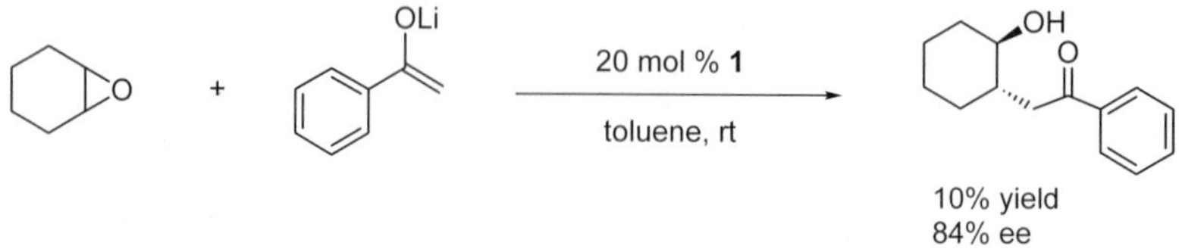

Scheme 10

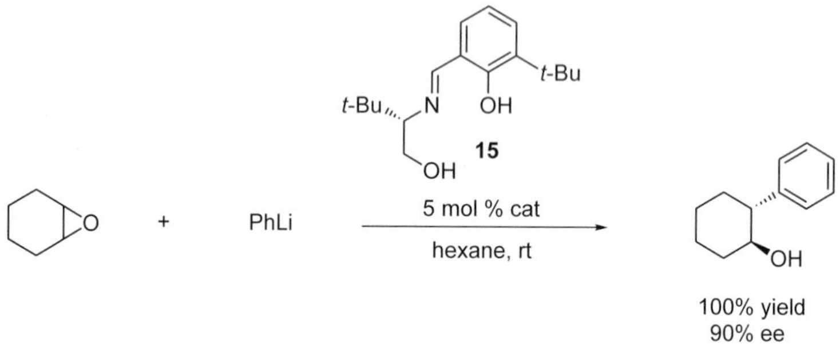

Scheme 11

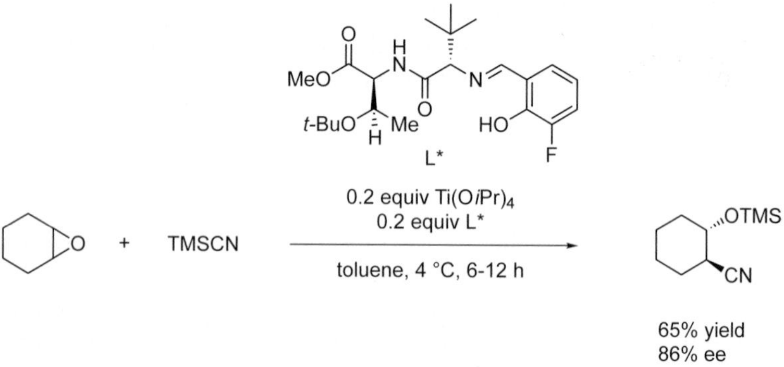

Scheme 12

under catalytic conditions, and the synthetic utility of the ring-opened products. The Ti(O-*i*-Pr)$_4$-catalyzed addition of TMSCN was discovered by Oguni to be dramatically accelerated by achiral tridentate Schiff base ligands [25]. Snapper and Hoveyda subsequently developed an asymmetric version of this reaction using chiral amino alcohol backbones. In addition, this work demonstrated the solid phase assembly of ligands for the rapid synthesis of potential catalysts (Scheme 12) [26].

2.5
Hydrogen and Halogen Nucleophiles

The asymmetric reduction of *meso*-epoxides with hydrogen or hydrides has been scarcely explored, despite the synthetic utility of the chiral secondary alcohol products. The lone example has been provided by Chan, who treated the disodium salt of epoxysuccinic acid with H_2 or MeOH as the reducing agent in the presence of a chiral rhodium catalyst (Scheme 13) [27]. Deuterium labeling experiments established that the reduction proceeded through direct cleavage of the epoxide C-O bond, rather than isomerization to the ketone followed by carbonyl reduction.

Desymmetrization with halogen nucleophiles was effectively demonstrated with two mechanistically-divergent chiral catalysts. Denmark disclosed a Lewis-base activated delivery of chloride that was catalyzed by the enantiopure phosphoramide 16. Binding of the phosphoramide was believed to induce dissociation of $SiCl_4$ into the chiral phosphorus/silicon cation and chloride anion, which subsequently ring-opened the activated epoxide. The best enantioselectivity was observed with *cis*-stilbene oxide, which was formed in 94% yield and 87% ee (Scheme 14) [28].

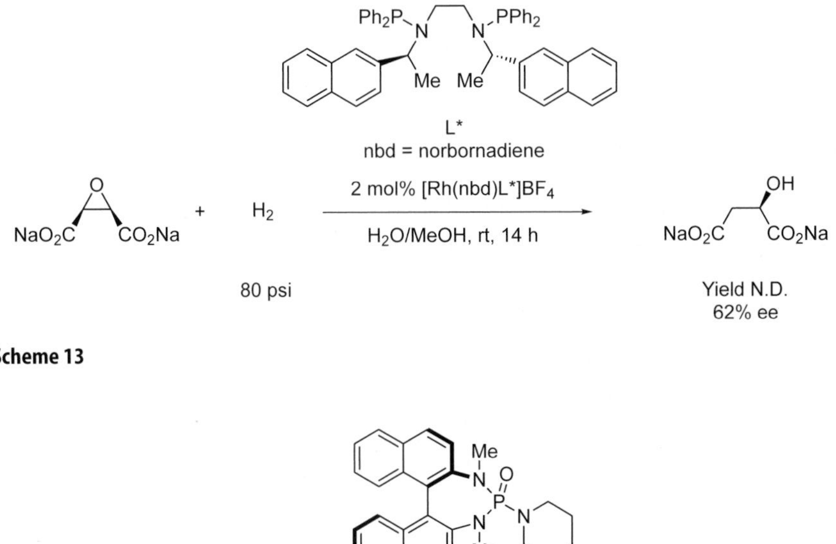

Scheme 13

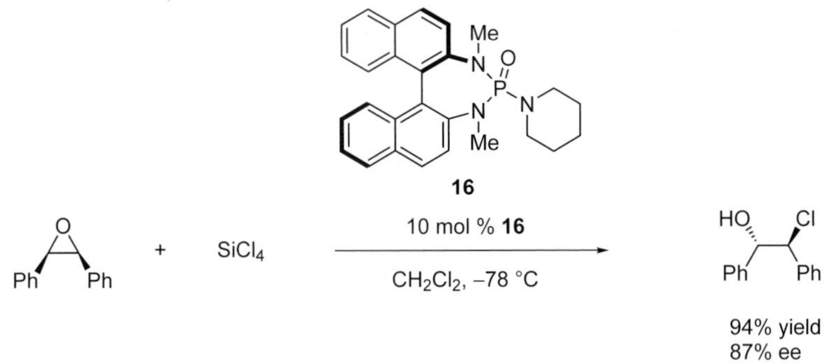

Scheme 14

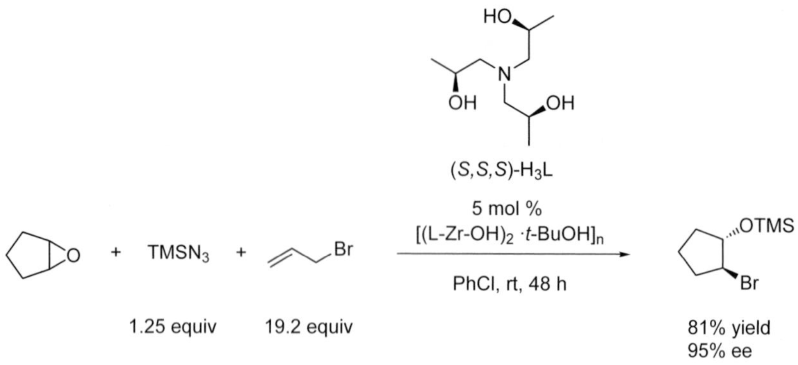

Scheme 15

Nugent adapted the zirconium trialkanolamine complex developed for the enantioselective addition of TMSN$_3$ (Scheme 1) to catalyze the addition of bromide, where substitution of bromide for azide at the metal center was proposed to account for the epoxide halogenation product (Scheme 15) [29]. When a large excess of the bromide source was used to suppress azido alcohol formation, a wide range of cyclic epoxides reacted with 5 mol % of the catalyst to afford good yields of the bromohydrins in 84–96% ee.

2.6
Enantioselective Deprotonation

While several stoichiometric chiral lithium amide bases effect the rearrangement of *meso*-epoxides to allylic alcohols [1], few examples using catalytic amounts of base have been reported. Asami applied a proline-derived ligand to the enantioselective deprotonation of cyclohexene oxide to afford 2-cyclohexen-

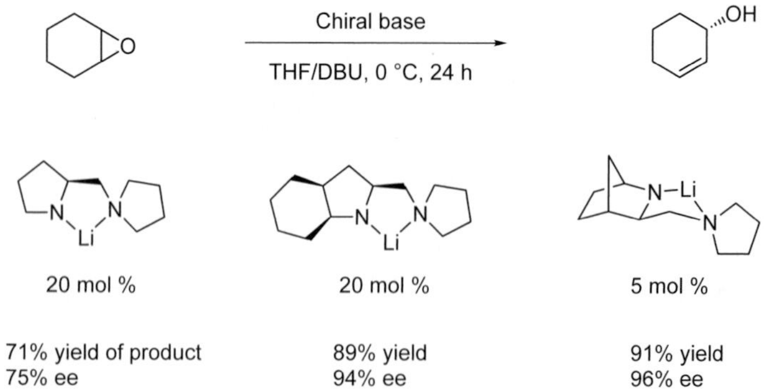

Scheme 16

1-ol in 71% yield and 75% ee [30]. Investigation of a related diamine ligand led to an improvement of product ee to 94% [31], but the lack of substrate generality and the limited availability of both enantiomers of the catalyst restrict its application (Scheme 16). A bicyclic variant was discovered by Andersson to display slightly broader substrate scope, as catalytic amounts of this lithium amide base afforded high levels of enantiopurity for cyclohexene and cycloheptene oxide [32]. In addition to accessing high levels of asymmetric induction, this diamine ligand has the additional advantage of being readily prepared as either enantiomer.

3
Kinetic Resolution of Racemic Epoxides

The use of kinetic resolution as a strategy in asymmetric synthesis has been reviewed extensively [33, 34]. The availability of epoxides in racemic form and the lack of effective enantioselective methods for preparing several important epoxide structural classes renders kinetic resolution by ARO a potentially powerful tool. For kinetic resolutions in which recovery of unreacted substrate is targeted, a cheap and easily handled reagent is desirable for effecting the resolution. Ideally, the ring-opened product would be of synthetic value, and each component of the kinetic resolution (unreacted substrate, product, and catalyst) would be easily isolated in pure form.

3.1
Kinetic Resolution with TMSN$_3$

Jacobsen successfully extended the Cr(salen)-catalyzed ARO of *meso*-epoxides with TMSN$_3$ to the reaction of terminal epoxides, generating valuable 1-amino-2-alkanol precursors [35]. The Cr(salen) complex 1 effectively distinguished between substrate enantiomers with k$_{rel}$ ranging from 44 to 230, reflecting a very high level of chiral recognition by this catalyst system. Most notable, simple aliphatic substrates such as propylene oxide in which the catalyst must differentiate between only a hydrogen and a methyl group were efficiently resolved (Table 2).

The Cr(salen) catalyst was shown to catalyze the resolution of 2,2-disubstituted epoxides, in which a methylene and a methyl group were distinguished by the chiral catalyst. The ARO of this difficult substrate class demonstrated a useful feature of kinetic resolution, in that the enantiopurity of the unreacted epoxide could be improved through higher substrate conversion (Scheme 17). Alternatively, allowing the reaction to proceed to only 40% conversion allowed production of the tertiary alcohol in 74% yield and 94% ee [36, 37, 38].

Table 2. Kinetic resolution of terminal epoxides with TMSN$_3$ catalyzed by **1**

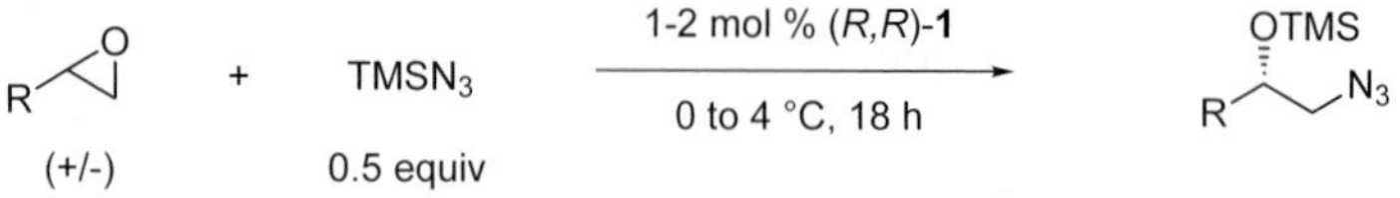

R	ee (%)	Yield (%)[a]	K$_{rel}$[b]
CH$_3$	97	98	230
CH$_2$CH$_3$	97	83	140
(CH$_2$)$_3$CH$_3$	97	89	160
CH$_2$Cl	95	94	100
CH$_2$C$_6$H$_5$	93	94	71
c-C$_6$H$_{11}$	97	84	140
(CH$_2$)$_2$CH=CH$_2$	98	94	280
CH(OEt)$_2$	89	96	44
CH$_2$CN	92	80	45

[a] Isolated yield based on TMSN$_3$
[b] krel=In[1−c(1+ee)]/In[1−c(1−ee)]

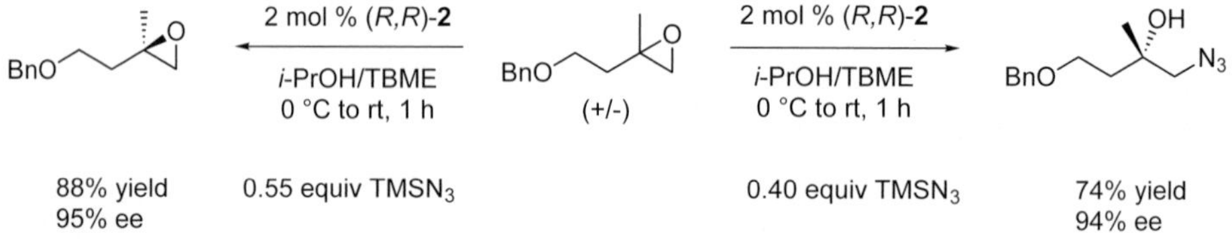

88% yield   0.55 equiv TMSN$_3$ 0.40 equiv TMSN$_3$ 74% yield
95% ee   94% ee

Scheme 17

3.2
Hydrolytic Kinetic Resolution

The ideal kinetic resolution would require no external resolving agent, so that
through an enantioselective isomerization or polymerization optically pure products are generated. In the absence of these methods [39], the possibility of an inexpensive nucleophile such as water to serve as the resolving agent could provide a very appealing alternative.

   Jacobsen disclosed a chiral Co(salen) catalyst that promoted the hydrolytic
kinetic resolution (HKR) of terminal epoxides [40]. Remarkably low levels of the
Co(salen)OAc complex **12** effected enantioselective epoxide hydrolysis to afford
mixtures of the unreacted epoxide and the ring-opened diol. Controlling the
amount of water in the HKR allowed either of these chiral products to be generated in high enantiopurity (Tables 3 and 4) [41]. Significant differences in vola-

Table 3. Hydrolytic kinetic resolution catalyzed by **12** to produce terminal epoxides in ≥99% ee

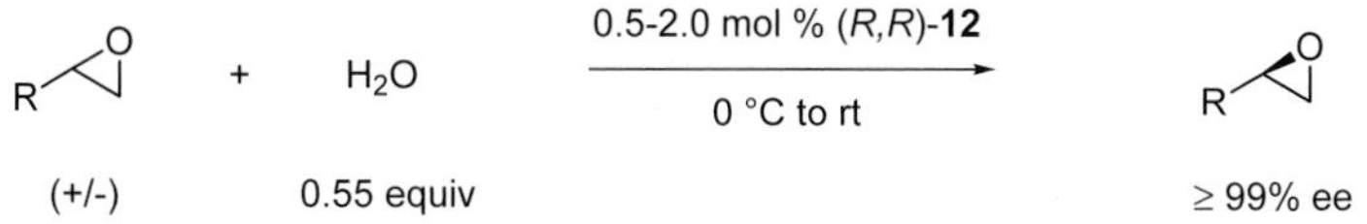

R	Catalyst (mol %)	Solvent	Time (h)	Yield (%)[a]
CH_3	0.2	–	18	92
$(CH_2)_3CH_3$	0.5	–	18	86
CH_2Ph	0.5	THF	18	92
$c\text{-}C_6H_{11}$	0.5	THF	18	87
t-Bu	2.0	1,2-hexanediol	48	82
CH_2Cl	0.5	THF	18	83
CF_3	0.5	–	18	75
CH_2CO_2Et	0.5	THF	18	92
CH_2NHBoc	2.0	THF	38	72
CO_2Me	2.0	THF	24	86
COMe	2.0	THF	24	80
CH_2OBn	0.5	THF	18	96
$m\text{-}Cl\text{-}C_6H_4$	0.8	THF	48	80

[a]Yield based on a maximum theoretical yield of 50%

Table 4. Hydrolytic kinetic resolution catalyzed by **12** to produce 1,2-diols

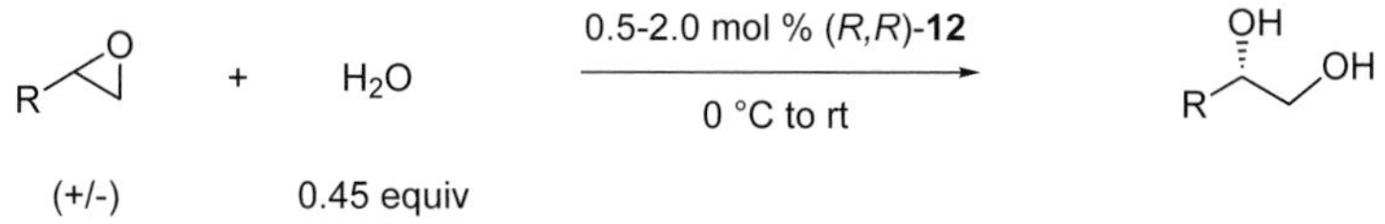

R	Catalyst (mol %)	Solvent	ee (%)	Yield (%)[a]
CH_3	0.2	–	99	89
$(CH_2)_3CH_3$	0.2	–	99	90
CH_2Ph	0.5	THF	95	81
$c\text{-}C_6H_{11}$	0.5	THF	99	82
t-Bu	2.0	1,2-hexanediol	95	80
CH_2Cl	2.0	THF	96	100
CF_3	0.5	–	99	82
CH_2CO_2Et	0.5	THF	96	80
OTBS	0.5	THF	98	83

[a]Yield based on a maximum theoretical yield of 50%

tility between the product diol and the epoxide, meanwhile, facilitated the purification of each component through fractional distillation.

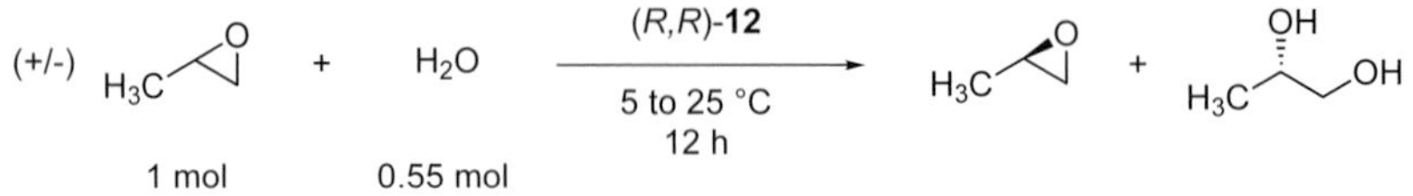

Cycle 1	44% yield	50% yield
	98.6% ee	98% ee
Cycle 2	46% yield	50% yield
	98.6% ee	98% ee
Cycle 3	48% yield	50% yield
	98.6% ee	98% ee

Scheme 18

The Co(salen) catalyst is remarkably insensitive to the steric properties of terminal epoxide substrates, as substituents ranging from methyl to cyclohexyl to *tert*-butyl groups are accommodated in the kinetic resolution. Propylene oxide presented an impressive illustration of catalyst enantiocontrol, where a k_{rel} exceeding 400 was estimated for this substrate. This epoxide served to further emphasize the synthetic utility of this process, as the HKR of 1 mole of propylene oxide proceeded efficiently with catalyst that had been recycled from previous kinetic resolutions (Scheme 18) The HKR of propylene oxide has also been effected on a multi-hundred kilogram scale in the pilot plant at ChiRex.

The hydrolytic kinetic resolution addressed a long-standing problem in enantioselective epoxide synthesis. The ability to access almost any terminal epoxide or 1,2-diol in high enantiopurity greatly expanded the chiral pool of compounds available for asymmetric synthesis. Equally important was the demonstration of practicality and efficiency that renders the ARO of a racemic mixture a synthetically viable approach.

3.3
Kinetic Resolution with Phenols

While its convenience and low cost seem to make water the ideal oxygen atom source for ARO reactions, the need for mono-protected 1,2-diols prompted the investigation of other oxygen nucleophiles in kinetic resolutions. Jacobsen reported that the Co(salen) complex 17 promoted the enantioselective addition of phenols and substituted phenols to racemic mixtures of terminal epoxides [42]. A wide range of aliphatic epoxides reacted effectively with *ortho*-, *meta*-, or *para*-substituted phenols bearing electron-donating or electron-withdrawing functionality (Scheme 19). This kinetic resolution offered an enantioselective route to β-aryloxy alcohols, which are important intermediates in pharmaceutical applications [43,44]. The dynamic kinetic resolution of epibromohydrin represented a particularly useful transformation, as the ring-opened product could be further elaborated to differentiated aryl glycidyl ethers or aryloxy propanolamines. An immobilized version of catalyst 17 was efficiently prepared that

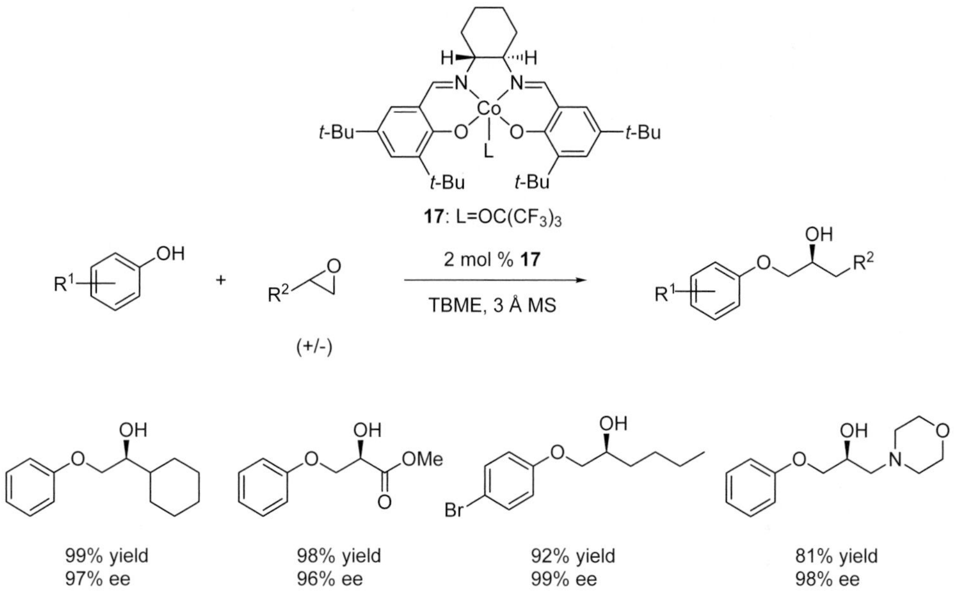
t-Bu
t-Bu
t-Bu
t-Bu
H
H
N
N
Co
O
O
L
17: L=OC(CF3)3
OH
R1
+
R2
O
(+/-)
2 mol % 17
TBME, 3 Å MS
OH
O
R2
R1
OH
O
OH
O
OMe
O
OH
O
Br
OH
O
N
O
99% yield
97% ee
98% yield
96% ee
92% yield
99% ee
81% yield
98% ee

Scheme 19

enabled the enantioselective synthesis of parallel libraries containing these compounds [45,46].

4 Conclusion

The development of asymmetric ring-opening reactions has given increased priority to the stereoselective reactions of epoxides in asymmetric synthesis. While a range of heteroatom and carbon-based nucleophiles has been explored with varying amounts of success, the list of nucleophiles to be discovered is certainly likely to continue growing. The potential to couple asymmetric ring-opening tranformations with asymmetric epoxidation methods by way of diastereo- and regioselective processes on chiral epoxides constitutes another important future challenge in ARO methodology.

References

1. Hodgson DM, Gibbs AR, Lee GP (1996) Tetrahedron 52:14361
2. Paterson I, Berrisford DJ (1992) Angew Chem Int Ed Engl 31:1179
3. Nugent WA (1992) J Am Chem Soc 114:2768
4. Yamashita H (1987) Chem Lett 525
5. Emziane M, Sutowardoyo KI, Sinou D (1988) J Organomet Chem 346:C7
6. Hayashi M, Kohmura K, Oguni N (1991) Synlett 774
7. Adolfsson H, Moburg C (1995) Tetrahedron Asymmetry 6:2023

8. Martínez LE, Leighton JL, Carsten DH, Jacobsen EN (1995) J Am Chem Soc 117:5897
9. Hansen KB, Leighton JL, Jacobsen EN (1996) J Am Chem Soc 118:10924
10. Konsler RG, Karl J, Jacobsen EN (1998) J Am Chem Soc 120:10780
11. Schaus SE, Larrow JF, Jacobsen EN (1997) J Org Chem 62:4197
12. Wu MH, Jacobsen EN (1997) Tetrahedron Lett 38:1693
13. Leighton JL, Jacobsen EN (1996) J Org Chem 61:389
14. Martínez LE, Nugent WA, Jacobsen EN (1996) J Org Chem 61:7963
15. Annis DA, Helluin O, Jacobsen EN (1998) Angew Chem Int Ed Engl 37:1907
16. Iida T, Yamamoto N, Sasai H, Shibasaki M (1997) J Am Chem Soc 119:4783
17. Wu MH, Jacobsen EN (1998) J Org Chem 63:5252
18. Wu J, Hou X-L, Dai L-X, Xia L-J, Tang M-H (1998) Tetrahedron Asymmetry 9:3431
19. Kolb HC, VanNieuwenhze MS, Sharpless KB (1994) Chem Rev 94:2483
20. Jacobsen EN, Kakiuchi F, Konsler RG, Larrow JF, Tokunaga M (1997) Tetrahedron Lett 38:773
21. Iida T, Yamamoto N, Matsunaga S, Woo H-G, Shibasaki M (1998) Angew Chem Int Ed Engl 37:2223
22. Wu MH, Hansen KB, Jacobsen EN (1999) Angew Chem Int Ed Engl 37: in press
23. Crotti P, Di Bussolo V, Favero L, Macchia F, Pineschi M (1997) Gazz Chim Acta 127:273
24. Oguni N, Miyagi Y, Itoh K (1998) Tetrahedron Lett 39:9023
25. Hayashi M, Tamura M, Oguni N (1992) Synlett 663
26. Cole BM, Shimizu KD, Krueger CA, Harrity JPA, Snapper ML, Hoveyda AH (1996) Angew Chem Int Ed Engl 35:1668
27. Chan ASC, Coleman JP (1991) J Chem Soc, Chem Commun 535
28. Denmark SE, Barsanti PA, Wong K-T, Stavenger RA (1998) J Org Chem 63:2428
29. Nugent WA (1998) J Am Chem Soc 120:7139
30. Asami M, Ishizaki T, Inoue S (1994) Tetrahedron Asymmetry 5:793
31. Asami M, Suga T, Honda K, Inoue S (1997) Tetrahedron Lett 38:6425
32. Södergren MJ, Andersson PG (1998) J Am Chem Soc 120:10760
33. Kagan HB, Fiaud JC (1987) Kinetic resolution. In: Eliel EL, Wilen SH (eds) Topics in stereochemistry. Wiley, New York, p 249
34. Noyori R, Tokunaga M, Kitamura M (1995) Bull Chem Soc Jpn 68:36
35. Larrow JF, Schaus SE, Jacobsen EN (1996) J Am Chem Soc 118:7420
36. Yields for all kinetic resolutions presented here are based on a maximum theoretical yield of 50%
37. Lebel H, Jacobsen EN (1998) J Org Chem 63:9624
38. Lebel H, Jacobsen EN (1999) Unpublished results
39. The intramolecular cyclization of epoxy alcohols provides one example of a reagent-less ARO transformation [22]
40. Tokunaga M, Larrow JF, Kakiuchi F, Jacobsen EN (1997) Science 277:936
41. Schaus SE, Brandes BD, Larrow JF, Tokunaga M, Hansen KB, Gould AE, Wu MH, Jacobsen EN (1999) submitted
42. Ready JM, Jacobsen EN (1999) J Am Chem Soc 121: in press
43. Wright JL, Gregory TF, Heffner TG, MacKenzie RG, Pugsley TA, Vander Meulen S, Wise LD (1997) Bioorg Med Chem Lett 7:1377
44. Baker NR, Byrne NG, Economides AP, Javeld T (1995) Chem Pharm Bull 43:1045
45. Annis DA, Jacobsen EN (1999) J Am Chem Soc 121:4147
46. Peukert S, Jacobsen EN (1999) submitted

Chapter 36   Polymerization Reactions

Chapter 36
Polymerization Reactions

Geoffrey W. Coates

Department of Chemistry, Baker Laboratory, Cornell University, Ithaca,
New York 14853-1301, USA
e-mail: gc39@cornell.edu

Keywords: Macromolecular stereochemistry, Main-chain chiral polymer, Atropisomeric polymer, Kinetic resolution, Anionic polymerization, Cationic polymerization, Insertion polymerization

1
Introduction

With the notable exceptions of natural rubber and gutta-percha, almost all naturally occurring polymers are optically active. Historically, interest in optically active synthetic polymers has focused on modeling natural polymers, interpreting the conformational properties of macromolecules in solution, and investigating the mechanism of polymerization reactions [1]. The synthesis of well-defined, optically active model polymers, and the understanding of their chiroptical properties, has made possible the conformational analysis of biopolymers using circular dichroism and optical rotatory dispersion [2, 3]. Isotactic polyolefins containing side-chain chirality often exhibit large, temperature-dependent optical rotations. Such characteristics provide clear experimental evidence that these polymers can maintain in solution the helical conformations that have been shown by X-ray studies to exist in the crystalline state [4, 5]. The stereoselective polymerization of racemic α-olefins has demonstrated that the chirality of the active catalytic site, not the chirality of the growing polymer chain, provides stereochemical control during polymerization using heterogeneous Ziegler-Natta catalysts [6].

Optically active synthetic polymers such as poly(trityl methacrylate) supported on silica gel [7, 8] as well as poly(ethylene glycol dimethacrylate) crosslinked in the presence of an optically active template [9] have found general use as chiral stationary phases for the optical resolution of various racemates by chromatography. A current area of investigation concerns the use of optically active polymers as reagents and catalysts for asymmetric synthesis [10, 11, 12].

Optically active polymers have been shown to exhibit different physical properties than their corresponding racemates [13]. Chiral polymers that exhibit two- or three-dimensional order [14] are valuable in applications that require piezoelectric, ferroelectric, and non-linear optical materials [15, 16]. It has been suggested that optically active polymers bearing photosensitive groups could find potential application as information storage materials, where data is preserved in the form of stereoisomeric structural variations of the polymer [17]. Polymers have several advantages over organic and inorganic materials for such applications since they exhibit mechanical stability, can be easily processed, and often permit a diverse range of functional group variation [16].

Due to the many important applications of optically active polymers, a significant goal is the development of new strategies for the synthesis of these polymers. Nature takes advantage of the ready availability of enantiopure monomers such as amino acids and sugars to construct its optically active polymers; e.g., proteins, nucleic acids and polysaccharides. For synthetic macromolecules, the analogous strategy of polymerizing optically active monomers has enjoyed considerable success, but suffers from the limited availability and/or high expense of enantiomerically enriched monomers. A far more efficient synthesis of chiral polymers is the polymerization of racemic or achiral monomers. For example, racemic monomers can be kinetically resolved using enantiopure polymerization catalysts, and achiral monomers can be enchained using enantioselective catalysts to give optically active polymers. In the field of polymer chemistry, there is considerable debate concerning the definition of catalysis as it pertains to polymer synthesis. Molecular species that construct only one polymer chain during a chemical reaction are called initiators, while those that make more than one chain are termed catalysts. This distinction between initiation and catalysis exists since the synthesis of an individual polymer chain is emphasized, not the

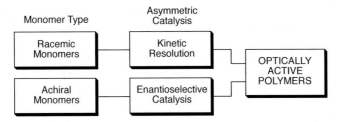

Fig. 1. Catalytic methods for the synthesis of optically active polymers from racemic and achiral monomers

addition of a monomer unit to the chain end. In contrast, if the repetitive process of monomer enchainment is the focus, then all molecular agents that produce polymer chains can be considered catalysts. The focus of this review is developments in the field of enantioselective polymerization that have occurred during the last decade [8, 18]. Therefore, all polymerization methods that involve the use of enantiopure catalysts or initiators for the direct, enantioselective synthesis of polymers from racemic or achiral monomers are reviewed (Fig. 1).

2
Chirality of Linear Macromolecules

The recognition of symmetry elements in stereoregular polymers is often much more difficult than in small molecules [19]. The identification of the symmetry properties of a conformationally flexible macromolecule is most easily carried out on the conformer exhibiting the highest symmetry. This form, most often the planar zig-zag, is conveniently represented as the Fischer projection to facilitate immediate recognition of the relative configuration of adjacent stereocenters (Fig. 2). Exactly opposite to the usual rules, horizontal lines represent bonds extending behind the plane of the paper while vertical lines depict bonds emerging from the paper [20].

Whereas small organic compounds are typically well defined with respect to their constitution, connectivity and stereochemistry, synthetic macromolecules are often inhomogeneous due to a distribution of molecular weights, defects in connectivity, different chain ends, and stereochemical defects. These impurities arise from the nature of the synthetic processes used to make the polymers, where successive steps are carried out without purification of the intermediate chain segments. In addition, it is impossible to separate polymers which differ only in minor details. For the analysis of the stereochemistry of these macromolecules, chemists are forced to use simplified models, where the polymers are assumed to have idealized structures and exist as infinitely long chains. Whereas the chirality of a finite chain can be determined using the normal criteria, an infinitely long chain contains the symmetry properties of a one-dimensional space group and translational symmetry operations must be considered. The two criteria for chirality of an infinitely long chain are (1) the absence of reflection elements of

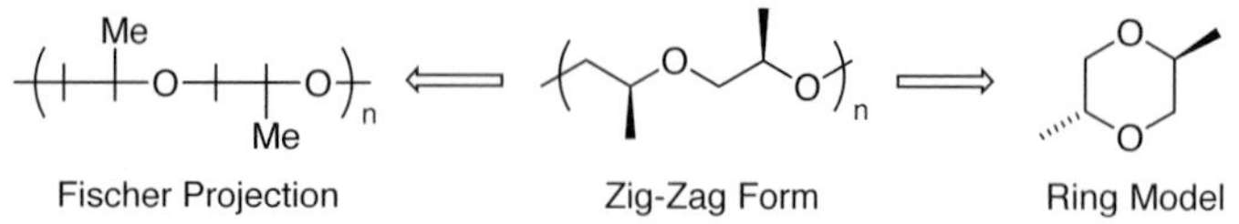

Fig. 2. Models of stereoregular polymers

R = Achiral Substituent (Me, Ph, CO_2Me, etc.)   X = Achiral Linker (C=O, O, NR, etc.)

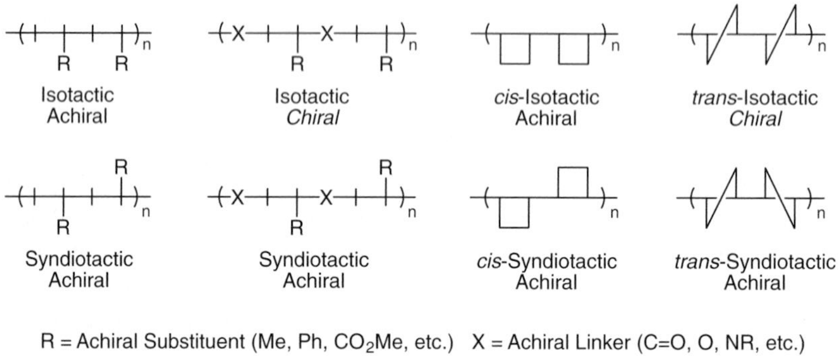

Fig. 3. Common architectures of stereoregular, linear polymers and cyclopolymers

symmetry (mirror planes) and (2) the absence of glide reflection elements (glide-mirror planes) [19].

An alternative method of identifying the chirality of a polymer modeled as an infinite chain is to draw a ring composed of the repeating unit of the polymer. If the ring has a center or plane of symmetry, then the chain is achiral (Fig. 2). For example, it is not intuitively obvious whether the polymer in Fig. 2 is chiral or achiral, but using the ring model its achirality is apparent due to the center of inversion.

The synthesis of chiral polymers from vinyl monomers requires complex architectures in order to circumvent the symmetry constraints of simple homopolymers. This was recognized in the early 1950s by Frisch [21] and Arcus [22] who examined the microstructures of stereoregular homo- and copolymers of achiral olefins. Shown above are Fischer projections for several possible stereochemical arrangements of chiral and achiral stereoregular polymers (Fig. 3). Note that simple vinyl homopolymers are achiral – only if the endgroups are different can an isotactic or syndiotactic polymer be chiral. However, the optical activity of such a polymer sample in the enantiomerically pure form will vanish as the molecular weight approaches infinity due to internal dilution [23, 24]. Such a molecule is described as being cryptochiral, where the model is chiral but the chirality of the polymer itself cannot be experimentally verified [25]. Only the structurally more complex polymers can exhibit chirality. Described in the following are synthetic strategies for the enantioselective synthesis of optically active polymers with these chiral microstructures.

3
Kinetic Resolution Polymerization of Racemic Monomers

A kinetic resolution polymerization [26] (also called 'stereoelective', 'enantio-asymmetric' and 'asymmetric-selective' polymerization) is a process where a single stereoisomer of a mixture of monomers is polymerized, giving polymers containing only one configuration of the repeating unit [27]. An example of such a process is when an enantiopure catalyst (C*) reacts with a racemic monomer (M_S, M_R) such that C* only reacts with M_S to make optically active poly(M_S), while M_R remains entirely unreacted (or vice-versa). In this ideal scenario, the ratio of the propagation rate constants ($k_S/k_R = R_{S/R}$) approaches infinity. The advantage of this procedure is that optically active polymers can then be synthesized directly from easily obtained racemic monomers. In addition, the remaining monomer, which has been resolved from the racemic mixture, can be used as a reagent for asymmetric synthesis. The disadvantage is that precious few catalytic systems (with the notable exception of biological systems) exhibit the ideal selectivity of the example described. As a result, the optical purity of the polymer formed is dependent on the degree of monomer conversion. Research in the area of kinetic resolution polymerization has focused on the polymerization of epoxides, episulfides, lactones, and olefins. Kinetic resolution in ring-opening polymerization has recently been reviewed [28, 29]. In general the selectivities have been poor, with the remaining monomer at half-conversion typically having an enantiomeric excess (ee) of less than 50% ($R_{S/R}<5$). Notable exceptions have recently been reported, though.

Sépulchre has reported the use of an easily prepared zinc-binaphthol (1) complex which gives a high degree of stereoelectivity in the polymerization of episulfides. In the polymerization of *rac*-ethylthiirane, the ee of the unreacted monomer is 66% at 46% conversion ($R_{S/R}=15$) (Scheme 1) [30]. Spassky and Sepulchre have previously reported the use of this compound for the highly selective of polymerization of *rac*-methylthiirane, where at 50% conversion, the optical purity of the unreacted monomer is 80% ($R_{S/R}=20$) [31].

Suda has reported the polymerization of *rac*-α-methylbenzyl methacrylate by a Grignard/binaphthyldiamine (2) initiator where the unreacted monomer at

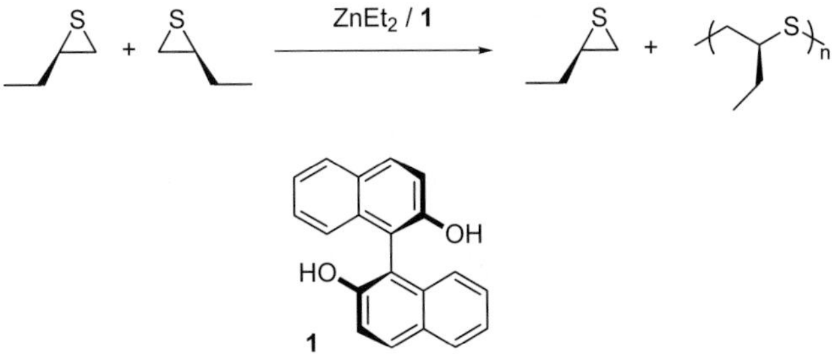

Scheme 1

57% conversion is 86% ee ($R_{S/R}=11$) (Scheme 2) [32]. Okamoto has previously shown that Grignard/(–)-sparteine (3) complexes can be used to resolve this monomer with slightly higher selectivity (monomer 94% ee at 77% conversion ($R_{S/R}=15$)) [33].

Recently, Spassky has demonstrated that the chiral Schiff base complex 4 is highly selective for the kinetic resolution polymerization of *rac*-lactide [34]. At 50% conversion, the residual monomer is approximately 80% ee ($R_{R/S}=20$) (Scheme 3). In addition, the polylactides exhibit narrow molecular weight distributions, consistent with the absence of significant transesterification.

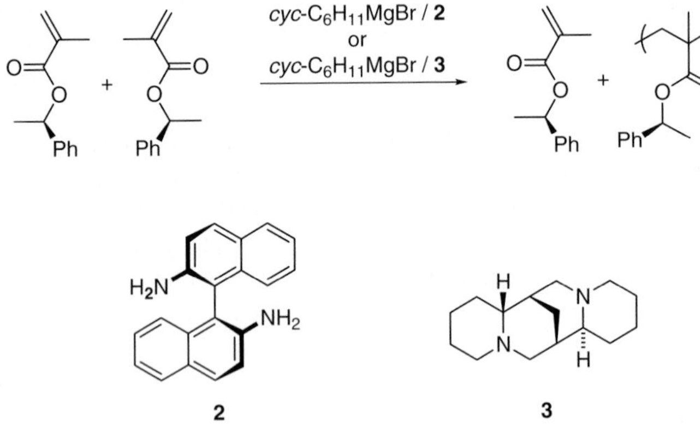

Scheme 2

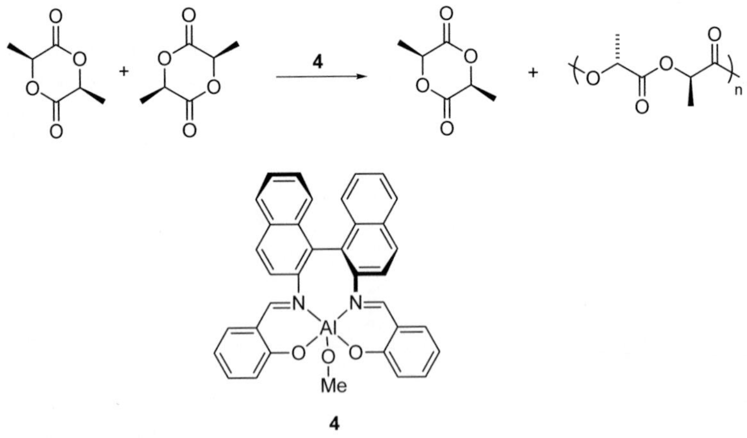

Scheme 3

4
Enantioselective Polymerization of Achiral Monomers

4.1
Atropisomeric Polymers

It is well known that isotactic polyolefins often exist as equimolar mixtures of right- and left-handed helices in the crystalline state [19, 35]. Upon dissolution they typically undergo rapid conformational changes due to a lack of rotational barriers [36]. In 1974, Drenth demonstrated that polymers bearing bulky side groups exist as stable helices in solution by resolving poly(*tert*-butyl isocyanide) into optically active fractions [37].

During the two decades after this important discovery, a tremendous amount of research has been directed toward the polymerization of sterically demanding achiral monomers with chiral initiators to create enantiomerically pure helical polymers (also known as 'helix-sense selective' or 'screw-sense-selective polymerization'). These polymers, known as atropisomers, are stable conformational isomers that arise from restricted rotation about the single bonds of their main chains. Key aspects of these reactions are enantiopure initiators that begin the polymerization with a one-handed helical twist, and monomers with bulky side-chains that can maintain the helical conformation due to steric repulsion. Notable examples of this fascinating class of polymers that are configurationally achiral but conformationally chiral include [8, 38, 39] poly(trityl methacrylate), polychloral, polyisocyanates, and polyisocyanides. Important advances in anionic and metal-based enantioselective polymerization methods have been reported in recent years.

4.1.1
Anionic Polymerization

One of the most studied polymerization systems employs alkyllithium initiators that are modified by chiral amine ligands for the polymerization of sterically bulky methacrylates [8, 38, 39, 40, 41], acrylates [42], crotonates [43], and acrylamides [44]. A primary example is the reaction of triphenylmethyl methacrylate with an initiator derived from 9-fluorenyllithium and (−)-sparteine (3) at −78 °C (Scheme 4). The resultant isotactic polymer is optically active, and is postulated to adopt a right-handed helix as it departs from the polymerization site. This polymer has been particularly successful as a chiral stationary phase for the chromatographic resolution of atropisomers [8]. Many modifications of the organolithium initiator/chiral ligand system have been explored. Recently, Okamoto has applied enantiopure radical initiators for the enantioselective polymerization of bulky methacrylate monomers [45].

Isocyanates can be polymerized by anionic initiators to give polymers that exhibit a rigid, helical conformation in solution. Elegant studies by Green and coworkers have revealed that the copolymerization of achiral isocyanates with small amounts (ca. 1%) of enantiopure chiral isocyanates yield polymers exhibiting unexpectedly high optical rotations [46]. Presumably the small amount of chiral side chains is sufficient to organize the dynamic helical polymer to a single screw-sense. Recently, Okamoto has polymerized bulky, achiral aliphatic and aromatic isocyanates using enantiopure lithium alkoxides and amides (5, 6) to produce optically active polymers (Scheme 5) [47, 48, 49]. The optical rotations of polymers produced from a given monomer were found to be highly sensitive to the nature of the initiator used, decreased with increasing polymer molecular weight, and are temperature dependent. Thus it is proposed that the chiral chain end controls the handedness of the helix despite facile intramolecular helical reversals.

The polymerization of bulky aldehydes such as chloral using enantiopure lithium alkoxides (7, 8) gives insoluble, isotactic polymers that exhibit optical activity in the solid state (Scheme 6) [8, 50, 51]. The solution and solid-state hel-

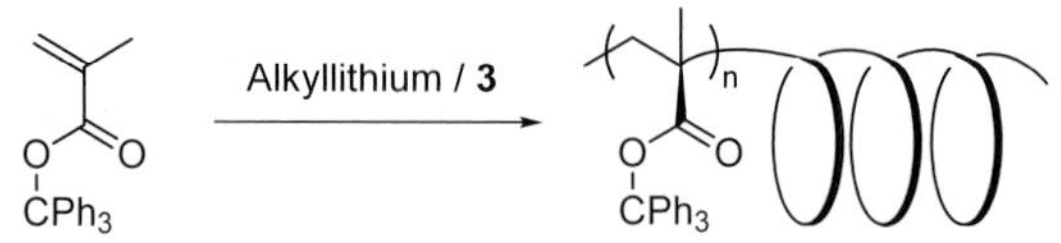

Scheme 4

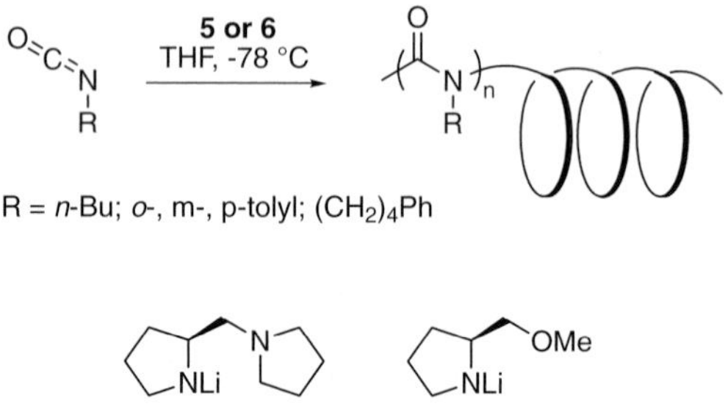

R = n-Bu; o-, m-, p-tolyl; (CH₂)₄Ph

Scheme 5

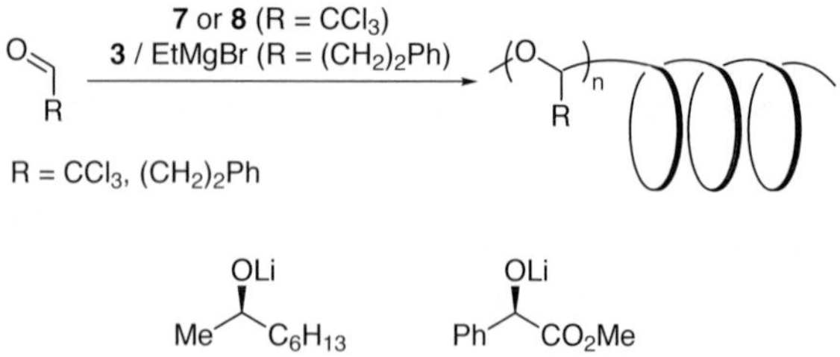

R = CCl₃, (CH₂)₂Ph

Scheme 6

ical conformation of optically active polychloral oligomers has been recently determined using NMR spectroscopic and crystallographic methods [52, 53, 54]. Okamoto has prepared optically active polymers of 3-phenylpropanal using a (–)-sparteine (3)/Grignard initiator [55]. Analysis of oligomers revealed that the growing chain-end reacts with the monomer in a Tischenko reaction to give an ester terminus.

4.1.2
Metal-Catalyzed Polymerization

Poly(isocyanides) typically exhibit stable helical conformations in solution since substituents on each main-chain atom restrict helix isomerization. Nolte and Drenth have combined Ni(II) complexes with optically active amines (9) to form enantioselective catalysts for the polymerization of *tert*-butyl isocyanide and other bulky isocyanides (Scheme 7) [56]. Nickel complexes containing 2-*tert*-butylphenyl isocyanide ligands produced polymers with exceptionally high optical rotation. Novak and Deming discovered a class of nickel carboxylate catalysts (10) for the enantioselective polymerization of *tert*-butyl and diphenylmethyl isocyanide (Scheme 7) [57]. Addition of two equivalents of cyanide increased the enantioselectivity of the polymerization. Takahashi has reported a bimetallic palladium-platinum complex (11) that produces high molecular weight, monodisperse polymers from achiral aromatic isocyanides in high yield (Scheme 7) [58]. When an enantiopure menthyl-substituted monomer is first oligomerized, the resulting single-handed helical oligomer complex can be used as an enantioselective catalyst for the polymerization of achiral isocyanides.

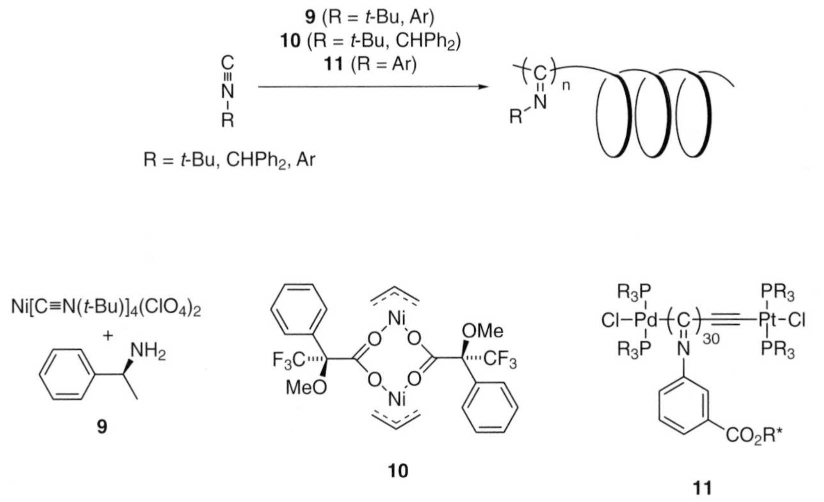

Scheme 7

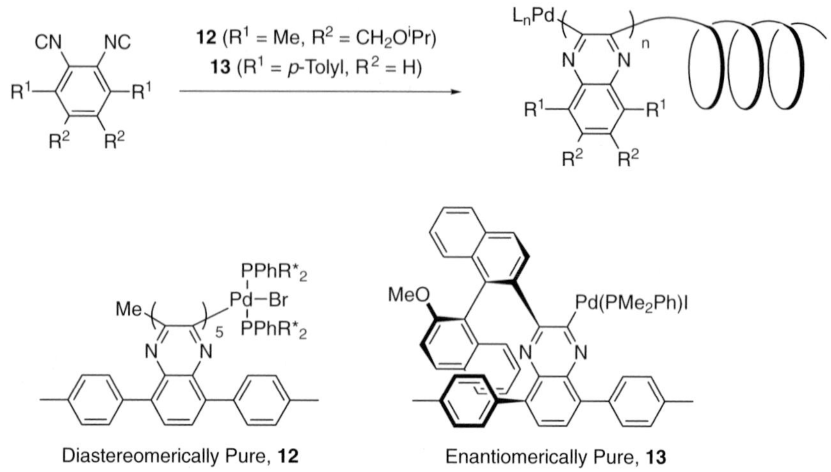

Scheme 8

Ito and coworkers have explored the use of enantiopure organopalladium complexes for the living cyclopolymerization of 1,2-diisocyanoarenes to poly(2,3-quinoxaline)s. Initial studies revealed that oligomers of sterically bulky diisocyanides were conformationally stable, and the diastereomerically pure oligomers could be resolved by chromatography with palladium complexes attached at the chain-end (12) [59, 60]. Addition of another diisocyanide monomer to the resolved oligomers yielded polymers with equal but opposite optical rotations upon cleavage of the palladium complex (Scheme 8). More recently, Ito has employed enantiopure, binaphthyl ligated palladium complexes (13) for the enantioselective cyclopolymerization of 1,2-diisocyano-3,6-di-p-tolylbenzene (Scheme 8) [61]. X-Ray structural analysis of a pentamer revealed the rigid, right-handed helical nature of the polymer.

4.2
Main-Chain Chiral Polymers

In 1961, Natta reported one of the first examples of enantioselective catalysis using a transition metal catalyst. In this reaction, an optically active polymer was formed from 1,3-pentadiene using a chiral organoaluminum/VCl_3 catalyst [62]. The optical activity of this polymer results from the main-chain chirality of polymer, where the methyl-substituted stereogenic centers are predominantly of one absolute configuration. Since this initial study, significant advances in the enantioselective synthesis of main-chain chiral polymers have been reported using ionic and metal-based techniques.

4.2.1
Metal-Catalyzed Polymerization

The discovery that group IV metallocenes can be activated by methylaluminoxane (MAO) for olefin polymerization has stimulated a renaissance in Ziegler-Natta catalysis [63]. The subsequent synthesis of well-defined metallocene catalysts has provided the opportunity to study the mechanism of the initiation, propagation, and termination steps of Ziegler-Natta polymerization reactions. Along with the advent of cationic palladium catalysts for the copolymerization of olefins and carbon monoxide [64, 65], these well-defined systems have provided extraordinary opportunities in the field of enantioselective polymerization.

4.2.1.1
Olefin Oligomerization

Isotactic polypropylene chains can be chiral only if their endgroups are different. Enantiomerically pure zirconocene/MAO catalysts have been employed to form isotactic polypropylene; as expected the high polymer in solution was optically inactive [66]. Anticipating that the oligomers from these catalysts should be optically active, Pino used (*R*)-**14**/MAO for the asymmetric oligomerization of propylene, 1-pentene, and 4-methyl-1-pentene using hydrogen as a chain transfer agent (Scheme 9) [67]. In the case of propylene, approximately 90% of the products (x<47) had a measurable optical activity. The oligomeric alkane fractions were characterized by polarimetry, and their absolute configurations were used to unambiguously determine the enantiofacial preference of the metallocene catalyst for the first time [23]. Fuhrmann has recently used $MgCl_2$-supported enantiopure titanium complexes for the synthesis of enantiomerically-enriched oligomers of 1-butene in up to 71% ee [68]. The aluminum-terminated oligomers can be hydrolyzed to the alkane, or oxidized to produce alcohols.

By raising the reaction temperature and lowering the olefin concentration, Kaminsky synthesized optically active olefin-terminated oligomers *via* β-hydrogen elimination chain transfer (Scheme 9) [24]. Propylene and 1-butene were oligomerized using (*S*)-**15**/MAO predominantly to products where 0<x<5. Although these functionalized alkene oligomers are of greater synthetic interest than the related saturated compounds, they are typically formed in lower percent enantiomeric excess (% ee) due to higher reaction temperatures.

$$(R)\text{-}\mathbf{14} / MAO$$
$$\text{Toluene, } H_2$$

R = Me, n-Pr, i-Bu

$$(S)\text{-}\mathbf{15} / MAO$$
$$\text{Toluene, 50 }^\circ C$$

R = Me, Et

(S)-**14**; L₂ = (S)-binaphtholate
(S)-**15**; L = O-acetyl-(S)-mandelate
(S)-**16**; L = biphenyl-2,2'-diolate

Catalyst	R	x	Temp. (°C)	ee (%)	Reference
(R)-**14**	Me	1	0	100	[67]
(R)-**14**	n-Pr	0	25	50	[67]
(R)-**14**	i-Bu	0	25	84	[67]
(S)-**15**	Et	1	50	27	[24]

Scheme 9

4.2.1.2
Diolefin Cyclopolymerization

Nonconjugated diolefins can be polymerized in an insertion and cyclization sequence, resulting in a polymer containing rings in the main chain. Whereas vinyl polymers have only two structures of maximum order (isotactic, syndiotactic), cyclopolymers are inherently more complicated. Fig. 3 shows that the *trans*-
isotactic microstructure contains no mirror planes of symmetry and is thus chiral by consequence of its main-chain stereochemistry. There are two criteria for
chirality of this polymer: (1) isotacticity (the same relative stereochemistry of
every other stereocenter); and (2) the presence of *trans* rings. The enantiofacial
selectivity of the first olefin insertion determines the tacticity of the cyclopolymer, and the diastereoselectivity of the cyclization step determines whether *cis*
or *trans* rings are formed. Using homogeneous Ziegler-Natta catalysts, Coates
and Waymouth have studied the effect of the catalyst geometry on the enantioselectivity and diastereoselectivity of ring formation with various α,ω-dienes [69,
70, 71]. Metallocene (R)-**14**/MAO produces an optically-active cyclopolymer of
1,5-hexadiene with $[\Phi]_{405}^{28}=+51.0^\circ$ (Scheme 10). Cyclopolymerization with (S)-
14 afforded the enantiomeric polymer, $[\Phi]_{405}^{28}=-51.2^\circ$. Microstructural analysis of the polymer by ^{13}C-NMR revealed a *trans* ring content of 72% and an enan-

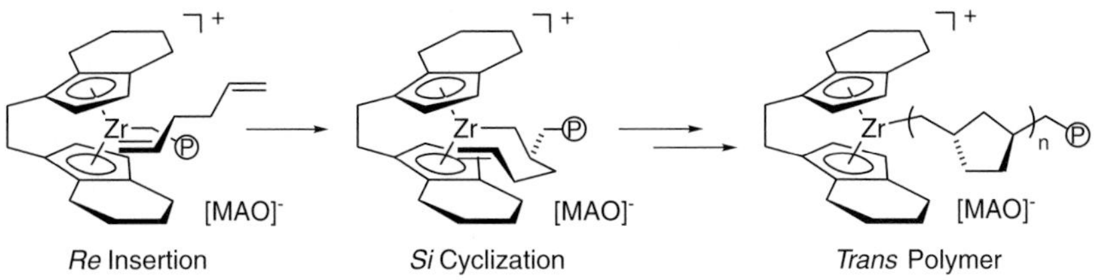

Re Insertion   *Si* Cyclization   *Trans* Polymer

Scheme 10

tiofacial selectivity for olefin insertion of 91%. The high degree of stereoregularity and predominance of *trans* rings are responsible for the optical activity of the polymer. Okamoto has recently carried out similar reactions using chromatographically resolved **16** (and the analogous hafnocene complex) and obtained comparable results [72, 73].

4.2.1.3
Alternating Copolymerization of Olefins and Carbon Monoxide

The synthesis of alternating copolymers from carbon monoxide (CO) and olefins using palladium catalysts is currently an area of intense research. In cases where α-olefins are used, the regiochemistry (head/tail orientations) and stereochemistry (tacticity) of olefin insertion have a strong influence on the physical and mechanical properties of the polymers. Unlike regioregular α-olefins homopolymers, these copolymers have a directionality along the polymer backbone due to the incorporation of CO. Therefore isotactic, regioregular CO/α-olefin polymers are chiral by virtue of their main-chain stereochemistry (Scheme 11).

In the early 1980s, it was discovered that cationic palladium catalysts with bidentate tertiary phosphines exhibited remarkable reaction rates for olefin/CO copolymerization [64, 65]. Although initial studies using bidentate arylphosphines produced CO/propylene polymers with poor regioregularity, it was later revealed that bidentate alkylphosphines and/or chiral phosphines produced polymers with a much higher degree of regioregularity. In the early 1990s, the first reports concerning the use of enantiopure, C_2-symmetrical ligated catalysts for the enantioselective copolymerization of α-olefins and CO began to appear.

In a 1990 patent, Wong briefly noted that palladium-based catalysts form optically-active propylene/CO copolymers when the enantiopure C_2-symmetrical phoshine ligands **17** and **18** are employed [74]. In 1992, Consiglio and coworkers published several papers concerning the use of enantiopure bidentate phosphine ligands in CO/propylene copolymerization [75, 76, 77]. The copolymers formed using ligands **19** and **20** were highly regioregular, and because of the simplicity of the ^{13}C-NMR spectra it was proposed that the polymers were isotactic. Proof of the isotactic microstructure (using ligand **20**) came when a circular dichroism spectrum of the copolymer revealed an intense band in the n-π*

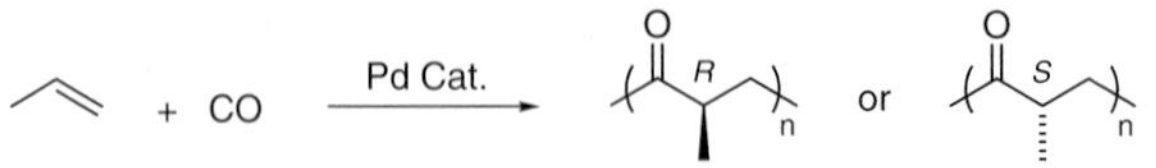

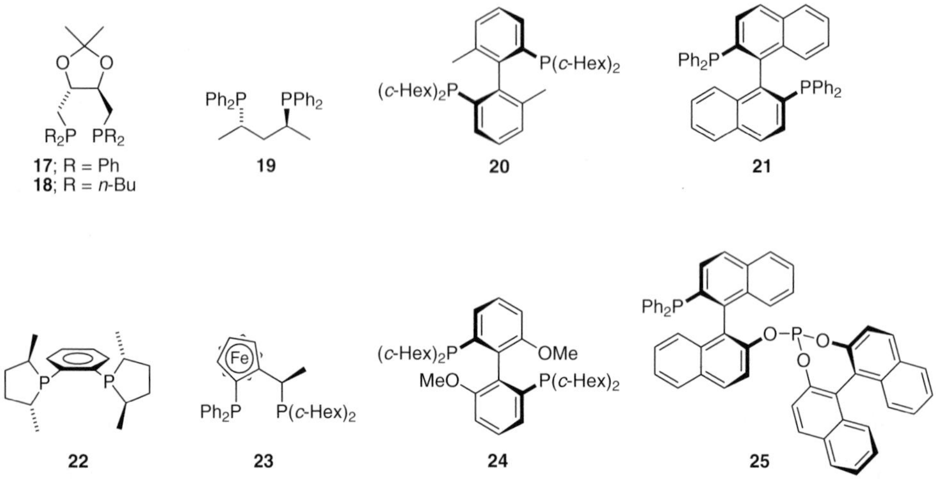

Catalyst[a]	Yield ($g_{poly}g_{Pd}^{-1}$)	Comment	Reference
17/A	93	$[\alpha]_D^{25}=+6.6°$ (HFIP)	[74]
18/A	703	$[\alpha]_D^{25}=+10.4°$ (HFIP)	[74]
20/A	NR[b]	Highly regioregular; $\Delta\varepsilon=-1.56$ L mol^{-1} cm^{-1}; $[\alpha]_D^{20}=+26°$; $T_m=245$ °C	[77, 78]
17/B	500	78% H-T linkages; $[\alpha]_D^{20}=-7°$ (CH$_2$Cl$_2$)	[80]
19/B	630	76% H-T linkages; $[\alpha]_D^{20}=-29°$ (CH$_2$Cl$_2$)	[80]
21/B	300	66% H-T linkages; $[\alpha]_D^{20}=+36°$ (CH$_2$Cl$_2$)	[80]
22/B	1462	$M_n=36,000$; $[\alpha]_D^{25}=-30°$ (HFIP); $[\alpha]_D^{25}=+70°$ (CHCl$_3$)	[84]
23/A	2975	$M_n=6300$; 99% H-T linkages; $\Delta\varepsilon=+1.84$ L mol^{-1} cm^{-1}	[85]
24/A	391	$M_n=6900$; 100% H-T linkages; $\Delta\varepsilon=+1.73$ L mol^{-1} cm^{-1}; $[\alpha]_D^{25}=-29.1°$ (HFIP); $T_m=237$ °C	[86, 83]
25/C	284	$M_n=65,000$; 100% H-T linkages; $[\alpha]_D^{24}=+57.2°$ (HFIP); $T_m=164$ °C	[82]

[a] A = Pd(OAc)$_2$, Ni(ClO$_4$)$_2$, naphthoquinone; B = [Pd(MeCN)$_2$](BF$_4$)$_2$; C = Pd(1,5-cyclooctadiene)(Cl)(Me), Na[B(3,5-(CF$_3$)$_2$C$_6$H$_3$)$_4$].
[b] Not reported.

Scheme 11

region (275 nm; $\Delta\varepsilon=-1.56$ L mol^{-1} cm^{-1}) [77]. It was later shown that this polymer exhibits a specific optical rotation of $[\alpha]_D^{20}=+26°$ [78]. Such chiroptical properties are only possible with an isotactic microstructure, where an excess of propylene units are enchained with the same absolute configuration. Interestingly, this polymer is isolated from the reaction in the spiroketal form, but is converted to the polyketone by dissolution in hexafluoroisopropanol (HFIP) followed by precipitation with methanol [79].

Sen and coworkers have synthesized optically active propylene/CO copolymers using more traditional chiral phosphines (**17, 19, 21**) [80]. Chien had previously carried out related polymerizations using **17** and **21** as ligands, but no chirooptical properties were reported [81]. Sen demonstrated that catalysts with ligands **17, 19**, and **21** gave moderately regioregular polymers with 66–78% head-to-tail linkages. The polymer formed using the atropisomeric binaphthyl ligand exhibited the highest optical rotation, although an enantioselectivity of the reaction was not reported.

In 1995, several new catalysts were reported to exhibit very high degrees of regioselectivity, stereoselectivity, and enantioselectivity for the synthesis of propylene/CO copolymers. From these studies, there is now good evidence that the regiochemistry results from a primary insertion of propylene into the Pd-acyl bond [78, 82]. There is agreement concerning the absolute configuration of these polymers. Based on the sign of the CD band [83] and from the isolation of oligomers of known absolute configuration [82], the copolymer that is dextrorotatory in HFIP is assigned the S-configuration. Note that the same polymer is *levorotatory* in chlorinated solvents [84]. Therefore it is difficult to measure and compare degrees of enantioselectivity since the specific optical rotations of these polyketones are also extremely sensitive to the sample concentration, temperature, and the molecular weight of the polymer [84]. In addition, chirooptical values for the pure polymers are unknown. Nevertheless, it is clear that these catalysts exhibit excellent selectivities. Sen has reported a highly enantioselective copolymerization using a Duphos-ligated (**22**) catalyst [84]. By measuring the ^{13}C-NMR spectrum of the copolymer in the presence of chiral shift reagents, an enantioselectivity of greater than 90% was determined. Consiglio has exploited ferrocene-based mixed aliphatic/aromatic phosphines (**23**) to produce optically-active copolymers with a purported enantioface selectivity of 97% [85]. Aliphatic phosphine **24** can be used to produce polymers where the % ee of olefin insertion is as high as 98% [83, 86]. Takaya has used a phosphine-phosphite bidentate ligand (**25**) to produce a copolymer with the highest reported specific optical rotation in HFIP ($[\alpha]_D^{24}$=+57.2°) [82].

In addition to propylene, other nonconjugated olefins have been copolymerized with CO using enantiopure palladium catalysts. Allylbenzene, 1-butene, 1-heptene, 4-methyl-1-pentene, and *cis*-2-butene [84, 85] as well as hydroxy- and carboxylic acid-functionalized monomers [87] have been polymerized to give optically active polymers. Waymouth, Takaya and Nozaki have recently reported the enantioselective cyclocopolymerization of 1,5-hexadiene and CO [88, 89].

The enantioselective copolymerization of styrenes and CO has also been achieved (Scheme 12). Using bidentate pyridine-imine ligands (**26**), Sen synthesized optically active styrene and 4-methylstyrene copolymers [80]. Based on a microstructural analysis, a 36% ee for olefin insertion was reported. Brookhart employed a C_2-symmetrical bisoxazoline complex (**27**) to produce styrene-based

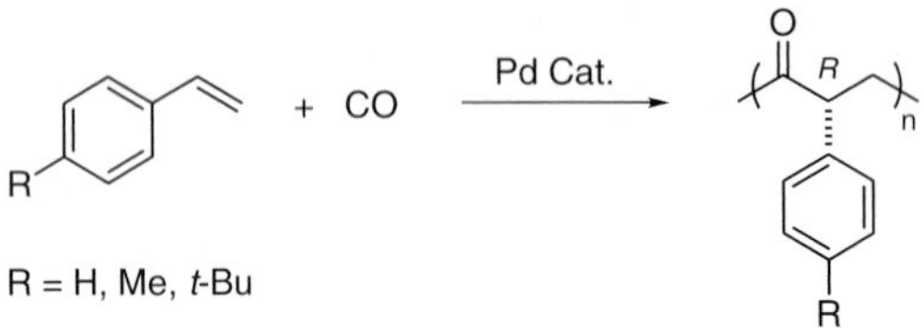

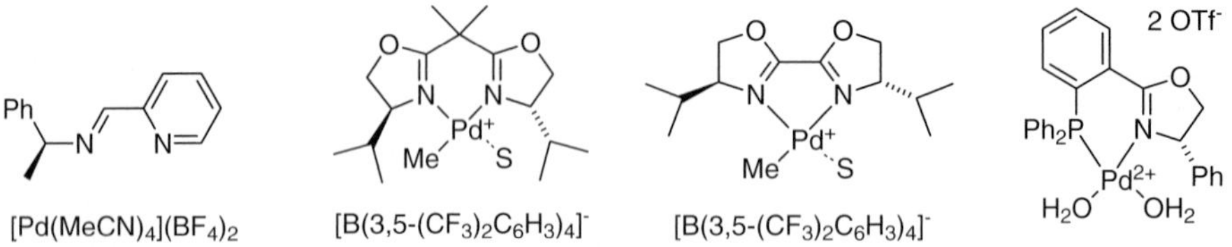

R = H, Me, t-Bu

[Pd(MeCN)$_4$](BF$_4$)$_2$	[B(3,5-(CF$_3$)$_2$C$_6$H$_3$)$_4$]$^-$	[B(3,5-(CF$_3$)$_2$C$_6$H$_3$)$_4$]$^-$	2 OTf$^-$
26	**27**	**28**	**29**

Catalyst[a]	R	Yield ($g_{poly}g_{Pd}^{-1}$)	Comment	Reference
26/A	H	NR[b]	$[\alpha]_D^{20}=-16°$ (CH$_2$Cl$_2$); M$_n$=5,600	[80]
26/A	Me	NR	$[\alpha]_D^{20}=-14°$ (CH$_2$Cl$_2$); M$_n$=14,000	[80]
27	t-Bu	179	$[\alpha]_D^{25}=-284°$ (CH$_2$Cl$_2$); M$_n$=26,000	[90]
28	H	13	$[\alpha]_D^{25}=-348°$ (Cl$_2$HCCHCl$_2$)	[92]
28	Me	66	$[\alpha]_D^{25} = -350°$ (CH$_2$Cl$_2$)	[92]
29	H	NR	$[\alpha]_D^{25}=-403°$ (HFIP/CHCl$_3$)	[93]
25/B	t-Bu	192	$[\alpha]_D^{23}=-261°$ (CH$_2$Cl$_2$); M$_n$=4,300	[82]

[a] A =[Pd(MeCN)$_2$](BF$_4$)$_2$; B=Pd(1,5-cyclooctadiene)(Cl)(Me), Na[B(3,5-(CF$_3$)$_2$C$_6$H$_3$)$_4$].
[b] Not reported

Scheme 12

copolymers that exhibited extremely high specific optical rotations [90]. After consideration of previous mechanistic studies and molecular models, it was proposed that propagation occurred by a 2,1-insertion mechanism to give an R-stereocenter. Brookhart has also devised a clever ancillary ligand exchange, where the chiral bisoxazoline ligand is replaced with an achiral bipyridine ligand during chain formation to create an optically active stereoblock polymer [91]. Musco [92] and Consiglio [93] have used palladium-oxazoline complexes (**28, 29**) to produce styrenic copolymers that exhibit high optical rotations. Takaya has briefly noted the use of his novel phosphine-phosphite bidentate ligands for the enantioselective production of aromatic polyketones [82].

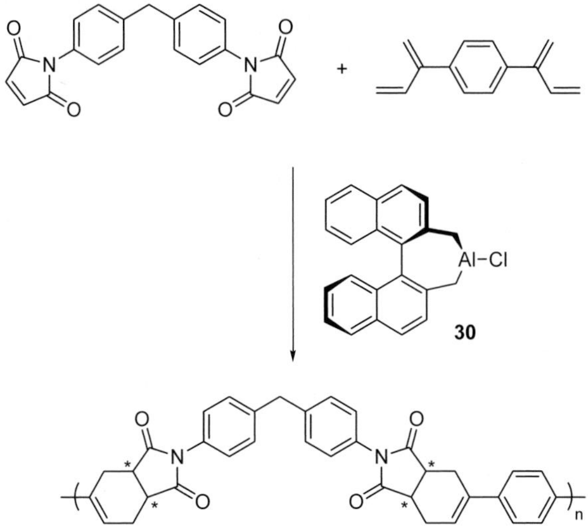

Scheme 13

4.2.1.4
Diels-Alder Polymerization

Recently, Itsuno described the first enantioselective Diels-Alder-based polymer synthesis using a chiral Lewis acid-mediated reaction between a bisdiene and a bisdienophile (Scheme 13) [94]. The bismaleimide and bisdiene were allowed to react with the enantiopure aluminum complex **30** (20 mol %) at −30 °C to give polymers with molar optical rotations ($[\Phi]_D$) as high as +243°. No information concerning the degree of stereochemical purity or absolute configuration of the polymer was given.

4.2.2
Anionic Polymerization

In 1960, Natta reported the first direct synthesis of an optically active polymer from an achiral monomer, where methyl sorbate was polymerized using (R)-2-pentyllithium [95]. Ozonolysis of the polymer (under conditions possibly allowing epimerization) produced (S)-methyl succinic acid in 5% ee, which provides evidence of asymmetric induction and absolute configuration of the polymer main chain. Since this initial report, a remarkable void in the literature exists concerning the synthesis of main-chain chiral polymers from achiral monomers using anionic initiators. Okamoto and Oishi have polymerized N-substituted maleimides with chiral anionic initiators (Scheme 14) [96, 97]. The polymer is assumed to have predominantly a *trans*-diisotactic microstructure, which adopts a secondary helical structure. The absolute configuration of the main chain has

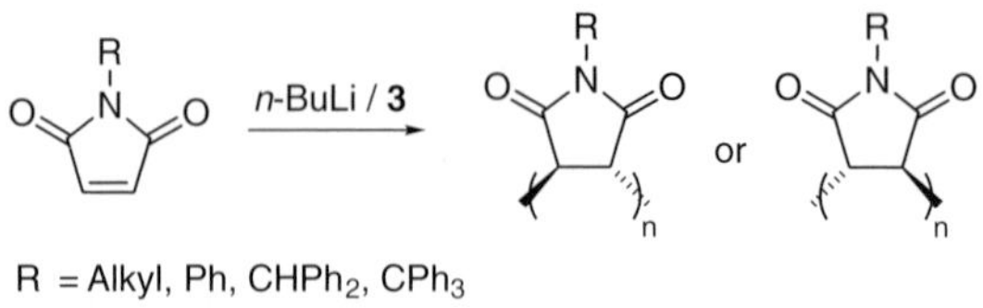

R = Alkyl, Ph, CHPh$_2$, CPh$_3$

Scheme 14

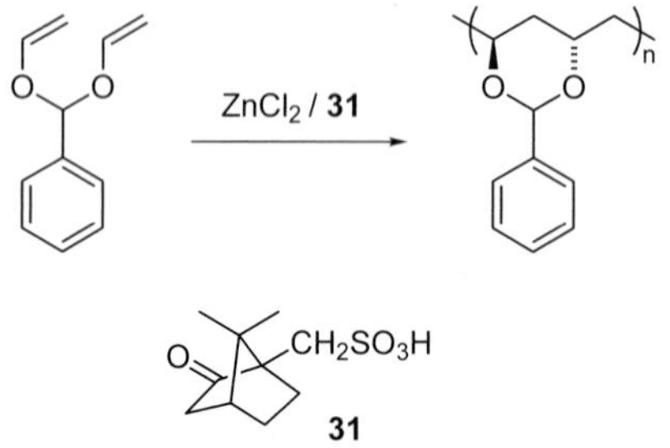

Scheme 15

not been assigned. Several studies have documented the influence of the N-substituent on the stereoselectivity of the polymerization [96, 98, 99]. More recently, Oishi has found that organolithium/bisoxazoline initiators produce polyimides that exhibit lower optical rotations [100].

4.2.3
Cationic Polymerization

Natta accomplished the first enantioselective cationic polymerization by reacting benzo- and naphthofurans with chiral cationic initiators [101, 102, 103]. Using aluminum trichloride modified with (S)-phenylalanine, the optical rotation of the polymer formed is independent of molecular weight, suggesting enantiomorphic site control. Autocatalytic behavior was also noted, as higher optical activities are obtained in polymerizations where a small amount of the optically active polymer is initially present. The absolute configuration of the polymer is unknown, although the microstructure is proposed to be *trans*-diisotactic.

Kakuchi and Yokota have reported the enantioselective cyclopolymerization of divinyl acetal and divinyl catechol using a chiral cationic initiator **31**/ZnCl$_2$ (Scheme 15) [104, 105]. The resultant polymer contains only cyclic units, however the relative ratio of *cis* and *trans* rings is not reported. By comparing the CD spectra of the polymer and a model compound, it was suggested that the *trans* rings of the polymer were predominantly of one absolute configuration.

5
Principal Alternatives

There are several alternative methods for the synthesis of optically active polymers from achiral or racemic monomers that do not involve polymerization catalysts. Optically active polymers have been formed from achiral dienes immobilized in a chiral host lattices [106]. In these reactions, the chiral matrix serves as a 'catalyst' and can be recovered following the reaction. For example, 1,3-pentadienes have been polymerized in perhydrotriphenylene and apocholic acid hosts, where asymmetric induction occurs via through-space interactions between the chiral host and the monomer [107, 108]. The resultant polymers are optically active, and the optical purities of the ozonolysis products are as high as 36%. In addition, achiral monomers have been found to pack in chiral crystals with the orientations necessary for topochemical solid-state polymerization [109]. In these reactions, the scientist is the 'enantioselective catalyst' who separates the enantiomeric crystals. The oligomers, formed by a $[2\pi+2\pi]$ asymmetric photopolymerization, can be obtained in the enantiomeric pure form [110].

Achiral polymers synthesized from achiral monomers have been modified after polymerization using asymmetric catalysts to yield optically active polymers. For example, enantioselective ketone reduction, hydrogenation, olefin epoxidation, and olefin hydroxylation have been carried on the functional groups of achiral polymers [111, 112]. Such functionalizations, however, are often incomplete or occur with a low degree of asymmetric control.

References

1. Ciardelli F (1987) Optically Active Polymers. In:Kroschwitz JI (ed) Encyclopedia of Polymer Science and Engineering. Wiley, New York, p 463
2. Ciardelli F, Pieroni O (1980) Chimia 34:301
3. Ciardelli F, Salvadori P (1985) Pure Appl Chem 57:931
4. Pino P (1965) Adv Polym Sci 4:393
5. Pino P, Ciardelli F, Zandomeneghi M (1970) Annu Rev Phys Chem 21:561
6. Pino P, Ciardelli F, Lorenzi GP, Natta G (1962) J Am Chem Soc 84:1487
7. Okamoto Y, Aburatani R (1989) Polym News 14:295
8. Okamoto Y, Nakano T (1994) Chem Rev 94:349
9. Wulff G (1991) Polym News 16:167
10. Han HS, Janda KD (1997) Tetrahedron Lett 38:1527
11. Herrmann WA, Kratzer RM, Blumel J, Friedrich HB, Fischer RW, Apperly DC, Mink J, Berkesi O (1997) J Mol Catal A:Chem 120:197
12. Nandanan E, Sudalai A, Ravindranathan T (1997) Tetrahedron Lett 38:2577
13. Jiang Z, Boyer M, Sen A (1995) J Am Chem Soc 117:7037
14. Moore JS, Stupp SI (1992) J Am Chem Soc 114:3429
15. Williams DJ (1984) Angew Chem Int Ed Engl 23:690
16. Wulff G (1989) Angew Chem Int Ed Engl 28:21
17. Ciardelli F, Pieroni O, Fissi A, Carlini C, Altomare A (1989) Br Polym J 21:97
18. Belfield K, Belfield J (1995) Trends in Polymer Science 3:180
19. Farina M (1987) Top Stereochem 17:1
20. IUPAC Macromolecular Nomenclature Commission (1981) Pure Appl Chem 53:733
21. Frisch HL, Schuerch C, Szwarc M (1953) J Poly Sci 11:559

22. Arcus CL (1955) J Chem Soc 2801
23. Pino P, Cioni P, Wei J (1987) J Am Chem Soc 109:6189
24. Kaminsky W, Ahlers A, Möller-Lindenhof N (1989) Angew Chem Int Ed Engl 28:1216
25. Mislow K, Bickart P (1976/77) Isr J Chem 15:1
26. Moss G (1996) Pure Appl Chem 68:2193
27. Kagan H, Fiaud J (1988) Top Stereochem 18:249
28. Tsuruta T, Kawakami Y (1989) Anionic ring-opening polymerization: stereospecificity for epoxides, episulfides and lactones. In: Allen G, Bevington J (ed) Comprehensive polymer science. Pergamon Press, Oxford, p 489
29. Spassky N, Dumas P, Le Borgne A, Momtaz A, Sepulchre M (1994) Bull Soc Chim Fr 131:504
30. Sépulchre M (1987) Makromol Chem 188:1583
31. Sépulchre M, Spassky N, Mark C, Schurig V (1981) Makromol Chem Rapid Commun 2:261
32. Kanoh S, Kawaguchi N, Suda H (1987) Makromol Chem 188:463
33. Okamoto Y, Ohta K, Yuki H (1977) Chem Lett 617
34. Spassky N, Wisniewski M, Pluta C, Leborgne A (1996) Macromol Chem Phys 197:2627
35. Vogl O, Jaycox GD (1987) Polymer 28:2179
36. Hoffman RW (1992) Angew Chem Int Ed Engl 31:1124
37. Nolte RJM, van Beijnen AJM, Drenth W (1974) J Am Chem Soc 96:5932
38. Okamoto Y (1996) Macromol Symp 101:343
39. Okamoto Y, Nakano T, Habaue S, Shiohara K, Maeda K (1997) J Macromol Sci, Pure Appl Chem A34:1771
40. Nakano T, Okamoto Y, Hatada K (1992) J Am Chem Soc 114:1318
41. Nakano T, Okamoto Y, Hatada K (1995) Polymer J 27:892
42. Habaue S, Tanaka T, Okamoto Y (1995) Macromolecules 28:5973
43. Ute K, Asada T, Nabeshima Y, Hatada K (1993) Macromolecules 26:7086
44. Okamoto Y, Hayashida H, Hatada K (1989) Polymer J 21:543
45. Nakano T, Shikisai Y, Okamoto Y (1996) Polymer J 28:51
46. Green M, Reidy M, Johnson R, Darling G, Oleary D, Willson G (1989) J Am Chem Soc 111:6452
47. Maeda K, Matsuda M, Nakano T, Okamoto Y (1995) Polymer J 27:141
48. Okamoto Y, Matsuda M, Nakano T, Yashima E (1993) Polymer J 25:391
49. Okamoto Y, Matsuda M, Nakano T, Yashima E (1994) J Poly Sci Part A:Poly Chem 32:309
50. Vogl O, Kratky C, Simonsick WJ, Xi F, Hatada K (1992) Makromol Chem Macromol Symp 53:151
51. Jaycox GD, Vogl O (1991) Polym J 23:1213
52. Vogl O, Xi F, Vass F, Ute K, Nishimura T, Hatada K (1989) Macromolecules 22:4658
53. Ute K, Oka K, Okamoto Y, Hatada K, Xi F, Vogl O (1991) Polymer J 23:1419
54. Ute K, Hirose K, Kashimoto H, Hatada K, Vogl O (1991) J Am Chem Soc 113:6305
55. Choi S, Yashima E, Okamoto Y (1996) Macromolecules 29:1880
56. Kamer PCJ, Nolte RJM, Drenth W (1988) J Am Chem Soc 110:6818
57. Deming TJ, Novak BM (1992) J Am Chem Soc 114:7926
58. Takei F, Yanai K, Onitsuka K, Takahashi S (1996) Angew Chem Int Ed Engl 35:1554
59. Ito Y, Ihara E, Murakami M, Sisido M (1992) Macromolecules 25:6810
60. Ito Y, Ihara E, Murakami M (1992) Angew Chem Int Ed Engl 31:1509
61. Ito Y, Ohara T, Shima R, Suginome M (1996) J Am Chem Soc 118:9188
62. Natta G, Porri L, Carbonaro A, Lugli G (1961) Chim Ind (Milano) 43:529 (Chem Abstr 61:727e)
63. Brintzinger HH, Fischer D, Mulhaupt R, Rieger B, Waymouth RM (1995) Angew Chem Int Ed Engl 34:1143
64. Sen A (1993) Acc Chem Res 26:303
65. Drent E, Budzelaar PHM (1996) Chem Rev 96:663
66. Kaminsky W, Möller-Lindenhof N (1990) Bull Soc Chim Belg 99:103
67. Pino P, Galimberti M, Prada P, Consiglio G (1990) Makromol Chem 191:1677

68. Fuhrmann H, Kortus K, Fuhrmann C (1996) Macromol Chem Phys 197:3869
69. Coates GW, Waymouth RM (1991) J Am Chem Soc 113:6270
70. Coates GW, Waymouth RM (1992) J Mol Catal 76:189
71. Coates GW, Waymouth RM (1993) J Am Chem Soc 115:91
72. Habaue S, Sakamoto H, Okamoto Y (1997) Polymer J 29:384
73. Habaue S, Sakamoto H, Baraki H, Okamoto Y (1997) Macromol Rapid Commun 18:707
74. Wong P (Shell) Eur. Pat. Appl. 0384517 (1990) (Chem Abstr 114:103079j)
75. Barsacchi M, Batistini A, Consiglio G, Suter UW (1992) Macromolecules 25:3604
76. Batistini A, Consiglio G, Suter UW (1992) Angew Chem Int Ed Engl 31:303
77. Batistini A, Consiglio G, Suter UW (1992) Polym Mater Sci Eng 67:104
78. Bronco S, Consiglio G, Hutter R, Batistini A, Suter UW (1994) Macromolecules 27:4436
79. Bastistini A, Consiglio G (1992) Organometallics 11:1766
80. Jiang Z, Adams SE, Sen A (1994) Macromolecules 27:2694
81. Xu FY, Zhao AX, Chien JCW (1993) Makromol Chem 194:2579
82. Nozaki K, Sato N, Takaya H (1995) J Am Chem Soc 117:9911
83. Bronco S, Consiglio G (1996) Macromol Chem Phys 197:355
84. Jiang Z, Sen A (1995) J Am Chem Soc 117:4455
85. Bronco S, Consiglio G, Di Benedetto S, Fehr M, Spindler F, Togni A (1995) Helv Chim Acta 78:883
86. Amevor E, Bronco S, Consiglio G, Di Benedetto S (1995) Macromol Symp 89:443
87. Kacker S, Jiang Z, Sen A (1996) Macromolecules 29:5852
88. Borkowsky SL, Waymouth RM (1996) Macromolecules 29:6377
89. Nozaki K, Sato N, Nakamoto K, Takaya H (1997) Bull Chem Soc Jpn 70:659
90. Brookhart M, Wagner MI, Balavoine GGA, Haddou HA (1994) J Am Chem Soc 116:3641
91. Brookhart M, Wagner MI (1996) J Am Chem Soc 118:7219
92. Bartolini S, Carfagna C, Musco A (1995) Macromol Rapid Commun 16:9
93. Sperrle M, Aeby A, Consiglio G, Pfaltz A (1996) Helv Chim Acta 79:1387
94. Itsuno S, Tada S, Ito K (1997) J Chem Soc Chem Commun 933
95. Natta G, Farina M, Donati M, Peraldo M (1960) Chim Ind (Milan) 42:1363 (Chem Abstr 55:24089a)
96. Oishi T, Yamasaki H, Fujimoto M (1991) Polym J 23:795
97. Okamoto Y, Nakano T, Kobayashi H, Hatada K (1991) Polym Bull 25:5
98. Liu W, Chen C, Chen Y, Xi F (1997) J Macromol Sci, Pure Appl Chem A34:327
99. Liu W, Chen C, Chen Y, Xi F (1997) Polym Bull 38:509
100. Onimura K, Tsutsumi H, Oishi T (1997) Polym Bull 39:437
101. Natta G, Farina M, Peraldo M, Bressan G (1961) Makromol Chem 43:68
102. Farina M, Bressan G (1963) Makromol Chem 61:79
103. Bressan G, Farina M, Natta G (1966) Makromol Chem 93:283
104. Haba O, Kakuchi T, Yokota K (1993) Macromolecules 26:1782
105. Haba O, Obata M, Yokota K, Kakuchi T (1997) J Poly Sci Part A:Poly Chem 35:353
106. Farina M (1984) Inclusion Polymerization. In: Atwood JL, Davies JED, MacNicol DD (eds) Inclusion Compounds. Academic Press, London, p 297
107. Farina M, Audisio G, Natta G (1967) J Am Chem Soc 89:5071
108. Mijata M, Kitahara Y, Takemoto K (1983) Makromol Chem 184:1771
109. Hasegawa M (1995) Adv Phys Org Chem 30:117
110. Addadi L, van Mil J, Gati E, Lahav M (1981) Macromol Chem Phys, Suppl 4:37
111. Sélégny E, Merle-Aubry L (1979) General methods of synthesis of optically active polymers. In: Sélégny E (ed) Optically Active Polymers. D. Reidel, Dordrecht, p 15
112. Cernerud M, Reina JA, Tegenfeldt J, Moberg C (1996) Tetrahedron: Asymmetry 7:2863

Chapter 37 Heterogeneous Catalysis

Chapter 37
Heterogeneous Catalysis

Hans-Ulrich Blaser · Martin Studer

Novartis Services AG, Catalysis & Synthesis Services, R 1055.6.28, CH-4002 Basel, Switzerland
e-mail: hans-ulrich.blaser@sn.novartis.com; martin.studer@sn.novartis.com

Keywords. β-Functionalized ketones, α-Keto acid derivatives, Cinchona modified Pt catalysts, Chiral imprints, Chiral metal surfaces, Chiral polymers, Cyanohydrin formation, Cyclic Dipeptides, Epoxidation catalysts, Heterogeneous catalysts, Hydrogenation catalysts, Modified metal oxides, Polypeptides, Tartrate-modified Nickel catalysts

1
Introduction/Overview

Among the various methods to selectively produce one single enantiomer of a chiral compound, enantioselective catalysis is arguably the most attractive method. At this time, homogeneous metal complexes with chiral ligands are the most widely used and versatile enantioselective catalysts [1]. From an industrial point of view, heterogeneous catalysts are of interest for a number of reasons, and

Table 1. Strong and weak points of homogeneous and heterogeneous catalysts

	Homogeneous	Heterogeneous
Strong points	defined on molecular level	separation, recovery
	scope, variability (design?)	handling
	(availability)	stability, re-use
Weak points	sensitivity	characterization
	activity, productivity	reproducibility
	(separation)	availability, preparation
		narrow scope

a comparison of the properties of homogeneous and heterogeneous catalysts is made in Table 1. Several recent reviews give a digest of the most relevant aspects of heterogeneous enantioselective catalysts [2, 3, 4, 5, 6]. This short overview describes the various types of chiral heterogeneous catalysts based on metals, metal oxides, and organic polymers. Special emphasis is placed on giving the synthetic chemist an impression of the present state of the art and of the substrate specificity of these heterogeneous catalytic systems. For detailed information on reaction conditions and other experimental details the cited references should be consulted.

2
Chirally Modified Hydrogenation Catalysts

Investigation of heterogeneous chiral hydrogenation catalysts started in the late fifties in Japan and has seen a renaissance in the last few years [2, 4, 6]. In spite of many efforts, only two classes of modified catalyst systems have been found that are of synthetic use at this time: nickel catalysts modified with tartaric acid [4] and platinum and, to a lesser degree, palladium catalysts modified with *Cinchona* alkaloids and analogs thereof [7, 8]. However, several laboratories are working to expand the scope of this interesting and potentially very versatile class of chiral catalysts.

2.1
Mechanism and Mode of Action

The influences of catalyst preparation, modifier and substrate structure, of various additives, and of reaction parameters have been reported for both the tartrate modified Ni catalysts [2, 4, 11] as well as for *Cinchona* modified Pt catalysts [2, 3, 7]. In addition, kinetic [2, 4, 12, 13] and molecular modeling studies [8] have been carried out. This experimental basis allows the discussion of their mode of action with some confidence and a mechanistic picture has been developed in the last few years that can explain the major observations, even though it is not accepted universally.

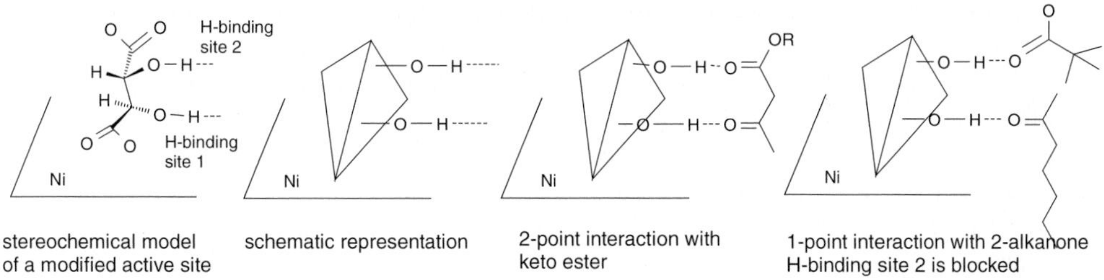

Fig. 1. Ni-tartrate catalyzed hydrogenation of β-keto esters and 2-alkanones. Important interactions between adsorbed tartrate and different ketones [14]

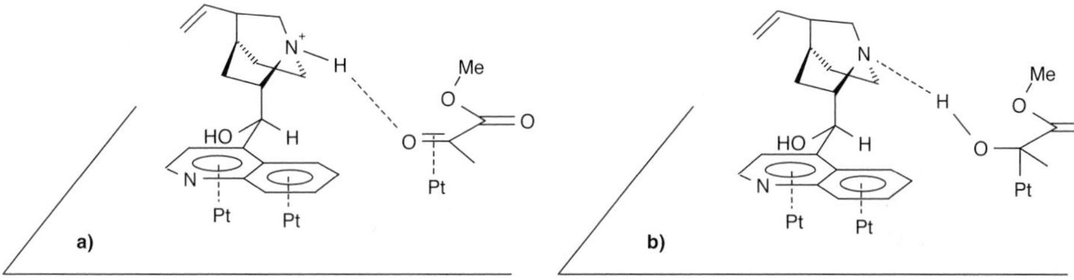

Fig. 2. Two intermediates proposed for Pt-cinchona catalyzed hydrogenation of α-keto esters: a) a protonated cinchonidine-ketone complex [8]; b) a stabilized half-hydrogenated ketone intermediate [3]

It is reasonably certain that a stepwise addition of hydrogen to the keto group takes place in the adsorbed state on the metal surface (Langmuir-Hinshelwood kinetics). While the reaction on an ordinary metal site leads to racemic alcohol, it is assumed that chiral active sites are formed by strong adsorption of a modifier molecule on the metal surface. Hydrogen bonding with the modifier as depicted in Fig. 1 and Fig. 2 not only controls the adsorption of the ketone but also facilitates and controls the addition of hydrogen. According to Sugimura et al. [14], this relatively simple model for the Ni catalysts is able to explain not only the observed absolute configuration of the product but also the steric effect of substituents in β-ketoesters and methyl ketones as well as the blocking effect of bulky organic acids. The Pt-*Cinchona* system has been modeled extensively, especially by the group of Baiker [8], and the modeling results are in agreement with the observed stereoinduction for a number of modifier-substrate combinations.

2.2
Tartrate-Modified Nickel Catalysts

The development of the nickel tartrate system and its successful preparative applications for the hydrogenation of β-functionalized and methyl ketones have been reviewed by Tai and Harada [4]. The preferred catalyst is freshly prepared

Raney nickel; the only useful modifier is *d*- or *l*-tartaric acid with bromide as co-modifier. The choice of the Ni alloy, the leaching and impregnation procedure (tartaric acid and NaBr concentrations, pH, T, time, ultrasound [15]) as well as the reaction conditions (solvent, p, T, acid co-modifiers) are crucial for getting good results. In general, the catalysts have relatively low activity and high pressures (up to 100 bar) and temperatures (usually 100 °C) are necessary [4]. Persistent and systematic optimization of the system parameters led to ee's up to 98% [14]. The Ni/SiO$_2$-tartrate system has recently been investigated in great detail (effect of impregnation and reaction parameters, catalyst characterization) and it looks as if modification by tartaric acid increases the specific activity by a factor of two [16].

2.3
Cinchona Modified Pt, Pd, and Related Catalysts

The *Cinchona* modified platinum catalysts are at the moment among the most selective catalytic systems known for the hydrogenation of α-keto acid derivatives and for other activated keto groups, whereas some Pd-cinchonidine catalysts give reasonable ee's for α,β-unsaturated acids [7, 8, 9]. For α-keto esters, the best catalysts are commercial 5% Pt/Al$_2$O$_3$ with low dispersion and a rather large pore volume. Carriers like SiO$_2$ or BaCO$_3$ and carbon supports are also suitable. The best modifiers are cinchonidine derivatives but some new types of modifiers [8, 10] also give good optical yields (see Fig. 3). The catalysts have to be pretreated in hydrogen at 300 to 400 °C before the reaction. The modifier can be directly added to the reaction solution but more elaborate modification procedures have also been reported [3]. The reaction conditions (solvent, modifier concentration, p, T) have a strong effect on rate and optical yield. Typical substrate/modifier ratios are in the range of 100 to 10,000, depending on substrate and solvent. Acetic acid is the solvent of choice for the hydrogenation of most α-ketoesters in order to get high enantioselectivity [17]. For the hydrogenation of α-ketoacid derivatives, the modified catalyst is 10 to 100 times more active than the unmodified Pt/Al$_2$O$_3$ (modifier accelerated catalysis) [18].

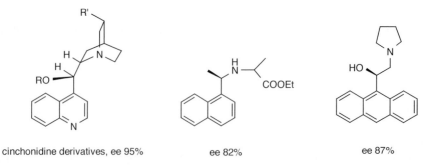

cinchonidine derivatives, ee 95%   ee 82%   ee 87%

Fig. 3. Structure of modifiers (best ee observed for the hydrogenation of ethyl pyruvate) [7, 8, 10]

2.4
New Catalytic Systems

As already pointed out, several research teams are actively looking for new catalytic systems and/or new applications. Until now, the results confirm the difficulty of this endeavor. In the area of the enantioselective hydrogenation of aromatic rings, Rh colloids stabilized and modified by chiral amines gave low but
significant ee's of 3 to 6% for the hydrogenation of o-substituted toluenes [19]. A
pyrazine derivative was hydrogenated with Pd/C in presence of camphor-10-sul-
fonic acid with 50% ee [20]. New results were also reported for the hydrogenation of pyruvic acid oxime (Pd/Al_2O_3-ephedrine, ee 26%) [21] and for an α,β-
unsaturated ketone (Pd/C-ephedrine, ee 36%) [22]. However, in all cases 0.5 to
more than 1 equivalents of modifier were necessary to give good results suggesting that 1:1 modifier-substrate adducts were the reactive species.

2.5
Synthetic Applications

2.5.1
Hydrogenation of Ketones

Enantioselectivities reported in Table 2 for aliphatic ketones and for α- and β-
ketoesters as well as for aliphatic methyl ketones are still among the highest for
hydrogenation reactions. The most outstanding results are 85% ee for tert-butyl
methyl ketone [11], 95% ee for ethyl pyruvate [17], and 98% ee for methyl 3-cy-
clopropyl-3-oxobutanoate [14]. While the preparation of the modified Raney-
nickel catalysts is not trivial, the Cinchona modifed Pt catalysts are easy to use
for preparative purposes. Both catalytic systems work well only at rather high
pressures, so special autoclaves are needed. Synthetic applications of the Ni-tartrate catalyst have been described for sex pheromones of the pine sawfly [25]; for
biologically active C_{10} to C_{16}-3-hydroxyacids [26]; for a diphosphine ligand [4,
27], and for an intermediate of tetrahydrolipstatin, a pancreatic lipase inhibitor
(ee 90 to 92%, 6 to 100 kg scale) [28]. Using a Pt Cinchona catalyst, the hydrogenation of ethyl 2-oxo-4-phenylbutyrate, an intermediate for the ACE inhibitor
benazepril, has been developed and scaled-up into a production process (10–
200 kg scale, chemical yield >98%, ee 79 to 82%) [29].

Several alternative methods with high ee's for various types of ketones are
known: reductions catalyzed by enzymes or baker's yeast [30] and microbial reagents [31], homogeneous hydrogenation (cf. Chapter 6.1), and stoichiometric
reductions with chiral metal hydrides [32].

2.5.2
Hydrogenation of C=C Bonds

Despite some recent progress as shown in Table 3, heterogeneous hydrogenation
of C=C bonds is not ripe for synthetic applications and certainly not competitive

Table 2. Hydrogenation of various ketones, best ee's and turnover frequencies

Substrate	R/R'	Catalyst	Modifier	ee (%)	tof (1/h)[a]	Ref.
CH_3COR	n-Alk	Ra-Ni	tartrate/NaBr/PvOH	71–80	<<1	[11]
CH_3COR	i-Pr	Ra-Ni	tartrate/NaBr/PvOH	85	<<1	[11]
$PhCOCF_3$		Pt/Al_2O_3	cinchonidine (Cd)	56	150	[23]
RCOCOOEt	n-Alk	Pt/Al_2O_3	O-methyl-dihydro-Cd	95	>50,000	[17]
$PhCH_2CH_2COCOOH$		Pt/Al_2O_3	O-methyl-dihydro-Cd	85	1000	[7]
$CH_3COCONHR$	var.	Pt/Al_2O_3	Cd	49–60	100	[24]
ketopantolactone		Pt/Al_2O_3	Cd	79	50	[8]
$RCOCH_2COOCH_3$	i-Pr	Ra-Ni	tartrate/NaBr	96–98	1	[14]
$RCOCH_2COOR'$	n-Alk	Ra-Ni	tartrate/NaBr	83–94	<1	[15]
$CH_3CO(CH_2)_3COOR$	Alk	Ra-Ni	tartrate/NaBr/PvOH	59–61	<1	[15]
$CH_3COCH_2SO_2CH_3$		Ra-Ni	tartrate/NaBr	71	n.a.	[4]
$CH_3COCH_2COCH_3$		Ra-Ni	tartrate/NaBr	91 (diol)	1	[15]

[a] Average turnover frequencies for complete conversion, rough estimates

Table 3. Hydrogenation of various C=C bonds, best ee's and turnover frequencies

Substrate	Catalyst	Modifier	ee (%)	tof (1/h)[a]	Ref.
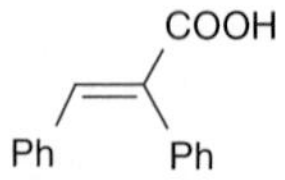	Pd/TiO_2	cinchonidine	72	400	[9]
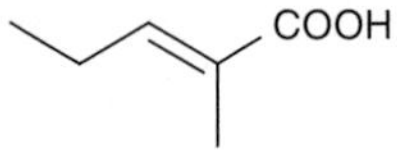	Pd/Al_2O_3	cinchonidine	52	2000	[8]
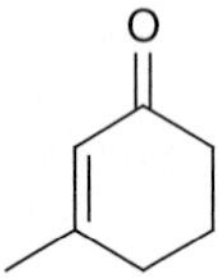	Pd black	dihydrovinpocetine	40	–	[7]

[a] Average turnover frequencies for complete conversion, rough estimates. Unmodified catalysts usually show higher activity.

with the homogeneous hydrogenation using Rh and Ru diphosphine complexes (cf. Chapter 5).

3
Modified Metal Oxides

Solid acids and bases are being increasingly applied for the catalytic synthesis of fine chemicals. However, chirally modified versions have not yet been developed to the point where their synthetic application seems feasible. Very little is known about their mode of action. In many cases, the preparation of the catalysts is not trivial and not always reproducible. Therefore, only a cursory overview is given here.

Titanium-pillared montmorillonites (Ti-PILC) modified with tartrates were described as heterogeneous Sharpless epoxidation catalysts [33] as well as for the oxidation of aromatic sulfides [34]. Metal oxides modified with histamine showed modest efficiencies for the kinetic resolution of activated amino acid esters ($k_R/k_S \approx 2$) [35]. Silica or alumina treated with diethylaluminium chloride and menthol catalyzed the Diels-Alder reaction between cylopentadiene and methacrolein with modest enantioselectivities of up to 31% ee [36]. Zeolite HY, modified with chiral sulfoxides had remarkable selectivities for the kinetic resolution of 2-butanol ($k_S/k_R = 39$) but unfortunately the catalyst is not very stable [37]. Clearly, this class of chiral catalysts, although of potential interest because of its variability, is not ready for synthetic applications.

4
Chiral Polymers

Because nature has provided us with so many chiral polymeric materials, their use for enantioselective synthesis is obvious. Indeed, some of the first successful attempts at enantioselective catalysis were carried out using aminocellulose by Bredig in 1932, metals on quartz by Schwab in 1932, and Pd supported on silk fibroin by Akabori in 1956 [2]. However, the catalysts were very difficult to reproduce and other strategies were pursued more successfully. In the mean time, impressive progress has been made for the synthetic application of polypeptides and of dipeptide gel catalysts. But even though good enantioselectivities are observed, there is very little understanding of their mode of action. For both systems there are indications that supramolecular interactions inside a polymeric aggregate might be important for catalysis and stereocontrol.

4.1
Polypeptides

Synthetic poly(amino acid) derivatives are highly selective catalysts for the asymmetric epoxidation of electron-deficient olefins with $NaOH/H_2O_2$ in a two-phase reaction system [38]. Parameters of importance for the catalytic performance are the type of amino acid, the degree of polymerization, the substituent at the terminal amino group (R in Fig. 4), catalyst pretreatments, and the organic solvent. In general, the catalysts are commercially available [39] or seem relatively easy to prepare even in larger quantities (>200 g) [40].

poly-L-alanine   poly-L-leucine

Fig. 4. Structure of most widely use polypeptide catalysts. R can be an alkyl or a polymeric residue

Table 4. Epoxidation of various activated C=C bonds using poly-peptide catalysts [39]

	R	R'	ee (%)
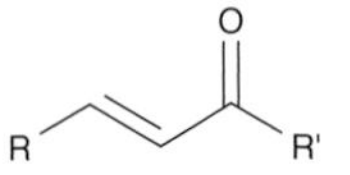	(subst)-Ph, 2-naphthyl 2'-styryl, cyclopropyl, 2-furyl, pyridyl	Ph, 2-naphthyl, 2-furyl, t-Bu, cyclopropyl	>90
		C(CH$_2$)$_2$OH, H	60
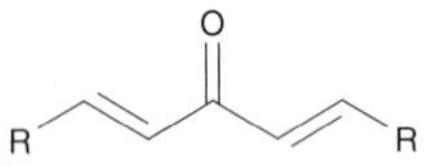	2-furyl, 2'-styryl (both C=C epoxidized)		80–90
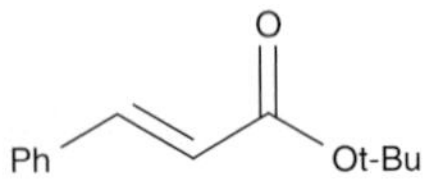			95
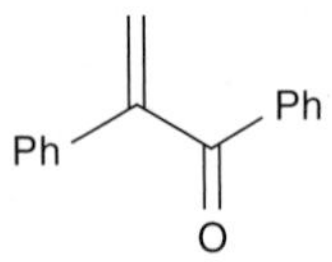			59

Recently, the scope of original Julia epoxidation was extended from the chalcones to a variety of activated C=C bonds and the catalytic system was broadened and extended to other oxidants, bases, and solvents [39]. In Table 4 we have summarized recently published examples that show the present scope and limitations. Generally, chemical yields are high but the activity of the catalysts is rather low (tof's ca. 0.1 h^{-1}) and relatively long reaction times and/or high catalyst loadings are necessary. Poly-L-alanine was used to prepare chiral intermediates for a leukotriene receptor antagonist with very high enantioselectivity [40].

4.2
Cyclic Dipeptides

Cyclic dipeptides, especially cyclo[(S)-phenylalanyl-(S)-histidyl], are efficient and selective catalysts for the hydrocyanation of aromatic aldehydes (Fig. 5) [41]. The catalysts are not available commercially but can be synthesized by conventional methods and their structure can be varied easily (Fig. 5) [41, 42, 43]. The catalysts are only selective in a particular heterogeneous state, described as a "clear gel" [41, 43]. It seems that their method of precipitation is crucial [41, 44] and that reproducing literature results is not always easy [42]. A recent study confirmed the importance of the aggregate formation and reported a second order rate dependence on the concentration of the cyclic dipeptide [45]. These findings indicate that the enantioselective catalytic species is not monomeric but either a dimer or polymer.

Because cyanohydrins are versatile intermediates, various methods for their asymmetric preparation have been worked out [46] (cf. Chapter 29). The chiral dipeptide catalysts allow the use of HCN but their application is restricted to aromatic aldehydes. Generally, 2% catalyst is needed and best optical yields are obtained at room temperature or below. As summarized by North [46], the pattern and the nature of substitution strongly affects the enantioselectivity (Table 5). Recently, the application to the synthesis of chiral side chains for liquid crystal polymers was described [44].

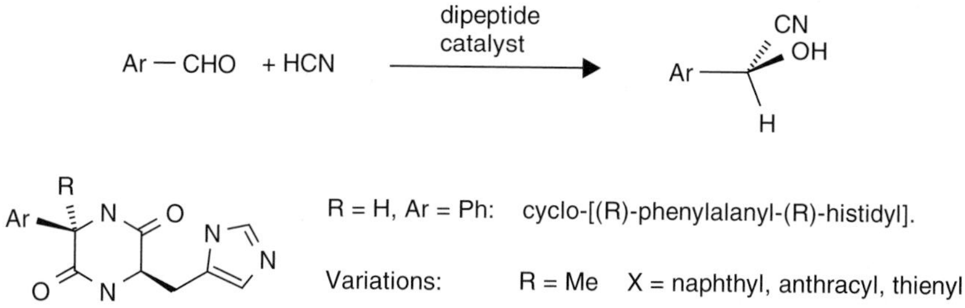

R = H, Ar = Ph: cyclo-[(R)-phenylalanyl-(R)-histidyl].

Variations: R = Me   X = naphthyl, anthracyl, thienyl

Fig. 5. Reaction scheme and structure of cyclo-[(R)-phenylalanyl-(R)-histidyl] and analogs

Table 5. Enantiomeric excesses for various aldehydes using cyclo-[(S)-phenylalanyl-(S)-histidyl]

Ar	ee (%)	Ref.
Ph, m-OR-Ph, m-tol, 2-naphthyl	91–97	[41]
p-OMe-Ph, o-OMe-Ph	78–84	[41]
heteroaromatic, aliphatic aldehydes	40–70	[41]
allyl-O-Ph	>98	[44]
$H_2C=CH-(CH_2)_nO$-Ph	72–90	[44]

5
Future Trends/Developments

Most groups working in the field of chiral heterogeneous catalysts use an empirical approach in order to expand the scope of known catalytic systems. There are, however, some attempts to try new principles and ideas. Even though the ee's are still low, there are some interesting first results:

- *Chiral metal surfaces.* Inherently chiral Ag metal surfaces were produced by cutting along the (643) surface to give Ag(643)^R and Ag(643)^S and characterization by LEED [47]. Not so surprisingly, no difference was detected for adsorption and decomposition of (R) and (S)-2-butanol.
- *Grafted auxiliaries.* Smith et al. [48] grafted chiral silyl ethers to a Pd surface through a Pd-Si bond. A borneoxysilyl-Pd catalyst was able to hydrogenate α-methylcinnamic acid with ee's up to 22%. Santini et al. [49] reported the preparation of a menthyl-Sn-Rh catalyst that hydrogenated ketopantolactone with 11% ee.
- *Chiral imprints.* The imprinting of organic [50] and inorganic materials [51] with transition state analog templates should, at least in principle, lead to what could be called artificial catalytic antibodies. Up to now, either the chiral recognition and/or the catalytic properties of such materials are still very poor. Some examples are a zeolite β, partially enriched in polymorph A [51], "chiral footprints" on silica surfaces [52], or several imprinted polymers [50].

References

1. Noyori R (1994) Asymmetric Catalysis in Organic Synthesis. John Wiley & Sons, Inc., Chichester; Ojima I (ed) (1993) Catalytic Asymmetric Synthesis. VCH, Weinheim
2. Blaser HU, Müller M (1991) Stud Surf Sci Catal 59:73
3. Webb G, Wells PB (1992) Catal Today 12:319
4. Tai A, Harada T (1986) In: Iwasawa Y (ed) Taylored Metal Catalysts. D Reidel, Dordrecht, p 265
5. Blaser HU, Pugin B (1995) In: Jannes G, Dubois V (eds) Chiral Reactions in Heterogeneous Catalysis. Plenum Press, New York, p 33
6. Baiker A, Blaser HU (1997) In: Ertl G, Knötzinger H, Weitkamp J (eds), Handbook of Heterogeneous Catalysis. p 2422
7. Blaser HU, Jalett HP, Müller M, Studer M (1997) Catalysis Today 37:441
8. Baiker A (1997) J Mol Catal A: Chemical 115:473
9. Nitta Y, Kobiro K (1996) Chem Lett 897
10. Pfaltz A, Heinz T (1997) Topics Catal 4:229
11. Osawa T, Harada T, Tai A (1997) Catal Today 37:465
12. Keane MA (1997) J Chem Soc Faraday Trans 93:2001
13. Blaser HU, Jalett HP, Garland M, Studer M, Thies H, Wirth-Tijani A (1997) J Catal 173:282
14. Sugimura T, Osawa T, Nakagawa S, Harada T, Tai A (1996) Stud Surf Sci Catal 101:231; Nakagawa S, Sugimura T, Tai A (1997) Chem Lett 859
15. Tai A, Kikukawa T, Sugimura T, Inoue Y, Osawa T, Fujii S (1991) J Chem Soc Chem Commun 795
16. Keane M (1997) Langmuir 13:41 and literature cited therein
17. Blaser HU, Jalett HP, Wiehl J (1991) J Mol Catal 68:215
18. Blaser HU, Garland M, Jalett HP (1993) J Catal 144:139

19. Nasar K, Fache F, Lemaire M, Beziat J-C, Besson M, Gallezot P (1994) J Mol Catal 87:107
20. Kikegawa K (1994) JP 07291943-A2 assigned to Koei Chem Ind Co Ltd; Chem Abstr (1996) 124:176150
21. Borszeky K, Mallat T, Aeschiman R, Schweizer WB, Baiker A (1996) J Catal, 161:451
22. Thorey C, Henin F, Muzart J (1996) Tetrahedron: Asymmetry 7:975
23. Mallat T, Bodmer M, Baiker A (1997) Catal Lett 44:95
24. Wang G-Z, Mallat T, Baiker A (1997) Tetrahedron: Asymmetry 8:2133
25. Tai A, Morimoto N, Yoshikawa M, Uehara K, Sugimura T, Kikukawa T (1990) Agric Biol Chem 54:1753
26. Nakahata M, Imaida M, Ozaki H, Harada T, Tai A (1982) Bull Chem Soc Jpn 55:2186
27. Bakos J, Toth I, Marko L (1981) J Org Chem 46:5427
28. Broger E, Hoffmann-LaRoche, Basel, personal communication
29. Sedelmeier G, Blaser HU, Jalett HP (1986) EP 206,993; assigned to Ciba-Geigy AG
30. Poppe L, Novak L (1992) Selective Biocatalysis; A Synthetic Approach, VCH, Weinheim
31. Simon H, Bader J, Günther H, Neumann S, Thanos J (1985) Angew Chem Int Ed Engl 24:539
32. Helmchen G, Hoffmann RW, Mulzer J, Schaumann E (eds) (1995) Houben-Weyl, 4th edn. Stereoselective synthesis vol E21d, chapter 2.3. Thieme, Stuttgart
33. Choudary BM, Valli VLK, Durga Prasad A (1990) J Chem Soc Chem Commun 1186
34. Choudary BM, Shobha Rani S, Narender N (1993) Catal Lett 19:299
35. Moriguchi T, Guo YG, Yamamoto S, Matsubara Y, Yoshihara M, Maeshima T (1992) Chem Express 7:625
36. Fraile JM, Garcia JI, Mayoral JA, Royo AJ (1996) Tetrahedron: Asymmetry 7:2263
37. Feast S, Rafiq M, Siddiqui H, Wells RPK, Willock DJ, King F, Rochester CH, Bethell D, Bulman Page PC, Hutchings GJ (1997) J Catal 167:533 and references cited therein
38. For a review see Aglietto M, Chiellini E, D'Antone S, Ruggeri G, Solaro R (1988) Pure & Appl Chem 60:415
39. Lasterra-Sanchez ME, Felfer U, Mayon P, Roberts SM, Thornton SR, Todd CJ (1996) J Chem Soc Perkin Trans I 343; Kroutil W, Lasterra-Sanchez ME, Maddrell SJ, Mayon P, Morgan P, Roberts SM, Thornton SR, Todd CJ, Tüter M (1996) J Chem Soc Perkin Trans I 2837; Bentley PA, Bergeron S, Cappi MW, Hibbs DE, Hursthouse MB, Nugent TC, Pulido R, Roberts SM, Wu LE (1997) J Chem Soc Chem Commun 739
40. Flisak JR, Gombatz KJ, Holmes MH, Jarmas AA, Lantos I, Mendelson WL, Novack VJ, Remich JJ, Snyder L (1993) J Org Chem 58:6247
41. Tanaka K, Mori A, Inoue S (1990) J Org Chem 55:181; Danda H (1991) Synlett 263
42. Hulst R, Broxterman QB, Kamphuis J, Formaggio F, Crisma M, Toniolo C, Kellog RM (1997) Tetrahedron: Asymmetry 8:1987
43. Noe CR, Weigand A, Pirker S (1996) Monatsh Chem 127:1081
44. Kim HJ, Jackson WR (1992) Tetrahedron: Asymmetry 3:1421; Kim HJ, Jackson WR (1994) Tetrahedron: Asymmetry 5:1541
45. Shvo Y, Gal M, Becker Y, Elgavi A (1996) Tetrahedron: Asymmetry 7:911
46. For a review see North M (1993) Synlett 807
47. McFadden CF, Cremer PS, Gellman AJ (1996) Langmuir 12:2483
48. Smith GV, Cheng J, Song R (1997) Catal Lett 45:73
49. Santini C, Cordonnier M-A, Candy JP, Basset JM (1995) Poster at the 3rd Paul Sabatier Conference, Strasbourg
50. Wulff G (1995) Angew Chem Int Ed Engl 34:1812; Shea KJ (1994) Trends in Polymer Sci 2:166
51. Davis ME (1993) Acc Chem Res 26:111; Davis ME, Katz A, Ahmad WR (1996) Chem Mater 8:1820
52. Morihara K, Kawasaki S, Kofuji M, Shimada T (1993) Bull Chem Soc Jpn 66:906 and references cited therein

Chapter 38   Catalyst Immobilization

Chapter 38.1
Catalyst Immobilization: Solid Supports

Benoît Pugin · Hans-Ulrich Blaser

Novartis Services AG, Catalysis and Synthesis Services, R-1055.6.29, CH-4002 Basel, Switzerland
e-mail: benoit.pugin@sn.novartis.com; hans-ulrich.blaser@sn.novartis.com

Keywords: Heterogenization, Immobilization, Support, Grafting, Copolymerization, Adsorption, Entrapment, Site isolation

1
Introduction

Control of stereoselectivity is easier with homogeneous than with heterogeneous catalysts. On the other hand, these soluble catalysts are more difficult to separate and to handle than the technically well-established heterogeneous catalysts. A promising strategy to combine the best properties of the two catalyst types is the heterogenization or immobilization of active metal complexes on insoluble supports or carriers [1, 2, 3]. Besides easy separation, immobilization opens opportunities like, e.g., the use of continuous flow reactors [4, 5, 6], site isolation [7], or the tuning of the catalyst environment [8, 9, 10] which in some cases can lead to improved catalytic performance. On the other hand, immobilization increases the complexity and the costs of the catalytic system.

To be of practical use, immobilized enantioselective catalysts should meet the following requirements:

(i) The preparation methods should be versatile, since it is not yet possible to predict which ligand and support will be the most suitable for a given sub-

strate and process. Modular systems that allow combinations of different support and ligand will therefore be preferred.
(ii) The selectivity, activity, and productivity of an immobilized catalyst should be comparable or better than that of the corresponding free analogs.
(iii) Separation should be achieved by simple filtration and at least 95% of the catalyst should be recovered.

Re-use is not mandatory, but would be a great advantage from an economic point of view.

First efforts in the field of catalyst immobilization showed the feasibility of the concept with respect to catalyst separation but in most cases lead to immobilized catalysts with very poor performance [1]. In the meantime, further efforts in catalyst preparation and the use of new supports have lead to several catalytic systems that meet these requirements. However, to our knowledge, none is applied on a technical scale yet.

2
Supports

The supports used for catalyst immobilization can be classified as follows:
- *Soluble polymers:* Non-crosslinked, linear polymers are soluble in suitable solvents. In the soluble state, high mobility of the bound catalyst and good mass transport are guaranteed and, therefore, catalytic properties will practically not be affected. However, separation of such catalysts is often problematic and costly since it is done either by ultrafiltration or precipitation.
- *Swellable polymers:* Crosslinked polymers are 3-dimensional networks and can easily be separated by sedimentation or filtration. Slightly crosslinked polymers such as, e.g., polystyrene crosslinked with 0.5--3% 1,4-divinylbenzene have to be used in solvents in which they swell, otherwise mass transport may be completely stopped.
- *Unswellable supports:* Highly crosslinked polymers (e.g., macroreticular polystyrenes or polyacrylates) and inorganic supports (metal oxides) are practically not swellable. In contrast to soluble or swellable polymers, these materials can therefore be used in a large variety of different solvents without changing their texture. To obtain immobilized catalysts with reasonable weights per mole of active sites these supports should have large specific surface areas (>100 m²/g). Also, to avoid mass transport problems, the pore size should be considerable larger than the size of the metal complex catalyst and the substrate [8]. The most frequently used insoluble support is silica gel. Different types, covering a large range of specific areas and pore sizes are commercially available.

3
Approaches for the Immobilization of Metal Complex Catalysts

A schematic view of the most important approaches that have so far been described in the literature is given in Fig. 1.

For all approaches it is important that the metal does not dissociate from the ligand. Since, with a few exceptions [12], the binding to monodentate ligands is usually insufficient [13], most immobilized catalysts are based on bi- or polydentate ligands.

For covalent binding or adsorption, the ligands have either to be attached to a linker or to an adsorbable moiety. In these cases, ligands with an additional function such as an NH or OH group have proven to be particularly useful, since they can be combined in a modular way with different linkers, supports, adsorbable moieties, or water-soluble groups. A highly versatile modular system has been developed that allows to bind various functionalized diphosphines to inorganic supports or to organic polymers via trialkoxysilane-isocyanate or di-isocyanate linkers [11, 14].

3.1
Immobilization via Covalently Bound Ligands

Covalent binding of ligands to supports is by far the most frequently used strategy for the heterogenization of metal complexes and, in general, the successful catalytic systems are sooner or later immobilized by this method. Covalent binding can be effected either by copolymerization or by grafting.

Copolymerization of functionalized ligands with a suitable monomer [e.g., 15, 16, 17] is being more frequently used than grafting. The polymerization is well

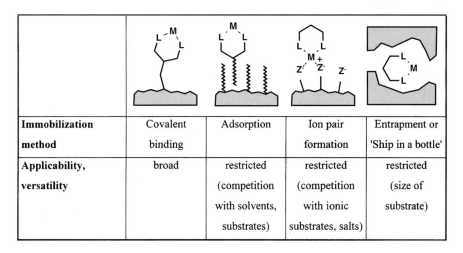

Immobilization method	Covalent binding	Adsorption	Ion pair formation	Entrapment or 'Ship in a bottle'
Applicability, versatility	broad	restricted (competition with solvents, substrates)	restricted (competition with ionic substrates, salts)	restricted (size of substrate)

Fig. 1

established, but it is difficult to predict and control the properties of the resulting polymer that is a random sequence of the original monomer units. In addition, the formation of inaccessible immobilized ligands cannot be excluded. Finally, unwanted polymerization could occur during the synthesis or storage of the functionalized ligands.

Grafting, i.e., reacting a functionalized ligand or metal complex with reactive groups of a preformed support [e.g., 11, 14, 18, 19] has several advantages. Different suitable supports are commercially available and many methods for introducing reactive groups onto non-functionalized polymers are described in the literature [20, 21, 22]. Their properties (e.g., solubility or swellabillity, particle size, separation, purity) can be checked before use. Finally, the ligands will preferentially bind at locations that are also accessible for the substrate during the catalytic reaction. A potential problem of the grafting method is the remaining reactive groups on the support that may interact with the immobilized catalysts.

The choice of the best method seems to depend on the catalytic system. A comparison of immobilized catalysts for enantioselective hydrogenation shows a clear superiority of the catalysts that were prepared by grafting. On the other hand, Itsuno [23] found that immobilized Lewis acid catalysts for Diels-Alder reactions showed better performance when they were prepared by copolymerization.

Table 1 gives some examples for catalysts that were immobilized by covalent binding and that are either typical or are promising with respect to potential future applications.

Most efforts have been made with catalysts for enantioselective hydrogenation and some have now catalytic performances that are of industrial interest.

Table 1. Examples of potentially useful immobilized catalysts prepared by covalent binding

reaction/**support**-linker-**catalyst**	preparation	ee[a]	tof[a)]	ref
hydrogenation of enamides				
	grafting	100	400	[18]
	grafting	97	60	[19]
	grafting	94.5	1500	[11, 24]
	grafting	95	2000	[14]
	grafting	97	200	[14]

Table 1. Continued

hydrogenation of imines

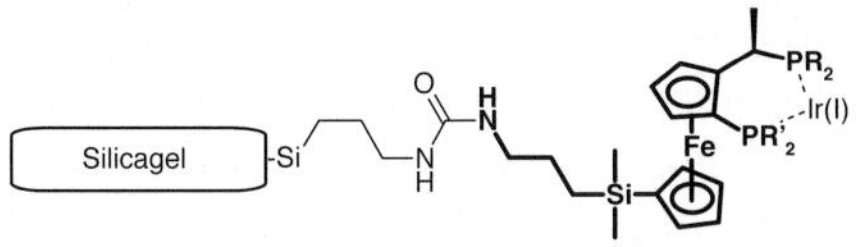

	grafting	79	10'000[b) [25]

addition of R_2Zn to aldehydes

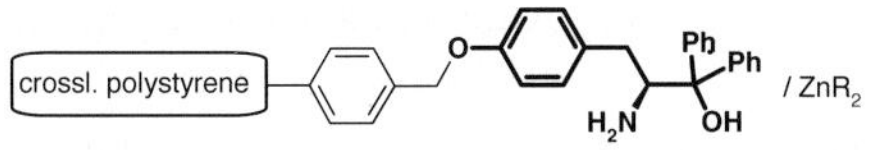

	copolym.	99	3	[4]

Diels Alder

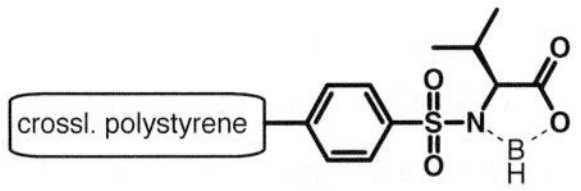

	copolym.	95	3	[6]

dihydroxylation of alkenes

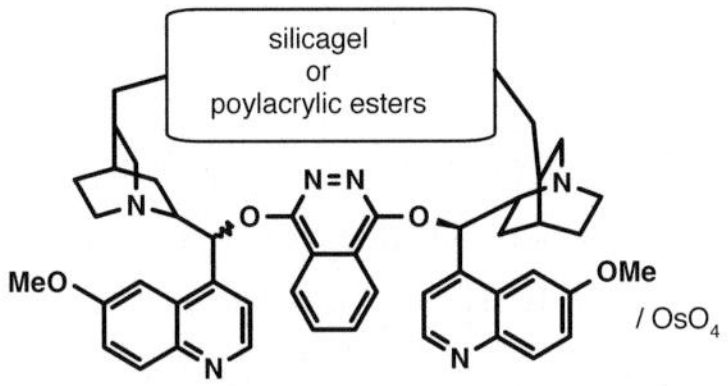

	grafting	99	6	[26]
	copolym.	99	6	[17]

epoxidation of alkenes

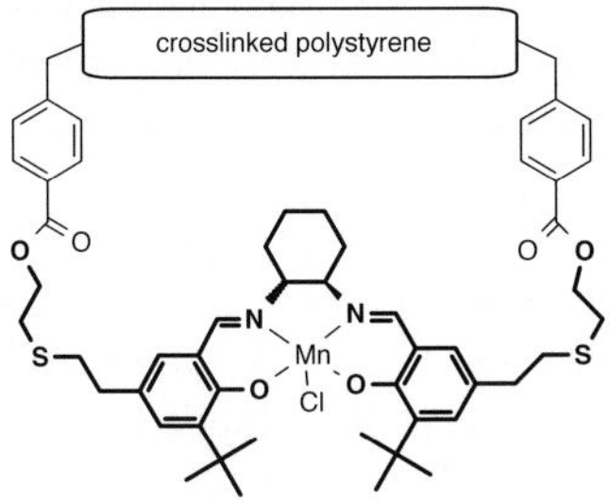

	copolym.	62	40	[27]

---

[a]Best ee and tof; tof=moles of substrate/moles of catalyst per h.
[b]Hydrogenation in presence of acid.

The recent activities directed towards the immobilization of the dihydroxylation and epoxidation catalysts have already led to systems with acceptable catalytic properties [17, 26, 27]. Both, immobilized Zn amino alcohol and oxazaborolidinone catalysts have been applied in continuous flow reactors [4, 6]. Interestingly, these immobilized oxazaborolidinone catalysts give better enantioselectivities than their soluble analogs [6].

3.2
Heterogenization via Adsorption and Ion Pair Formation

This approach has not often been used so far, but seems promising with respect to practical applications. It relies on various adsorptive interactions between a carrier and a metal complex. The advantage is the easy preparation of the heterogenized catalyst by a simple adsorption procedure, very often without the need to functionalize the ligand. Electrostatic forces were used to bind cationic Rh diphosphine complexes to anionic resins [28, 29]. The resulting catalysts could be recycled 20 times with very little leaching. Toth et al. [30] functionalized diphosphine ligands with amino groups and adsorbed the corresponding Rh complexes on Nafion. Another approach by Inoue relied on the interaction of lipophilic ligands with surface methylated silica [31]. Further examples were described by Brunner [32, 33] and Inoue [34]. With few exceptions [28], adsorbed complexes have lower enantioselectivities than covalently bound catalysts. It is to be expected that the applicability of this elegant immobilization strategy is limited by strong solvent and salt effects.

3.3
Heterogenization via Entrapment

This method relies on the size of the metal complex rather than on a specific adsorptive interaction. There are two different preparation strategies: One, often called the 'ship-in-a-bottle' approach, is based on building up catalysts in well defined cages of porous supports. Recently, enantioselective Mn epoxidation catalysts with different salen ligands have been assembled in zeolites. In zeolite EMT [35] ees up to 88% and in zeolite Y [36] ees up to 58% were obtained with cis-β-methylstyrene. However, both entrapped catalysts were much less active than their homogeneous counterparts. Rh diphosphine complex were entrapped in the interlayers of Smectite [37]. The resulting catalyst was active for the enantioselective hydrogenation of N-acetamidoacrylic acid (ee 75%).

The other approach is to build up a polymer-network around a preformed catalyst. Using this method, Jacobs et al. [38] occluded Jacobsen's Mn-Salen epoxidation catalysts and Noyori's Ru-Binap-catalyst in poly-dimethylsiloxane and demonstrated that leaching strongly depends on the size and the solubility of the metal complex and the swelling of the polymer [39].

4
Effect of Immobilization on Catalytic Performance

Unfortunately, heterogenization of a homogeneous catalyst often leads to a change of its catalytic properties. Most of the time the consequences are negative but sometimes the performance is improved. Even though most effects are poorly understood and ill-defined, several factors may be distinguished.
(i) Interactions between functional groups on the surface of the support and the metal center can influence the catalytic performance. Stille [8, 9, 10] dem-

onstrated that the enantioselectivity of a chiral hydrogenation catalyst bound to a support with additional chiral groups significantly changes with the change of the configuration of these groups.

(ii) Restricted conformational flexibility through geometrical confinement can be positive as described by Corma et al. [40, 41, 42] for proline amide Rh complexes that were attached in modified USY zeolites or negative as observed by Pugin and Müller [11] for a diphosphine Rh complex that was immobilized on silica gels with different pore sizes. Itsuno et al. [6] studied an asymmetric oxazaborilidinone Diels-Alder catalysts bound to polystyrene. They were able to raise the enantioselectivity from 65% to 95% by increasing the length and the polarity of the crosslinking agent.

(iii) Attaching a complex to a rigid support via a covalent bond can lead to so-called site isolation, i.e., different active centers no longer interact with each other. The most remarkable positive effects are observed for complexes that are prone to form inactive dimers [7]. To our knowledge, no example for the alternative situation, the so-called 'site cooperation', where site interactions are important for good catalytic performance, has been described yet.

(iv) Mass transport can be hindered or even completely stopped if unsuitable solvents are used with swellable supports [43] or if solid supports with too large particle size or with too narrow pores are used. It is conceivable that this may in some cases lead to a large population of highly reactive catalytic intermediates that, instead of reacting with the substrate, deactivate the system by undesired side-reactions.

5
Conclusions and Outlook

Immobilized catalysts are a fascinating topic and currently several research groups both in industry and at university are active. There are many examples demonstrating that efficient separation by filtration or sedimentation can be achieved. In a few cases re-use of the recovered catalysts is possible without loss of performance. Immobilized catalysts can be applied in continuous flow reactors. This technique has advantages in cases where a high concentration of catalyst is required to obtain good selectivities or where degradation of the support particle by stirring is a problem. Also, in a few cases, catalytic properties could be improved by the effect of site isolation or by tuning the catalyst's surroundings.

However, immobilized catalysts are much more complex and expensive than their homogeneous counterparts and immobilization usually goes along with changes of the catalytic property that cannot be predicted. For these reasons there is now a strong tendency to avoid heterogenization and to develop soluble catalysts that can be separated by extraction [44], by ultrafiltration [45], or that can be precipitated by changing the temperature [46] or the pH [47]. Promising results have also been obtained recently with 'supported aqueous phase' (SAP) catalysts [48]. These are based on a thin film of a water-soluble catalyst in water or another polar solvent that is adsorbed on a hydrophilic support and that reacts with a water-insoluble solution of the substrate. All these approaches have their strong and weak points and all of them require a functionalization of the ligand.

The chances for the application of immobilized catalysts will improve with the number of chiral ligands that have an anchoring group, the number of efficient immobilization methods, better understanding of the interactions between supports and catalysts and the increasing experience in manufacturing well-defined solid supports. Also, in the near future it is to be expected that combinatorial solid phase synthesis and screening of chiral ligands [49] will become an important issue that will give another push to the development of this still young field.

References

1. Hartley FR (1986) in: Hartley FR (ed) The Chemistry of Metal Carbon Bond, Vol 4, Wiley, p 1163.
2. Selke R, Häupke K, Krause HW (1989) J Mol Catal 56:315
3. Shuttleworth SJ, Allin SM, Sharma PK (1997) Synthesis 1217
4. Itsuno S, Sakurai Y, Ito K, Maruyma T, Nakahama S, Fréchet JMJ (1990) J Org Chem (1990) 55:304
5. Gamez P, Fache F, Lemaire M (1994) Bull Soc Fr 131:600
6. Kamahori K, Ito K, Itsuno S (1996) J Org Chem 61:8321
7. Pugin B (1996) J Mol Catal 107:273
8. Matsuda T, Stille JK (1978) J Am Chem Soc 100:268
9. Baker GL, Fritschel SJ, Stille JK (1981) J Org Chem 46:2960
10. Deschenaux R, Stille JK (1985) J Org Chem 50:2299
11. Pugin B, Müller M (1993) In: Guisnet et al. (ed) Heterogeneous Catalysis and Fine Chemicals III, Elsevier. p 107
12. Eisen M, Korpal T, Blum J (1990) J Mol Catal 61:19; Eisen M, Bernstein T, Blum J, Schumann H (1987) J Mol Catal 43:199
13. Kinting A, Krause H, Capka M (1985) J Mol Catal 33:215
14. Pugin B (1996) Patent EP 728768 (96.08.28) assigned to Ciba-Geigy
15. Achiwa K (1978) Heterocycles 9:1539
16. Baker GL, Fritschel SJ, Stille JR (1981) J Org Chem 46:2954
17. Song CE, Yang JW, Ha HJ, Lee Sg (1996) Tetrahedron Asymmetry 7:645
18. Nagel U, Kinzel E (1986) J Chem Soc Chem Commun 1098
19. Nagel U, Leipold J (1996) Chem Ber 129:815
20. Frechet JMJ, Darling GD, Itsuno S, Lu PZ, de Meftahi V, Rolls WA (1988) Pure Appl Chem 60:353
21. Lieto J, Milstein D, Albright RL, Minkiewics JV, Gates BC (1983) Chemtech 46
22. Soutif JC, Brosse JC (1990) Reactive Polymers 12:3
23. Kamahori K, Tada S, Ito K, Itsuno S (1995) Tetrahedron Asymmetry 6:2547
24. Pugin B, Spindler F, Müller M (1992) Patents EP 496699 (92.01.16) and EP 496700 (92.07.29) assigned to Ciba-Geigy
25. Pugin B (1996/1997) Patents EP 729969 (96.09.04), WO 9632400 (96.10.17), WO 9702232 (97.01.23)
26. Song CE, Yang JW, Ha HJ (1997) Tetrahedron Asymmetry 8:841
27. Minutolo F, Pini D, Petri A, Salvadori P (1996) Tetrahedron Asymmetry 7:2293
28. Selke R, Häupke K, Krause HW (1989) J Mol Catal 56:315
29. Selke R, Capka M (1990) J Mol Catal 63:319

30. Toth I, Hanson BE (1992) J Mol Catal 71:365
31. Ishizuka N, Togashi M, Inoue M, Enomoto S (1987) Chem Pharm Bull 35:1686
32. Brunner H, Bielmeier E, Wiehl J (1990) J Organomet Chem 384:223
33. Brunner H, Bublak P, Helget M (1997) Chem Ber/Recueil 130:55
34. Inoue M, Ohta K, Ishizuka N, Enomote S (1983) Chem Pharm Bull 31:3371
35. Ogunwumi SB, Bein T (1997) J Chem Soc Chem Commun 901
36. Sabater MJ, Corma A, Domenech A, Fornés V, Garci H (1997) J Chem Soc Chem Commun 1285
37. Mazzei M, Marconi W, Riocci M (1980) J Mol Catal 9:381
38. Vankelecom IFJ, Tas D, Parton RF, Van de Vyver V, Jacobs PA (1996) Angew Chem Int Ed Engl 35:1346
39. Janssen KBM, Laquire I, Dehaen W, Parton RF, Vankelekom IFJ, Jacobs PA (1997) Tetrahedron Asymmetry 8:3481
40. Corma A, Iglesias M, del Pino C, Sanchez F (1991) J Chem Soc Chem Commun 1253
41. Corma A, Iglesias M, del Pino C, Sanchez F (1992) J Organometal Chem 431:233
42. Corma A, Iglesias M, Martin MV, Rubio J, Sanchez F (1992) Tetrahedron Asymmetry 3:845
43. Dumont W, Poulin JC, Dang TP, Kagan HB (1973) J Am Chem Soc 95:8295
44. Joó F, Kathó Á (1997) J Mol Catal 116:3
45. Felder M, Giffels G, Wandrey C (1997) Tetrahedron Asymmetry 8:1975
46. Bergbreiter, DE (1987) Chemtech 686
47. Bergbreiter DE, Liu YS (1997) Tetrahedron Lett 38:3703
48. Wan KT, Davis ME (1995) J. Catal 152:25
49. Cole BM, Shimizu KD, Krueger CA, Harrity JPA, Snapper ML, Hoveyda AH (1996) Angew Chem, Int Ed Engl 35:1668

Chapter 38.2
Catalyst Immobilization: Two-Phase Systems

Günther Oehme

Institut für Organische Katalyseforschung an der Universität Rostock e.V.,
Buchbinderstr. 5–6, D-18055 Rostock, Germany
e-mail: goehme@chemie1.uni-rostock.de

Keywords: Biphasic catalysis, Asymmetric phase-transfer catalysis, Micellar catalysis, Vesicles, Microheterogeneous systems

1
Introduction

Most asymmetric-catalytic reactions are carried out in homogeneous organic media. The well known advantages are high activity and selectivity and a good reproducibility under mild conditions; however, catalyst recycling can be difficult. From a practical point of view it is desirable to separate the catalytic system from the reactants using two immiscible liquid phases. After conversion the phases can be separated and the catalyst phase can be used in the next cycle. In accordance to Southern [1] we will use the following terms for two-phase systems:
- The term biphasic catalysis will be used for liquid-liquid systems containing two immiscible liquids without any phase-transfer agent,
- the term phase-transfer catalysis will be used for biphasic liquid-liquid systems containing a phase-transfer agent for transportation of a part of catalyst or reactant into one favored phase, and
- thirdly (in a free extension), the term micellar or vesicular catalysis will be used for microheterogeneous liquid-liquid systems containing colloidal assemblies of self-organized amphiphiles. The surfactant simultaneously is the phase-transfer agent and the organic phase [2].

Liquid-solid phase-transfer systems and other systems with immobilized catalysts are not mentioned here.

2
Biphasic Catalytic Systems

Two important industrial processes are based on biphasic systems: the Shell higher olefin process (SHOP) [3] and the hydroformylation developed by Ruhrchemie/Rhône Poulenc [4]. Prerequisite was the synthesis of water-soluble ligands, especially water-soluble phosphines. Scheme 1 shows a selection of optically active phosphines for asymmetric reactions under biphasic conditions.

Some excellent reviews collect the early and recent literature with respect to typical complex-catalyzed reactions [5, 6, 7, 8, 9, 10, 11, 12]. Most of the water-soluble phosphines were synthesized by direct sulfonation of the phenyl group leading to mixtures of products. Sinou et al. [13] investigated the asymmetric hydrogenation of precursors of α-amino acids [14] and even of dehydropeptides [15]. In comparison to homogeneous systems in organic solvents activity and

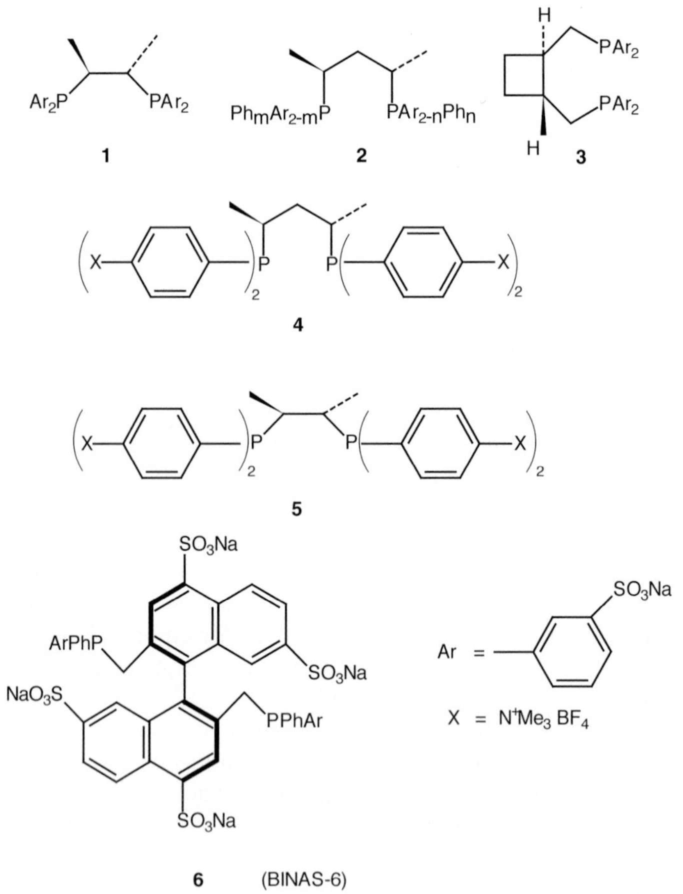

Scheme 1

enantioselectivity decrease in the presence of water [16] and Table 1 shows some characteristic results in biphasic systems.

A favored ligand can be derived from BDPP, [(S,S)-1.3-bis(diphenylphosphino)pentane]. Bakos [17] showed that two quite different asymmetric catalytic reactions, the hydrogenation of acetophenonbenzylimine and the hydrogenolysis of sodium cis-epoxysuccinate [18] by rhodium(I) complexes depend on the degree of sulfonation of the ligand 2 (Scheme 2).

Table 1. Hydrogenation in the biphasic system water/AcOEt (1/1)

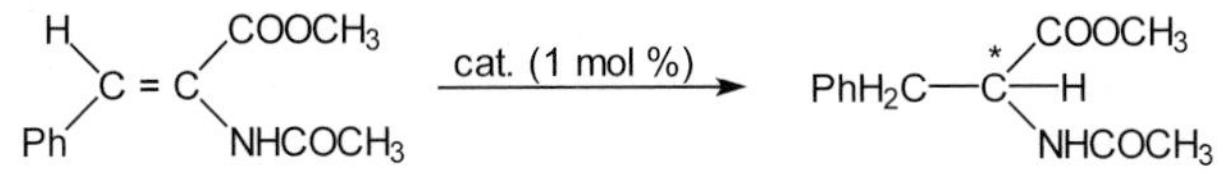

catalyst	pressure in bar	optical yield in % ee	Ref.
Rh(cod)Cl]$_2$+2.2 **1**	10	88 (R)	[14]
[Rh(cod)Cl]$_2$+2.2 **2**	15	67 (R)	[14]
[Rh(cod)Cl]$_2$+2.2 **3**	1	67 (S)	[14]
[Rh(nbd) **4**] [BF$_4$]$_5$[a]	14	44 (R)	[19]

[a]Biphasic system water/AcOEt-benzene (2/1–1); cod=cis,cis-1.5-cyclooctadiene; nbd = norbornadiene

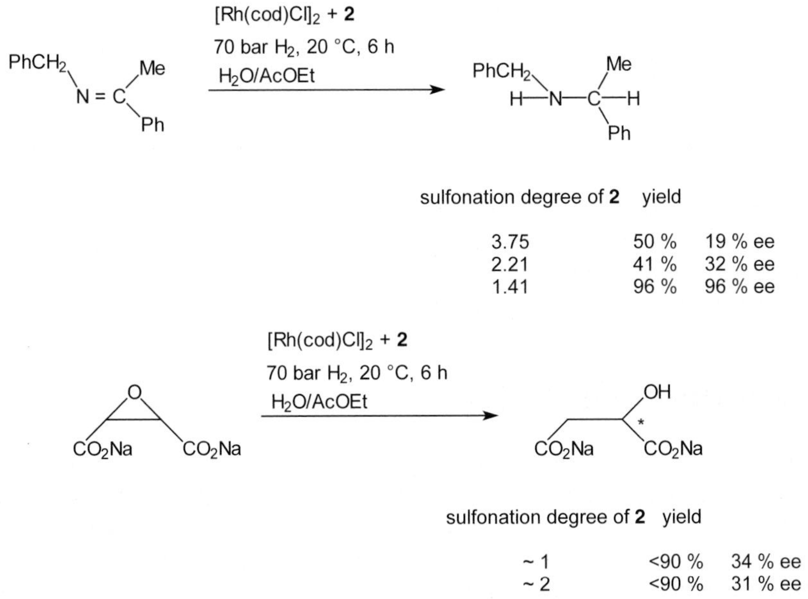

sulfonation degree of **2**	yield	
3.75	50 %	19 % ee
2.21	41 %	32 % ee
1.41	96 %	96 % ee

sulfonation degree of **2**	yield	
~ 1	<90 %	34 % ee
~ 2	<90 %	31 % ee

Scheme 2

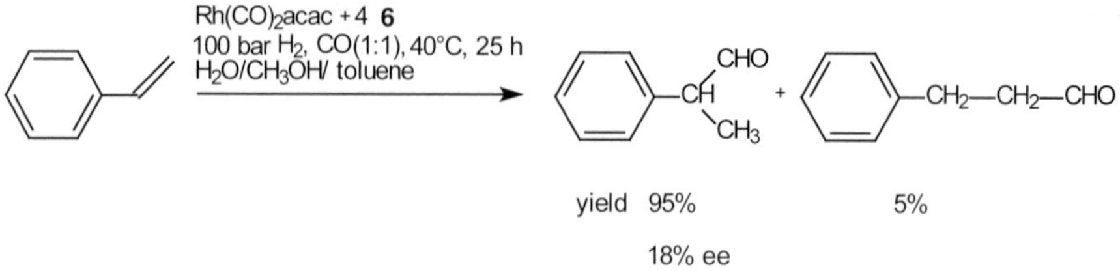

Scheme 3

Best results were observed with a sulfonation degree between one and two.

Also, outstanding enantioselectivities could be achieved with the rhodium complex of ligand 4 in the hydrogenation of a DOPA precursor [19]. In the case of a H_2O-slurry of the substrate the reaction yielded 95% ee.

A very simple experiment, the adsorption of a water-soluble catalyst on a silica gel prepared with organic solvent or on a controlled pore glass treated with an organic solvent, led to a new generation of catalysts, SAP (supported aqueous phase) catalysts [20,21]. In a comprehensive study Wan and Davis [22] investigated the use of Ru-BINAP-4 SO_3Na as catalyst in the homogeneous, biphasic, and SAP asymmetric hydrogenation of the naproxen precursor 2-(6'-methoxy-2'-naphthyl)acrylic acid. The best enantioselectivities were 96% ee for the homogeneous, 83% ee for the biphasic. and 77% ee for the SAP catalysis, the relative activities were 1:0.0026:0.14.

The low activity and enantioselectivity of the biphasic systems seems to be a characteristic feature. Eckl et al. [23] described the use of NAPHOS and its sulfonated derivative BINAS (6) as ligands in the asymmetric hydroformylation of styrene (Scheme 3). NAPHOS yielded 34% ee (in toluene), BINAS 18% ee (in toluene/water) of phenylmethylacetaldehyde.

The achievements in asymmetric biphasic catalysis are not very encouraging except for some hydrogenation reactions. Nevertheless, the simplicity of catalyst-product separation in biphasic systems should be a challenge to improve activity and stereoselectivity. New developments of biphasic systems containing a hydrocarbon and a perfluorinated hydrocarbon which are miscible at slightly enhanced temperature and separable at ambient temperature have not yet been used in asymmetric catalysis [24, 25].

3
Phase-Transfer Catalytic Systems

Phase-transfer catalysis (PTC) has been developed to a very important laboratory method in organic and organometallic chemistry. Numerous books [26, 27, 28, 29, 30] and reviews [31, 32, 33] give a good insight into methods and results. In principle PTC is a two-phase liquid-liquid system with an amphiphilic PT catalyst for equilibration of the reactants between the phases. Actually, the asymmetric potential of PTC is rather low and only a few examples with taylor-made catalysts gave spectacular results. All successful chiral catalysts are ammonium salts or crown ethers. A selection of PT catalysts is summarized in Scheme 4.

Some highlights will be discussed here. Schiff bases of glycine esters were alkylated under PT conditions with benzylated *Cinchona* bases (9) as catalysts in the presence of 50% NaOH and dichloromethane with a maximum of 64% ee [34]. Outstanding enantioselectivities could be observed by alkylation of a special phenylindanone with modified benzylcinchoninium salt according to Scheme 5.

The enantioselectivity of the alkylation depends on the electrophile. Best results were surprisingly obtained with methyl chloride [35].

The Michael addition is one of the favored reactions in enantioselective PTC. For instance, the reaction of an indanone similar to those in Scheme 5 with methyl vinyl ketone in the presence of catalyst **8** in a toluene system (50% NaOH) gave the Michael product in 95% yield and 80% ee [36].

The best result (99% ee) was achieved by Michael addition of an indanone derivative to methyl vinyl ketone in the presence of a chiral crown ether at −78 °C in toluene (Scheme 6) [37].

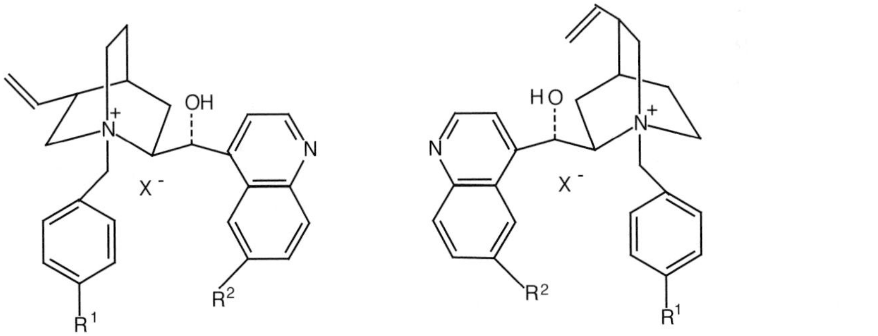

7   R^1 = H

   R^2 = OCH$_3$

   Quininium salt

8   R^1 = CF$_3$,   R^2 = H

9   R^1 = R^2 = H

   Cinchoninium salt

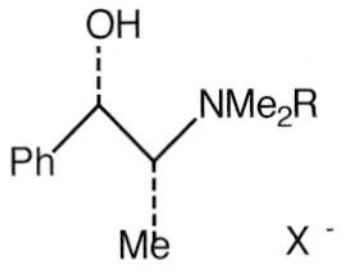

N-methyl-benzyl-ephedrinium salt **10**

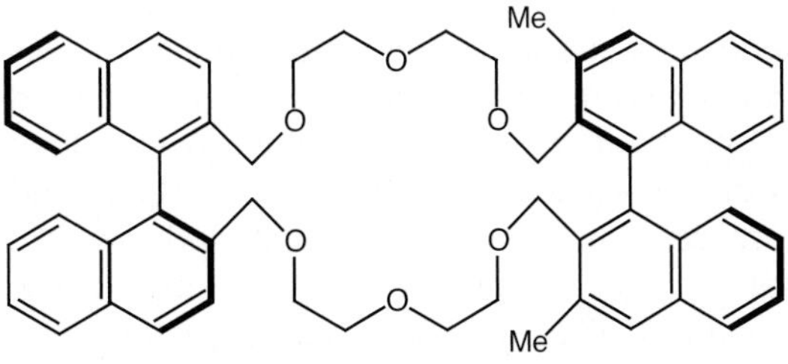

chiral crown ether **11**

Scheme 4

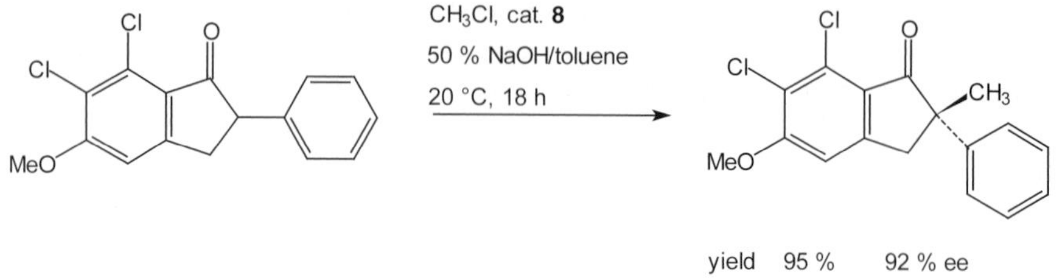

CH₃Cl, cat. **8**

50 % NaOH/toluene

20 °C, 18 h

yield   95 %   92 % ee

Scheme 5

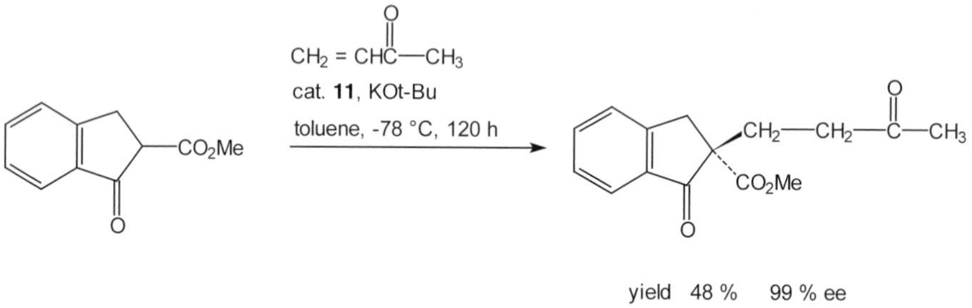

$CH_2 = CHC(O)-CH_3$

cat. **11**, KOt-Bu

toluene, -78 °C, 120 h

yield   48 %   99 % ee

Scheme 6

In a more recent paper an enhancement of the rate by sonication was reported [38]. Benzylmethylephedrinium bromide **10** was used in this reaction.

Many experiments have been described for the PT reduction of ketones and imines with $NaBH_4$ as reagent. The catalysts were ammonium salts of *Cinchona* alkaloids and ephedrine and even chiral crown ethers. Only medium enantiose-lectivities could be achieved.

Ding, Hanson, and Bakos [39] used an amphiphilic chiral rhodium-phosphine complex in the catalytic asymmetric two-phase hydrogenation of methyl (Z)-α-acetamidocinnamate in ethyl acetate/water and obtained enantioselectivities of up to 69% ee.

In the field of asymmetric oxidation reactions the epoxidation of α,β-unsaturated carbonyl compounds was investigated. In the case of 1,4-naphthoquinone derivatives and *tert*-butyl hydroperoxide as reagents enantioselectivities up to 78% ee were observed with quininium and quinidinium salts as PT catalysts [40].

Important seems to be also the epoxidation of chalcones with hydrogen per-oxide as reagent leading to 48% ee for the epoxide [41] (Scheme 7).

Masui et al. [42] reported a high optical yield (up to 79% ee) in the autoxida-tion of cyclic ketones in the presence of 4-trifluoromethylbenzyl-cinchoninium bromide **8**, 50% NaOH/toluene, oxygen and triethyl phosphite.

With the same type of PT catalyst Aires-de-Sousa et al. [43] described recently an enantioselective synthesis of N-arylaziridines starting from N-acylarylhydrox-ylamines in a PT system of NaOH/toluene. High conversion needs highly con-centrated NaOH (≥33%) and yielded only up to 50% ee. Decreasing the NaOH

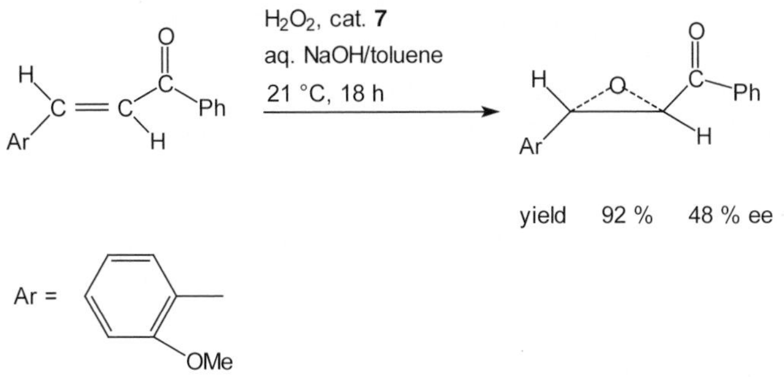

Scheme 7

concentration to 9% decreased the conversion from 79% to 12% and enhanced the enantioselectivity to 61% ee.

Only few examples of asymmetric catalytic reactions by PTC are really successful but in all cases it was the result of an acribic development. The potential of PTC in asymmetric catalytic reactions will be a topic of future application.

4
Catalysis in Micelles and Vesicles

This term is connected with assemblies of typically structured surfactants above a critical micelle concentration (cmc) or a critical vesicle concentration (cvc). The aggregates have colloidal dimensions and are spherically shaped [44].

The core of a micelle and the bilayer of a vesicle are comparable with a liquid-crystalline phase and can influence the stereoregularity of asymmetrically catalyzed reactions. Self-organization and the neighborhood of hydrophilic and hydrophobic regions are close to those of natural systems and we designate this as membrane mimetic or enzyme mimetic chemistry [45]. The large field of artificial enzymes was recently reviewed by Murakami et al. [46].

The enhancement of reaction rate as well as the stereoselectivity of hydrolytic reactions were studied by several authors [47]. Typical substrates were hydrophobized activated esters of amino acids and typical catalysts were surface active peptides with histidine as active component. The kinetic resolution of racemic esters was determined. Brown [48] and Moss [49] gave explanations for the stereoselectivity. Ueoka et al. [50] reported one example where non-functional amphiphiles as cosurfactants can enormously improve the stereoselectivity: the saponification of D,L-p-nitrophenyl N-dodecanoylphenylalaninate with the tripeptide Z-PheHisLeu-OH as catalyst in assemblies of ditetradecyldimethylammonium bromide yielded practically pure L-N-dodecanoylphenylalanine upon the addition of between 7 to 20 mol % of the anionic surfactant sodium dodecyl sulfate (SDS).

As models for metalloenzymes, Scrimin, Tonellato, Tecilla and coworkers [51] describe in a series of papers amphiphilic chiral complexes as catalysts for the hydrolysis of α-amino acid esters. In some cases moderate kinetic resolution of the racemic esters was observed [52]. Best results were obtained with bilayer forming cosurfactants in the gel state [53].

An enantioselective oxidation of 3,4-dihydroxy-L-phenylalanine catalyzed by an N-lauroyl-L- or -D-histidine-Cu(II) complex in the presence of cetyltrimethyl-ammonium bromide as cosurfactant was described by Yamada et al. [54].

One of the most successful asymmetric catalytic reactions is the asymmetric hydrogenation of amino acid precursors by means of optically active rhodium(I)phosphine or phosphinite complexes [55]. Usually, the reaction is carried out in methanol as solvent. When water is used the activity and enantioselectivity decrease significantly [16], but the addition of micelle forming surfactants leads to a solubilization of catalyst and substrate and increases activity and enantioselectivity. The results are somewhat better than the ones obtained with methanol as solvent [56]. Table 2 shows the effect with different types of surfactants.

All types of surfactants promote the reaction but only the hydrogen sulfate was active in the case of the cationic amphiphiles. There is no need to work with water-soluble complexes. However, surfactant could be substituted by polymer bound amphiphiles [57, 58].

Less successful was the use of achiral catalysts in chiral micelles. The induced enantioselectivity in the resulting α-amino acid derivatives was in all cases below 10% ee depending on the type of amphiphile [59]. Other asymmetric reac-

Table 2. Hydrogenation[a]

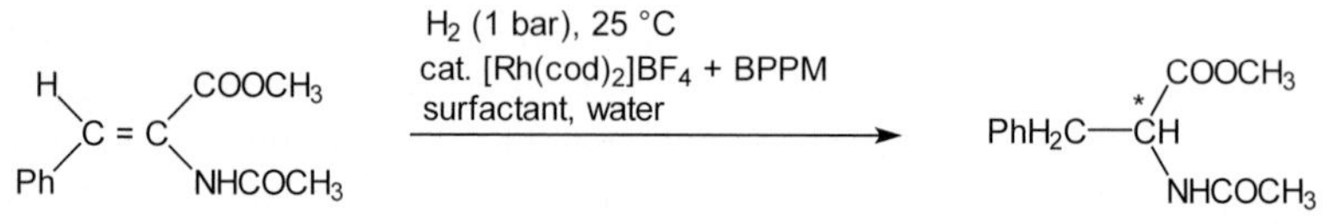

surfactant	t/2 in min	optical yield in % ee (R)
none in water (methanol)	90 (2)	78 (90)
sodium dodecyl sulfate (SDS)	6	94
cetyltrimethylammonium hydrogen sulfate	5	95
N-dodecyl-N,N-dimethyl-3-ammonio-1-propanesulfonate	5	93
decaoxyethylene-hexadecyl ether	7	95

[a]Rh : surfactant : substrate = 1:20:100;

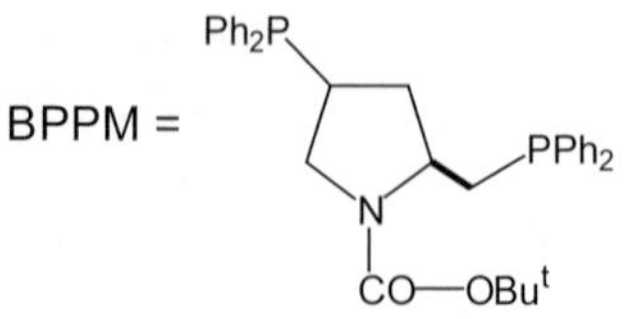

BPPM =

cod = cis, cis-1.5-cyclooctadiene

tions in optically active surfactant assemblies gave distinctly higher inductions. Examples are the conversion of aromatic aldehydes with chloroform and ammonia in presence of N-hexadecyl-N-methylephedrinium bromide [60] (28% ee), the reduction of phenyl ethyl ketone with sodium borohydride in the presence of dodecyl-β-D-glucopyranoside [61] (98% ee) and the hydroxylation of olefins via acetoxymercuration [62] (up to 96% ee) in the presence of N-hexadecyl-N-methylephedrinium bromide, and finally the Michael addition of nitromethane to chalcone with piperidine as catalyst and N-dodecyl-N-methylephedrinium bromide as chiral surfactant [63] (17% ee).

Recently, Zhang and Sun [64] reported the reduction of a series of phenyl alkyl ketones with NaBH$_4$ to optically active alcohols in reverse micelles of different ephedrinium bromides. Addition of sugars like D-fructose and D-glucose enhanced the stereoselectivity up to 27% ee. Other asymmetric catalytic reactions in reverse micelles have been investigated by Nozaki et al. [65] and Buriak and Osborn [66].

In summary, the influence of micellar and vesicular media on asymmetric catalytic reactions could be of general interest in the future.

References

1. Southern TG (1989) Polyhedron 8:407
2. Kunitake T, Shinkai S (1980) Adv Phys Org Chem 17:435
3. Keim W (1984) Chem Ing Techn 56:850
4. Wiebus E, Cornils B (1994) Chem Ing Techn 66:916
5. Joó F, Tóth Z (1980) J Mol Catal 8:369
6. Sinou D (1987) Bull Soc Chim Fr 480
7. Barton M, Atwood JD (1991) J Coord Chem 24:43
8. Kalck P, Monteil F (1992) Adv Organomet Chem 34:219
9. Herrmann WA, Kohlpaintner CW (1993) Angew Chem Int Ed Engl 32:1524
10. Sinou D (1994) Trends Organomet Chem 1:15
11. Roundhill DM (1995) Adv Organomet Chem 38:155ó
12. Joó F, Katho A (1997) J Mol Catal A 116:3
13. Sinou D (1995) Organometallic Catalysis in Water and in a Two-Phase System In: Horvàth IT, Joó F (eds) Aqueous Organometallic Chemistry and Catalysis, Kluwer, Dordrecht p 215
14. Amrani Y, Lecomte L, Sinou D, Bakos J, Tóth I, Heil B (1989) Organometallics 8:542
15. Laghmari M, Sinou D, Masdeu A, Claver C (1992) J Organomet Chem 438:213
16. Lecomte L, Sinou D, Bakos J, Tóth I, Heil B (1989) J Organomet Chem 370:277
17. Bakos J (1995) Chiral Sulfonated Phosphines in Enantioselective Catalysis In: Horvath IT, Joó F (eds) Aqueous Organometallic Chemistry and Catalysis, Kluwer, Dordrecht p 231
18. Bakos J, Orosz A, Cserepi S, Tóth I, Sinou D (1997) J Mol Catal A 116:85
19. Tóth I, Hanson BE, Davis ME (1990) Catal Lett 5:183
20. Arhancet JP, Davis ME, Merola JS, Hanson BE (1989) Nature (London) 339:454
21. Wan KT, Davis ME (1993) Tetrahedron Asymmetry 4:2461
22. Wan KT, Davis ME (1994) J Catal 148:1
23. Eckl RW, Priermeier T, Herrmann WA (1997) J Organomet Chem 532:243
24. Horváth IT, Rabái J (1994) Science 266:72
25. Cornils B (1997) Angew Chem 119:2147; Angew Chem Int Ed Engl 36:2057
26. Weber WP, Gokel GW (1977) Phase Transfer Catalysis in Organic Synthesis, Springer, Berlin

27. Starks CM, Liotta C (1978) Phase Transfer Catalysis: Principles and Techniques, Academic Press, Orlando
28. Dehmlow EV, Dehmlow SS (1993) Phase Transfer Catalysis, 3rd edn. VCH, Weinheim p 80 ff.
29. Goldberg Y (1992) Phase Transfer Catalysis: Selected Problems and Applications, Gordon and Breach, Philadelphia, Chapter 6
30. Noyori R (1994) Asymmetric Catalysis in Organic Synthesis, Wiley, New York, Chapter 7, p 333
31. Goldberg Y, Alper H (1996) Phase-Transfer Catalysis and Related Systems. In: Cornils B, Herrmann WA (eds) Applied Homogeneous Catalysis with Organometallic Compounds, VCH, Weinheim p 844
32. O'Donnel MJ (1993) Asymmetric Phase Transfer Reactions. In: Ojima I (ed) Catalytic Asymmetric Synthesis, VCH, New York, Chapter 8, p 389
33. Chiral Phase Transfer Catalysts (1997) In: Sasson Y, Neumann R (eds) Handbook of Phase Transfer Catalysis, Blackie, London
34. O'Donnel MJ, Bennett WD, Wu S (1989) J Am Chem Soc 111:2353
35. Dolling U-H, Hughes DL, Bhattacharya A, Ryan KM, Karady S, Weinstock LM, Grabowski EJJ (1987) In: Starks CM (ed) Phase Transfer Catalysis, ACS Symposium Series: 326, Washington, p 67
36. Conn RSE, Lovell AV, Karady S, Weinstock LM (1986) J Org Chem 51:4710
37. Cram DJ, Sogah GDY (1981) J Chem Soc, Chem Commun 625
38. Mirza-Aghayan M, Etemad-Moghadam G, Zaparucha A, Berlan J, Loupy A, Koenig M (1995) Tetrahedron Asymmetry 6:2643
39. Ding H, Hanson BA, Bakos J (1995) Angew Chem 107:1728; Angew Chem Int Ed Engl 34:1645
40. Harigaya Y, Yamaguchi H, Onda M (1981) Heterocycles 15:183
41. Wynberg H, Greijdanus B (1978) J Chem Soc, Chem Commun 427
42. Masui M, Ando A, Shioiri T (1988) Tetrahedron Lett 29:2835
43. Aires-de-Sousa J, Lobo AM, Prabhakar S (1996) Tetrahedron Lett 37:3183
44. Fendler JH (1987) Chem Rev 87:877
45. Fendler JH (1982) Membrane Mimetic Chemistry, Wiley, New York
46. Murakami Y, Kikuchi J, Hiseada Y, Hayashida O (1996) Chem Rev 96:721
47. Bunton CA, Savelli G (1986) Adv Phys Org Chem 22:213
48. Brown JM, Bunton CA (1974) J Chem Soc, Chem Commun 969
49. Moss RA, Chiang Y-CP, Hui Y (1984) J Am Chem Soc 106:7506
50. Ueoka R, Matsumoto Y, Yoshino T, Watanabe N, Omura K, Murakami Y (1986) Chem Lett 1743
51. Scrimin P, Tonellato (1991) Ligand Surfactants: Aggregation, Cations Binding and Transport and Catalytic Properties. In: Mittal KL, Sha DD (eds) Surfactants in Solution, Plenum Press, New York, Vol. 11, p 339
52. Scrimin P, Tecilla P, Tonellato U (1994) J Org Chem 59:4194
53. Cleij MC, Scrimin P, Tecilla P, Tonellato U (1996) Langmuir 12:2956
54. Yamata K, Shosenji H, Otsubo Y, Ono S (1980) Tetrahedron Lett 21:2649
55. Oehme G, Paetzold E, Selke R (1992) J Mol Catal 71:L1
56. Grassert I, Paetzold E, Oehme G (1993) Tetrahedron 49:6605
57. Flach HN, Grassert I, Oehme G (1994) Macromol Chem Phys 195:3289
58. Flach HN, Grassert I, Oehme G, Capka M (1996) Colloid Polym Sci 274:261
59. Grassert I, Vill V, Oehme G (1997) J Mol Catal A 116:231
60. Zhang, Y, Li W (1988) Synth Commun 18:1685
61. Hui Y, Yang C (1988) Huaxue Xuebao 46:239, CA 109:190679
62. Zhang Y, Bao W, Dong H (1993) Synth Commun 23:3029
63. Zhang Y, Fang X, Wang S (1992) Youji Huaxue 12:488, CA 118:59364
64. Zhang Y, Sun P (1996) Tetrahedron Asymmetry 7:3055
65. Nozaki K, Yoshida M, Takaya H (1994) J Organomet Chem 473:253
66. Buriak JM, Osborn JA (1996) Organometallics 15:3161

Chapter 39   Combinatorial Approaches

Chapter 39
Combinatorial Approaches

Ken D. Shimizu[1] · Marc L. Snapper[2] · Amir H. Hoveyda[2]

[1]Department of Chemistry and Biochemistry, University of South Carolina, Columbia, South Carolina 29208, USA
[2]Department of Chemistry, Merkert Chemistry Center, Boston College, Chestnut Hill, Massachusetts 02467, USA
e-mail: shimizu@psc.sc.edu; marc.snapper@bc.edu; amir.hoveyda@bc.edu;

Keywords: Chiral catalysts, Asymmetric synthesis, High-throughput screening, Ligand diversity, Solid phase chemistry

1
Introduction

Combinatorial chemistry has emerged as an important and formidable strategy in the search for effective therapeutic agents [1]. The practicality, efficiency, and productivity of diversity-based protocols for drug discovery have already had a notable impact on the pharmaceutical industry. Although the major applications of combinatorial libraries remain in the search for biologically active compounds, it has also been recognized that such methods might be effective in the identification of compounds with any attractive properties. Combinatorial and related strategies have indeed been recently utilized in investigations in materials science [2], molecular recognition [3, 4], polymer chemistry [5], and asymmetric catalysis [6].

This article provides a brief overview of the recently developed diversity-based approaches in the screening and identification of effective metal-based chiral catalysts for enantioselective synthesis. This first wave of reports indicates that high throughput technologies might represent a viable and efficient route for the development of useful chiral catalysts. In a few instances, it is likely that the more traditional screening approaches, often based on *a priori* mechanistic bias, would have been less successful, at least within the same time span.

2
Diversity-Based Approach for Drug vs Catalyst Discovery

The searches for therapeutic agents and asymmetric catalysts share a number of facets and can reap similar benefits from diversity-based protocols. Traditionally, both fields have relied on iterative approaches wherein a single compound is designed, synthesized, and tested (Scheme 1). The cycle is repeated until a compound is obtained with the desirable levels of enantioselectivity or activity. In contrast, as illustrated in Scheme 1, a high throughput strategy enables the chemist to simultaneously generate and test notably larger numbers of candidates, potentially reducing the entire search cycle to one or two iterations. A more comprehensive search can thus be completed more efficiently.

Combinatorial chemistry is not an irrational method; rather, it brings together rational design and high throughput evaluation and is rooted in empirical observations and logical deduction. Structure-selectivity observations remain the basis for propagating molecular features from one generation of catalysts to the next. High throughput strategies therefore permit more initial "guesses" and a greater allowance for failure. Combinatorial chemistry is particularly well suited toward optimizing novel and previously unexamined reactions for which little mechanistic data are available. Such strategies can be viewed as the chemist's attempt to address the notion that mechanistic subtleties that often differentiate the selectivity and reactivity from one substrate or catalyst to another may not be generalized – these mechanistic intricacies can vary unpredictably. Such a broad-based approach relieves the chemist from the risk of following a relatively narrow path that is selected on the basis of fickle mechanistic parameters. It is

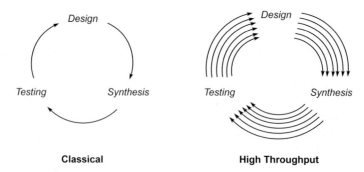

Classical   **High Throughput**

Scheme 1. The diversity-based approach can provide a wealth of reactivity and selectivity data in an efficient manner perhaps fair to state that almost all successful asymmetric catalysts have benefited, at some point in their development, from serendipitous observations. Combinatorial chemistry integrates this aspect of catalyst discovery into the overall search process, increasing the rate at which "advantageous mutations" can occur. Nevertheless, a combinatorial approach can significantly promote the mechanistic studies of new asymmetric processes, as it can put forth a large structure-selectivity database from which data can be rapidly generated.

Three important issues need to be addressed for the successful implementation of combinatorial asymmetric catalyst design: (i) sources of diversity, (ii) high throughput synthesis of catalysts, and (iii) high throughput catalyst screening.

(i) *Sources of diversity.* Two distinct approaches have been adopted in introducing the element of diversity into asymmetric reactions. The first is analogous to that developed for drug design, where a modular catalyst composed of changeable subunits is used. Variation of the modules generates an exponential number of catalysts with myriad steric and electronic attributes. When organometallic complexes are employed as reaction initiators, metal centers can be modified as well. The second route for introducing diversity into asymmetric reactions involves variation of the reaction conditions. Solvent, concentration, temperature, and reaction times are potential parameters that can be altered. These protocols towards introducing diversity are not mutually exclusive.

(ii) *High throughput catalyst synthesis.* To retain the benefits of a diversity-based method, it is imperative that the catalyst is easily and efficiently synthesized. At the same time, reaction initiators must be of high purity due to the potential deleterious effects of impurities in metal-catalyzed reactions. These stringent requirements can therefore limit the types of catalysts that are amenable to combinatorial and related techniques. Most "combinatorial catalysts" to date have been made through parallel syntheses. The modularity of the chiral ligands has been critical to accelerate their preparation by making the fabrication of each catalyst identical, regardless of structural variations. In many cases, hundreds of compounds can be manually synthesized by a single researcher. A high level of efficiency requires that lengthy separations be avoided by utilizing only high-yielding reactions that join various modules.

(iii) *High throughput catalyst screening.* Currently, the greatest bottleneck is in assaying each catalyst for asymmetric induction. A successful approach in drug design has been to test mixtures of compounds and then, by means of a deconvolution strategy, to identify the active component. This approach is not easily amenable to asymmetric catalytic processes where two opposing and neutralizing outcomes (R and S enantiomers) can exist; a mixture of chiral complexes may yield a racemic product even with a mixture of selective catalysts. Accordingly, recent reports generally involve testing individual systems. It merits mention that the parallel screening strategy has also been applied in therapeutic discovery efforts, due to the difficulties involved in the accurate deconvolution of various mixtures and in order to remove any synergism that may exist between several active compounds. As such, combinatorial chemistry does not necessarily have to involve the generation of mixtures of compounds; it may be better characterized by the modular nature of the constituent compounds that in different combinations provide large numbers of molecular ensembles.

3
Diversity-Based Investigations on the Discovery of Catalysts for Enantioselective Synthesis

3.1
Catalytic Enantioselective Aldehyde Alkylation

One of the earliest demonstrations of the viability of a combinatorial approach in the discovery of asymmetric reactions was reported by Ellman in 1995 [7]. The Berkeley team's selection of the dialkylzinc addition to aldehydes (Scheme 2) to gauge the potential utility of a combinatorial approach was based on several factors:

(i) Previous studies had clearly demonstrated a direct correlation between ligand structure and enantioselectivity. Thus, variations of the 2-pyrrolidinemethanol ligand framework could lead to significant improvements in asymmetric induction.

(ii) Although extensive work had been done with aromatic aldehydes as substrates, the levels of enantioselectivity with aliphatic systems left ample room for improvement.

Thirteen different ligands were synthesized on a solid support from a 4-hydroxyproline precursor and attached to the support (commercially available Merri-

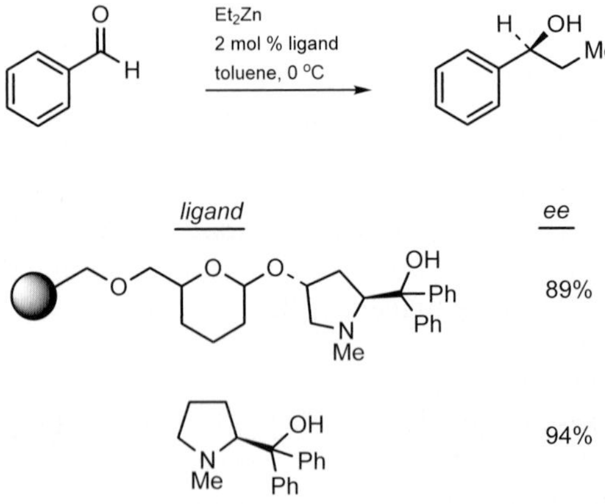

Scheme 2. In catalytic alkylation of aldehydes, similar but slightly lower enantioselectivity is attained when the chiral ligand is anchored to a solid support

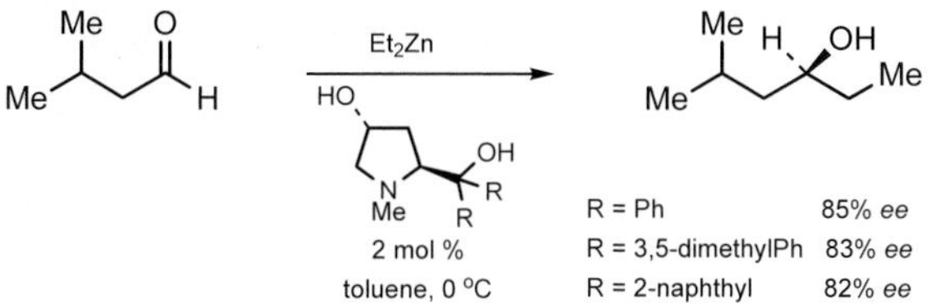

Scheme 3. Influence of various chiral ligands on the enantioselective addition of Et$_2$Zn to an aliphatic aldehyde

field resin) through the 4-hydroxy substituent. The ligands were initially screened while still covalently anchored to the support. As depicted in Scheme 2, the levels of enantioselectivity for reactions initiated by the resin-bound ligands proved to be high, but slightly lower than those obtained with the free ligands in solution (e.g., 89% vs 94 % ee).

Screening was carried out with unbound chiral ligands, synthesized on solid supports and subsequently freed from the resin. Representative data are shown in Scheme 3. It is important to note that, subsequent to cleavage from the solid support, little or no purification of the ligands was required to maintain excellent enantioselectivity. This is a tribute to the efficient multistep synthesis carried out on the Merrifield resin and to the benefits of ligand-accelerated catalysis [8]. The effects of the chiral pyrrolidinone ligand are sufficiently dominant so as not to allow adventitious side products from ligand synthesis to catalyze product formation and thus lower overall selectivity or hinder accurate analysis of the screening process.

3.2
Catalytic Enantioselective Hydrogenation

Chiral phosphines have been used on numerous occasions in conjunction with various transition metals to effect enantioselective bond formation [9]. As with many other metal-catalyzed asymmetric reactions, it is nearly impossible to predict which phosphine ligands or metal centers will afford the most desirable outcomes. Gilbertson and coworkers therefore set out to prepare a sixty-three member library of chiral phosphines that could be screened for catalytic and enantioselective olefin hydrogenation. A particular feature of these phosphines is that they were built within a helical peptide scaffold. Thus, a variable sequence of four to five amino acids was inserted into the peptide Ac-Ala-Aib-Ala-[···]-Ala-Aib-Ala-NH$_2$ to yield a range of chiral ligands. These researchers argued that folding of the polypeptide backbone would bring together two different donor phosphine units to present effectively a bisphosphine system and an appropriate chiral environment to the transition metal (Scheme 4) [10]. The terminal Ala-Aib-Ala sequences were designed to promote helix formation by bringing the two synthetic amino acids with phosphine side chains in close proximity when they are positioned at spacings of (i, i+1) and (i, i+4).

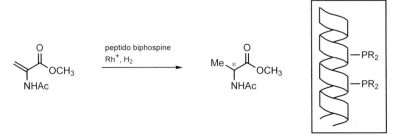

Scheme 4. Phosphine units attached to helical peptide scaffolds have been screened as catalysts for enantioselective hydrogenation

The modular peptidic ligands are amenable to well-established solid-phase synthesis protocols, enabling ready access to a diverse array of structures. Sixty-three different peptides were synthesized in parallel on pins and tested for asymmetric induction while still attached to the solid-phase. As depicted in Scheme 4, Rh(I) was selected as the metal center (based on ample precedence) and the enantioselective hydrogenation of an α-amino acid was examined. Although relatively low levels of enantioselectivity were observed (≤19% ee), this study demonstrated that a combinatorial protocol can efficiently provide the chemist with a wealth of data. It is not clear whether any reliable mechanistic information can be gleaned from these initial results because of the low levels of stereodifferentiation. However, future developments of these ligands, along the lines demonstrated by Gilbertson, could well afford more practical levels of selectivity: the helical peptidic bisphosphines have been recently reported to effect the hydroformylation of styrene with 40% ee [11].

3.3
Catalytic Enantioselective Addition of TMSCN to Meso Epoxides

As part of a program directed towards the development of new catalytic and enantioselective reactions, we have utilized diversity-based protocols to introduce variations within a modular peptide-based ligand as the means to identify effective chiral ligands for enantioselective TMSCN addition to *meso* epoxides (Scheme 5). These peptidic ligands had previously been used in other enantioselective catalytic processes [12] and, as depicted in Scheme 5, are composed of three subunits: Schiff base (SB), amino acid 1 (AA1), amino acid 2 (AA2) [13]. Such peptide-based systems offer various attractive features: (i) A number of chiral amino acids are available in the non-racemic form, thus allowing a gamut of optically pure chiral ligands to be readily accessed. (ii) Peptidic systems can be prepared efficiently, in parallel, and by established solid phase protocols. (iii) There is ample precedent that polypeptides effectively associate with a range of transition metal centers.

In principle, 8000 (20^3) different chiral catalysts could be made from the 20 natural amino acids and 20 different aldehydes. However, to control the num-

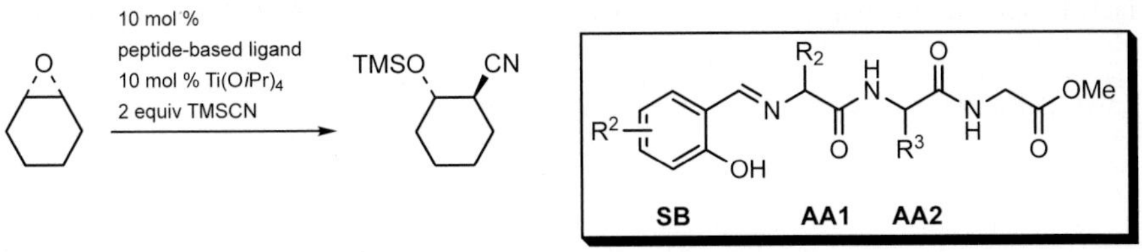

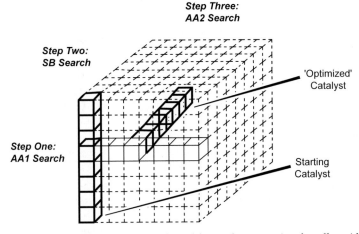

Scheme 5. Peptidic Schiff bases may be screened for identification of an effective chiral ligand for catalytic enantioselective addition of TMSCN to *meso*-epoxides

Fig. 1. Representation of a search strategy adopted for catalyst screening that allows identification of effective ligands without examination of all possibilities

bers of compounds synthesized and screened, a representational search strategy was employed (illustrated in Fig. 1). Each of the three subunits in the modular ligand was successively optimized, such that the first amino acid 1 (AA1 - shown in red) was varied and the other two subunits were kept constant. *tert*-Leucine was found to be optimal at position AA1 and this structural element was retained in successive generations. The second position (AA2) was then altered, and *O-tert*-butylthreonine was identified as the best AA2. Finally, from a pool of salicylic aldehydes, 3-fluorosalicylaldehyde was selected as the best Schiff base (SB). In the end, only a representative sampling of sixty (20×3) catalysts was necessary to identify one that affords nearly a 95:5 ratio of enantiomers (89% ee). The initial catalyst provided the addition product with only 26% ee (cyclohexene oxide as substrate); successive modifications of the ligand structure enhanced the level of selectivity in three steps to afford finally a synthetically attractive level of enantioselectivity (with 3-fluorosalicylaldehyde-*tert*-leucine-*O*-*tert*-butylthreonine-glycine-OMe as the catalyst). It is unlikely that any mechanistic considerations would have pointed to this peptidic complex as being one of the most suitable species.

Ken D. Shimizu · Marc L. Snapper · Amir H. Hoveyda

Table 1. Optimized ligands for catalytic enantioselective addition of TMSCN to *meso*-epoxides[a]

Entry	Substrate	Product	ee [%]	Yield [%]	Optimized Ligand
1	(cyclopentene oxide)	NC (R)··· OTMS (S) (cyclopentane)	83	72	(naphthyl Schiff-base ligand)
2	(cyclohexene oxide)	NC··· OTMS (cyclohexane)	87	65	(fluoro-substituted Schiff-base ligand)
3	(cycloheptene oxide)	NC··· OTMS (cycloheptane)	84	68	(naphthyl Schiff-base ligand)
4	*n*-Pr, *n*-Pr (epoxide)	*n*-Pr NC··· OTMS *n*-Pr	78	69	(phenyl Schiff-base ligand)

[a]Conditions: 20 mol % Ti(O*i*Pr)$_4$, 20 mol % ligand, 4 °C, toluene, 6–20 h

The aforementioned strategy for catalyst screening raises an intriguing question: Is the "optimal catalyst" identified by this process truly the very best catalyst? And if it is not, does the attendant improvement in enantioselectivity (>99% ee), assuming there will be no notable difference in efficiency, justify the additional effort that would be required to achieve it? In the approach described above, we have made certain assumptions about the additivity and absence of cooperativity between the three subunits. At least for this small sampling, these assumptions seem to hold true, but without testing every combination we cannot definitively answer this important question. Examination of every possibility would tax and detract from the efficiency of the general screening method. An important practical advantage of the above approach is that, in a relatively short amount of time, it allowed us to identify a selective catalyst for an entirely new asymmetric process. That is, the search strategy is not an open-ended odyssey but a well-structured, program that allows a fairly comprehensive assessment of a ligand framework with respect to a specific asymmetric reaction in a finite and predictable period of time.

When we applied the above search strategy to various other *meso*-epoxide substrates, a number of important observations were made [14]. One significant trend that emerged from these studies was that for each epoxide substrate, as de-

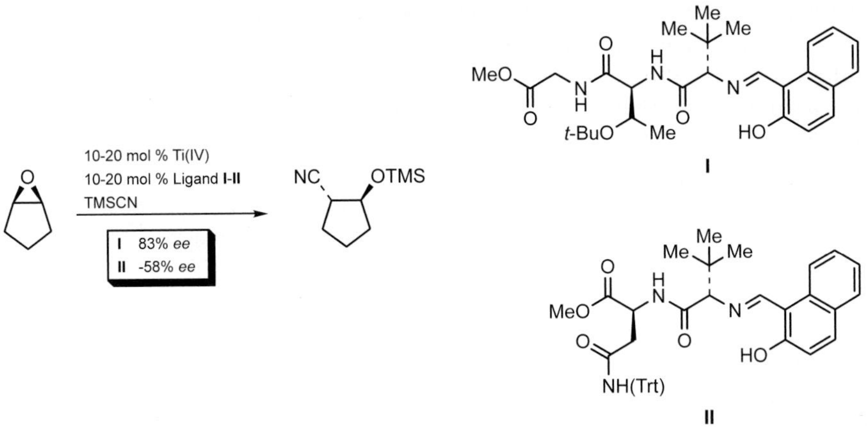

Scheme 6. Subtle modifications in the structure of a peptide ligand may unexpectedly lead to significant variations in selectivity

picted in Table 1, a similar but unique chiral catalyst was identified. This type of catalyst/substrate selectivity is akin to that observed in Nature where many reactions have their own unique enzymes. The high levels of selectivity observed with enzymatic reactions are often accompanied by a lack of substrate generality. In this instance, however, because ligand modification is relatively straightforward, substrate specificity does not necessarily imply lack of generality. Another important issue raised by the data in Table 1 is that the search for a "truly general catalyst" is perhaps unrealistic: catalysts that afford exceptional selectivity do so because they associate with specific structures with great fidelity. To expect high specifity and broad range generality may be somewhat contradictory.

An important factor that emerged from our studies is that the above method of catalyst identification increases the frequency with which unexpected observations are made. For example, as illustrated in Scheme 6, a subtle alteration in the structure – and not the stereochemical identity – of the peptide ligand leads to inversion of stereochemistry in the epoxide-opening reaction (compare reaction with ligands **I** and **II**). These observations validate our choice of individually synthesizing and testing each catalyst, as mixtures of catalysts can lead to racemic products.

3.4
Catalytic Asymmetric Carbene Insertion

In the above examples, the modularity was contained within the chiral ligand. Burgess and Sulikowski adopted an alternate approach by matching an array of chiral ligands with a range of metal centers [15]. A third dimension of diversity was introduced by changing the reaction conditions through variation of the solvent systems. The approach is thus an amalgamation of the diversity strategies outlined earlier. All told, five chiral ligands coupled with six different metal salts were examined in four different solvents. Ninety-six of the possible one hundred and twenty different combinations were examined in less than a week for their ability to direct the asymmetric carbene C-H insertion, Eq (1). The most effective catalyst was found to be a Cu(I)·(bis)oxazoline ligand complex which was optimized to give a 3.9:1 diastereomeric ratio, compared to the previously reported 2.3:1 selectivity. The unprecedented catalysis of carbene insertion by Ag(I) was also observed, underlining an additional strength of the high throughput approach.

(1)

4
Conclusions and Outlook

The studies discussed in this brief overview represent the chemist's initial attempts to establish a reasonably general protocol for the identification and discovery of metal-based catalysts that effect bond formation in an enantioselective manner. These research activities are perhaps the result of the appreciation of the principle that a "rational" approach has its shortcomings: mechanisms may not be general and can be unpredictable. Such principles can vary with subtle changes in reaction conditions or substrate structure. Development of high throughput protocols is based on the realization that, even within a single class of substrates, the identity of the "optimum catalyst" may change. Perhaps this area of research has its deepest roots in the history of asymmetric catalysis: it is more than often the unanticipated "hit" that becomes the key factor that fuels a successful investigation. If so, why not carry out research in a manner that enhances the probability of making positive chance observations? Combinatorial and related strategies likely arise from the scientist's desire to harness and effectively utilize serendipity, the factor that has been most instrumental in allowing us to achieve our most impressive successes. The interesting and enlightening observation is often the unpredicted; diversity-based approaches allow us to uncover the often convoluted and hidden path to success.

This line of research does not advocate that we abandon rational or rigorous investigations of detailed mechanisms of important processes. Elements of design and *a priori* decisions are still required in determining what collection of catalysts need be prepared; the framework is simply broader and thus initial bias, that may be based on few initial observations, has less of a chance to direct us in the wrong direction. Diversity-based strategy will allow us to base our mechanistic hypotheses on a much wider pool of data points – it will discourage us from making naive generalities, which are more than often revised soon after a few additional experiments.

The above investigations are the first steps in the exciting road that lies ahead of us. It is likely that we will soon be able to screen significantly larger catalyst collections. A recent report by Jacobsen's group [16] in connection with an impressive library of chiral metal complexes represents an important first step in this direction. The high throughput approach to enantioselective reaction discovery will likely present us with a more complete picture, where the hidden subtleties are highlighted, where the exciting exceptions, as well as the more useful generalities, become more apparent.

Acknowledgments: We are most grateful to our colleagues Dr. Bridget Cole, Mr. Clinton Krueger, Mr. Kevin Kuntz, and Dr. Joseph Harrity for making numerous invaluable intellectual and experimental contributions to the project discussed in this article. Research in our laboratories was generously supported by the National Science Foundation (CHE-9632278 to A. H. H.), Johnson and Johnson (Focused Giving Grant to A. H. H.), Pfizer (Young Faculty Award to A. H. H.), Massachusetts Department of Public Health (Breast Cancer Scholar Award to M. L. S.), and the National Institutes of Health (postdoctoral fellowship to K. D. S.; F32-GM-17821).

References

1. For recent reviews on combinatorial chemistry, see special issues of: (a) Chem Rev (1997) 97; (b) Acc Chem Res (1996) 26
2. Briceño G, Change H, Sun X, Schultz PG, Xiang X-D (1995) Science 270:273
3. Cheng Y, Suenaga T, Still WC (1996) J Am Chem Soc 118:1813
4. Goodman MS, Jubian B, Linton B, Hamilton AD (1995) J Am Chem Soc 117:11610
5. Brocchini S, Tangpasuthadol JV, Kohn J (1997) J Am Chem Soc 119:4553
6. Borman S (1996) Chem Eng News 4:37
7. Liu G, Ellman JA (1995) J Org Chem 60:7712
8. Berrisford DJ, Bolm C, Sharpless KB (1995) Angew Chem Int Ed Engl 34:1059
9. Ojima I (1993) Catalytic Asymmetric Synthesis. VCH, Weinheim
10. Gilbertson SR, Wang X (1996) Tetrahedron Lett 37:6475
11. Gilbertson SR (1996) US patent 5,541,289
12. Nitta H, Yu D, Kudo M, Mori A, Inoue S (1992) J Am Chem Soc 114:7969
13. Cole BM, Shimizu KD, Krueger CA, Harrity JPA, Snapper ML, Hoveyda AH (1996) Angew Chem, Int Ed Engl 35:1668
14. Shimizu KD, Cole BM, Krueger KA, Kuntz KW, Snapper ML, Hoveyda AH (1997) Angew Chem, Int Ed Engl 36:1704
15. Burgess K, Lim H-L, Porte AM, Sulikowski GA (1996) Angew Chem, Int Ed Engl 35:220
16. Francis, MB, Finney NS, Jacobsen EN (1996) J Am Chem Soc 118:8983

Chapter 40   Catalytic Antibodies

Chapter 40
Catalytic Antibodies

Paul Wentworth Jr. · Kim D. Janda

Department of Chemistry, The Scripps Research Institute and The Skaggs Institute for
Chemical Biology, 10550 North Torrey Pines Road, La Jolla, CA 92037, USA
e-mail: paulw@scripps.edu; kdjanda@scripps.edu

Keywords: Catalytic antibody, Hapten, Enantiofacial, Enantioselective, Diels-Alder cycloaddition, Cationic reactions, Aldol condensation, Disfavored cyclization

Abbreviations

ER   Enhancement ratio, equivalent to k_{cat}/k_{uncat}
IgG   Immunoglobulin-G
K_m Michaelis-Menten constant, equivalent to the substrate concentration at which an enzyme or catalytic antibody is 50% saturated
k_{cat} Catalytic rate constant, equivalent to the maximal rate of turnover by an enzyme or catalytic antibody when saturated with substrate
k_{uncat} The rate constant for a non-catalyzed reaction
TS   transition state

1
Introduction

The *de novo* design and synthesis of new catalysts for organic synthesis and biochemistry has long been the goal of chemists and biochemists alike. This chapter documents the rise of antibodies as a new class of such catalysts and describes their increasing scope and utility in asymmetric chemical processes.

For the purposes of this treatise, the definition of asymmetric synthesis is a modification of that proposed by Morrison and Mosher [1] and as such will be applied to stereospecific reactions in which a prochiral unit in either an achiral or a chiral molecule is converted, by utility of other reagents and/or a catalytic antibody, into a chiral unit in such a manner that the stereoisomeric products are produced in an unequal manner. As such, the considerable body of work devoted to antibody-catalysis of stereoselective reactions including chiral resolutions, isomerizations and rearrangements are considered to be beyond the scope of this discussion. For information regarding these specific topics and more general information regarding the catalytic antibody field the following papers [2,3,4,5,6] and reviews [7,8,9,10] are recommended.

1.1
Antibody Structural Features

Antibodies, termed more correctly immunoglobulins, are important components of the mammalian defense mechanism. Structurally, they are symmetrical protein molecules composed of two pairs of polypeptide chains, two heavy chains (M_R 50,000) and two light chain (M_R 25,000), interlinked by disulfide bonds [11]. Sequence comparison of different monoclonal immunoglobulin G (IgG) molecules reveals that the carboxy-terminal half of the light chain and approximately three-quarters of the heavy chain show little sequence variation [12]. By contrast, the N-terminal regions of both the antibody light and heavy chains show high degrees of structural variability. These regions of variability compose the binding pockets, or antigen-binding regions, of the antibodies. Antibody recognition of its antigen (hapten) is highly specific and, in general, comprises noncovalent interactions leading to high affinity constants, typically $K_d > 10^{-4}$ to 10^{-14} M^{-1} [13]. If linked to catalysis, this binding may supply up to 20 kcal·M^{-1} of free energy, sufficient to catalyze most chemical reactions. The mammalian immune system comprises *ca.* 10^{10} different antibody molecules, each with a distinct variable region sequence and hence binding site specificity, the net result of this vast protein library is that immunoglobulins possess high affinity and unmatched structural specificity towards virtually any molecule. It is this vast library of chiral binding sites and binding energies that the field of catalytic antibodies is perusing in its quest for novel biocatalysts.

2
Enantiotopic and Enantiofacial Selective Reactions

One of the first demonstrations of antibodies as chiral determinants in an organic reaction involved the enantioselective perturbation of a *meso*-substrate [14]. For this case phosphonate 1 was used as a hapten and the *meso*-diacetate 2 as the substrate for antibodies elicited to this hapten (Scheme 1). This strategy engages

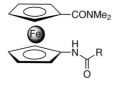

33 R = (CH$_2$)$_3$CO$_2$H
34 R = 4-carboxyphenyl

Scheme 1. Antibody 37E8, elicited to the phosphonate hapten **1** catalyzes the enantiotopic hydrolysis of the *meso*-diacetate **2** (>98% ee) at pH 8.0 (ATE [0.1 M *N*-(2-acetamido)-2-aminoethanesulfonic acid (ACES), 0.052 M Tris, 0.052 M ethanolamine] and 37 °C

1 as a transition state (TS) mimic; however, some significant extensions to preceding work were implicit in this study.

First, the lack of an aromatic component in **1** meant that decreased immunogenicity was observed. Second, initial screening assays for highlighting catalytic clones were not readily susceptible to a UV assay. Enantiotopic hydrolysis of **2** to the hydroxyacetate **3** was best accomplished by antibody 37 EB with a turnover number (catalytic rate constant), k_{cat}=0.007 min^{-1}, and with excellent enantioselectivity, 86% ee.

One of the first enantiofacially selective processes catalyzed by an antibody involved the hydrolysis of enol esters [15]. Hapten **4** was used to elicit antibodies for the hydrolysis of enol ester **5** (Scheme 2). This reaction proceeds *via* a putative enolate intermediate **6** and the key asymmetric induction step involves antibody-catalyzed enantiofacial protonation of one of the prochiral faces of **6**. Antibody 27B5 catalyzes the hydrolysis of the enol ester **5** with a turnover number k_{cat}=0.01 min^{-1} corresponding to an enhancement ratio (ER), k_{cat}/non-catalyzed rate (k_{uncat})=300 and provides an optically enriched mixture of the *R*-ketone product **7** (42% ee). Although the asymmetric induction is lower than that achievable by natural enzymes for certain substrates [16], it was a successful demonstration, at entry level, for catalytic antibodies and asymmetric induction.

The enantioface selective protonation of prochiral enol derivatives is a simple and attractive route for the preparation of optically active carbonyl derivatives. Reports of stoichiometric protonation of metal enolates by a chiral proton source at low temperature leads to optical yields from 20 to 85% ee and yeast esterase catalyzes the hydrolysis of 1-acetoxycycloalkenes with enantioselectivities between 41 and 96% for enol protonation [17,18]. These reactions involve enolates under basic conditions. Hydrolysis of enol ethers under acidic conditions proceeds *via* a rate-determining carbon protonation and is catalyzed by carboxylic acids [19,20]. Reymond et al. [21] reasoned that a complementary

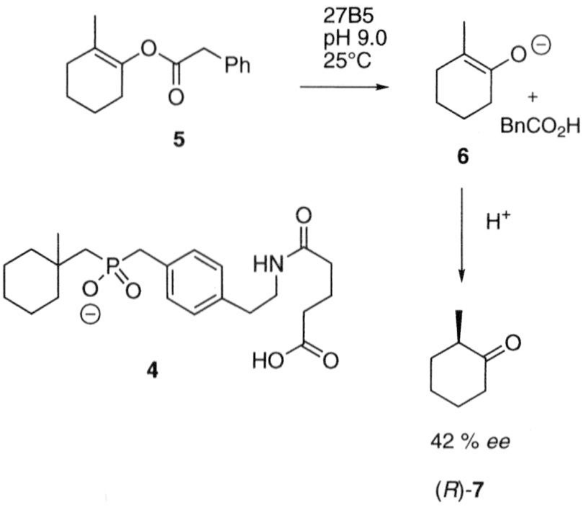

Scheme 2. Antibody 27B5 elicited to the phosphonate hapten **4** catalyzes the enantiofacial protonation of the enolate intermediate **6** during the acid-catalyzed hydrolysis of the enol ester **5** to yield the chiral ketone **7** with 42% ee

carboxylic acid elicited in an antibody binding site to either of the diastereomers of the N-methylpiperidinium salt **8a** or **8b**, both of which are 'bait-and-switch' haptens, would be in an optimal position for enantiofacial protonation of enol ethers (Scheme 3). Furthermore, in its conjugate base form this carboxylate may subsequently serve to stabilize the intermediate oxocarbenium ion **9**. The tetrahedral geometry of the ammonium ion of **8a** and **8b** was seen as a critical factor for supplying a binding pocket for the asymmetric pyamidalization of the carbon undergoing protonation.

The most active antibody generated, 14D9, elicited to **8a** catalyzes the enantioselective hydrolysis of the enol ether substrates **10** and **11** to the chiral aldehyde **12** and the enol ether **13** to ketone **14** (Scheme 3) [22, 23]. The process is readily scaled up for the cyclic substrate **13** and leads to the ready production of gram quantities of optically active ketone (−)-(S)-**14** in 65% yield (based on starting material) and 86% ee, demonstrating the potential of this catalytic antibody in organic synthesis [24].

In an alternative approach, Nakayama and Schultz [25] have successfully achieved the enantiofacial reduction of prochiral ketones. By utilizing the phosphonate hapten **15** catalytic antibodies were elicited which catalyze a highly stereospecific reduction of ketone **16** with sodium cyanoborohydride as a cofactor (Scheme 4). The most active antibody, A5, was found to have a pH optimum at acidic pH, consequently the reductions were performed in aqueous buffer at pH 5.0. The reaction was followed for multiple turnovers (>25) without any decrease in activity or stereoselectivity highlighting the utility of this catalytic system.

Lineweaver-Burke analysis of the steady-state kinetic data for A5 revealed a $k_{cat}=0.1$ min^{-1}, equivalent to a rate enhancement of 290. The background reac-

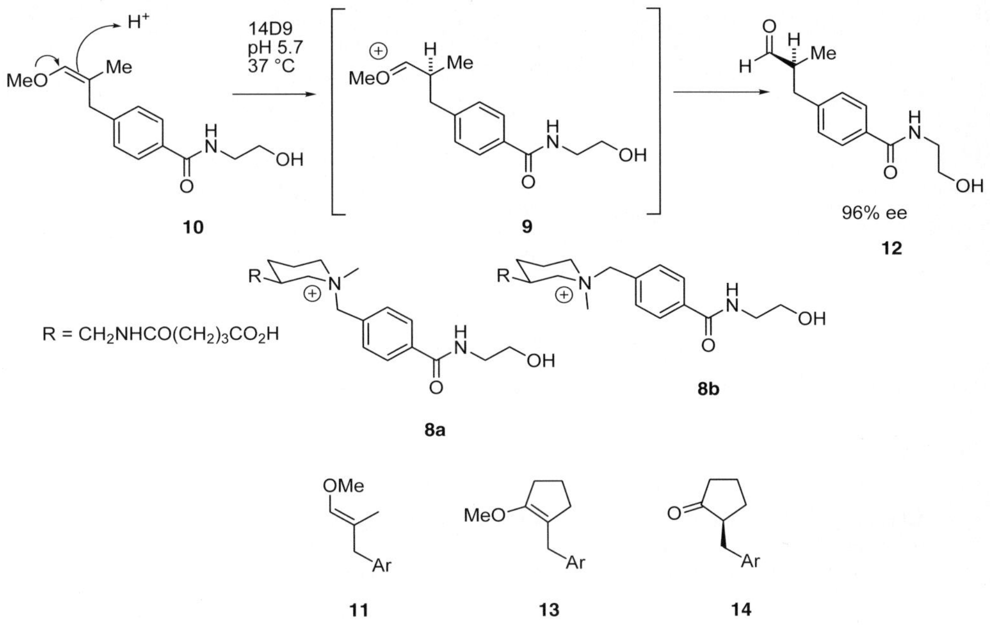

R = CH₂NHCO(CH₂)₃CO₂H

Scheme 3. Enantiofacial protonation of the enol ethers **10, 11, 13** by antibody 14D9, elicited to hapten **8a**, yields chiral products **12, 14** with upto 96% ee

Substrate[a] (conditions)	conv. [%]	ee [%]	configuration
10 (5 days, 37 °C)[b]	56	97	(S)
11 (7 days, 37 °C)	20	93	(S)
13 (1 h, 20 °C)[c]	100	93	(S)

[a]Solutions containing 1 mM substrate and 30 μM antibody
[b]In 50 mM MES buffer, 10 mM NaCN, and 90 mM NaCl, pH 5.5
[c]In 50 mM bis-tris buffer and 100 mM NaCl, pH 7.0

tion yields the α-hydroxy amide (R)-**17** with a 56% de. In contrast, the antibody-catalyzed process completely switches the stereospecificity of this reaction by generating the (S)-**17** product with a diastereomeric excess of >99%. This result emphasizes the fact that antibodies, even when elicited to achiral haptens, can provide binding pockets which discriminate between enantiomeric transition states with a high degree of selectivity.

Epoxides are key chiral synthetic intermediates and their enantioselective preparation by oxidation of achiral alkenes is a key reaction in many synthetic strategies. Sharpless' asymmetric epoxidation is suitable for most allylic alcohols [26, 27], but few general procedures exist for unfunctionalized olefins. Jacobsen's manganese salen-mediated epoxidation is suitable for and gives good selectivities with Z-olefins (85 to 90% ee) [28]. The enzyme chloroperoxidase

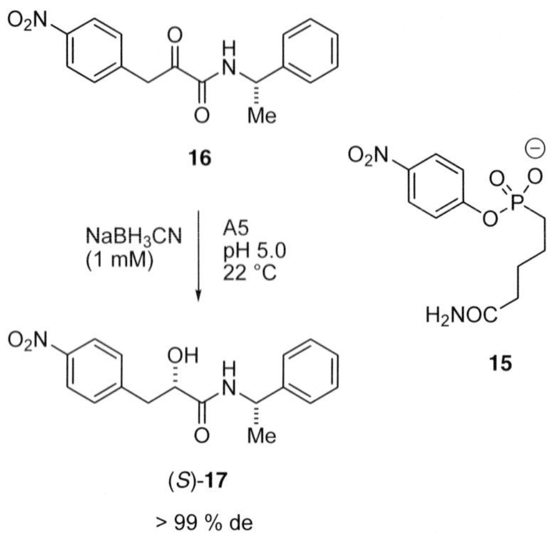

Scheme 4. Antibody A5, elicited to hapten **15**, catalyzes the highly enantiospecific reduction of α-ketoamide **16** (>99% de) with sodium cyanoborohydride (NaCNBH₃) as a cofactor. This is a good example of an achiral hapten generating a catalyst possessing exquisite chiral discrimination

has recently been shown to catalyze the epoxidation of unfunctionalized alkenes with hydrogen peroxide as the oxidant with 66 to 97% ee [29]. Reymond et al. [30] sought to expand the role of antibody-catalysis into the realm of catalytic asymmetric epoxidation. Mechanistic investigations into the origin of the catalytic power of enol ether hydrolysis supplied by antibody 14D9 (vide supra) revealed an equal contribution of general acid catalysis supplied by a carboxyl group and pyramidalization of the enol ether's β-carbon by hydrophobic contacts. It was reasoned that antibodies possessing the latter effect, coupled with a suitable oxidizing agent may be capable of catalyzing an enantioselective epoxidation of alkenes.

Based on this reasoning, the library of monoclonal antibodies elicited to the bait-and-switch haptens **8a** and **8b**, vide supra, were rescreened for their ability to catalyze the asymmetric epoxidation of the alkene substrate **18** (Scheme 5).

After a survey of oxidizing reagents the previously reported combination of hydrogen peroxide and acetonitrile [31,32] in aqueous buffer, under neutral conditions, effected clean epoxidation without damaging the antibodies. Testament to the successful mimicry of the transition-state for epoxidation by **8a** and **8b**, nine anti-**8a** and six anti-**8b** antibodies were found to catalyze the epoxidation reaction. One antibody, 20B11 elicited to **8a**, was studied in detail for its ability to catalyze the epoxidation of a range of alkenes **18** to **24** (Scheme 5). Alkenes **18** to **22** were indeed substrates, but **23** and **24** were not, revealing the importance both of double substitution at the benzylic carbon and proper localization of the double bond in the binding pocket. As shown in Scheme 5, the asymmetric induction supplied by 20B11 ranges from 67 to >98% ee for alkenes **18**, **19**, and **20**.

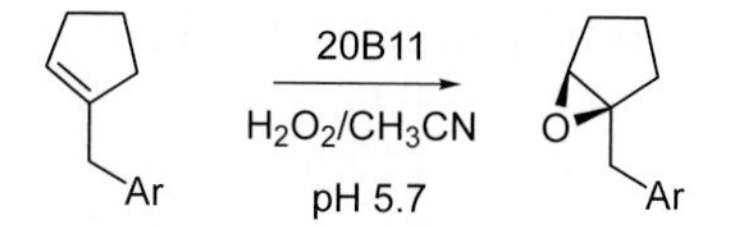

Scheme of alkene structures 18-24 and reaction of 18 with 20B11, H$_2$O$_2$/CH$_3$CN, pH 5.7.

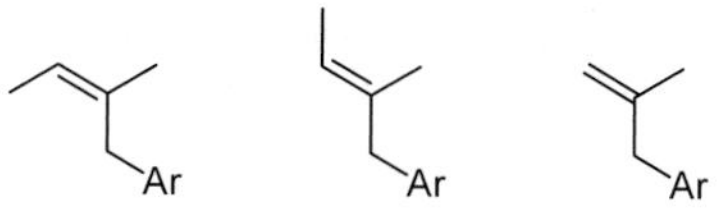

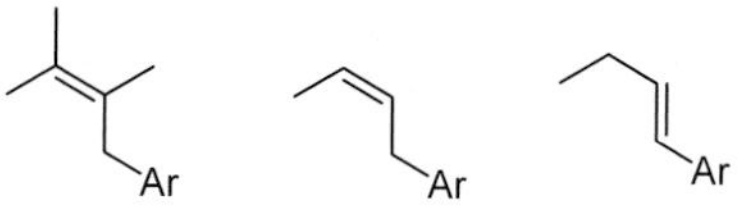

Alkene	K_m	k_{cat}, × 10^5 s^{-1}	ER	ee[a], [%]
18	85	5.0	40	>98
19	120	6.4	125	67
20	140	3.0	50	>98
21	260	1.4	60	nr[b]
22	60	3.6	15	nr

[a]Absolute configuration not determined
[b]Not recorded

Scheme 5. Alkenes **18** to **24** were utilized as substrates for antibody 20B11 mediated chiral epoxidation. An oxidation system of H$_2$O$_2$/CH$_3$CN was found to be sufficiently mild so as not to damage the biocatalyst

It should be noted, however, that the competing background reaction (which yields racemic products) is of a sufficient rate to result in optical purities of only 47, 64 and 71%, respectively.

3
C-C Bond Forming Processes

3.1
Diels-Alder Cycloaddition Reaction

The Diels-Alder reaction is one of the most useful carbon-carbon bond-forming reactions in organic chemistry and can lead to the rapid assimilation of complex molecules containing a high degree of asymmetry. It is a bimolecular process and is a classic example of a reaction that demands control of translational entropy. It is accelerated by both high pressure and ionic solutions (8 M LiCl) and proceeds through an entropically disfavored, highly ordered transition state, showing large activation entropies: -30 to -40 cal·mol^{-1} K^{-1} [33, 34].

While it is one of the most important and versatile transformations available to the organic chemist, the reaction between an unsymmetrical diene and dienophile can generate up to eight stereoisomers [35]. By increasing the electron-withdrawing character of the substituent on the dienophile Danishefsky [36] has shown that the regioselectivity of the Diels-Alder reaction can be controlled such that only the four *ortho*-adducts are produced (Scheme 6).

However, complete stereochemical control of the Diels-Alder reaction to yield only disfavored *exo*-products in enantiomerically pure form has proven to be very difficult by chemical means. Furthermore, only recently has a potentially enzymatic Diels-Alder reaction been reported [37]. Therefore, attempts to generate antibodies which can catalyze stereoselective Diels-Alder reactions is seen as an ongoing major target in the field.

Of particular difficulty when attempting to elicit catalytic antibodies for a bimolecular reaction, where by necessity the transition-state is very similar to products, is choosing a suitable hapten design that does not lead to strong product recognition and hence inhibition. Three strategies have been developed to circumvent this problem. In the first, the reaction is chosen to generate an unstable bicyclic intermediate which can either spontaneously rearrange or eliminate to furnish the product [38]. A more general strategy engages a highly constrained bicyclic hapten, which elicits a binding pocket that juxtaposes the diene and dienophile in an 'entropic trap', but with the additional feature of locking the developing cyclohexene product into a high energy pseudo-boat conformation [39,40,41,42,43,44]. This product destabilization, while perhaps reducing the turnover number of any catalyst generated, serves to aid the release of the product and minimize product inhibition. The final strategy replaces the rigid

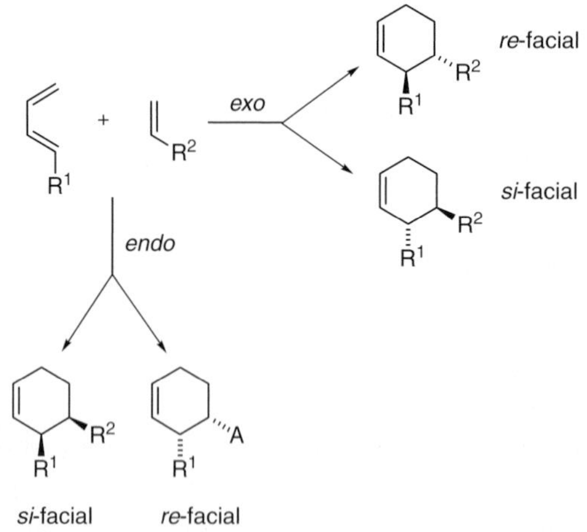

Scheme 6. Diastereo- and enantioselectivity in the Diels-Alder reaction between a substituted diene and dienophile

bicyclic core, common to both of the above strategies, with a freely rotating metallocene. One then relies on the antibodies being able to freeze out a conformer of the hapten which mimics the Diels-Alder transition state during the evolution of the immune response to elicit catalysts [45].

Gouverneur et al. [44] were interested in using catalytic antibodies to control the stereochemical outcome of the reaction between diene 25 and *N,N*-dimethylacrylamide 26 (Scheme 7).

The uncatalyzed reaction leads to the formation of only two diastereomers; the *ortho-endo*-27 and the *ortho-exo* (*trans*)-28, in a ratio of 85:15. Two bicyclic haptens 29 and 30 were designed, one to mimic the *exo*-31 and one a mimic of the *endo*-32 transition-state structures. Kinetic studies showed that two monoclonal antibodies, 13D4 and 7D4, derived from immunization with hapten 29 and four antibodies, 22C8, 27R4, 14F2, and 8B11 derived from immunizations with hapten 30, catalyze exclusively the formation of either the *exo*-28 or *endo*-27 adducts, respectively. Antibodies 7D4 (*exo*) and 22C8 (*endo*) provided the

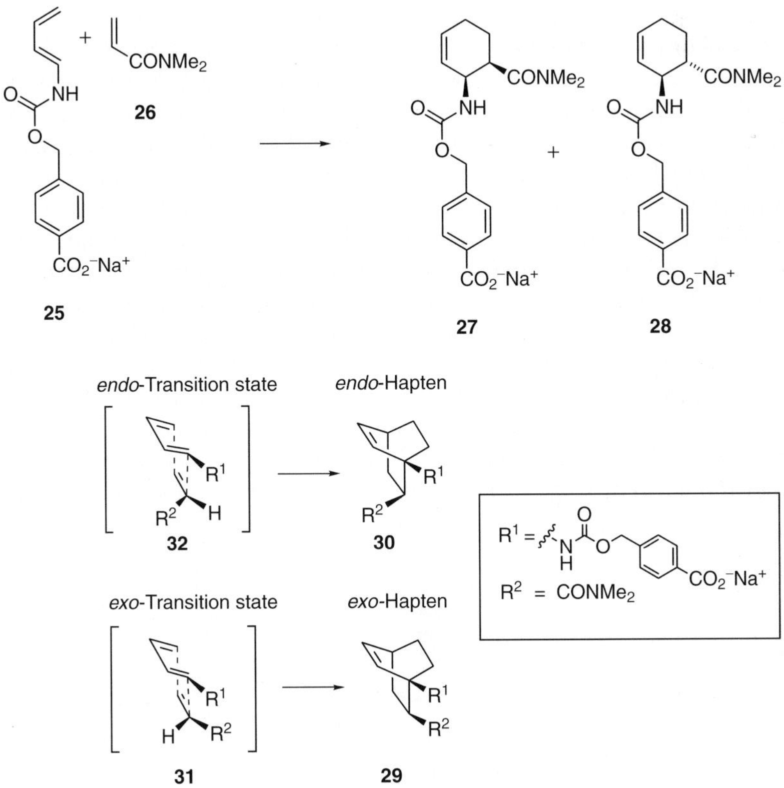

Scheme 7. Highly constrained haptens **30** and **31** were utilized to elicit catalytic antibodies for the diastereo- and enantioselective Diels-Alder reaction between diene **25** and dimethylacrylamide **26**

best rate enhancements and were studied in some detail. The turnover numbers for these antibodies were k_{cat}=3.44×10^{-3} and 3.17×10^{-3} min^{-1}, respectively, equivalent to effective molarities (k_{cat}/k_{uncat}) of 4.8 M (7D4) and 18 M (22C8). These rate enhancements were slightly lower than for previous examples of Diels-Alderase antibody and were rationalized as being a result of a less than ideal transition state representation [38,39]. Both the bicyclic haptens **29** and **30** are mimics of a synchronous cycloaddition transition state, whereas *ab initio* studies had revealed that the reaction between **25** and **26** actually proceeds with considerable asynchronicity [44,46].

Each antibody catalyzes its respective processes not only with high diastereoselectivity (>98% de), but also with excellent enantioselectivity (>98% ee), such that the antibody-catalyzed Diels-Alder reaction gives essentially optically pure *endo*-**27** or *exo*-**28** adducts.

A more recent enterprise has focused on the compounds **33** and **34** perceived as freely rotating haptens for the same Diels-Alder cycloaddition between diene **25** and dienophile **26** (Fig. 1) [45]. Critical to the success of this enterprise is antibody recognition and freezing out of conformers which resemble either the Diels-Alder *exo*-**31** or *endo*-**32** transition states.

From a library of antibodies which recognize hapten **33** seven were found to be catalysts, (1 *endo* and 6 *exo*) and from the antibodies that recognized **34**, eight antibody catalysts (7 *endo* and 1 *exo*) were found. From these sublibraries, the most efficient *endo* (4D5, k_{cat}=3.43×10^{-3} min^{-1}, EM 5 M) and *exo* (13G5, k_{cat}= 3.17×10^{-3} min^{-1}, EM 18 M) catalysts were studied in detail. Both undergo multiple turnovers without evidence of product inhibition and the reaction occurs with complete regio- and high diastereoselectivity (>98% de) and enantioselectivity (>98% ee). X-Ray crystal structure analysis of antibody 13G5, in complex with a ferrocenyl inhibitor containing the essential haptenic core which elicited it, revealed that the antibody does indeed bind the the hapten with the ring substitutents in an eclipsed conformation [47]. In addition, three antibody residues have been implicated as being key, both in terms of the catalytic rate enhancement and the marked stereochemical control. Tyrosine-L36 acts as a Lewis acid activating the dienophile for nucleophilic attack, and asparagine-L91 and aspartic acid-H50 form hydrogen bonds to the carboxylate side chain that substitutes

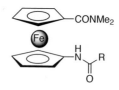

33 R = (CH$_2$)$_3$CO$_2$H
34 R = 4-carboxyphenyl

Fig. 1. The freely rotating metallocenes **33** and **34** were utilized in a new strategy for the elicitation of enantioselective Diels-Alderase antibodies for the carbamate diene substrate **25**. It is this hydrogen bonding network that is directly responsible for the pronounced stereoselectivity imparted by 13G5.

3.2
Cationic Reactions

Carbocations can be difficult to generate and are such highly reactive intermediates that it is not easy to predict or control their reaction pathways [48]. This is of particular relevance in biochemical systems where the nucleophilic nature of the carbenium ion makes it susceptible to attack from peptidic side chain residues and/or backbone peptide bond components, thereby nullifying the required activity and essentially alkylating the protein catalyst. Nonetheless, nature has set us an impressive target in the way it deals with these intermediates [49,50,51]. A number of enzymes utilize cationic processes, one of the most remarkable being 2,3-oxidosqualene cyclase which catalyzes the formation of a highly complex tetracyclic triterpenoid with the concurrent generation of seven asymmetric centers [52,53].

For more than three decades chemists have been attempting to mimic cationic cyclization reactions [54,55]. This work has lead to the realization that cationic cyclization reactions can be divided into three distinct steps: initiation, propagation and termination. Each of these steps must be rigidly controlled if one wishes to precisely organize the reaction outcome.

For antibodies to be successful catalysts for the initiation and control of cationic cyclization, they must be able to simultaneously stabilize point charges, overcome entropic barriers, and provide a chiral environment to elicit asymmetry [56]. In essence the problem reduces to that of generating a carbocation in an environment that stabilizes its formation and controls its subsequent reaction pathways.

The primary approach was a development of the classical system investigated by Johnson [57], which involves initiation of carbocation formation following solvolysis of a sulfonate ester. In this scenario, once cyclization has occurred, the newly formed carbocation can be captured by either elimination or attack of a nucleophile. At an entry level, in an attempt to catalyze the cyclization of the acyclic sulfonate ester **35** two haptens, **36** and **37**, were utilized in a 'bait-and-switch' strategy (Scheme 8) [58]. HPLC assay revealed that four antibodies (4C6, 16B5, 1C9, and 6H5) elicited to the N-oxide hapten **36** and one antibody, 87D7, elicited to the N-methylammonium hapten **37** were 'initiation' catalysts, i.e., they catalyzed the solvolysis of the sulfonate ester bond of **35**. A remarkable feature of antibody catalysis of this reaction is the narrow product distribution observed. Of all the possible products **38** to **42** inferred from the work of Johnson, only the cyclized products **38** and **39** were detected, a testimony to the antibodies exquisite binding of a putative cyclic transition state as programmed by the haptens **36** and **37**.

Antibody 4C6, elicited to the N-oxide hapten **36** yields cyclohexene **38** (2%) and *trans*-2-(dimethylphenylsilyl)cyclohexanol **39** (98%). Whereas 18G7, elicit-

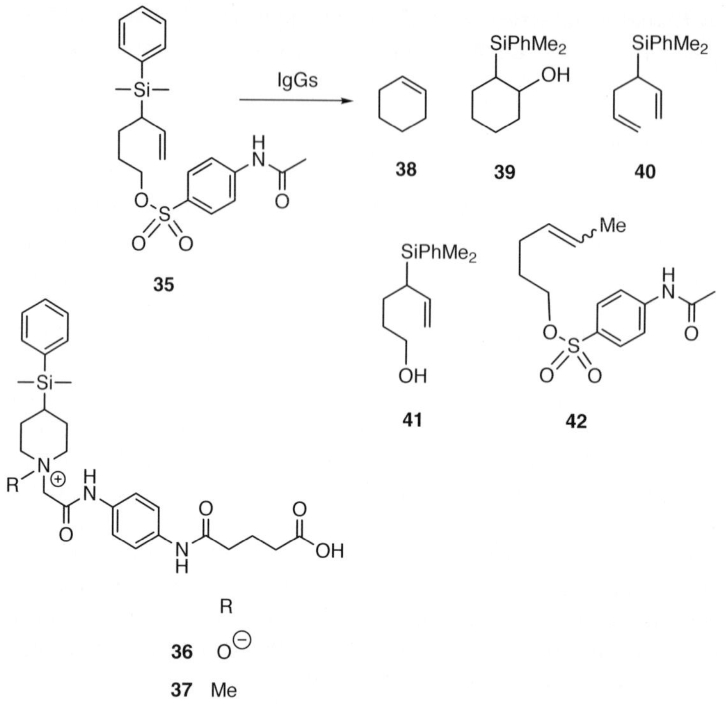

Scheme 8. Antibodies elicited to the *N*-oxide hapten **36** catalyze the cationic cyclization of the arenesulfonate **35** to cyclohexene **38** and the diastereomeric alcohol **39**. The background reaction generates a much broader product spectrum **38** to **42**

ed to the quaternary ammonium hapten **37**, gives a complete reversal of this product distribution, **38** (90%) and **39** (10%). In terms of the absolute rates, both these are efficient catalysts, for 4C6 k_{cat}=0.02 min^{-1} and K_m=230 µM, and for 18G7 k_{cat}=0.02 min^{-1}, K_m=25 µM. The ratio of cyclohexene to cycloxehanol is entirely a reflection of the antibody's ability to exclude water from the binding site and hence prevent it from acting in the termination step. It is still unclear if the chiral induction at C2, leading to formation of exclusively the *anti*-cyclohexanol derivative **39**, is a result of direct antibody control or simply due to the steric constraints imposed by the bulky C1 moiety, leading to antarafacial quenching by a water molecule [59].

Subsequent studies with this antibody catalyst have involved discrete substrate modification experiments in order to probe and modify the reaction course and hence product outcome. Thus, the silicon atom of the phenyldimethylsilyl group was replaced with a carbon atom so that a potential β-effect would no longer bias the reaction course. In addition, a methyl group was added to the terminal olefin appendage for the purpose of both eliminating the chemical advantage of a reaction route involving a secondary rather than a primary carbocation and also to increase the potential asymmetry of the products. This led to substrate modification experiments with the olefins **43** to **45** (Scheme 9) [60,61].

When olefin **43** was incubated with antibody 87D7, clean conversion to a single diastereomeric alcohol **46** was observed in 60% yield (k_{cat}=0.013 min^{-1}, K_m 58 μM), with no observable product inhibition. With olefin **44**, where the silyl moiety had been replaced with carbon, a single diastereomeric exocyclic alcohol **47** was formed in 80% yield. The most interesting result however, was obtained with *trans*-olefin **45** which lead to the chiraly defined cyclopropane **48** in 63% yield (k_{cat}=0.021 min^{-1}, K_m 102 μM). Even under harsh Johnson-like conditions (formic acid/sodium formate, 80 °C, 2 h) this product could not be detected in the uncatalyzed reaction, revealing the exquisite power of antibody-catalysis not only to generate optically pure products from cationic cyclizations, but also to reroute cation reactivity to elicit novel homochiral products. Furthermore, the rate accelerations for this antibody catalyst is within an order of magnitude of those of natural enzymes that catalyze similar processes [62].

A unified reaction pathway invoking a protonated cyclopropane **49** was formulated to rationalize formation of the reaction products **46** to **48**. Thus, for substrate **43**, addition of water to **49** at the α-carbon generates the cyclohexanol **46**. For the substrates **44** and **45**, which both contain an electron-donating methyl group, products **47** and **48** are formed by either addition of water to the β-carbon of intermediate **49**, or loss of a proton from **49**. The observed product distribution and asymmetry can thus be ascribed to the direct control of the central carbocation intermediate **49** by the antibody catalyst.

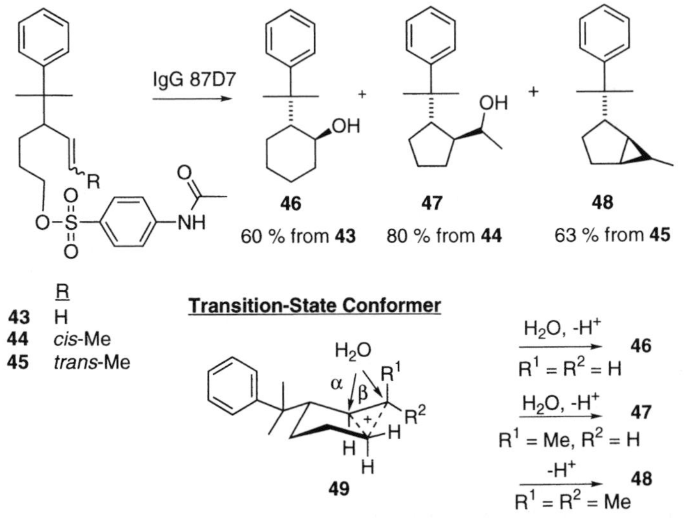

Scheme 9. Substrate modification experiments with alkenes **43, 44, 45** and antibody 87D7 revealed the intermediacy of a protonated cyclopropane **49** in the antibody-catalyzed cationic cyclization process

More recent efforts in this area have moved away from bait-and-switch haptens and focused on cationic transition state mimics, such as the amidinium ion species 50, used to generate antibodies for the cationic cyclization of the arenesulfonate 51 (Scheme 10) [63]. Antibody 17G8, elicited to 50, catalyzes the solvolysis of the terpenoid 51 (K_m 35 μM, k_{cat} 3.6∞10^{-3} min^{-1}) and re-routes the product distribution from the diastereomeric pair of cyclohexanols 52 and 53 and 1,2-cyclohexene 54, into the cyclic products 55 and 56. However, the enantioselectivity of the 17G8-catalyzed process is poor, carbocycles 55 and 56 being formed in a 24 and 37% ee, respectively. This poor stereoselectivity is attributed in part to the planar nature of the transition state analog 50 at the 2-position, a critical locus for asymmetric induction by the antibody during cyclization of the chair-like transition structure 57.

Arguably the most demanding antibody-catalyzed cationic cyclization thus far reported has involved the formation of the decalins 58, 59, and 60 (Scheme 11) [64]. The trans-decalin epoxide 61 hapten, a TS mimic, was immunized as a diastereomeric mixture. Monoclonal antibody HA5-19A4 emerged as the best catalyst for the cyclization of arenesulfonate 62. The olefinic fraction (70%) was predominantly a mixture of 58, 59, and 60 with an enantiomeric excess of 53, 53, and 80%, respectively, with a significant proportion (30%) of cyclohexanols.

Kinetic investigations revealed that the antibody first catalyzes the ionization of the arenesulfonate 62 to generate the first carbocation, this process has an ER of 3.2∞10^3 and a K_m=320 μM. The resulting cation can then either cyclize to decalins 58, 59, 60 in a concerted process (as via the transition structure 63) or in a stepwise fashion. The formation of significant amounts of cyclohexanols

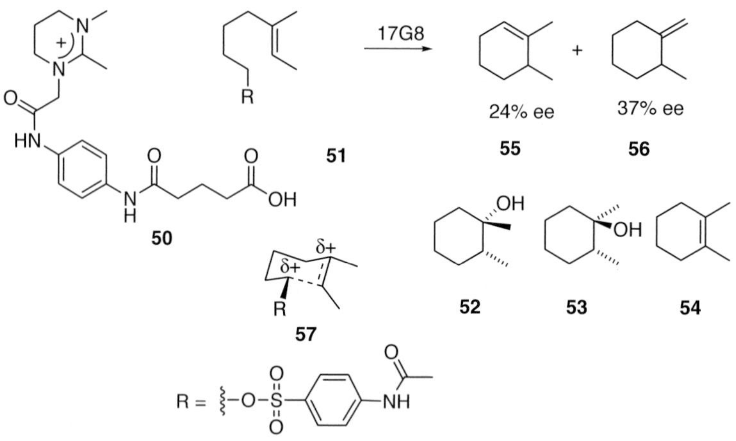

Scheme 10. The amidinium 50 was engaged as a transition state analog hapten to generate antibodies for the cationic cyclization of the arenesulfonate 51. Antibody 17G8 catalyzes this cyclization process to yield two cyclohexene derivatives 55 and 56, albeit with low enantioselectivity 24 and 37% ee, respectively. The remarkable feature the 17G7-catalyzed process is a complete change in the product outcome from non-catalyzed process which yields 52, 53, 54

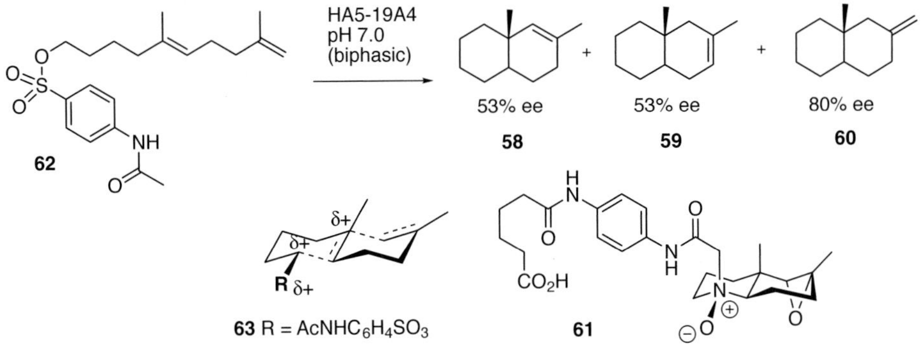

Scheme 11. Antibody HA5–19A4, elicited to that transition state analog hapten, diastereomeric epoxide 61, catalyzed the formation of decalins 58, 59, 60 in a tandem cyclization reaction with moderate to good ees from the arenesulfonate 62

seems to indicate that the latter may be the case. Most interestingly, inhibition studies with 61 strongly suggest that the isomer of the haptenic mixture of that elicited this antibody has an axial representation of the leaving sulfonate group, which would indicate a formal reversal of the Stork-Eschenmoser [65, 66] concept of equatorial leaving groups and presents an interesting challenge for future study.

3.3
Aldol Condensation

The aldol condensation is one of the most utilized C-C bond forming reactions in both organic chemistry and nature. A variety of efficient reagents have been developed to control the stereochemical outcome of this reaction, but they are required in stoichiometric amounts and, in general, require preformed enolates and involve extensive protecting group strategies [67,68,69]. More recently, catalytic aldol reactions have been explored [70,71]. In addition a number of enzymes is known to catalyze the aldol condensation and much mechanistic information has been gleaned about their modes of action [72,73].

Class I aldolases utilize the ε-amino group of a lysine (Lys) residue in the active site to form a Schiff base with one of the substrates, which thus activates the substrate to an aldol donor. In an attempt to mimic this mechanism, the β-diketone 64 was used as a hapten in the hope of elicitating of a Lys residue in an antibody binding site (Scheme 12) [74].

The hapten was designed to trap the requisite Lys residue in the active site and then form the essential enamine intermediate 65 by dehydration of the tetrahedral carbinolamine intermediate 66. The trapping of a nucleophile in antibody-binding sites for enhanced efficiency of antibody catalysis had previously been reported by Wirsching and co-workers [75] and has been dubbed 'reactive immunization'. By utilizing this reactive immunization strategy, two antibodies,

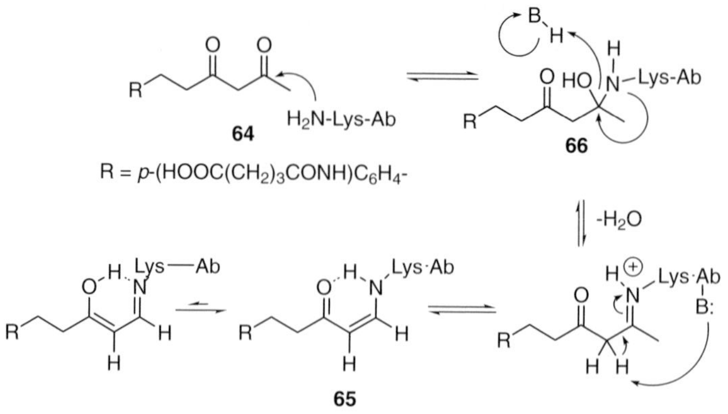

Scheme 12. Diketone hapten **64** was utilized as a reactive immunization hapten and trapped a lysine residue in an antibody binding site by the mechanism outlined

38C2 and 33F12, were found that possessed a Lys residue in their binding sites and that catalyze the aldol reaction between a variety of aliphatic ketones and aldehydes.

Reaction of the branched 3-phenylpropionaldehyde **67** acceptor with acetone is the most efficient process yet observed and shows no product inhibition (Scheme 13). In addition, only 1 mol % of catalyst is required to achieve high conversion of substrate.

The background reaction at pH 7.5 under identical conditions gives a $k_{uncat}=$ $2.28\times10^{-7}\,M^{-1}\,min^{-1}$ [76]. For antibody 38C2 and 33F12, this gives a $(k_{cat}/K_m)/$ k_{uncat} of ca. 10^9. The proficiency [77] of this process was attributed in a large part to the entropic advantage gained from the juxtaposition of the bi-reactant system in the antibody binding site reflected also in the high effective molarity (EM) of this process ($>10^5$ M). In fact, the catalytic efficiency of these antibodies is only 3 orders of magnitude lower than that of the most studied type-I aldolase enzyme, 2-deoxyribose 5-phosphate aldolase [78].

The stereoselectivity of the antibody-catalyzed addition of acetone to aldehyde **67** revealed that the ketone was added to the *re*-face of **67** regardless of the stereochemistry at C2 of this substrate. The aldol process follows a classical Cram-Felkin mode of attack on (*S*)-**67** to generate the (4*S*,5*S*)-**68** diastereomer and the anti-Cram-Felkin mode of attack on the (*R*)-**67** to yield the (4*S*,5*R*)-**69** diastereomer. The products are formed at a similar rate and yield, therefore there is no concomitant kinetic resolution of the racemic aldehyde. The two antibodies differ in their diastereofacial selectivity, reflecting the ability of the antibodies to orient the **67** on opposite sides of the prochiral faces of the nucleophilic antibody-enamine complex of acetone. Heathcock and Flippin [79] have shown that the chemical reaction of the lithium enolate of acetone with (*S*)-**67** yields the (4*S*,5*S*)-**68** diastereomer a 5% de for this Cram-Felkin product. The generation of the (4*S*,5*R*)-**69** and (4*R*,5*R*)-**70** products in a ratio of 11:1 by the

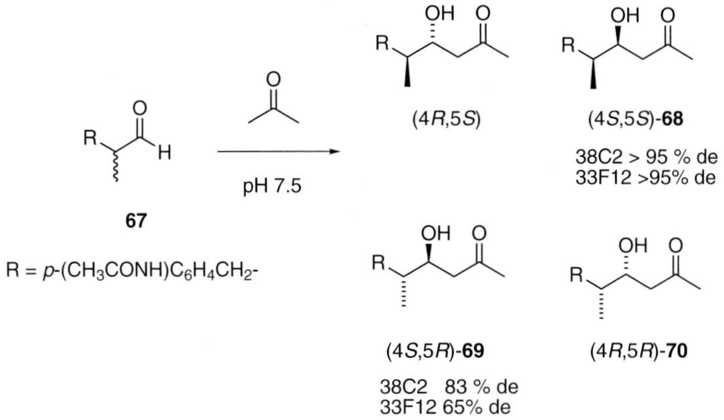

Scheme 13. Diastereoselectivity of the aldol reaction between racemic aldehyde **67** and acetone catalyzed by antibodies 38C2 and 33F12

38C2-catalyzed process is a remarkable reversal of this typical Cram-Felkin stereoselectivity of the aldol, to a disfavored and energetically more demanding anti-Cram-Felkin model.

The crystal structure of unliganded 33F12, shows a lysine residue, Lys-H93, at the bottom of a hydrophobic well, but linked *via* a hydrogen-bonded water to a tyrosine moiety, TyrL41 [80]. The high catalytic rates imparted by this catalyst are now rationalized as being a composite of both imine formation with LysH93 and general base catalysis supplied by the TyrL41 residue.

One normally expects antibodies to have a low tolerance to substrate modifications, however an ongoing feature of these aldolase antibodies is their wide scope. They accept a remarkable range of aldol donors and acceptors and perform crossed-, intramolecular- and retro-variants of this reaction, with high yields, rates, and stereospecificities [81,82,83]. Substrate modification experiments have revealed that when acetone is the aldol donor in a ketone-aldehyde crossed aldol reaction, stereoinduction is linked to attack of the *si*-face of a prochiral aldehyde with typically >95% ee and when hydroxyacetone is the donor substrate, attack occurs preferentially at the *re*-face of the aldehyde leading to a diastereomeric α,β-dihydroxy ketones with the two stereogenic centers having an α-*syn* configuration. This reaction leads to stereospecificities of typically 70 to >99% ee.

4
Disfavored Cyclization

For reactions were there are several possible outcomes, the final product distribution reflects the relative free energies of each transition state when the reaction is under kinetic control, this is in fact the empirical basis behind enantio-

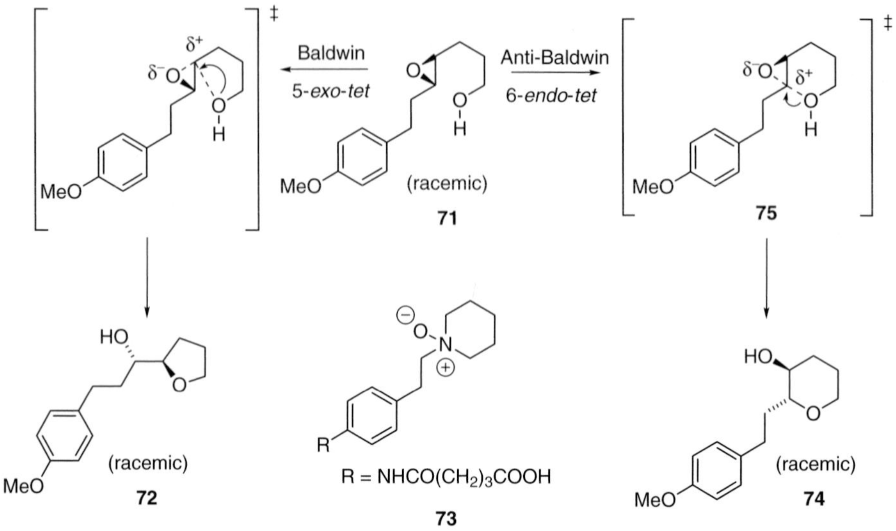

Scheme 14. *N*-Oxide hapten **73** elicited an antibody, 26D9, which re-routes the cyclization of the hydroepoxide **71** to the disfavored product **74**

and diastereospecific reactions [84]. Baldwin's rules predict that for the acid-catalyzed ring closure of the hydroepoxide **71** the tetrahydrofuran **72** arising from 5-*exo-tet* attack will be preferred (Scheme 14) [85,86]. Janda et al. [87] raised antibodies to the cyclic hapten **73** and generated a catalytic antibody, 26D9, which reverses the kinetic outcome of the reaction and produces exclusively the tetrahydropyran product **74** in optically pure form.

The hapten was designed to elicit antibodies which would strategically place a negatively charged amino acid adjacent to the epoxide moiety in a position to selectively stabilize the disfavored the 6-*endo-tet* TS structure **75**. A remarkable feature of 26D9 catalysis of this reaction is its stereoselectivity, accepting only the (S,S)-enantiomer of **71** as a substrate and leading to a kinetic resolution of racemic **71**. The turnover number for this catalyst is $k_{cat}=4.6\times10^{-6}$ min^{-1}. The comparison of k_{cat}/k_{uncat} was not possible for this process because no 6-*endo-tet* product **74** could be detected. This process is an unprecedented achievement in *de novo* catalyst generation, as catalytic antibodies were elicited to a reaction for which there is no enzymatic or synthetic equivalent.

5
Conclusions

Since the first reports of antibody catalysis appeared just over a decade ago [88, 89], >50 different chemical reactions have been catalyzed by these remarkable proteins and their scope and application is expanding at a rapid pace [7]. Improved hapten design strategies such as reactive immunization and refinements of the more classical transition state analog and bait-and-switch hapten approaches are resulting in improvements in both the catalytic power and applicability of these remarkable enzyme-mimics. From the perspective of asymmetric catalysis (vide supra), initial attempts focusing on relatively trivial problems have been superseded by more complex systems targeting asymmetric C-C bond forming reactions with real synthetic viability. However, despite these notable successes, many problems still remain to be solved. These include limitations in the scope of the reactions that have been achieved and much more work still remains to optimize catalyst stereospecificity and performance under the rigorous conditions of organic synthesis, especially in large-scale reactions. Nonetheless, continued exploratory approaches with catalytic antibodies is undoubtedly going to bear fruit in the field of asymmetric chemistry for many decades to come.

References

1. Morrison JD, Mosher HS (1976) Asymmetric Organic Reactions, 2nd edn. ACS, Washington DC
2. Lo C-HL, Wentworth Jr. P, Jung KW, Yoon J, Ashley JA, Janda KD (1997) J Am Chem Soc 119:10251
3. Janda KD, Benkovic SJ, Lerner RA (1989) Science 244:437
4. Tanaka F, Kinoshita K, Tanimura R, Fujii I (1996) J Am Chem Soc 118:2332
5. Kitazume T, Lin JT, Takeda M, Yamazaki T (1991) J Am Chem Soc 113:2123
6. Kitazume T, Lin JT, Yamamoto T, Yamazaki T (1991) J Am Chem Soc 113:8573
7. Wentworth Jr. P, Janda KD (1998) Curr Opin Chem Biol 2:138
8. Blackburn GM, Datta A, Denham H, Wentworth Jr. P (1998) Adv Phys Org Chem in press:
9. Janda KD, Shevlin CG, Lo C-HL (1996) Catalytic antibodies: Chemical and biological approaches In: Comprehensive Supramolecular Chemistry. (Yakito M, ed), Pergamon, London, p 43
10. Schultz PG, Lerner RA (1995) Science 269:1835
11. Burton DR (1990) TIBS 15:64
12. Kabat EA, Wu TT, Perry HM, Gottesman KS, Foeller C (1991) Sequences of Proteins of Immunological Interest, 5th edn. US Department of Health and Human Services, Public Health Service, NIH,
13. Stryer L (1988) Molecular Immunology. In: Biochemistry. Freeman, New York, p 889
14. Ikeda S, Weinhouse MI, Janda KD, Lerner RA, Danishefsky SJ (1991) J Am Chem Soc 113:7763
15. Fujii I, Lerner RA, Janda KD (1991) J Am Chem Soc 113:8528
16. Matsumoto K, Tsutsumi S, Ihori T, Ohta H (1990) J Am Chem Soc 112:9614
17. Matsumoto K, Ohta H (1991) Tetrahedron 32:4729
18. Potin D, Williams K, Rebek Jr. J (1990) Angew Chem Int Ed Engl 29:1420
19. Kresge AJ, Segatys DS, Chen HL (1977) J Am Chem Soc 99:7228
20. Kresge AJ, Chiang Y (1991) Science 253:395
21. Reymond J-L, Janda KD, Lerner RA (1992) J Am Chem Soc 114:2257
22. Reymond J-L, Jahangiri GK, Stoudt C, Lerner RA (1993) J Am Chem Soc 115:3909
23. Jahangiri GK, Reymond J-L (1994) J Am Chem Soc 116:11264
24. Reymond J-L, Reber J-L, Lerner RA (1994) Angew Chem Int Ed Engl 33:475
25. Nakayama GR, Schultz PG (1992) J Am Chem Soc 114:780
26. Sharpless KB, Verhoeven TR (1979) Aldrichimica Acta 12:63
27. Hoyeda AH, Evans DA, Fu GC (1993) Chem Rev 93:1307
28. Jacobsen EN, Zhang W, Muci AR, Ecker JR, Deng L (1991) J Am Chem Soc 113:7063
29. Allain EJ, Hager LP, Deng L, Jacobsen EN (1993) J Am Chem Soc 115:3909

30. Koch A, Reymond J-L, Lerner RA (1994) J Am Chem Soc 116:803
31. Payne GB, Deming PH, Williams PH (1961) J Org Chem 26:659
32. Bach RD, Knight JW (1991) Org Synth 60:80
33. Ciobanu M, Matsumoto K (1997) Liebigs Ann Chem 623
34. Sauer J (1966) Angew Chem Int Ed Engl 5:211
35. March J (1992) Addition to Carbon-Carbon Multiple Bonds. In: Advanced Organic Chemistry. John Wiley & Sons, New York, p 839
36. Danishefsky S, Hershenson FM (1979) J Org Chem 44:1180
37. Oikawa H, Katayama K, Suzuki Y, Ichihara A (1995) J Chem Soc Chem Commun 1321
38. Hilvert D, Hill KW, Nared KD, Auditor M-TM (1989) J Am Chem Soc 111:9261
39. Braisted AC, Schultz PG (1990) J Am Chem Soc 112:7430
40. Suckling CJ, Tedford MC, Bence LM, Irvine JI, Stimson WH (1992) Bioorg & Med Chem Lett 2:49
41. Suckling CJ, Tedford C, Bence LM, Irvine JI, Stimson WH (1993) J Chem Soc Perkin Trans 1 1925
42. Meekel AAP, Resmini M, Pandit UK (1995) J Chem Soc Chem Commun 571
43. Resmini M, Meekel AAP, Pandit UK (1996) Pure & Appl Chem 68:2025
44. Gouverneur VE, Houk KN, De Pascual-Teresa B, Beno B, Janda KD, Lerner RA (1993) Science 262:204
45. Yli-Kauhaluoma JT, Ashley JA, Lo C-H, Tucker L, Wolfe MM, Janda KD (1995) J Am Chem Soc 117:7041
46. Houk KN, Gonzalez J, Li Y (1995) Acc Chem Res 28:81
47. Heine A et al. (1998) Science 279:1934
48. Olah GA, Schleyer R (1992) Carbonium ions, Wiley Interscience, New York
49. Abe I, Rohmer M, Prestwich GD (1993) Chem Rev 93:2189
50. Cane DE (1990) Chem Rev 90:1089
51. Woodward RB, Bloch K (1953) J Am Chem Soc 75:2023
52. Schroefer Jr. GJ (1982) Annu Rev Biochem 51:555
53. Griffin JH, Buntel CJ (1992) J Am Chem Soc 114:9711
54. Johnson WS, Chenera B, Tham FS, Kullnig RK (1993) J Am Chem Soc 115:493
55. Johnson WS (1968) Acc Chem Res 1:1
56. Li T, Lerner RA, Janda KD (1997) Acc Chem Res 30:115
57. Johnson WS, Bailey DM, Owyang R, Bell RA, Jacques B, Crandall JK (1964) J Am Chem Soc 86:1959
58. Li T, Janda KD, Ashley JA, Lerner RA (1994) Science 264:1289
59. Li T, Hilton S, Janda KD (1995) J Am Chem Soc 117:3308
60. Li T, Janda KD, Lerner RA (1996) Nature 379:326
61. Lee JK, Houk KN (1997) Angew Chem Int Ed Engl 36:1003
62. Lewinsohn E, Gijzen M, Croteau R (1992) Arch Biochem Biophys 293:167
63. Hasserodt J, Janda KD, Lerner RA (1996) J Am Chem Soc 118:11654
64. Hasserodt J, Janda KD, Lerner RA (1997) J Am Chem Soc 119:5993
65. Eschenmoser A, Ruzicka L, Jeger O, Arigoni D (1955) Helv Chim Act 38:1890
66. Stork G, Burgstahler AW (1955) J Am Chem Soc 77:5068
67. Heathcock CH (1990) Aldrichimica Acta 23:99
68. Heathcock CH (1981) Science 214:395
69. Evans DA (1988) Science 240:420
70. Bach T (1994) Angew Chem Int Ed Engl 33:417
71. Carreira EM, Lee W, Singer RA (1995) J Am Chem Soc 117:3649
72. Gijsen HJM, Qiao L, Fitz W, Wong C-H (1996) Chem Rev 96:443
73. Wong C-H, Whitesides GM (1994) Enzymes in Synthetic Organic Chemistry, Pergammon, Oxford
74. Wagner J, Lerner RA, Barbas III CF (1995) Science 270:1797
75. Wirsching P, Ashley JA, Lo C-HL, Janda KD, Lerner RA (1995) Science 270:1775
76. Reymond J-L, Chen Y (1995) Tetrahedron Lett 36:2575
77. Radzicka A, Wolfenden R (1995) Science 267:90

78. Lai CY, Nakai N, Chang D (1974) Science 183:1204
79. Heathcock CH, Flippin LA (1983) J Am Chem Soc 105:1667
80. Barbas CF III et al. (1997) Science 278:5346
81. Lerner RA, Barbas III CF (1996) Acta Chem Scand 50:672
82. Hoffmann T et al. (1998) J Am Chem Soc in press
83. Zhong G, Hoffmann T, Lerner RA, Danishefsky S, Barbas III CF (1997) J Am Chem Soc 119:8131
84. Schultz PG, Lerner RA (1993) Acc Chem Res 26:391
85. Baldwin JE (1982) J Chem Soc Chem Commun 734
86. Baldwin JE, Wsik MS (1982) Tetrahedron 38:2939
87. Janda KD, Shevlin CG, Lerner RA (1993) Janda KD, Shevlin CG, Lerner RA (1993) Science 259:490
88. Tramontano A, Janda KD, Lerner RA (1986) Science 234:1566
89. Pollack SJ, Shultz PG (1986) Science 234:1570

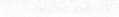

Chapter 41   Industrial Applications

Chapter 41.1
The Chiral Switch of Metolachlor

Hans-Ulrich Blaser · Felix Spindler

Novartis Services AG, Catalysis & Synthesis Services, R 1055.6.28, CH-4002 Basel, Switzerland
e-mail: hans-ulrich.blaser@sn.novartis.com; felix.spindler@sn.novartis.com

Keywords: Imine hydrogenation, Ir diphosphine complexes, Ir ferrocenyldiphosphine complexes, Rh diphosphine complexes, Technical process, Industrial application, MEA imine, Ir ferrocenyldiphosphine complexes, (*S*)-Metolachlor, Chiral switch

1
Introduction

Up to now, relatively few enantioselective catalysts are used on an industrial scale. One reason for this is the fact that enantioselective catalysis is a relatively young discipline: Up to 1985, only few catalysts affording enantioselectivities up to 95% were known [1]. This has changed dramatically in recent years and now many chiral catalysts are known that catalyze a variety of transformation with ee's >98% [2]. Another reason is that the application of enantioselective catalysts on a technical scale presents some very special challenges and problems [3]. Some of these problems are due to the special situation for manufacturing chiral products, others are due to the nature of the enantioselective catalytic process.

Enantiomerically pure compounds will be used above all as pharmaceuticals and vitamins [4], as agrochemicals [5], and as flavors and fragrances [6]. Pharmaceuticals and agrochemicals usually are multifunctional molecules that are produced via multistep syntheses. Compared to basic chemicals, they are relatively small scale products with short product lives, produced in multipurpose batch equipment. The time for development of the production process is often very short since "time to market" affects the profitability of the product.

1.1
Critical Factors for the Application of Enantioselective Catalysts

Whether a synthetic route containing an enantioselective catalytic step can be considered for a particular product is usually determined by the answer to two questions:
- Can the costs for the over-all manufacturing process compete with alternative routes?
- Can the catalytic step be developed in the given time frame?

The following critical factors determine the technical feasibility of an enantioselective process:
- The *enantioselectivity* of a catalyst should be >99% for pharmaceuticals unless further enrichment is easy (via recrystallization or at a later stage via diastereomer separation). Ee's >80% are often acceptable for agrochemicals.
- The *catalyst productivity*, given as turnover number (ton) or as substrate/catalyst ratio (s/c), determines catalyst costs. Ton's should be >1,000 for high value products and >50,000 for large scale or less expensive products (catalyst re-use increases the productivity).
- The *catalyst activity* (turnover frequency, tof, h^{-1}), affects the production capacity. Tof's should be >500 h^{-1} for small and >10,000 h^{-1} for large scale products.
- *Availability and cost of the catalyst*: chiral ligands and many metal precursors are expensive and/or not easily available. Typical costs for chiral diphosphines are 100 to 500 $/g for laboratory quantities and 5,000 to >20,000 $/kg on a larger scale (only few ligands are available commercially). Chiral ligands used for early transition metals are usually cheaper.
- The *development time* can be a hurdle, especially if an optimal catalyst has to be developed for a particular substrate (substrate specificity) and/or when not much is known on the catalytic process (technological maturity).

For most other aspects such as catalyst stability and sensitivity, handling problems, catalyst separation, space time yield, poisoning, chemoselectivity, process sensitivity, toxicity, safety, special equipment, etc., enantioselective catalysts have similar problems and requirements as nonchiral catalysts.

Which of these criteria will be critical for the development of a specific process will depend on the particular catalyst and transformation. In the following section we describe the various steps that were necessary to develop a technical process for the production of (*S*)-metolachlor.

2
(*S*)-Metolachlor: The Problem

Metolachlor was first described in 1972 [7]; it is an *N*-chloroacetylated, *N*-alkoxy-alkylated *ortho* disubstituted aniline (Fig. 1). The unusual functionalization pattern renders the amino function extremely sterically hindered. As a consequence, *metolachlor* has two chiral elements: a chiral axis (atropisomerism, due to hindered rotation around the C_{Ar}-N axis) and a stereogenic center, leading to four stereoisomers. In 1982 it was found that the two 1'*S*-stereoisomers provide most of the biological activity [8]. This was the driving force for finding a production process for the biologically active stereoisomers – a formidable task due to the very special structure and properties of this molecule, the large production volume, and also because of the extremely efficient production process for the racemic product [9]. During the course of the development efforts, the following minimal requirements evolved for a technically viable catalytic system: ee ≥80%, substrate to catalyst ratio >50,000 and tof >10,000 h^{-1}.

2.1
Outline of Synthetic Routes

Of the many possible approaches to synthesize enantiomerically pure compounds, enantioselective catalysis is arguably the most elegant method. In such a case, the overall synthesis is usually designed around the enantioselective catalytic transformations. The reason for this is that only a limited number of effective catalytic enantioselective transformations are available. In addition, it is often quite difficult to transfer the results obtained for a particular substrate to even a close analogue due to the high substrate specificity of many catalysts (low tolerance for structure variation even within a class of substrates). Moreover, the assessment of the a technical feasibility of enantioselective catalysts is also hampered because there is little information on catalyst activity or other aspects

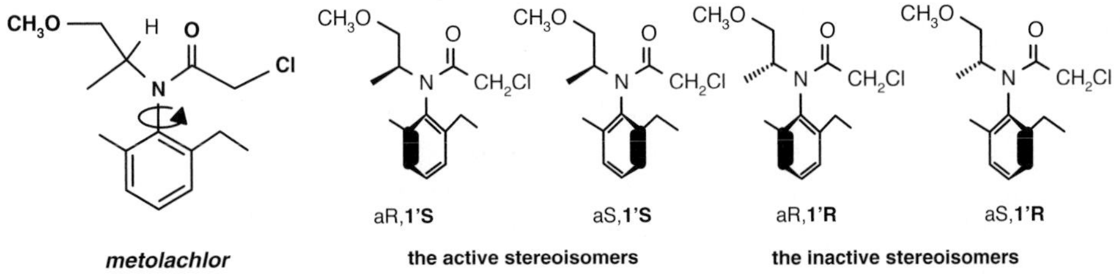

Fig. 1. Structure of metolachlor and its individual stereoisomers

available (in the literature enantioselectivity is the dominant criterion) and because few applications with "real" substrates exist (usually simple model reactions are studied).

Many possibilities exist for the enantioselective preparation of (S)-metol-
achlor, four synthetic routes were considered in some detail:

Enamide hydrogenation (Fig. 2). This idea clearly was inspired by the successful
L-dopa process of Monsanto [10]. At that time, little was known on the effects of
the substituents at the C=C bond and the amide nitrogen. A selective synthesis
of one of the enamides looked difficult.

Nucleophilic substitution of a (R)-methoxyisopropanol derivative (Fig. 3). Here,
the proposed key step was the enantioselective hydrogenation of methoxyacetone in analogy to the Pt-*Cinchona* catalyzed hydrogenation of α-ketoesters [11]
(the Ru-binap system was not yet known at that time). The nucleophilic substitution with clean inversion was expected to be difficult.

Hydrogenation of MEA imine (Fig. 4). Because the racemic metolachlor is produced via a reductive alkylation, it was obvious to try to hydrogenate the imine

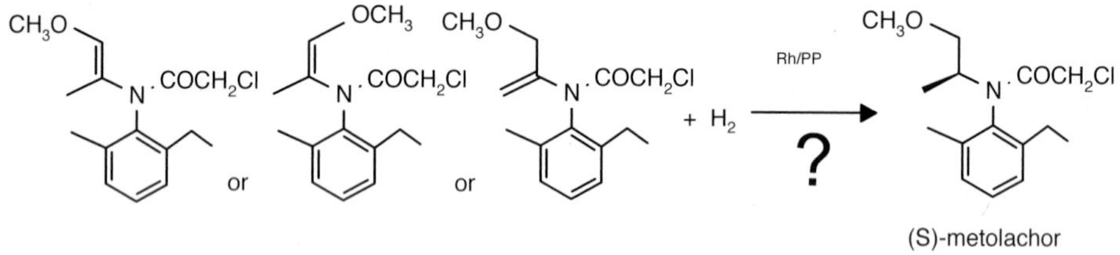

Fig. 2. Enamide hydrogenation: Structures of tested enamides

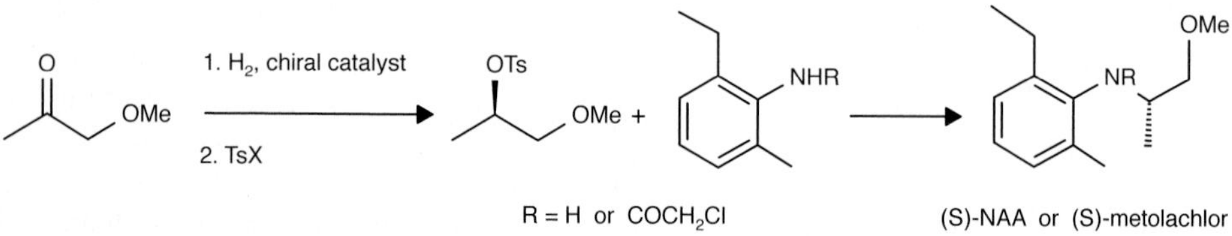

Fig. 3. Enantioselective hydrogenation of methoxyacetone and nucleophilic substitution with
a MEA derivative

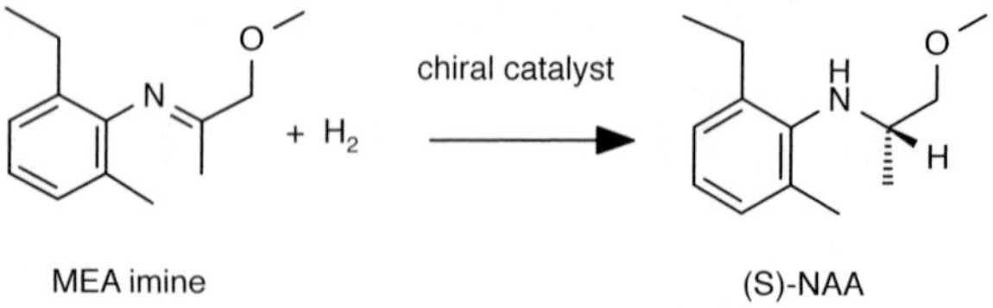

Fig. 4. Imine hydrogenation: Structures of MEA imine and (S)-N-alkylated aniline

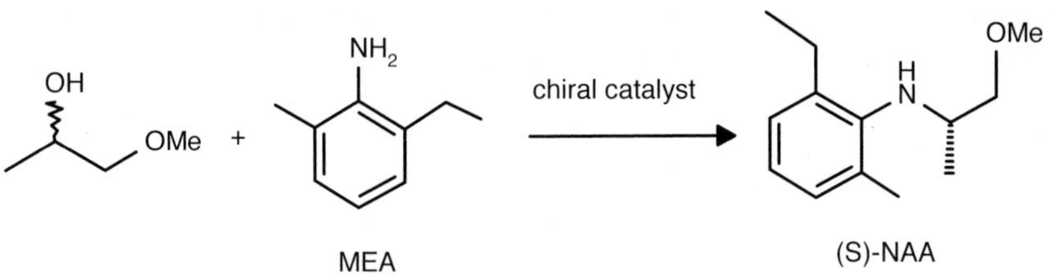

Fig. 5. Alkylation of MEA with methoxyisopropanol

intermediate, either isolated or formed in situ. Unfortunately, at that time only one single imine hydrogenation was described in the literature with 22% ee [12].

Direct catalytic alkylation with racemic methoxyisopropanol (Fig. 5). This idea was based on an alternative process developed for the racemic product with heterogeneous catalysts in the gas phase [13] and some results of the N-alkylation of aliphatic amines with primary alcohols using homogeneous Ru phosphine catalysts [14].

2.2
Assessment and Screening of Proposed Routes

2.2.1
Assessment Criteria

When assessing proposed routes the following criteria are important:
– chances of success for the catalytic step according to precedents, i.e., is there a closely related, efficient catalytic transformation,
– number and perceived difficulty of the non-catalytic steps, and
– first approximations for costs and ecology of the over-all synthesis

In Table 1 the four proposed routes are classified according to these criteria. The overall ranking was used for setting priorities to carry out practical work. Because the enantioselective catalysis is usually considered to be the most difficult

Table 1. Comparison of possible routes for the synthesis of (S)-metolachlor

Route	Catalytic step	Other steps	Cost (ecology)	Priority
Enamide	close analogy ee >90%	enamide synthesis difficult	high (medium)	1
Substitution	weak analogy ee >80%	substitution difficult	high (bad)	2
Imine	weak analogy ee <30%	as in current process	medium (good)	3
Direct alkylation	no precedent	as in current process	low (very good)	4

step, its chances of success very often dominate the decision and accordingly, the enamide and the substitution route were tested first.

2.2.2
Screening Results for Routes 1 and 2

Enamide route. The preparation of the three MEA enamides proved to be rather difficult. Disappointingly, we did not succeed to hydrogenate any of the three isomers using seven different Rh diphosphine complexes at normal pressure and temperatures up to 50 °C.

Substitution route. The hydrogenation of methoxyacetone was somewhat more successful: Using a Pt/C catalyst modified with cinchonidine as described by Orito et al. [14] produced (R)-methoxyisopropanol, but ee's were never higher than 12%. The *direct alkylation* was not tested experimentally, because chances for success were considered to be too low.

3
Imine Hydrogenation

The results of the route screening left the hydrogenation of the MEA imine as the only realistic possibility. As a matter of fact, it took more than 10 years (Table 2) and the collaboration of an untold number of research and development chemists, technicians, engineers, and workmen until the production plant for making (S)-metolachlor was opened on November 16, 1996. The most time-consuming part was of course to find the right catalyst: metal, ligand, and additives. Many of the classical strategies were used: screening of various metal-diphosphine combinations; use of different metal precursors; synthesis of novel ligands; ligand fine tuning; screening and optimization of solvents, additives and reaction

Table 2. Milestones in the history of *S-metolachlor*

1970	Discovery of the biological activity of rac-*metolachlor* (patent for product and synthesis)
1978	Full-scale plant for the production of rac-*metolachlor* in operation (capacity >10,000 t/y)
1982	Synthesis and biological tests of the four stereoisomers of *metolachlor*
1983	First unsuccessful attempts to synthesize S-*metolachlor* via enantioselective catalysis
1985	Rhodium/cycphos catalyst gives 69% *ee* for the imine hydrogenation (UBC Vancouver)
1986	Discovery of new iridium diphosphine catalysts that are more active and selective than Rh catalysts for the hydrogenation of *MEA* imine
1993	Ir/ferrocenyl-diphosphine catalysts and acid effect are discovered. Process development starts
1993/4	Patents for rac-*metolachlor* expire
1995/6	Pilot results for S-*metolachlor*: ee 79%, ton 1,000,000, tof >200,000 h^{-1}, first 300 t produced
1996	Full-scale plant for production of >10,000 t/y S-*metolachlor* starts operation

conditions. The development of the technical process, the design of the high pressure equipment and the construction of the plant were carried out in less than 3 years, a very remarkable achievement in itself.

3.1
Finding the Right Metal-Ligand Combination

The history of the development of a technically feasible catalyst for the enantioselective hydrogenation of MEA imine has been described [15]. Very important were collaborations, initially with a research team of the University of British Columbia at Vancouver [16] and later with the group of J.A. Osborn [17] of the University of Strasbourg.

Screening of Rh diphosphine complexes (Fig. 6). First positive results were obtained by trying to adapt Rh diphosphine catalysts originally developed for the hydrogenation of olefins. An extensive ligand screening led to [Rh(nbd)Cl]$_2$/cycphos as the best catalyst: 69% ee were achieved at –25 °C, the best tof was 15 h^{-1} at 65 bar, r.t., far too low for an industrial application [16]. Nevertheless, these results represented a remarkable progress for the enantioselective hydrogenation of *N*-arylimines.

Screening of Ir diphosphine complexes. The next breakthrough was obtained when iridium was used instead of rhodium. This idea was inspired by results of Crabtree et al. [18] who described an extraordinarily active Ir/tricyclohexylphosphine/pyridine catalyst that was able to hydrogenate even tetra-substituted C=C bonds. The highest ee's were observed with an Ir-bdpp catalyst in the presence of additional iodide ions (ee 84% at 0 °C), but the activity was disappointing; ton's up to 10,000 and tof's of 250 h^{-1} (100 bar and 25 °C) with somewhat lower ee's were obtained for Ir-diop-iodide catalysts [17, 19]. A major problem of these new Ir diphosphine catalysts was an irreversible catalyst deactivation.

These results, especially the good enantioselectivities, were very promising and represented by far the best catalyst performance for the enantioselective hydrogenation of imines at that time. Nevertheless, it was also clear that we could probably not reach our ambitious goals using Ir complexes with "classical" diphosphine ligands. Even though Ir/diop and Ir/bdpp catalysts showed much higher activities than the best Rh complexes for MEA imine, they were still far below the requirements: A new approach was clearly required.

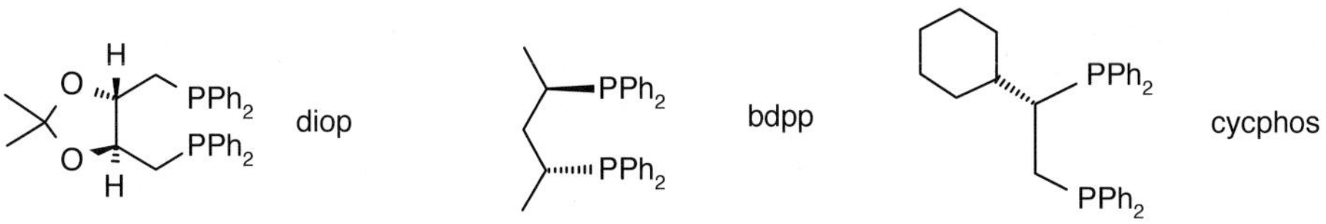

Fig. 6. Imine hydrogenation: Structure of important ligands

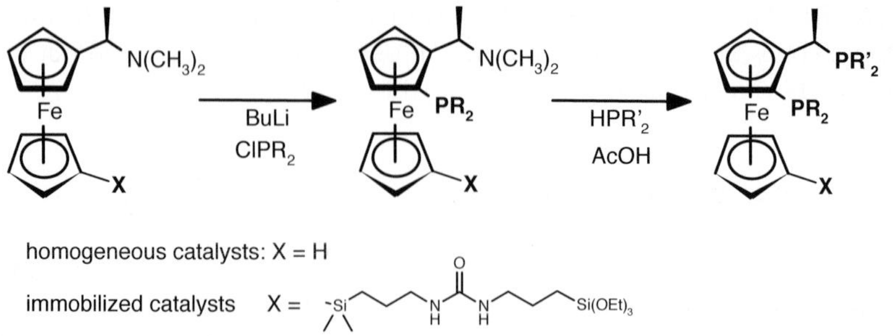

Fig. 7. Preparation and structure of ferrocenyl diphosphine ligands

Synthesis and screening of a new ligand class. Since we could not get stable catalysts with the known diphosphine ligands, new types were tested, among others, novel ferrocenyl-diphosphines (PPF) developed by Togni and Spindler [20]. Their mode of preparation (Fig. 7) allows an efficient fine tuning of the electronic and steric properties of the two phosphino groups, something that is often very difficult with other ligand classes. Indeed, the Ir complexes of such diphosphines proved to be very efficient. Especially PPF-P(3,5-$(CH_3)_2C_6H_3)_2$ (R=Ph, R'=3,5-xylyl), named xyliphos, turned out to give an exceptionally active catalyst and, even more important, it did not deactivate!

3.2
Optimization of Reaction Medium and Conditions

Using xyliphos as ligand, a screening of solvents and additives as well as an optimization of the reaction conditions were carried out. Most remarkable was the effect observed when 30% of acetic acid were added to the reaction mixture of MEA imine and Ir-xyliphos-NBu$_4$I: we observed a rate increase by a factor of 5 while the time for 100% conversion was more than 20 times shorter than without additives. The effect of pressure and temperature was investigated in the presence of acid and iodide. The reaction rate was approximately proportional to the hydrogen pressure, ee's decreased from 81% at −10 °C to 76% at 60 °C. Using optimized conditions, the isolated imine can be hydrogenated at a hydrogen pressure of 80 bar and 50 °C with a substrate to catalyst ratio (s/c) of 1,000,000. Complete conversion is reached within 4 h with an enantioselectivity of 79% with an initial tof exceeding 1,800,000 h^{-1} [21]. These results set a new standard concerning catalyst activity and productivity for a homogeneous enantioselective hydrogenation.

3.3
Ligand Fine Tuning

As described above, the Ir-xyliphos catalysts showed extremely high catalyst activities and productivities. On the other hand, the enantioselectivity to the desired S-enantiomer just barely meets the requirement. Therefore, we tried to improve the ee's by tuning of the electronic and steric properties of the new ferrocenyl ligands. As shown in Table 3, we could indeed improve the selectivity of the catalyst [21]. However, as observed before with other ligands, an improvement in selectivity was always offset by a loss in activity and often productivity. In the end, xyliphos was the best compromise regarding activity and selectivity for a technical process.

3.4
Alternative Catalytic Systems

During the course of the development work two alternative variants based on the very active xyliphos were investigated in some detail.

Immobilized Ir diphosphine catalysts. Immobilized catalysts were investigated for two reasons. First, because many Ir catalysts deactivate via dimerization it was attempted to prevent this by site isolation. Indeed, several of immobilized Ir diphosphine complexes showed improved activities and stabilities [21]. Best results were obtained with ferrocenyl diphosphine ligands immobilized on SiO_2 via the X substituent (see Fig. 7): ee 79%, ton up to 120,000. These are at the moment the most active heterogeneous catalysts known. The second reason for immobilization was their expected better separation properties, allowing different work up strategies. Because the catalytic performance of the homogeneous analogs was so much better (see Table 3), this approach was not investigated further.

Reductive alkylation using Ir ferrocenyl diphosphine catalysts. Because the synthesis, isolation, and purification of the MEA imine are cost factors, the most attractive method would be an enantioselective reductive alkylation, in analogy to the existing process for *rac*-metolachlor. At that time, enantioselective reduc-

Table 3. MEA imine hydrogenation with Ir-ferrocenyldiphoshine complexes: Comparison of catalyst performances of (a) homogeneous catalysts, (b) immobilized on silica gel, and (c) for reductive alkylation in a two-phase system (formulas see Fig. 7)

R	R'	ton	tof (h^{-1})	ee	type	comments
Ph	3,5-xylyl	1,000,000	>200,000	79	(a)	production process
p-CF$_3$C$_6$H$_4$	3,5-xylyl	800	400	82	(a)	ligand screening
Ph	4-t-Bu-C$_6$H$_4$	5000	80	87	(a)	low temperature
Ph	4-(n-Pr)$_2$N-3,5-xyl	100,000	28,000	83	(a)	optimized conditions
Ph	3,5-xylyl	120,000	12,000	79	(b)	optimized conditions
Ph	3,5-xylyl	10,000	700	78	(c)	optimized conditions

tive alkylations were not known. Nevertheless, we tried the direct reductive alkylation of MEA with methoxyacetone using our most active Ir-ferrocenyl diphosphine catalysts. In a two-phase system, we indeed achieved ton's up to 10,000 and ee's up to 78% [21]. Again, due to the superior performance of the MEA imine hydrogenation this approach was abandoned (see Table 3).

3.5
Technical Process

Once a catalyst system with the required performance was found and confirmed, attention was turned to finding a technically feasible overall process. Without going into too much detail, the following aspects had to be dealt with. The technical preparation of methoxyacetone and 2-methyl-6-ethylaniline as well as the chloroacetylation step were already established in the existing process for *rac*-metolachlor. However, the production of the MEA imine in the required quality remained to be worked out. This proved to be not trivial, since most high performance catalysts are quite sensitive to all kinds of impurities. Another problem to be solved was the availability and the technical handling of the ligand and the organometallic catalyst precursor. In the end, a large-scale synthesis for the ligand was developed, while the Ir precursor is supplied by a commercial manufacturer. Scale up from the 50 mL screening autoclave via 6.3 L and 50 L stirred tanks, and a 1,000 L loop reactor to the final 10 m^3 production autoclave was carried out without much problems. The catalyst and the process are very well behaved if all starting materials have the required quality. Indeed, the production plant in Kaisten has been producing since about one year without major problems.

4
Conclusions

The case of (S)-metolachlor allows some generalized conclusions.
- The chiral switch from the racemate to an enriched form is attractive not only for pharmaceuticals but also for agrochemicals. Enantioselective hydrogenation is an especially attractive and technically feasible technology to allow this. The activity of the catalyst and not so much its enantioselectivity is often the major problem to be solved.
- The selection of the catalytic system is especially difficult when the required catalyst performance is very high. In addition, every enantioselective catalytic reaction must be treated individually. Besides establishing the technical feasibility of the catalytic key step, it is important to evaluate the entire reaction sequence to the final product.
- The time for process development obviously depends very much on the state of the art of a given catalytic technology. When one has to start almost at point zero as for the enantioselective imine hydrogenation, it may take many years to reach the goal. In our experience, an empirical approach is the fastest

way to find or develop a catalytic system for a problem that has no close precedent. Mechanistic information is especially helpful in later stages of process development or for trouble shooting.

References

1. Morrison JD (ed) (1985) Asymmetric Synthesis. vol 5, Academic Press, New York
2. Brunner H, Zettlmeier W (1993) Handbook of Enantioselective Catalysis. VCH, Weinheim
3. Sheldon RA (1993) Chirotechnology. Marcel Decker Inc, New York; Collins AN, Sheldrake GN, Crosby J (eds) (1992 and 1997) Chirality in Industry. vol I and vol II, John Wiley, Chichester
4. Stinson SC (1997) Chem Eng News, October 20, p 38
5. Ramos Tombo GM, Bellus D (1991) Angew Chem, 103:1219; Fischer HP, Buser HP, Chemla P, Huxley P, Lutz W, Mirza S, Ramos Tombo GM, Van Lommen G, Sipido V (1994) Bull Soc Chim Belg 103:565
6. Noyori R (1992) Chemtech 22:366
7. Vogel C, Aebi R (1972) DP 23 28 340, assigned to Ciba-Geigy AG
8. Moser H, Ryhs G, Sauter HP (1982) Z Naturforsch 37b:451
9. Bader R, Flatt P, Radimerski P (1992) EP 605363-A1 assigned to Ciba-Geigy AG
10. Vineyard D, Knowles W, Sabacky M, Bachmann G, Weinkauf D (1977) J Am Chem Soc 99:5946
11. Orito Y, Imai S, Niwa S, (1979) J Chem Soc Jpn 1118; Orito Y, Imai S, Niwa S, (1980) J Chem Soc Jpn 670; Orito Y, Imai S, Niwa S, (1982) J Chem Soc Jpn 137
12. Levi A, Modena G, Scorrano G (1975) J Chem Soc, Chem Commun 6
13. Rusek M (1991) Stud Surf Sci Catal 59:359
14. Watanabe Y, Tsuji Y, Ige H, Ohsugi Y, Ohta T (1984) J Org Chem 49:3359 and references cited therein
15. Spindler F, Pugin B, Jalett HP, Buser HP, Pittelkow U, Blaser HU (1996) In: Malz RE (ed) Catalysis of Organic Reactions (Chem Ind). vol 68, Marcel Dekker Inc, New York, p 153
16. Cullen WR, Fryzuk MD, James BR, Kutney JP, Kang G-J, Herb G, Thorburn IS, Spogliarich R (1990) J Mol Catal 62:243; Becalski AG, Cullen WR, Fryzuk MD, James BR, Kang G-J, Rettig SJ (1991) Inorg Chem 30:5002
17. Ng Cheong Chan Y, Osborn JA (1990) J Am Chem Soc 112:9400
18. Crabtree R, Felkin H, Fellebeen-Khan T, Morris G (1979) J Organomet Chem 168:183
19. Spindler F, Pugin B, Blaser HU (1990) Angew Chem Int Ed Engl 29:558
20. Togni A, Breutel C, Schnyder A, Spindler F, Landert H, Tijani A (1994) J Am Chem Soc 116:4062
21. Jalett HP, Pugin B, Spindler F, Novartis Services AG, unpublished results

Chapter 41.2
Process R&D of Pharmaceuticals, Vitamins, and Fine Chemicals

Rudolf Schmid · Michelangelo Scalone

Process Research and Catalysis Department, Pharmaceuticals Division,
F. Hoffmann-La Roche AG, CH-4070 Basel, Switzerland
e-mail: rudolf.schmid@roche.com; michelangelo.scalone@roche.com

Keywords: Pantothenic acid, Pantolactone, Dextromethorphan, Cilazapril, Mibefradil, Vitamin E, (R,R,R)-α-Tocopherol, Orlistat, Tetrahydrolipstatin, β-Ketoester hydrogenation, α-Ketolactone hydrogenation, Enamide hydrogenation, Imine hydrogenation, α,β-Unsaturated acid hydrogenation, Allylic alcohol hydrogenation, γ-Oxo-olefin hydrogenation, Ene reaction, Glyoxylate-ene reaction, Rh(BPPM), Ru(MeOBIPHEP), Raney-Ni/tartaric acid, Ti(BINOL), Enantioselective hydrogenation, Chiral drugs, Ruthenium diphosphine catalysts, Rhodium diphosphine catalysts, Iridium diphosphine catalysts

List of Additional Abbreviations

s/c	molar substrate to catalyst ratio used in the reaction
TON	turnover number (moles of substrate produced per moles of catalyst used)
TOF	turnover frequency (turnover numbers per time unit (h^{-1}))

tfa   trifluoroacetate
ox   oxidation
res   resolution
cy-Hex   cyclohexyl
aqu   aqueous

1
Introduction

Enantioselective catalysis, due to its enormous potential to install the chirality
in enantiomerically pure compounds selectively, cost-effectively, and with minimal environmental impact has justifiably attracted great attention in the life-
science and related fine chemical industries. At Roche, the evaluation and application of enantioselective catalytic methodology was prompted by the desire to
replace labor- and equipment-intensive resolution processes for established commercial products such as, e.g., pantothenic acid and dextromethorphan and to
perform chiral switches for commercial racemic products, namely for α-tocopherol (vitamin E). Subsequently, applications towards pharmaceuticals have
become more and more important as single enantiomer drugs have predominated. Specifically, the development of short routes to chiral drug building blocks is
now the dominant motive. In the following, selected Roche Process R&D examples of enantioselective catalysis applications, most of which have reached pilot
or pre-pilot stage in development, are presented and discussed [1].

2
Pantothenic Acid: an α-Ketolactone Hydrogenation

Pantothenic acid (**4**), a water-soluble vitamin, is currently manufactured from
(R)-pantolactone ((R)-**1**) and β-alanine (Scheme 1) Commercial syntheses for

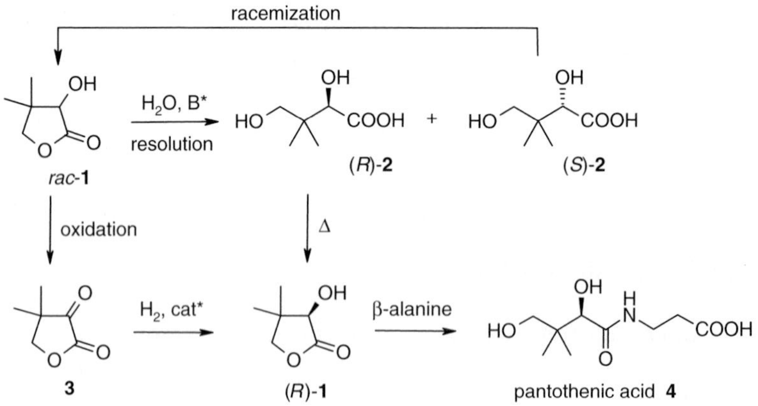

Scheme 1

(*R*)-1 involve resolutions at the stage of pantoic acid (**2**), the undesired enantiomer being recycled back into *rac*-**1**. The enantioselective hydrogenation of ketopantolactone (**3**; easily accessible from *rac*-**1** by oxidation) by means of a Rh(BPPM)Cl catalyst to afford (*R*)-**1** was discovered by Ojima, Kogure, and Achiwa [2] (Table 1, entry 1). The catalyst, however, due to the low turnover number (TON), appeared not to be sufficiently active for technical application [3].

Table 1. Development and state of the art of the pantolactone hydrogenation, **3**→(*R*)-**1**

Entry	Cat* [a]	s/c	Solvent	T [°C]	p [atm]	t[b] [h]	ee [%]	Ref.
1	Rh(BPPM)Cl	1,000	THF	50	50	>48	72	[2b]
2	Rh(BPPM)Cl	50,000	toluene	70	30	6	83	[4]
3	Rh(*m*-Tol-POPPM)Cl	50,000	"	45	30	7	81	[5]
4	Rh(*m*-Tol-POPPM)(tfa)	50,000	"	40	40	1	91	[5]
5	Rh(*m*-Tol-POPPM)(tfa)	200,000	"	40	40	13	90	[5]
6	Rh(BCPM)Cl	10,000	THF	50	50	45	90	[7b]
7	Rh (Cy-oxoProNOP)(tfa)	70,000	toluene	40	40	48	96	[8]

[a] For ligand structures and abbreviations see Fig. 1
[b] Time required to achieve complete conversion

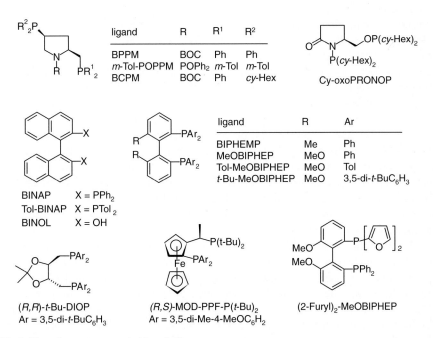

Fig. 1. Ligand structures and abbreviations

Investigations in our laboratories which included a) elaboration of a suitable purification protocol for **3**, b) solvent screening, and c) optimization of reaction conditions and of experimental techniques demonstrated that molar substrate to catalyst ratios (s/c) up to 50,000 and 83% ee could be achieved with the Achiwa catalyst (Table 1, entry 2) [4]. Subsequent ligand fine tuning (cf. Fig. 1) and, even more importantly, variation of the anionic ligand at Rh, led to catalysts with even higher activity and selectivity (Table 1, entries 3–5) [5]. The best catalyst, a Rh(*m*-Tol-POPPM) trifluoroacetato complex, afforded turnover frequencies (TOF) up to 50,000 h^{-1} and 91% ee. In conjunction with this work, a new continuous gas-phase dehydrogenation of *rac*-**1** was developed to prepare the hydrogenation substrate [6]. In the final process, (*R*)-**1** of 91% ee is isolated by distillation and upgraded to 99.9% ee by crystallization. The almost racemic material from the mother liquor is recycled back into the oxidation step. Rhodium is recovered from the distillation residue. The overall process proved suitable for technical implementation, although optimization of the oxidation step is still required. More recently, Achiwa [7] and Mortreux [8] have reported new highly active and enantioselective catalysts for the pantolactone hydrogenation (Table 1, entries 6 and 7). The potential of these catalysts towards technical application is still to be realized.

3
Dextromethorphan: Enamide and Imine Hydrogenations

Dextromethorphan (**10**), a commercial antitussive agent that is widely used in cough-relieving medications, is manufactured by a resolution/recycling route (Scheme 2: **6**→*rac*-**7**→(*S*)-**7**→(*S*)-**9**→**10**). Noyori and the Takasago group demonstrated that *N*-formyl-enamide (*Z*)-**8a**, accessible by formylation of **6**, can be enantioselectively hydrogenated with Ru(BINAP) type catalysts to provide (*S*)-**9a** of 98% ee (Table 2, entry 1) [9]. Investigation at our laboratories proved the *N*-acetyl-enamide (*Z*)-**8b** to be an even better substrate: its hydrogenation with a Ru(MeOBIPHEP)(tfa)$_2$ catalyst at 120 °C and s/c 20,000 afforded (*S*)-**9b** of 98% ee in 97% yield after distillation (entry 2) [10]. Very high space-time yields are possible due to the high substrate concentration of 60%. Both the synthesis of (*Z*)-**8b** and the conversion of (*S*)-**9b** into **10** were developed into efficient processes [11]. Thus, overall an efficient and competitive process is at hand which, however, deviates substantially from the established resolution route by going through new intermediates and therefore would require further process development before full scale-up in the plant.

Recently, the attractive direct enantioselective hydrogenation of **6** to (*S*)-**7** has met with some success. Thus, the Lonza group showed that the dihydrogen phosphate salt of the chemically rather labile imine **6** can be hydrogenated with Ir-ferrocenyldiphosphine catalysts in a two-phase solvent system with up to 89% ee (Table 2, entry 3) [12]. Similar results were achieved in our laboratories with the hydrogen sulfate salt of **6** and an Ir catalyst derived from a sterically bulky *t*-Bu-DIOP ligand in a monophasic solvent system (Table 1, entry 4) [13]. Chemose-

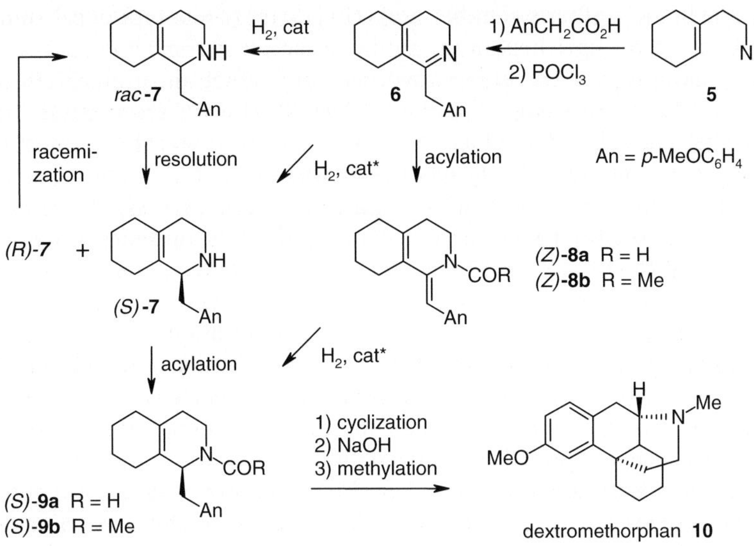

Scheme 2. Resolution and enantioselective hydrogenation routes to dextromethorphan

Table 2. Enantioselective hydrogenation approaches to dextromethorphan (**10**)

Entry	Substrate	Cat* [a]	s/c	Solvent	T [°C]	p [atm]	ee [%]	Ref.
1	(Z)-**8a**	[Ru((S)-Tol-BINAP)(SnCl$_6$)]$_2$NEt$_3$	1,000	MeOH	75	35	98	[9b]
2	(Z)-**8b**	Ru((S)-MeO BIPHEP)(tfa)$_2$	20,000	MeOH	120	35	98	[10]
3	**6**-H$_3$PO$_4$	Ir((R,S)-MOD-PPF-P(t-Bu)$_2$)(cod)BF$_4$	1,500	[b]	rt	70	89	[12]
4	**6**-H$_2$SO$_4$	Ir((R,R)-t-Bu-DIOP)Cl/NBu$_4$I	1,000	[c]	60	100	84	[13]

[a] For ligand structures and abbreviations see Fig. 1
[b] toluene/aqu NaOH, NBu$_4$Cl
[c] MeOH/toluene/i-Pr$_2$NEt

lectivity problems, in particular overhydrogenation, and the relatively low s/c ratios and ee's are issues which remain to be resolved in this direct approach.

4
Cilazapril and Mibefradil: α,β-Unsaturated Acid Hydrogenations

Further replacements of resolution by enantioselective hydrogenation routes have been developed in the syntheses of cilazapril (**15**) [14] (an angiotensin converting enzyme inhibitor; the active ingredient of the antihypertensive *Inhib-*

ace^{TM}) and in the synthesis of mibefradil (**19**) [15, 16] (a new type of calcium antagonist; the active ingredient of the antihypertensive *Posicor*^{TM}).

In the cilazapril case (Scheme 3), hydrogenation of tetrahydropyridazinecarboxylic acid **14** was realized with a Ru(Tol-MeOBIPHEP)(OAc)$_2$ catalyst to produce the key chiral building block (*S*)-**13** [17]. The hydrogenation substrate **14** was prepared from **12**, an intermediate in the current resolution route, by isomerization and transesterification. In the optimized process, **14** was hydrogenated at s/c 40,000 and at 40 bar/100 °C in MeOH in the presence of 1 equiv of Et$_3$N, to afford (*S*)-**13** of 95 to 97% ee. A lower pressure/lower temperature version afforded material with 99% ee. In both variants, enantiomerically pure (*S*)-**13** was obtained in >95% yield after a single crystallization. Of critical importance for the high catalyst productivity was the optimization of the **12** to **14** conversion in terms of reaction conditions and purification steps. It is noteworthy that this required far more effort than the demonstration, optimization, and scale-up of the enantioselective hydrogenation itself.

In the mibefradil case, (*S*)-**17**, the key chiral building block for establishing the tetralin skeleton is currently produced by a 5-step resolution/recycling process starting from **16** (Scheme 4). An efficient enantioselective route has been developed which involves a high-yield 2-step conversion of **16** into the trisubstituted acrylic acid **18** and its hydrogenation with a Ru(MeOBIPHEP)(OAc)$_2$ catalyst [18]. In this process – which constitutes one of the rare examples of a hydrogenation of a tetrasubstituted olefin – the ee increased strongly with pressure (35% at 5 bar, 95% at 250 bar). In the optimized process at 180 bar (*S*)-**17** was obtained with 94% ee. Scale-up was performed in a continuous stirred tank reactor system (CSTR) at 270 bar, s/c 1,000 and 30 °C to provide (*S*)-**17** of 93.5% ee [19]. Upgrading to 98% ee was achieved by crystallization of the sodium salt. Compared to batch mode the CSTR system allows for a higher space-time yield and

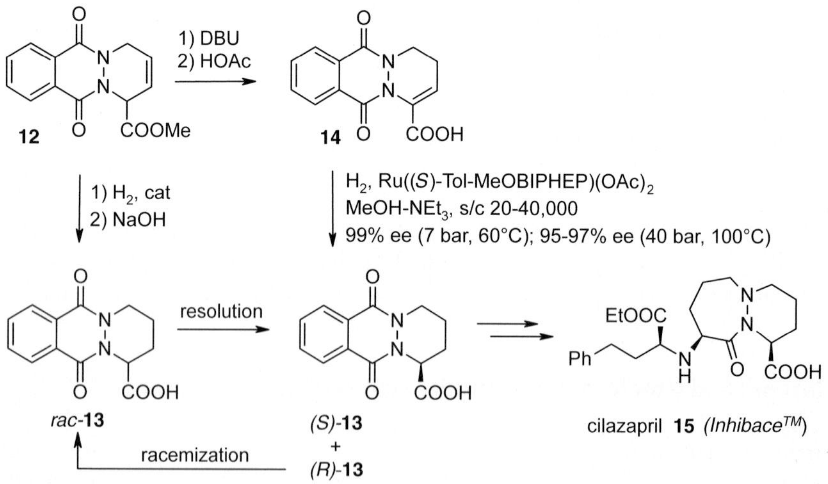

Scheme 3. Resolution and enantioselective hydrogenation in the synthesis of cilazapril

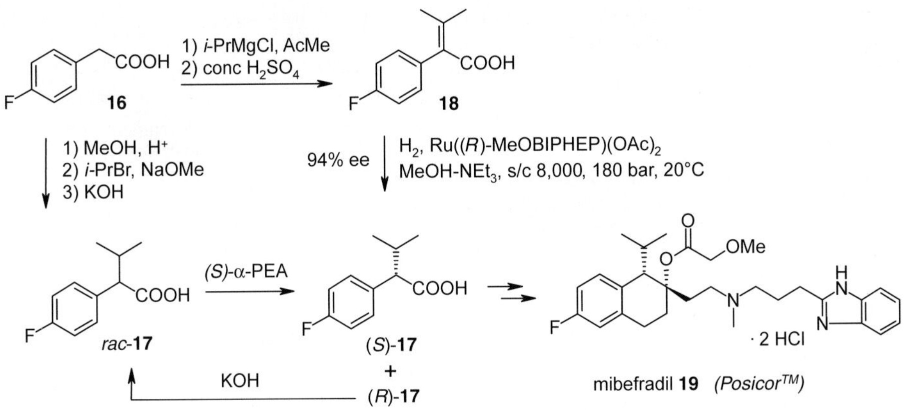

Scheme 4. Resolution and enantioselective hydrogenation in the synthesis of mibefradil

requires less reactor volume; both are factors which translate into lower investment costs particularly when working at high pressure. Overall, the enantioselective route to (S)-**17** is shorter (2 vs 5 steps) and higher-yielding (80 vs 70%) than the resolution route. The resulting significant cost advantage renders the new process attractive for implementation in the plant.

5
Vitamin E: the Quest for an Economic Total Synthesis of (R,R,R)-α-Tocopherol

The total synthesis of α-tocopherol in the naturally occurring (R,R,R)-configuration (**26**, Scheme 5) was a potent promoter for establishing enantioselective catalytic methodologies at Roche. Based on the fundamental work of Noyori and Takaya on the Ru(BINAP)-catalyzed hydrogenation of allylic alcohols [20] the synthesis of the C_{15} side-chain alcohol **25** was realized in our laboratories as outlined in Scheme 5. The hydrogenations of **20** and **24** proceeded extremely well with Ru(MeOBIPHEP)(tfa)$_2$ catalysts: TON's of 20,000 and 100,000 and ee's of 98.5 and 98%, respectively, were achieved, thus providing **25** of >98% (R,R)-content. A major difficulty on the larger scale remains the E/Z isomer separation (by fractional distillation) of **24** or its acetate.

The hydrogenation of **28**, which is readily accessible from dehydrolinalol (**27**), was investigated as an alternative entry into the side chain chemistry. Ru(BINAP)- or Ru(MeOBIPHEP)-catalyzed hydrogenation of this γ-oxo-substituted olefin afforded **23** with only 64 and 77% ee, respectively. Remarkably, MeOBIPHEP diphosphines containing P(2-furyl)$_2$ moieties afforded ee's >90%, the best ligand found so far being the unsymmetrical ligand (2-furyl)$_2$-MeOBIPHEP [21].

Numerous approaches were pursued to establish the quaternary chroman center in the correct stereochemistry, several of them involving catalytic methodology, e.g. Sharpless asymmetric epoxidation or enantioselective cyclization, but none of them proved economically viable [22]. At present, an economic total

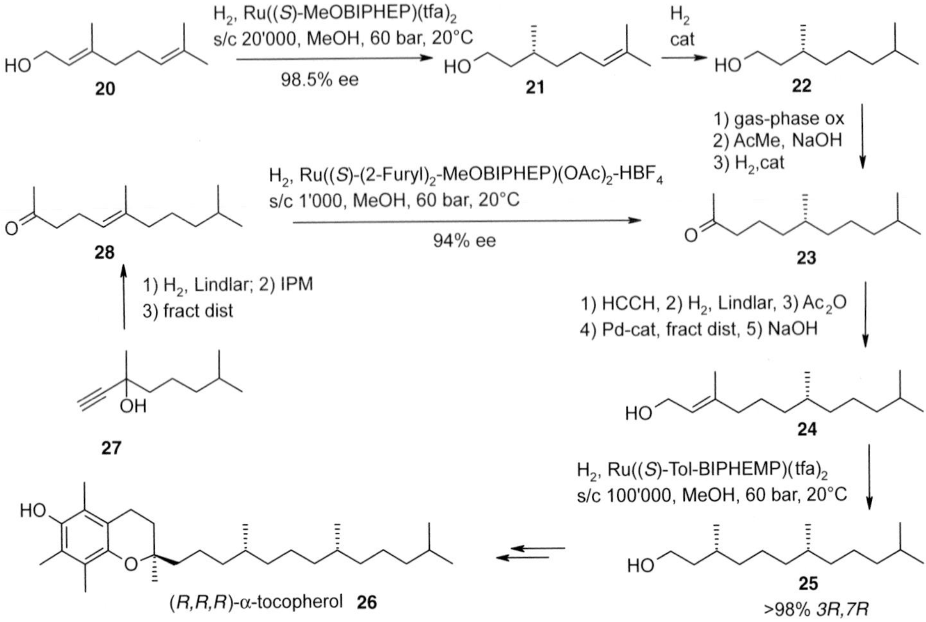

Scheme 5. (R,R,R)-α-Tocopherol side-chain building blocks

synthesis of **26** remains elusive. Although enantioselective catalysis was highly successful to establish the stereogenic centers, especially those of the side-chain, the chemistry involved in synthesizing the substrates for the catalytic reactions as well as the coupling of the building blocks is by far too lengthy and expensive. As of today, partial synthesis of **26** by permethylation of mixtures of naturally occurring α-, β-, γ-, and δ-tocopherols remains the only competitive route to produce **26**.

6
Orlistat: a β-Ketoester and an α-Pyrone Hydrogenation

Orlistat (**32**; tetrahydrolipstatin, *Xenical*^{TM}) is a potent inhibitor of pancreatic lipase [23] which has been launched for the treatment of obesity in 1998. Large amounts of **32** required for clinical development were obtained using a route based on the enantioselective reduction of β-ketoester **29** to provide β-hydroxyester (R)-**30** followed by diastereoselective elaboration strategies (via (S,S,R)-**31**, Scheme 6) [24]. For the reduction a heterogeneous version based on the use of the Raney-Ni/tartaric acid catalyst [25] and a homogeneous process with an Ru(BI-PHEMP)Cl$_2$ catalyst were investigated. The heterogeneous process, although being less enantioselective, was chosen for scale-up due to its shorter development time. The catalyst could be recycled up to 15 times, whereby the ee slightly dropped, but crystallization afforded enantiomerically pure material in >80% yield.

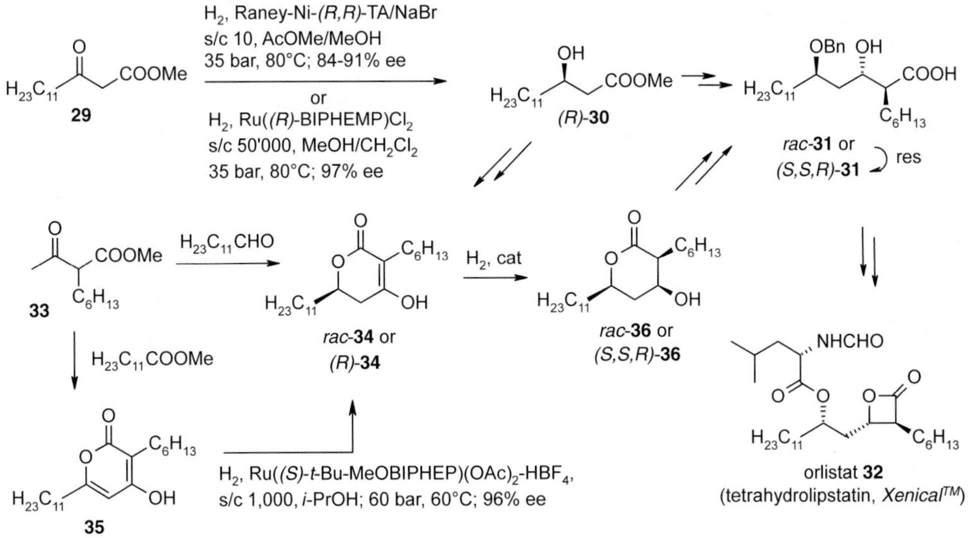

Scheme 6. Enantioselective and resolution routes to orlistat

Subsequently, the enantioselective route was replaced by a resolution process (33→rac-34→rac-36→rac-31→(S,S,R)-31→32) [26]. This synthesis proved superior due to the reduced number of steps, although the undesired stereoisomer (R,R,S)-31 could not be recycled. Ecological considerations led to the investigation of another enantioselective approach which is based on the hydrogenation of α-pyrone 35 to dihydropyrone (R)-34 [27]. Ee-values up to 96% were achieved with a cationic Ru catalyst derived from the electron-rich and sterically bulky t-Bu-MeOBIPHEP ligand. The relatively low TON of 1,000 calls for further catalyst improvement in this hitherto unprecedented type of reduction.

7
A Glyoxylate-Ene Reaction for a Collagenase Inhibitor Intermediate

The chemical development of the collagenase-selective inhibitor 39 [28] provides an example for a partially concurrent, partially sequential evaluation and scale-up of various conceptual approaches to establish the stereogenic center of an α-hydroxyester building block (Scheme 7). In one approach, hydroxyester rac-37, obtained in 3 steps from inexpensive acid chloride 36, was resolved by enzymatic kinetic resolution to afford (R)-37 in high ee [29]. This approach, although suffering from a low overall yield, served well to produce the first tens-of-kg of product rapidly. Another approach to produce (R)-37 or the corresponding ethyl or methyl esters by Ru(MeOBIPHEP)-catalyzed hydrogenation of the corresponding α-ketoesters suffered from the relative inaccessibility of the substrates and insufficient enantioselectivities of 80–90% ee.

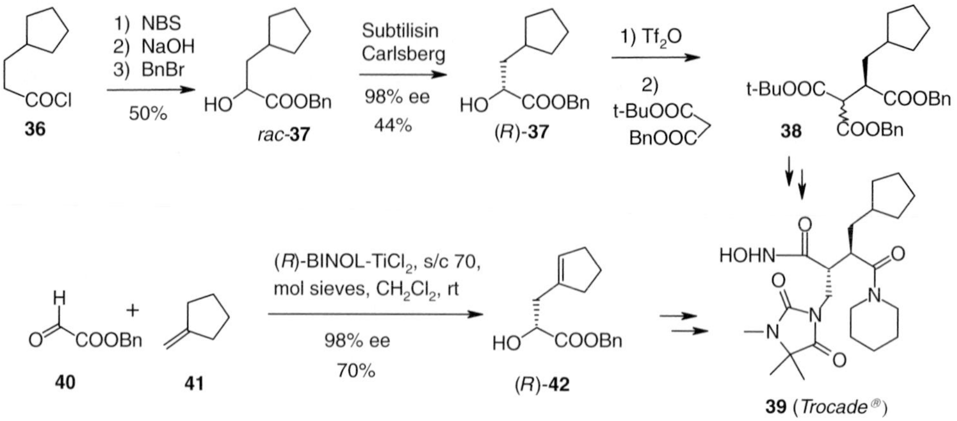

Scheme 7. Enantioselective glyoxylate-ene synthesis of an α-hydroxyester intermediate for a collagenase-selective inhibitor

Most remarkably, a Mikami type enantioselective glyoxylate-ene reaction [30] of **40** and **41** catalyzed by (R)-BINOL-TiCl₂ provided the α-hydroxyester (R)-**42** in excellent 98% ee and in good 70% yield [31]. This approach constitutes a very efficient and short entry into the optically active series. It was successfully developed and scaled-up into a reliable process to produce (R)-**42** on a large scale.

8
Concluding Remarks

The enantioselective catalytic processes described above have produced over 10 tons of chiral intermediates in piloting studies and over 4 tons of chiral building blocks for final bulk drug substances. With regard to batch-scale, the largest ones performed so far were 150 kg in the hydrogenation of pantolactone and 250 kg in the glyoxylate-ene reaction. Some of the processes described have served well to produce bulk drug material required for clinical development.

With respect to methodologies, enantioselective hydrogenation has proven to be highly versatile and powerful in our applications. Ee's often were very high, occasionally even at high reaction temperatures. Therefore, the challenges to be met were more frequently: a) to achieve high catalyst productivity (TON) and – sometimes intimately related to that – b) to secure an appropriate quality of the hydrogenation substrate, and finally c) to develop processes within given short time frames. Enantioselective hydrogenation remains the tool most widely applicable to our current problems, but we foresee other methodologies such as enantioselective C-C bond formation and oxidation to gain in importance in the future.

References

1. For a review covering part of these and other applications see: Schmid R, Broger EA, Cereghetti M, Crameri Y, Foricher J, Lalonde M, Müller RK, Scalone M, Schoettel G, Zutter U (1996) Pure & Appl Chem 68:131
2. (a) Achiwa K, Kogure T, Ojima I (1977) Tetrahedron Lett 18:4431; (b) Ojima I, Kogure T, Terasaki T, Achiwa K (1978) J Org Chem 43:3444; (c) Ojima I, Kogure T, Yoda Y (1985) Organic Synthesis 63:18
3. (a) Morimoto T, Takahashi H, Fujii K, Chiba M, Achiwa K (1986) Chem Lett: 2061; (b) Ojima I, Clos N, Bastos C (1989) Tetrahedron 45:6901
4. Broger EA, Crameri Y, F. Hoffmann-La Roche AG, unpublished results
5. (a) Broger EA, Crameri Y (1983) US 4539411 and US 4620013 to F. Hoffmann-La Roche AG; (b) Broger EA, Crameri Y (1985) EP 158875 to F. Hoffmann-La Roche AG; (c) Broger EA, Crameri Y (1987) EP 218970 to F. Hoffmann-La Roche AG
6. Nösberger P (1983) EP 86322 to F. Hoffmann-La Roche AG
7. (a) Takahashi H, Hattori M, Chiba M, Morimoto T, Achiwa K (1986) Tetrahedron Lett 27:4477; (b) Takahashi H, Morimoto T, Achiwa K (1987) Chem Lett: 855
8. Roucoux A, Thieffry L, Carpentier J-F, Devocelle M, Méliet C, Agbossou F, Mortreux A (1996) Organometallics 15:2440 and references cited therein
9. (a) Kitamura M, Sho T, Noyori R, Takatani H (1987) JP 1034964 to Takasago Perfumery Co Ltd; (b) Sayo N, Takemasa T, Kumobayashi H (1988) EP 307168 to Takasago Int Corp
10. (a) Broger EA, Crameri Y, Heiser B (1988) EP 315886 to F. Hoffmann-La Roche AG; (b) Broger EA, Heiser B (1990) EP 397042 to F. Hoffmann-La Roche AG
11. Wehrli C (1996) EP Appl 96115782.3 and 96115783.1 to F. Hoffmann-La Roche AG
12. Werbitzky O (1995) WO Appl 97/03052 to Lonza AG
13. Scalone M, Broger EA, Wehrli C (1996) EP Appl 96120844.4 to F. Hoffmann-La Roche AG
14. Attwood MR, Hassall CH, Kröhn A, Lawton G, Redshaw S (1986) J Chem Soc Perkin Trans I, 1011
15. Hengartner U, Ramuz H (1985) EP 177960 to F. Hoffmann-La Roche AG; (b) Branca Q, Jaunin R, Märki HP, Marti F, Ramuz H (1987) EP 268148 to F. Hoffmann-La Roche AG
16. For a recent review see Casas A, Graul A, Castaner J (1997) Drugs Future 22:1091
17. Broger EA, Crameri Y, Imfeld M, Montavon F, Widmer E (1993) EP 570764 to F. Hoffmann-La Roche AG
18. Crameri Y, Foricher J, Scalone M, Schmid R (1997) Tetrahedron: Asymmetry 8:3617 and references cited therein
19. Wang S, Kienzle F (1998) Organic Process Res & Dev 2:226
20. Takaya H, Ohta T, Sayo N, Kumobayashi H, Akutagawa S, Inoue S, Kasahara I, Noyori R (1987) J Am Chem Soc 109:1596
21. (a) Broger EA, Müller RK (1993) EP 565975 to F. Hoffmann-La Roche AG, (b) see also ref. [1]
22. For a review on the Roche work on (R,R,R)-α-tocopherol and stereoisomers see: Netscher T (1996) Chimia 50:563
23. Weibel EK, Hadvary P, Hochuli E, Kupfer E, Lengsfeld H (1987) J Antibiotics 40:1081
24. (a) Barbier P, Schneider F (1988) J Org Chem 53:1218 and references cited therein; b) Zutter U, Karpf M, F. Hoffmann-La Roche AG, unpublished results
25. Harada T, Tai A, Yamamoto M, Ozaki H, Izumi Y (1981) Stud Surf Sci Catal 7:364
26. Karpf M, Zutter U (1991) EP 443449 to F. Hoffmann-La Roche AG
27. Broger EA, Karpf M, Zutter U (1994) EP 643052 to F. Hoffmann La Roche AG; see also ref. [1]
28. Broadhurst MJ, Brown PA, Johnson WH (1995) EP 684240A1 to F. Hoffmann-La Roche AG
29. Wirz B, Brown P, Hilpert H, F. Hoffmann-La Roche AG, unpublished results
30. Mikami K (1996) Pure Appl Chem 68:639 and references cited therein
31. Brown P, Hilpert H (1996) EP Appl 96110814.9 to F. Hoffmann-La Roche AG

Chapter 41.3
Cyclopropanation

Tadatoshi Aratani

Organic Synthesis Research Laboratory, Sumitomo Chemical Co., Ltd., Takatsuki, Osaka 569-1093, Japan
e-mail: aratani@sc.sumitomo-chem.co.jp

Keywords: Cyclopropanation, Diazoacetate, Amino alcohols, Schiff base, Copper complexes

1
Cyclopropanes in Agro and Pharma Chemicals

Certain kinds of cyclopropanecarboxylic acids are important in the production of pyrethroid, an insecticide with low mammalian toxicity [1]. For example, chrysanthemic acid is an acid component of allethrin (Fig. 1). Various kinds of alcohols have been developed to produce pyrethroids for special application [2]. Chrysanthemic acid has two chiral centers and there are four optical isomers. There is a close correlation between the chirality of a molecule and its biological activity [3]. In the case of chrysanthemic acid, the most effective isomer is shown to be the *d-trans* isomer, which is followed by the *d-cis* isomer whereas

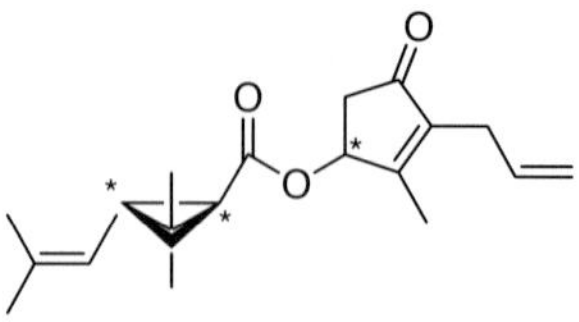

Fig. 1

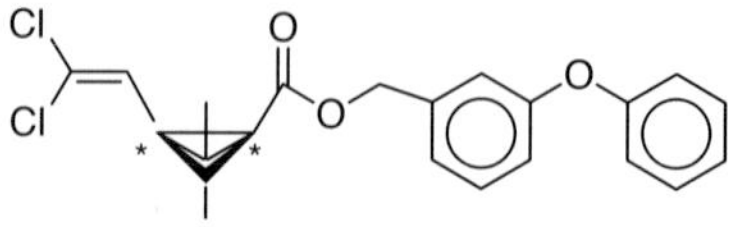

Fig. 2

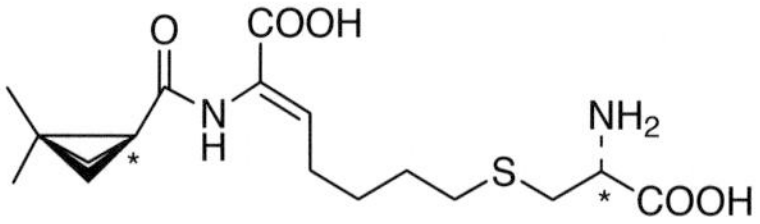

Fig. 3

the *l-trans* and *l-cis* isomers are almost ineffective [4]. The naturally occurring chrysanthemic acid found in the pyrethrum flower also has the *d-trans* or 1*R*,3*R* configuration.

Permethrinic acid, 3-(2,2-dichlorovinyl)-2,2-dimethylcyclopropanecarboxylic acid, is another kind of cyclopropanecarboxylic acid producing insecticides of higher performance and stability [5]. The structure of permethrin, a totally synthetic pyrethroid, is shown in Fig. 2. The most effective isomer of permethrinic acid is shown to be the *d-cis* isomer rather than the *d-trans* isomer [6].

Some cyclopropanes have proved to be useful as pharmaceutical intermediates. The compound, (+)-*S*-2,2-dimethylcyclopropanecarboxylic acid, is a component of cilastatin (Fig. 3), which is administrated in combination with imipenem, a carbapenem antibiotic [7]. In spite of its high and wide antibacterial activity, imipenem is found to be easily decomposed in the kidneys. This metabolism is suppressed by cilastatin, an enzyme inhibitor for dehydropeptidase I.

Certain kinds of cyclopropylamines have been developed as a component of so-called new quinolone antibacterials [8].

2
Chiral Copper Carbenoid Reaction

In Kyoto in 1966, Nozaki and Noyori and their coworkers discovered that reaction of ethyl diazoacetate with styrene in the presence of a chiral copper catalyst gives the cyclopropanated product in optically active form (Scheme 1). This experiment showed that the carbene derived from diazoacetate is not free but is

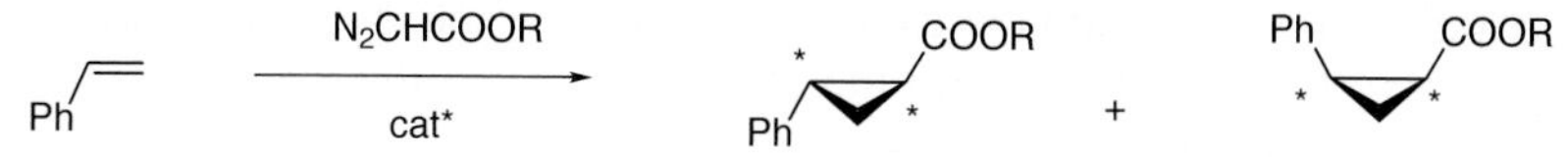

Scheme 1

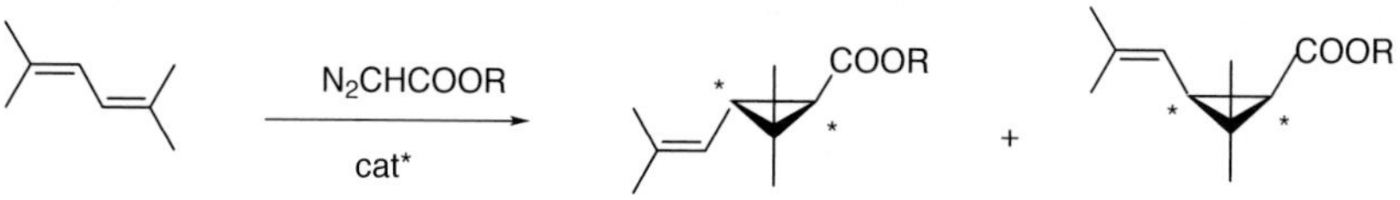

$$Ph-CH=CH_2 \xrightarrow[\text{cat*}]{N_2CHCOOR} Ph \overset{*}{\triangle}^{*}COOR + Ph \overset{*}{\triangle}^{*}COOR$$

Scheme 1

Scheme 2

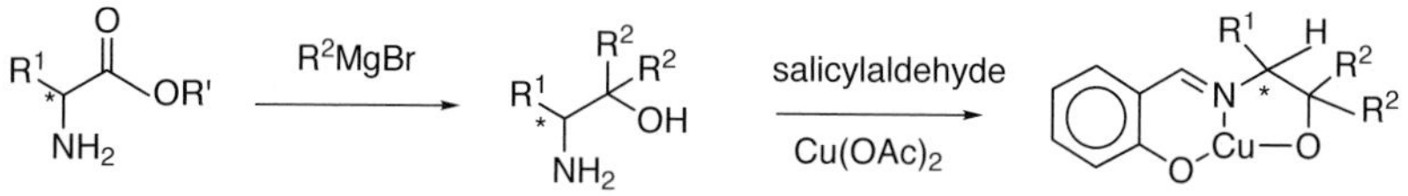

$$\xrightarrow[\text{cat*}]{N_2CHCOOR} \quad \overset{*}{\triangle}^{*}COOR \quad + \quad \overset{*}{\triangle}^{*}COOR$$

Scheme 2

Scheme 3

$$R^1\underset{NH_2}{\overset{O}{\|}}OR' \xrightarrow{R^2MgBr} R^1\underset{NH_2}{\overset{R^2}{\underset{OH}{|}}}R^2 \xrightarrow[\text{Cu(OAc)}_2]{\text{salicylaldehyde}}$$

Scheme 3

Scheme 4

$$\xrightarrow[\text{cat*}]{N_2CHCOOR} \quad \overset{*}{\triangle}COOR$$

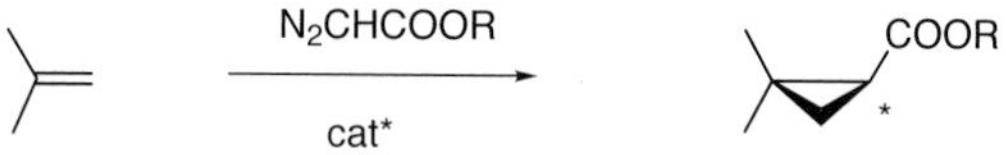

Scheme 4

combined to the catalyst to form a carbene-copper complex, which is responsible for the asymmetric induction [9].

In 1971, Sumitomo started to apply this reaction to the chrysanthemic acid synthesis (Scheme 2). The first problem was how to choose a suitable catalyst which would achieve the highest ee of the product. Here we describe our approach to this problem [10, 11, 12, 13, 14, 15, 16, 17, 18]. Other effective catalysts [19, 20, 21, 22, 23, 24, 25] are discussed by Pfaltz [26].

Reaction of optically active α-amino esters with an excess of Grignard reagent gives optically active amino alcohols with complete retention of configuration [27, 28]. The amino alcohol (Scheme 3) has two substituents, R^1 and R^2. The R^1 comes from the amino acid and R^2 comes from the Grignard reagent. The amino alcohol was reacted with salicylaldehyde to give a Schiff base, whose treatment with cupric acetate followed by alkaline work-up afforded a copper(II) complex, in which the Schiff base was incorporated as a tridentate ligand [29, 30].

In the chrysanthemic acid synthesis (Scheme 2), the ee of the product increased with the bulkiness of the R^2 group [11]. The highest ee achieved was 70%, when R^1 was methyl (the amino acid was alanine) and R^2 was 2-octyloxy-5-tert-butylphenyl. The catalyst with the R-configuration (from D-amino acid) favored the formation of d-trans and d-cis isomers to that of l-trans and l-cis isomers, respectively.

Further improvement of the stereoselectivity was achieved by selection of the alkyl group of the diazoacetate [12]. Both the preference for the trans isomer and the ee of the trans isomer increased in proportion to the bulkiness of the alkyl group. Reaction of *l*-menthyl diazoacetate with 2,5-dimethyl-2,4-hexadiene in the presence of the *R*-catalyst gave the product in which the most effective isomer, *d-trans*, predominated by as much as 92%.

The catalyst was successfully applied to the production of *S*-2,2-dimethylcyclopropanecarboxylic acid (Scheme 4). Reaction of ethyl diazoacetate with isobutylene in the presence of the *R*-catalyst gave the corresponding ethyl ester in 92% ee [10, 22, 31].

3
Copper Complexes of *N*-Salicylideneamino Alcohols

The catalyst was a copper(II) chelate having the chiral Schiff base, *N*-salicylideneamino alcohol, as a tridentate ligand. The Schiff base occupies three of the four coordination sites leaving the fourth site vacant. In the absence of another ligand, the coordination number four is satisfied by dimerization. The binuclear structure (Fig. 4) was suggested by measurements of the magnetic susceptibility and confirmed by X-ray crystallography [14]. The two copper atoms are bridged by two phenolic oxygen atoms of salicylaldehyde.

In the presence of donating ligands, the dimer collapses into two equivalents of the mononuclear complex. For example, addition of pyridine gave rise to the pyridine adduct (Fig. 5) in which pyridine occupies the fourth coordination site of copper.

In an attempt to prepare another kind of mononuclear complex (Fig. 6) having the Schiff base as a bidentate ligand, the amino alcohol was reacted with bis(salicylaldehydato)copper [32]. The product was again a mononuclear complex (Fig. 7), in which the same Schiff base resides as a tridentate ligand and the

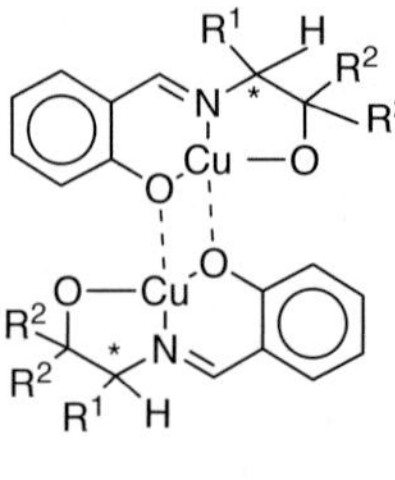

Fig. 4

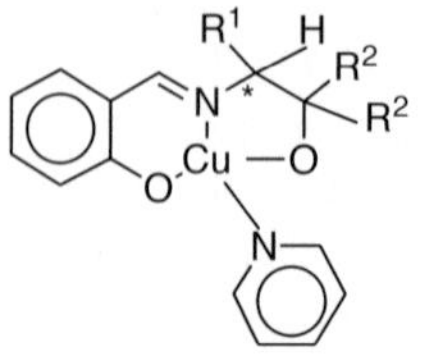

Fig. 5

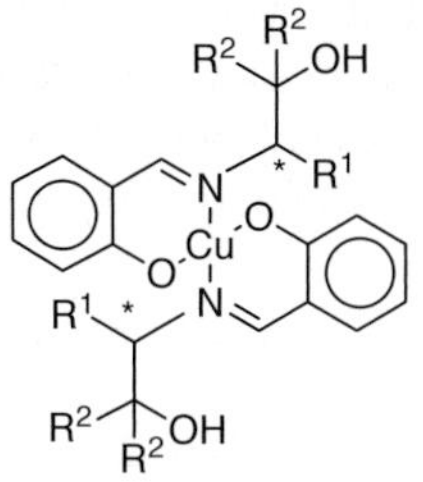

Fig. 6

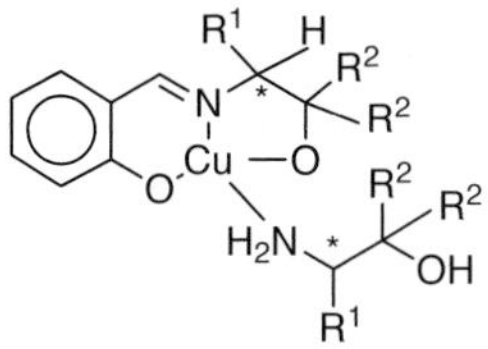

Fig. 7

last coordination site is occupied by the amino group of the amino alcohol. The structure was established by X-ray crystallography [15].

4
Experimental Section

4.1
R-2-Amino-1,1-di-(2-methoxyphenyl)-3-phenylpropanol (Scheme 3) [33]

To a solution of Grignard reagent prepared from 2-bromoanisole (46.8 g, 250 mmol) and magnesium turnings (6.08 g, 250 mmol) in diethyl ether (150 mL), a solution of D-phenylalanine ethyl ester (9.66 g, 50 mmol) in ether (50 mL) was added in the course of 5 h. The reaction mixture was heated under reflux for 5 h and was then carefully added to dilute hydrochloric acid under cooling. The precipitate was filtered and treated with an aqueous ammonia to give the amino alcohol as white crystals (yield: 12.4 g, 68%). A sample recrystallized from cyclohexane showed mp 102–103 °C, $[\alpha]_D$ +42.5 ° (c 1.00, chloroform).

4.2
Mononuclear Copper Complex of R-N-Salicylidene-2-amino-1,1-di(2-methoxyphenyl)-3-phenylpropanol (Fig. 7) [34]

To a suspension of bis(salicylaldehydato)copper [32] (0.43 g, 1.4 mmol) in methanol (10 mL), a solution of R-2-amino-1,1-di-(2-methoxyphenyl)-3-phenylpropanol (1.1 g, 3.0 mmol) in methanol (10 mL) was added in the course of 1 h. The addition was carried out with vigorous stirring at room temperature. After the reaction mixture had been stirred for another 1 h, the solid product was filtered, washed with methanol, and dried in vacuo. Recrystallization from benzene-methanol gave grayish blue crystals (yield: 0.75 g, 58%), mp 205–206 °C (dec), $[\alpha]_{546}$ +860 ° (c 0.065, benzene). The magnetic susceptibility μ was 1.78 B.M. For X-ray crystallography, see [15].

4.3
Binuclear Copper Complex of R-N-Salicylidene-2-amino-1,1-di-(2-butoxy-5-tert-butylphenyl)-3-phenylpropanol (Fig. 4) [34]

The corresponding amino alcohol, R-2-amino-1,1-di-(2-butoxy-5-tert-butylphenyl)-3-phenyl-propanol, was prepared by the reaction of a Grignard reagent from 2-bromo-4-tert-butylphenyl butyl ether [35] with D-phenylalanine ethyl ester. In this case, neither the amino alcohol nor its hydrochloride was crystalline. The reaction mixture was treated with aqueous hydrochloric acid and the inorganic phase was discarded. The separated organic layer was neutralized with aqueous ammonia and concentrated in vacuo to give a crude mixture of the amino alcohol and butyl 4-tert-butylphenyl ether, a hydrolysate of the Grignard reagent. The yield was estimated to be 67% (LC) based on a pure sample, $[\alpha]_D$ +43.5 ° (c 1.00, chloroform), isolated by column chromatography on silica gel.

An equimolar mixture of the crude amino alcohol and salicylaldehyde in toluene was heated under reflux for 2 h. Evaporation of the toluene followed by distillation of butyl tert-butylphenyl ether (bp 100 °C/3 torr) gave the corresponding Schiff base as a yellow viscous residue. The yield was estimated to be 65% (LC) based on phenylalanine ethyl ester. A pure sample, $[\alpha]_D$ –8.00 ° (c 1.00, chloroform), was isolated by column chromatography on silica gel [36].

An equimolar mixture of the Schiff base and cupric acetate monohydrate in ethanol was heated under reflux for 2 h. The ethanol was evaporated in vacuo and the dark green residue was dissolved in toluene. The toluene solution was treated with aqueous sodium hydroxide to complete complex formation. Removal of the toluene followed by trituration in methanol induced crystallization of the copper complex as a bluish green mass. Filtration of the solid followed by drying in vacuo gave a pure sample, $[\alpha]_{546}$ +1040 ° (c 0.087, benzene), mp 186–188 °C (dec), in 77% yield based on the Schiff base. The magnetic susceptibility μ was 0,86 B.M. For X-ray crystallography, see [14].

4.4
Ethyl S-2,2-Dimethylcyclopropanecarboxylate (Scheme 4) [37]

To a solution of the binuclear copper complex of R-N-salicylidene-2-amino-1,1-di-(2-butoxy-5-tert-butylphenyl)-3-phenylpropanol (0.40 g, 0.55 mmol copper) in toluene (50 mL), was dissolved isobutylene (14 g) under a nitrogen atmosphere. Addition of a toluene solution of phenylhydrazine (10%, 0.30 mL, 0.28 mmol) induced an instant color change of the solution from green to pale yellow indicating the reduction of copper valency from +2 to +1.

Ethyl diazoacetate (purified by distillation, 16.15 g, 142 mmol) in toluene (40 g) was added dropwise at 40 °C in the course of 7 h. During the addition, isobutylene gas (33 g) was continuously blown into the solution. Evolution of nitrogen gas started as soon as the addition was started and, at the end of addition, a quantitative amount of nitrogen gas (3.4 L) was observed.

The reaction mixture was heated to 80 °C to remove an excess of isobutylene. GC analysis of the reaction mixture (105.2 g) showed that the yield of ethyl 2,2-dimethylcyclopropanecarboxylate was 82% based on ethyl diazoacetate. Distillation gave a pure sample, bp 80 °C /60 torr, $[\alpha]_D$ +105.6 ° (c 2.0, chloroform). The ee was 92% by GC analysis of the corresponding d-octyl ester.

4.5
Binuclear Copper Complex of *R-N*-Salicylidene-2-amino-1,1-di-(5-*tert*-butyl-phenyl-2-octyoxy)propanol (Fig. 4) [34]

The amino alcohol, *R*-2-amino-1,1-di-(2-octyoxy-5-*tert*-butyl-phenyl)-propanol, was prepared by the reaction of a Grignard reagent from 2-bromo-4-*tert*-butylphenyl octyl ether [35] with D-alanine ethyl ester. The copper complex was prepared in the similar manner as above. In this case, neither the amino alcohol, the Schiff base, nor the copper complex was crystalline.

The Schiff base was purified by column chromatography on silica gel (65% yield based on alanine ethyl ester). An equimolar mixture of the Schiff base and cupric acetate monohydrate in ethanol was heated under reflux for 1 h. After evaporation of the ethanol in vacuo, the dark green residue was dissolved in toluene and treated with aqueous sodium hydroxide. Removal of the toluene and drying in vacuo gave the copper complex as a dark green viscous oil, $[\alpha]_{546}$ +730 ° (c 0.076, benzene), which is used in the subsequent reaction.

4.6
l-Menthyl Diazoacetate [38]

A mixture of *l*-menthyl glycine [39] (19.7 g, 92 mmol), isoamyl nitrite (12.0 g, 100 mmol), and acetic acid (1.6 g, 27 mmol) in chloroform (400 mL) was heated under reflux for 25 min [40]. The reaction mixture was washed with 1M sulfuric acid, with a saturated aqueous solution of sodium bicarbonate, and finally with water. After removal of the solvent, the residue was purified by column chromatography on silica gel (160 g, methylene chloride) to give yellow waxy crystals (15.0 g, 73%), $[\alpha]_D$ –86.8 ° (c 1.0, chloroform), IR υ 2125 cm^{-1} (film) and ^{1}H-NMR δ 5.29 ppm (chloroform-d, TMS).

4.7
l-Menthyl *d*-Chrysanthemate (Scheme 2) [38]

To a solution of the binuclear copper complex of *R*-*N*-salicylidene-2-amino-1,1-di(5-*tert*-butyl-2-octyloxyphenyl)-3-phenylpropanol (0.30 g, 0.40 mmol copper) in 2,5-dimethyl-2,4-hexadiene (17.6 g, 160 mmol), a solution of *l*-menthyl diazoacetate (4.5 g, 20 mmol) in the diene (4.4 g, 40 mmol) was added at 40 °C in the course of 7 h. At the beginning of the addition, the reaction mixture was heated to 70 °C to facilitate decomposition of the diazoacetate and thereafter the temperature was maintained at 40 °C. At the end of the addition, a nearly quantitative amount of nitrogen gas had been evolved.

Removal of the unreacted diene, bp 45 °C/20 torr, followed by distillation, bp 123 °C/0.2 torr, gave *l*-menthyl chrysanthemate (4.7 g, 76%) whose GC analysis showed the following composition: *d-trans*, 89.9%; *l-trans*, 2.7%, and the *cis* isomers, 7.4%. The *cis/trans* ratio was 7/93 and ee of the *trans* isomer was 94%. Complete hydrolysis of the *l*-menthyl ester with potassium hydroxide in aqueous ethanol followed by esterification with *d*-2-octanol in the presence of thionyl chloride and pyridine gave the corresponding *d*-2-octyl ester. GC analysis revealed the following composition: *d-trans*, 90.4%; *l-trans*, 4.7%; *d-cis*, 3.6%, and *l-cis*, 1.3%. The ee was calculated to be 90% for the *trans* isomer and 47% for the *cis* isomer [41].

5
Conclusions

The discovery of the chiral copper carbenoid reaction in Kyoto has led to the introduction of a chiral cyclopropane production in Sumitomo. However, at the present time, the catalysts in our hands are not as selective as those of natural origin. The *Pyrethrum* flower is still much more skillful and beautiful in chrysanthemic acid synthesis. Further endeavors should be exerted to complete the "man-made asymmetric catalysis" [42,43,44,45].

Acknowledgments. The author would like to thank all the people who were involved in the course of this work for their help and encouragement.

References

1. Matsui M, Yamamoto I (1971) Pyrethroids. In: Jacobson M, Crosby DG (eds) Naturally Occurring Insecticides. Marcel Dekkar, New York, chap 1
2. Matsuo N, Miyamoto J (1997) Development of synthetic pyrethroids with emphasis on stereochemical aspects. In: Hedin PA, Hollingworth RM, Masler EP, Miyamoto J, Thompson DG (eds) Phytochemicals for pest control. Am Chem Soc, Washington DC, chap 14
3. Collins AN, Sheldrake GN, Crosby J (eds) (1992) Chirality in industry: the commercial manufacture and applications of optically active compounds. Wiley, New York
4. For the racemization of chrysanthemic acid, see Suzukamo G, Fukao M, Nagase T (1984) Chem Lett 1799
5. Elliott M, Janes NF (1978) Chem Soc Rev 7: 473

6. A stereoselective synthesis of this carboxylic acid was reported: Martin P, Greuter H, Rihs G, Winkler T, Bellus D (1981) Helv Chem Acta 64:2571 and ref 13

7. Budavari S (ed) (1996) The Merck Index, 12th edn. Merck & Co., Inc., Whitehouse Station NJ, references 2310, 2331 and 4954

8. For example, a stereoselective synthesis of 1*R*-*cis*-2-fluorocyclopropylamine for DU-6859: Tamura O, Hashimoto M, Kobayashi Y, Katoh T, Nakatani K, Kamada M, Hayakawa I, Akiba T, Terashima S (1992) Tetrahedron Lett 24:3483, 3487

9. Nozaki H, Moriuti S, Takaya H, Noyori R (1966) Tetrahedron Lett 5239

10. Aratani T (1985) Pure Appl Chem 57:1839

11. Aratani T, Yoneyoshi Y, Nagase T (1975) Tetrahedron Lett 16:1707

12. Aratani T, Yoneyoshi Y, Nagase T (1977) Tetrahedron Lett 18:2599

13. Aratani T, Yoneyoshi Y, Nagase T (1982) Tetrahedron Lett 23:685

14. Yanagi K, Minobe M (1987) Acta Cryst C43:1045. The 3D file, R7644.vrml, is attached in the CD-ROM version. Information about VRML plug-ins is available at http://www.sdsc. edu/vrml/.

15. Yanagi K, Minobe M (1987) Acta Cryst C43:2060. The 3D file, R7601.vrml, is attached in the CD-ROM version.

16. Aratani T, Yoshihara H, Ageha T, Inukai Y, Yanagi K (1987) Absolute configuration of 2,3-*cis*-disubstituted cyclopropanecarboxylic acids. In: Greenhalgh R, Roberts (eds) Pesticide Science and Biotechnology. Blackwell Scientific Publications, Oxford, p 49

17. Yanagi K, Aratani T, Minobe M (1986) Acta Cryst C42:745.

18. Yanagi K, Aratani T (1987) Acta Cryst C43:263.

19. Nakamura A, Konishi A, Tatsuno Y, Otsuka S (1978) J Am Chem Soc 100:3443, 6544

20. Fritschi H, Leutenegger U, Pfaltz A (1986) Angew Chem Intern Ed Engl 25:1005

21. Lowenthal RE, Abiko A, Masamune S (1990) Tetrahedron Lett 31:6005

22. Evans DA, Woerpel KA, Hinman MM, Faul MM (1991) J Am Chem Soc 113:726

23. Doyle MP (1991) Recl Trav Chim Pays-Bas 110:305; Doyle MP, Forbes DC (1998) Chem Rev 98:911

24. Ito K, Katsuki T (1993) Tetrahedron Lett 34: 2661

25. Nishiyama H, Itoh Y, Sugawara Y, Matsumoto H, Aoki K, Itoh K (1994) J Am Chem Soc 116:2223

26. Jacobsen EN, Pfaltz A, Yamamoto H (eds) Comprehensive asymmetric catalysis. Springer, Berlin Heidelberg New York, chap xx.x

27. Mckenzie A, Roger R, Wills GO (1926) J Chem Soc 779

28. Benjamin BM, Schaeffer CJ, Collins CJ (1957) J Am Chem Soc 79: 6160

29. Holm RH, Everett, GW Jr, Chakravorty A (1966) Metal complexes of Schiff bases and β-ketoamines. In: Cotton FA (ed) Progress in inorganic chemistry. Interscience Publishers, New York, vol 7, p 83

30. Houghton RP, Pointer DJ (1965) J Chem Soc 4214

31. An enzymatic resolution of this carboxylic acid has been developed: Meyer HP (1993) Chimia 47:123; CEN (1994) Sept 19, p 61

32. Kotowski A (ed) (1966) Gmelin Handbook of Inorganic Chemistry, 8th edn. Verlag Chemie, Weinheim, Copper, Part B4, p 1568

33. For the amino alcohols synthesis, see Sumitomo Chem (1976) JP Patent 839405

34. For the preparation of copper complexes, see Sumitomo Chem (1977) US Patent 4029683, 4029690

35. Prepared by bromination of 4-*tert*-butylphenol with bromine followed by etherification with the corresponding alkyl bromide in the presence of a phase transfer catalyst

36. Another way to isolate the pure Schiff base was hydrolysis of the purified copper complex in the presence of aqueous EDTA

37. For the cyclopropanation procedure, see Sumitomo Chem (1985) US Patent 4552972, 4603218

38. For the alkyl chrysanthemate synthesis, see Sumitomo Chem (1980) US Patent 4197408

39. Prepared by ammonolysis of *l*-menthyl chloroacetate in DMF

40. Takamura N, Mizoguchi T, Koga K, Yamada S (1975) Tetrahedron 31:225
41. Murano A (1972) Agr Biol Chem 36:2203
42. Kagan HB (1985) Chiral ligands for asymmetric catalysis. In: Morrison JD (ed) Asymmetric synthesis. Academic Press, Orlando FL, vol 5, p 1
43. Parshall GW, Nugent WA (1988) Chemtech 18:184, 376
44. Hartmuth CK, VanNieuwenhze MS, Sharpless KB (1994) Chem Rev 94:2483
45. Noyori R (1994) Asymmetric catalysis in organic synthesis. Wiley, New York

Chapter 41.4
Asymmetric Isomerization of Olefins

Susumu Akutagawa

Takasago International Corporation, Nissay Aroma Square, 5-37-1, Kamata Ohta-ku,
Tokyo 144-8721, Japan
e-mail: akutag@bni.co.jp

Keywords: Allylamine, BINAP, Catalyst recycle, Citronellal, Enamine, Ene reaction, Isomerization, Menthol, Rhodium complex, Telomerization, Terpenoids

1
Introduction

With its characteristic oriental note and cooling effect, (–)-menthol is in daily use among many consumer products including tobacco flavors, mouth-cares, toothpaste, plasters, and in pharmaceuticals. Currently, its world market is approaching 12,000 tons annually, with the selling price in the range of $30–45/kg. About 70% of the market is supplied from natural products isolated from the essential oil of *Mentha arvensis* cultivated mainly in India and China. Among the eight stereoisomers of menthol, only the (1*R*,3*R*,4*S*)-configuration exhibits genuine biological properties. Thus, the major synthetic problem is the control of stereoisomers. There are two commercialized synthetic processes, one is a resolution method and the other is an asymmetric methodology, equally sharing the remaining 3,500-ton market.

Since 1984 Takasago has been producing (–)-menthol based on Rh-BINAP catalysts (see Chapter 23). The catalysts can convert *N,N*-diethylgeranylamine **1** to citronellal enamine **2** enantioselectively, Eq. (1) [1, 2, 3]. Both chemical and enantioselectivities are extremely high at 99% yield and 98.5% ee, respectively. Besides, the enantioselectivity is independent of the reaction temperature in the range of 25 to 100 °C which is favorable for obtaining a high TON. This article summarizes the process development in the practical use of the Rh-BINAP catalyst for the production of (–)-menthol **3** and related enantiopure terpenoids in a total amount of over 2,300 tons.

2
Process Development

In the practice of homogeneous asymmetric catalysts, we must solve such problems as high cost of chiral auxiliaries and difficulties in the handling of sensitive catalysts. For industrial applications, the consumption of chiral auxiliaries should be kept to a minimum by improving TON as much as possible. When first discovered, the TON of the Rh-BINAP catalysis in Eq. (1) was only 100 as usual laboratory works. A feasibility study indicated that the TON must be more than 100,000 for the profitable manufacturing of **2** on a 2,500-ton scale. It was also necessary to complete the synthetic scheme not only before substrate production but also after the asymmetric process to complete the target molecules. The studies on process development described below have realized this criterion and enabled the commercial operation.

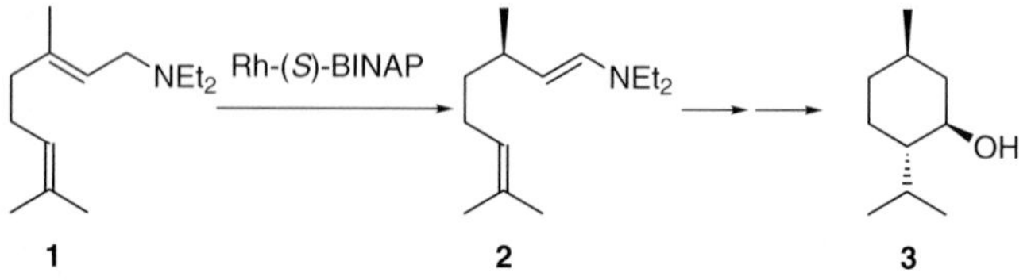

$$(1)$$

1   2   3

2.1
Substrate Production

During the 1970's, the lithium diethylamide catalyzed anionic telomerizations of myrcene **4**, Eq. (2) [4] and isoprene, Eq. (3) [5] with secondary aliphatic amines were discovered. These reactions are highly chemo- and regioselective and opened the way for the production of various useful terpenoids. The selective formation of *N,N*-diethylnerylamine **5** from isoprene is noteworthy, because this reaction is only one example hitherto known that can effect isoprene coupling in the natural fashion.

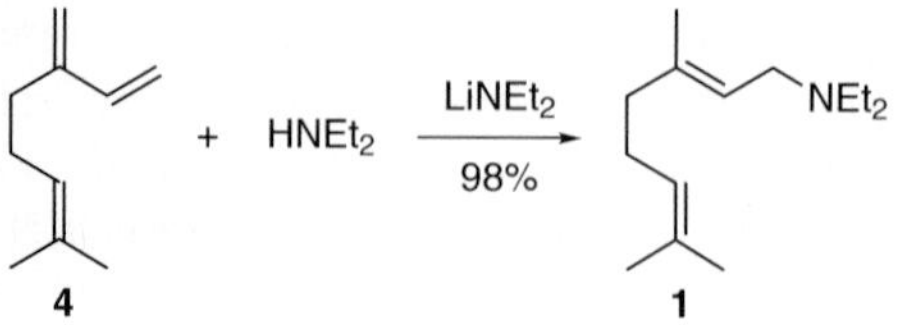

$$(2)$$

4   1

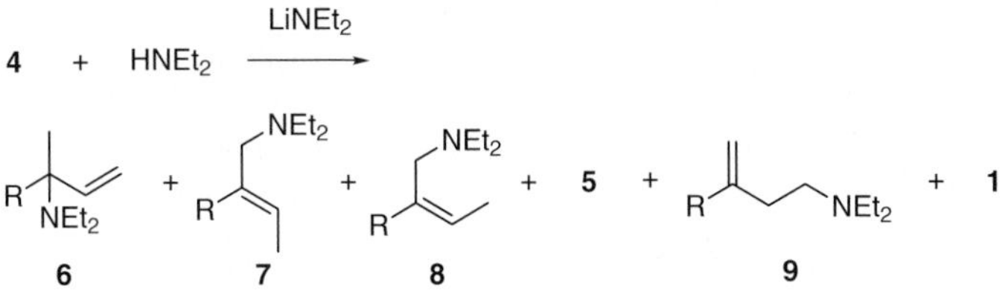

Scheme 1

Products	Adduct of	Ratio [%]	Relative volatility
6	3-4	0.03	0.75
7	*cis* 4-1	0.20	0.80
8	*trans* 4-1	0.50	0.85
5	*cis* 1-4	1.20	0.90
9	1-2	0.07	0.98
1	*trans* 1-4	98.00	1.00

$$(3)$$

We are now producing 3,000 tons of *N*,*N*-diethylgeranylamine **1** annually according to Eq. (2) [6]. In the industrial operation, where drastic conditions such as the higher reaction temperature of 120 °C and the lower catalyst ratio (1 to 100) are required, the regioselectivity drops to 92%. The crude telomer consists of six regioisomers, which were formed by all possible modes of addition between diethylamine and the conjugated diene of **4** (Scheme 1) [3].

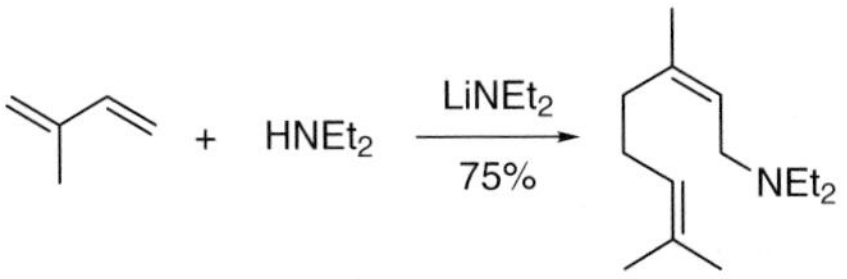

It is essential to remove isomers **5** and **9** from **1** for the asymmetric reaction, because the former reduces the enantiomeric purity while the latter acts as a strong catalyst poison (see Chapter 23). As the volatility of **9** is very close to that of **1**, a distillation column with 80 theoretical plates is applied to furnish **1** in the purity of 99.98% for the Rh-BINAP catalysis.

2.2
Catalyst Preparation

Industrially, we have been using Tol-BINAP **10** instead of the prototype BINAP (Fig. 1). The merit is its higher crystallization properties both in the resolution and as rhodium complexes.

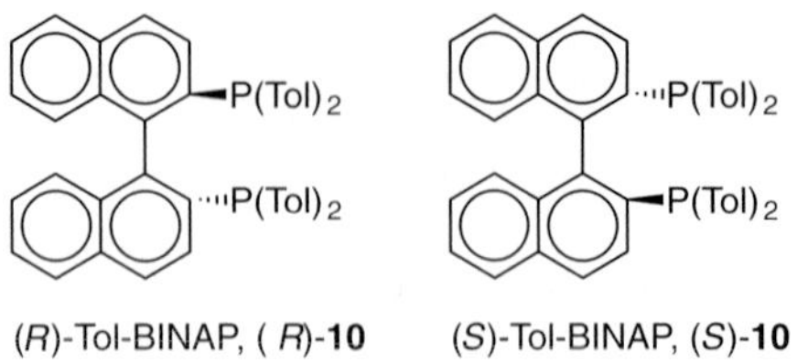

(R)-Tol-BINAP, (R)-10   (S)-Tol-BINAP, (S)-10

Fig. 1

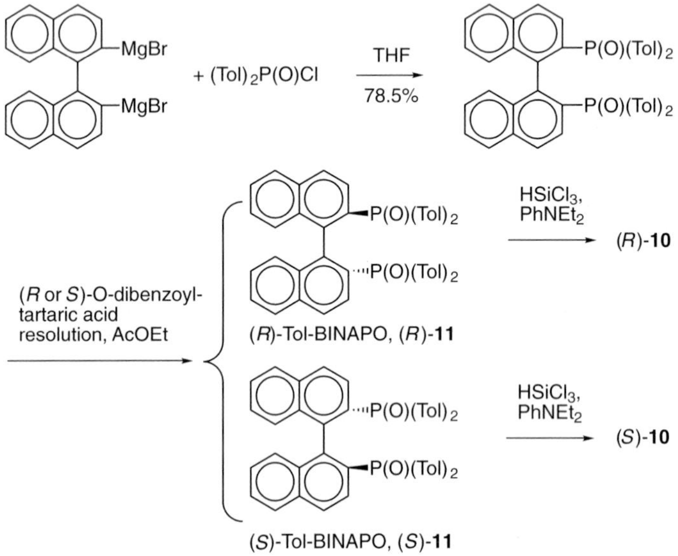

Scheme 2

Compared to the recent publication [7], our synthetic scheme consists of a sequence of classical organic syntheses suitable for 100 kg scale production (Scheme 2) [8,9]. Generally, it is difficult to introduce diphenylphosphino groups at the sterically hindered 2,2'-positions of binaphthyl in a high yield. The combination of organomagnesium bromide with diphenylphosphinyl chloride gives Tol-BINAPO 11 in a yield of 87.5%, the value is much higher than that obtained from the organolithium and diphenylphosphinous chloride (37%). The resolution of 11 can be carried out perfectly according to the following procedure. When 45% mol equivalent of (R)-O-dibenzoyltartaric acid is added to racemic 11 in ethyl acetate at 15 °C, the (R,R)-diastereomeric salt precipitates quantitatively. The optical purity of (R)-11, liberated by sodium hydroxide, is sufficient for the trichlorosilane reduction to 10 without further purification. Similarly, the (S)-isomer is obtainable by applying 45% mol equivalent of (S)-O-dibenzoyltartaric acid to the mother liquor containing (S)-rich 11.

$$[\text{Rh(cod)Cl}]_2 + \text{NaClO}_4 + (R)\text{-}\mathbf{10} \xrightarrow{\text{H}_2\text{O-CH}_2\text{Cl}_2,\ \text{PhCH}_2\text{NMe}_3\text{Br}} [\text{Rh(cod)}\{(R)\text{-}\mathbf{10}\}]^+\text{ClO}_4^-$$

$$\mathbf{12}$$

$$\mathbf{12} + (R)\text{-}\mathbf{10} + \text{H}_2 \xrightarrow{\text{THF}} [\text{Rh}(\{(R)\text{-}\mathbf{10}\}_2]^+\text{ClO}_4^-$$

$$\mathbf{13}$$

Scheme 3

In our asymmetric process development, the discovery of the thermally stable rhodium bis-BINAP complex was outstanding as it enabled the repeated use of catalyst [10]. In Scheme 3, the synthesis of the cationic rhodium bis-Tol-BINAP complex **13** is illustrated. Its precursor **12** can be prepared quantitatively using cheap sodium perchlorate in a binary system in the presence of a phase transfer catalyst. It is possible to convert **12** to **13** by monitoring the reaction either by the volume of hydrogen absorbed or the color change from orange (**12**) to deep red (**13**).

2.3
Improvement of TON

In the sense of coordination chemistry, almost all catalyst inhibitors were removed to attain high a TON. First, we introduced the treatment of substrate **1** by distillation over Vitride, a toluene solution of $\text{NaAlH}_2(\text{OCH}_2\text{CH}_2\text{OCH}_3)_2$, to remove donor substances such as oxygen, moisture, carbon dioxide, and in particular sulfur-containing impurities which originated from turpentine. Thus, the TON was raised to 1,000 from the original 100. Second, the removal of an amine isomer, **9** in Scheme 1, by fractional distillation increased the TON to 8,000. The coordination order of substrates and products to a rhodium complex is supposed to be proportional with the basicity order of the simple tertiary amine **9**, the allylic amine **1**, and the enamine **2**. This phenomenon enables the fast replacement of the enamine formed from the metal by a substrate molecule that allows the smooth catalytic cycle. As the coordination of **9** to rhodium is too strong, it acts as a strong catalyst poison even in a small amount.

2.4
Catalyst Recycle System

Finally, we have established the practical asymmetric isomerization process as follows: In a 15-m^3 batch reactor, a mixture of 7 tons of **1** and 6.71 kg of the catalyst **13** (the molar ratio of catalyst to substrate is 1 to 8,000) in THF (3 m^3) are charged. The isomerization is completed in 18 h at 100 °C, providing **2** in 99% yield and 98.5% ee [3, 11, 12]. After the reaction, the whole products (THF, enamine, and catalyst) are subjected for distillation under reduced pressure (initially 400 torr, finally 2 torr) to recover THF and **2**. The distillation residue is an orange-brown solid mass containing mainly **12** and free ligand **10**. The reverse coordination of these two components to **13** is slow in the solid, however, the complex **13** can be precipitated in a pure form by the addition of *n*-heptane to the residue.

During the early stage of the manufacturing, the recovery of **13** was 90%, by which we assumed 10% of the catalyst was decomposed during the reaction. Hence the additional 0.671 kg of fresh catalyst **13** was required for the following batch to keep the same operation. Thus, the total TON was 80,000. Now, as a result of total quality control, the catalyst recovery has become 98%. Consequently the next batch requires 2% of the fresh catalyst in additon to the recovered amount thus exemplifying the total TON of 400,000. It is also possible to recover rhodium metal and **11** from the *n*-heptane solution, thus making the total mass balance of precious materials higher than 99.9%.

2.5
Enamine to Menthol

After the isomerization, the production of (–)-menthol is carried out according to Scheme 4. On usual hydrolysis, **2** gives (*R*)-citronellal **14** quantitatively in 98.5% ee. The cyclization of **14** to (–)-isopulegol **15**, an intramolecular ene reaction, is promoted by various acid catalysts. Whereas ordinary Lewis acid gives a

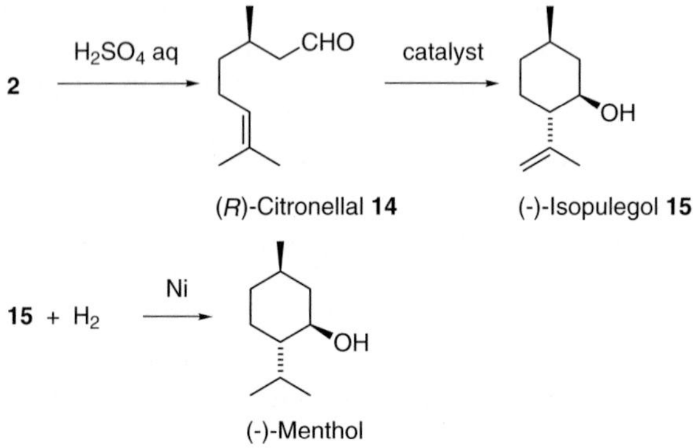

(*R*)-Citronellal **14**   (-)-Isopulegol **15**

(-)-Menthol

Catalyst	Conversion [%]	15 [%]
SiO$_2$	100	62
Rh(PPh$_3$)$_3$Cl	25	85
ZnCl$_2$	15	70
ZnBr$_2$	12	99
ZnI$_2$	8	100
ZnBr$_2$, calcined at 160 °C	97	98

Scheme 4

mixture of conformers of **15**, zinc bromide (calcined at 160 °C) catalyzed the reaction stereoselectively. The 100% enantiomeric purity of **15** can be accomplished
by the crystallization of isopulegol (98.5% ee) at –50 °C from *n*-heptane solution. Finally, the last step is a simple hydrogenation.

3
Application

In the isomerization, one favorable feature is the desirable stereochemical correlation between substrate geometries, product configurations, and ligand chi-
ralities, as shown in Scheme 5. Since both enantiomers of the ligand as well as
the substrates are easily obtainable, this stereochemical relation provides economical advantages in the option of taking the starting material either from natural resource (renewable turpentine) or petroleum. It is also possible to produce
both enantiomers of citronellal from a single intermediate only by changing the
ligand chirality.

The present asymmetric technology has enabled the manufacturing of enantiomeric pairs of aroma chemicals that is a strategically powerful means in the
fragrance business. Both enantiomers of citronellol **16** (Fig. 2) are precious fragrances inaccessible before this technology. We are supplying a pair of isomers

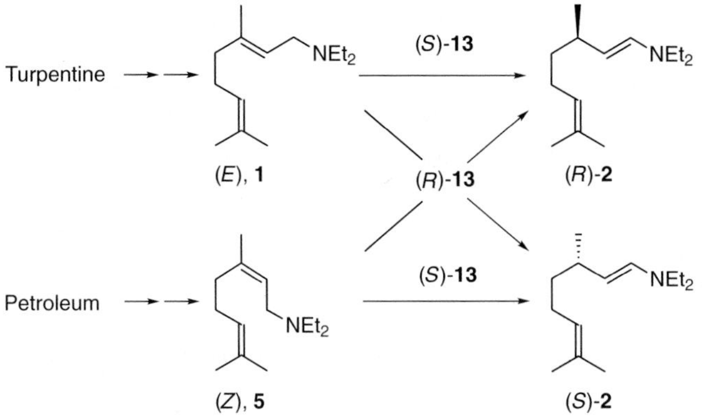

Scheme 5

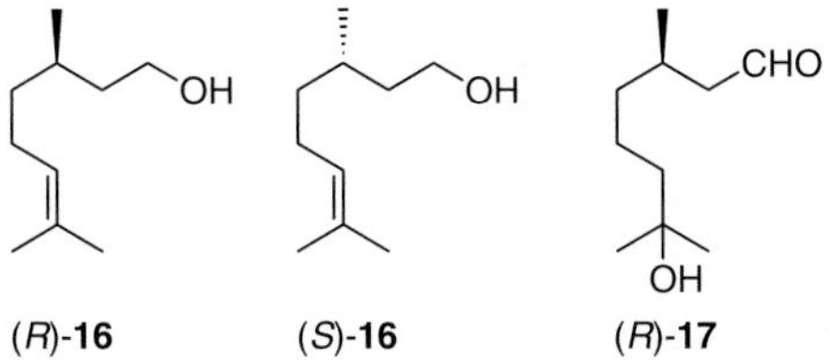

Fig. 2

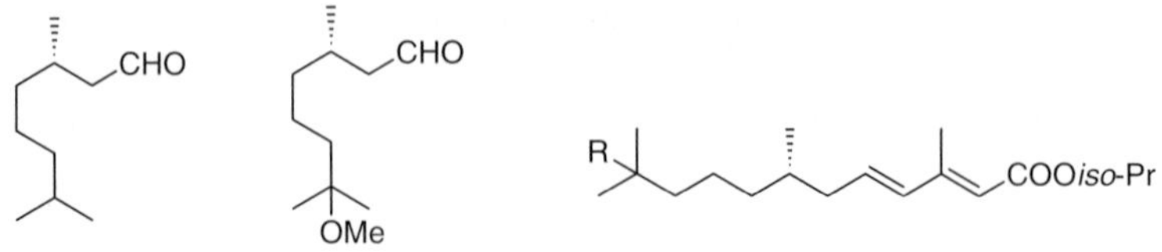

Fig. 3

of **16** on a 200 ton scale, produced by the copper-chromite catalyzed hydrogenation of **14**. A lily of the valley fragrance, 7-hydroxycitronellal **17**, where the (R)-form is less skin irritant compared to its (S)-enantiomer, is produced for the perfumery industry in 300 ton amounts.

Besides aroma chemicals, we are supplying key intermediates for the synthetic insect growth regulators, (S)-3,7-dimethyl-1-octanal **18a** and (S)-7-methoxy-citronellal **18b**, on 100 ton scales each. Hydropren **19a** is effective for mosquitoes and Methopren **19b** is used for controlling cockroach, where only (S)-forms are active (Fig. 3).

4
Conclusion

As a pioneer, Takasago started the synthesis of menthol in the early 1960's. Originally, our raw material was citronellal obtained from Indonesian and Taiwanese citronella oil. The enantiomeric purity of natural citronellal was only 82%. On a glance at the formula of N,N-diethylnerylamine in the literature, an idea of asymmetric isomerization spontaneously came to me that was driven by the major need for enantiopure citronellal. Besides, the first synthesis of BINAP by the late Prof. Takaya was quite timely to realize this idea. We believe that our menthol process has become a milestone in the progress of homogeneous asymmetric catalysis. The rhodium-BINAP catalysts, though using very extensive components, have become one of the cheapest catalysts in the chemical industry by extensive process development. During the period 1983 to 1996, we have produced 22,300 tons of menthol, for which the consumption of Tol-BINAP was only 125 kg. Thus one part of the chiral ligand has multiplied its chirality to 180,000 parts of the product. This value has enabled the precious metal catalyst to become an economically feasible reagent.

References

1. Akutagawa S (1992) A practical synthesis of (–)-menthol with the Rh-BINAP catalyst. In: Collins AN, Sheldrake GN, Crosby J (eds) Chirality in industry. John Wiley & Sons, London, p 313
2. Akutagawa S (1992) Practical asymmetric syntheses of (–)-menthol and related terpenoids. In: Noyori R (ed) Organic synthesis in Japan past, present, and future. Tokyo Kagaku Dojin, Tokyo, p 75

3. Akutagawa S (1995) Applied Catalysis A; General 128:171
4. Fujita T, Suga K, Watanabe S (1973) Chem Ind (London) 231
5. Takabe K, Katagiri T, Tanaka J (1972) Tetrahedron Lett 4009
6. Takabe K, Katagiri T, Tanaka J, Fujita T, Watanabe S, Suga K (1989) Addition of di-alkylamines to myrcene: *N,N*-diethylgeranylamine. In: Bruce E, Smart BE (eds) Organic Synthesis vol 67. John Wiley & Sons, New York, p 44
7. Cai D, Payack JF, Bender DR, Hughes DL, Verhoeven TR, Reider PJ (1994) J Org Chem 59:7180
8. Takaya H, Mashima K, Koyano K, Yagai M, Kumobayashi H, Taketomi T, Akutagawa S, Noyori R (1986) J Org Chem 51:629
9. Takaya H, Akutagawa S, Noyori R (1989) (*R*)-(+)- and (*S*)-(−)-2,2'-bis(diphenylphosphi-no)-1,1'-binaphthyl (BINAP). In: Bruce E, Smart BE (eds) Organic Synthesis vol 67. John Wiley & Sons, New York, p 20
10. Tani K, Yamagata T, Tatsuno T, Tomita K, Akutagawa S, Kumobayashi H, Otsuka S (1985) Angew Chem Int Ed Engl 24:217
11. Tani K, Yamagata T, Akutagawa S, Kumobayashi H, Taketomi T, Takaya H, Miyashita A, Noyori R, Otsuka S (1984) J Am Chem Soc 106:5208
12. Tani K, Yamagata T, Otsuka S, Kumobayashi H, Akutagawa S (1989) (*R*)-(−)-*N,N*-Diethyl-(*E*)-citronellal enamine and (*R*)-(+)-ctronellal via isomerization of *N,N*-diethylgeran-ylamine or *N,N*-diethylnerylamine. In: Bruce E, Smart BE (eds) Organic Synthesis vol 67. John Wiley & Sons, New York, p 33

Chapter 42   Future Perspectives in Asymmetric Catalyis

Chapter 42
Future Perspectives in Asymmetric Catalysis

Eric N. Jacobsen

Department of Chemistry and Chemical Biology, Harvard University, Cambridge, MA 02138, USA
e-mail: jacobsen@chemistry.harvard.edu

1
Introduction

Taken together, the chapters in this collection highlight the remarkable progress made in asymmetric catalysis since the inception of the field three decades ago. Indeed, it is difficult to think of a transformation that involves the reaction of achiral starting materials to give chiral products that has *not* been subjected to asymmetric catalysis with some degree of success. And the progress continues at an accelerating pace: during the period that this collection was being written and edited (1998-Spring 1999), significant breakthroughs were made in many topics that were too poorly developed to merit coverage at the beginning of the project. For example, several enantioselective catalytic systems were developed for the venerable Strecker reaction [1–5] – the addition of hydrogen cyanide to imines – whereas only one example had been described prior to 1998 [6]. Important progress was also made during 1998–9 in conjugate addition catalysis [7], and significant breakthroughs were achieved with asymmetric catalysis of the brand-new ring-closing metathesis reaction [8]. As such, this collection is clearly just a snapshot of a rapidly growing field. It is due to this fact that the publishers at Springer have committed themselves to keeping *Comprehensive Asymmetric Catalysis* current by regular updates in a CD-ROM format, and we hope that this will make this collection most useful.

Given this high level of ongoing activity in the field, it is interesting to consider what new advances are likely during the coming years. This exercise, of course, requires a notion of what it is that remains to be done. If one defines the ultimate goal of the field to be the successful development of general and practical asymmetric catalysts for every synthetically interesting transformation, then much future effort will need to be directed not only toward new reaction discoveries, but also to process research and development. From a more fundamental perspective, one could add the goal of the elucidation of the mechanism of action of every one of those catalyst systems, including stereochemical models with complete predictive values. An even more ambitious goal would be the attainment of a level of mechanistic understanding that would allow rational design of new chiral catalyst systems. Regardless of which of those is set as the ideal, it is quite clear that the field is still in its relative infancy, and a great deal of work remains ahead.

2
New Reaction Design

Only partial solutions have been provided thus far to many of the most important transformations amenable to asymmetric catalysis. For example, no generally effective methods exist yet for enantioselective epoxidation or aziridination of terminal olefins, or for hydroxylation of C-H bonds of any type. Despite the enormous advances in asymmetric hydrogenation catalysis, highly enantioselective reduction of dialkyl ketones remains elusive [9]. And as far as asymmetric C-C bond-forming reactions are concerned, the list of successful systems is certainly shorter than the list of reactions waiting to be developed.

The methods by which new asymmetric catalytic reactions will be discovered in the future will most likely be as interesting as the reactions themselves. Certainly, we will continue to see the tried-and-true method of taking a known metal-catalyzed reaction and rendering it asymmetric by incorporating either known or new chiral ligands and optimizing enantioselectivity through a combination of design, intuition, persistence, and good fortune. However, the recognition that an effective asymmetric catalyst relies on the successful combination of a large number of interrelated variables – not all of which are necessarily well-understood – renders this exercise largely empirical. As a result, the possibility of using high-throughput screening methods in asymmetric catalysis research has been widely recognized to be extremely desirable. The challenge of identifying enantioselective catalysts from mixtures of possible catalysts is significant to say the least, and many obstacles must be overcome before the full potential of combinatorial chemistry can be realized. Nonetheless, at the time of writing in 1999, nearly all of the leading research laboratories in the field make use of GC and HPLC autosamplers for screening asymmetric reactions, and it is certain that the level of automation will only increase. Perhaps of even greater significance, we are already seeing the first successes in the application of combinatorial strategies to the discovery and/or optimization of new chiral catalysts [1, 10–11].

3
Development of Practical Catalyst Systems

The very first genuine success in the field of asymmetric catalysis, the hydrogenation of dehydroamino acids developed by Knowles at Monsanto [12], helped set an awesome standard for future work. The Knowles reaction became a commercial process for the synthesis of an important pharmaceutical, and it found continuous use in that context for decades. The Monsanto L-dopa process made it clear that the ultimate test of practicality – commercialization – was an attainable standard for asymmetric catalysis, and this has colored subsequent research in the field ever since. In that light, it has become clear that it takes more than just high yields and ee's in a catalytic reaction to attain practicality, and some of the very best process research in the world has been done in the context of asymmetric catalysis. Especially notable examples can be found in the Takasago menthol process (Chapter 23), the CIBA-Geigy (Novartis) imine hydrogenation [13], and the Sharpless epoxidation (Chapter 18.1). In the first two cases, extremely precious metals and relatively complex synthetic ligands are required for the synthesis of high-volume, low-margin chiral products. In the case of the Sharpless epoxidation, an inherently unstable catalyst system is employed that is sensitive to moisture, concentration, temperature, and aging. The fact that such obstacles have been overcome and that these processes have each been used for manufacture on a multi-ton scale is a testament to how concerted effort in process research can be rewarded with dramatic success, and it certainly bodes well for the future development of commercial processes using asymmetric catalysis.

Yet, as noted in the introduction, the number of truly useful enantioselective catalysts is still limited, and there are only a handful of systems that have been used in a commercial context. With notable exceptions, even the number of asymmetric catalytic reactions that have seen application in academic target-oriented synthesis research is relatively small. There is no doubt that a major emphasis in future research will be placed on rendering known reactions more practical.

This is of course more easily said than done, and the factors that determine practicality vary greatly according to the system at hand. One of the most obvious is the issue of catalyst turnover number. Loadings of 5 mol % of a non-recyclable catalyst are acceptable in the commercial process for the Sharpless epoxidation, whereas millions of turnovers are needed to render the CIBA-Geigy imine hydrogenation system viable. The difference, in this case, is that the Sharpless reaction employs inexpensive titanium with tartrate ester ligands, while the CIBA-Geigy Josiphos catalyst uses precious iridium and an expensive phosphine ligand. Other issues are less straightforward yet. For example, the need for high dilution or low reaction temperature in a catalytic process can render scale-up extremely difficult or expensive. The seemingly simple problem of catalyst removal from the product can prove critical, especially in the case of toxic heavy-metal-catalyzed reactions. Given that many of these issues fall outside the scope of what is generally considered "academic" research, there is no doubt that at least part of the responsibility for advancing the field will continue to be assumed by the industrial sector. It is certainly noteworthy in that context that several companies (e.g. ChiRex, Chirotech, Catalytica) are committing increasing resources to commercialization of asymmetric catalytic technologies discovered in academic laboratories.

4
Mechanism

It is easy to forget that the first publication on the topic of asymmetric catalysis described a mechanistic study [14]. From the very origins of the field, it was recognized that enantioselectivity can provide useful insights into a catalytic process that are otherwise difficult to attain. Ultimately, the degree and sense of asymmetric induction in a reaction can help shed light on the most difficult and important mechanistic questions, including the precise geometry of the selectivity-determining transition state.

However, the now-classic work on the mechanism of the Rh-catalyzed asymmetric hydrogenation of dehydroamino acids ended up serving a dual role [15]. On one hand, it helped establish that some of the very best work in mechanistic chemistry could be done in the context of asymmetric catalysis. On the other, it provided a stark example of the Curtin-Hammett Principle at work, and thereby highlighted the dangers associated with trying to devise stereochemical models for even the most selective reactions. While "Halpern's Law" (various versions exist, but all follow along the lines of "if you can detect an intermediate, it probably is not an intermediate!") may be a somewhat cynical view of the situation, it is a simple fact that even a very highly enantioselective reaction requires only a small (<5 kcal/mol) difference in activation barrier energies leading to the enantiomeric products. As a result, the construction of stereochemical models is typically more art than science, and very few reactions are understood at the level of detail that allows complete predictability, much less rational design.

Will research in asymmetric catalysis always be an empirical endeavor, such that it will never be possible to carry out completely rational design of new systems? It is likely that one might get as many different answers to this question as there are researchers in the field. However, no one would argue that research activity in asymmetric catalysis will continue to grow, and the collection of more and more information about selectivity in catalysis will certainly help guide future work and facilitate the discovery of more effective systems.

References

1. Sigman MS, Jacobsen EN (1998) J Am Chem Soc 120:4901
2. Sigman MS, Jacobsen EN (1998) J Am Chem Soc 120:5315
3. Ishitani H, Komiyama S, Kobayashi S (1998) Angew Chem Int Ed Engl 37:3186
4. Krueger CA, Kuntz KW, Dzierba CD, Wirschun WG, Gleason JD, Snapper ML, Hoveyda AH (1999) J Am Chem Soc 121:4284
5. Corey EJ, Grogan MJ (1999) Org Lett (in press)
6. Iyer MS, Gigstad KM, Namdev ND, Lipton M (1996) J Am Chem Soc 118:4910

7. Takaya Y, Ogasawara M, Hayashi T, Sakai M, Miyaura N (1998) J Am Chem Soc 120:5579
8. La DS, Alexander JB, Cefalo DR, Graf DD, Hoveyda AH, Schrock RR (1998) J Am Chem Soc 120:9720
9. Jiang Q, Jiang Y, Xiao D, Cao P, Zhang X (1998) Angew Chem Int Ed Engl 37:1100
10. Cole BM, Shimizu KD, Krueger CA, Harrity JPA, Snapper ML, Hoveyda AH (1996) Angew Chem Int Ed Engl 35:1668
11. Francis MB, Jacobsen EN (1999) Angew Chem Int Ed Engl 38:937
12. Knowles WS, Sabacky MJ (1968) Chem Comm 1445
13. Spindler F, Blaser H-U (1997) Chimia 51:297
14. Nozaki H, Moriuti S, Takaya H, Noyori R (1966) Tetrahedron Lett 5239
15. Halpern J (1982) Acc Chem Res 15:332

Subject Index

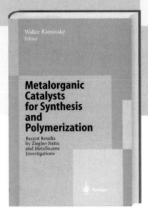

Walter Kaminsky
Editor

**Metalorganic
Catalysts
for Synthesis
and
Polymerization**

Recent Results
by Ziegler-Natta
and Metallocene
Investigations

Springer

W. Kaminsky (Ed.)

Metalorganic Catalysts for Synthesis and Polymerization

Recent Results by Ziegler-Natta and Metallocene Investigations

1999. XIII, 674 pp. 302 figs.
Hardcover DM 298*
£ 114.50 / FF 1123 / Lit. 329.120
ISBN 3-540-65813-0

45 years after the discovery of transition metals and organometallics as cocatalysts for the polymerization of olefins and for organic synthesis, these compounds have not lost their fascination. The birthday of Karl Ziegler, the great pioneer in this metalorganic catalysis, is now 100 years ago. Polyolefins and polydienes produced by Ziegler-Natta catalysis are the most important plastics and elastomers. New impulses for the polymerization of olefins have been brought about by highly active metallocenes and other single site catalysts. Just by changing the ligands of the organometallic compounds, the structure of the polymers produced can be tailored in a wide manner. In invited lectures and posters, relevant aspects of the metalorganic catalysts for synthesis and polymerization are discussed in this book. This includes mechanism and kinetics, stereochemistry, material properties, and industrial applications.

**Please order from
Springer-Verlag
P.O. Box 14 02 01
D-14302 Berlin, Germany
Fax: +49 30 827 87 301
e-mail: orders@springer.de
or through your bookseller**

* This price applies in Germany/Austria/Switzerland and is a recommended retail price. Prices and other details are subject to change without notice. In EU countries the local VAT is effective. d&p · 64336/1 SF · Gha

Springer

Topics
in Organometallic
Chemistry

Volume 1

A. Fuerstner (Ed.)

Alkene Metathesis
in Organic Synthesis

With contributions by numerous experts

The tremendous progress in olefin meta-thesis achieved during the last decade can hardly be overestimated. Due to the devel-opment of a new generation of well defined and high-performance organometallic cata-lysts, this reaction is rapidly evolving into one of the most versatile tools for advanced organic chemistry, natural product synthe-sis, fine chemical production and polymer sciences. Written by some of the leading experts in this field, this monograph intends to familiarize the reader with the most exciting developments and frontiers in this flourishing field of chemistry research.

1999. XI, 231 pp.
Hardcover DM 169*
£ 65 / FF 637 / Lit. 186.640
ISBN 3-540-64254-4

Volume 2

S. Kobayashi (Ed.)

Lanthanides: Chemistry
and Use in Organic Synthesis

With contributions by numerous experts

The use of lanthanides in organic chemistry and in organic synthesis has attracted broad interest recently because of the unique reactivities and selectivities exhibit-ed by compounds of these rare earth ele-ments. In particular, several major advances have been made in the last five years. This volume of *Topics in Organometallic Chemistry* is designed to provide the spe-cialist and non-specialist alike with a much needed overview of recent developments. Contributions by leaders in the field focus on stereoselective organic synthesis using lanthanides, the principles of lanthanide chemistry, lanthanide-based chiral cata-lysts, low-valent lanthanide compounds, polymer synthesis using lanthanide cata-lysts, and polymer-supported lanthanide catalysts used in organic synthesis.

1999. IX, 307 pp.
Hardcover DM 179*
£ 69 / FF 675 / Lit. 197.690
ISBN 3-540-64526-8

**Please order from
Springer-Verlag
P.O. Box 14 02 01
D-14302 Berlin, Germany
Fax: +49 30 827 87 301
e-mail: orders@springer.de
or through your bookseller**

* This price applies in Germany/Austria/Switzerland and is a recommended
  retail price. Prices and other details are subject to change without notice.
  In EU countries the local VAT is effective. d&p · 64336/2 SF · Gha

Springer

Topics in Organometallic Chemistry

Volume 3
S. Murai (Ed.)

Activation of Unreactive Bonds and Organic Synthesis

With contributions by numerous experts

In this volume, areas ranging from the activation of C-H bonds to the chemical transformation of dinitrogen are authoritatively discussed by leading experts in the field. The cleavage and formation of chemical bonds is fundamental to organic synthesis; these new activation methodologies make hitherto infeasible reactions extremely easy and create new opportunities for innovative organic transformations, for both industry and academia. This is the first book that provides a thorough and timely coverage of both inorganic and organic synthetic aspects of bond activation.

1999. IX, 272 pp.
Hardcover DM 198*
£ 76 / FF 746 / Lit. 218.680
ISBN 3-540-64862-3

Volume 4
J.M. Brown, P. Hofmann (Eds.)

Organometallic Bonding and Reactivity

Fundamental Studies

Written by experts and pioneers in the field, the volume addresses state-of-the-art theoretical and experimental methodologies applicable to fundamental problems of structure and reactivity of organometallic compounds. The principles of ab initio and density functional theory, as well as integrated force field/quantum chemistry approaches are outlined, with particular emphasis on their applicability to transition metal organometallic molecules and their reactions. Specific case studies, spanning a range from static structural aspects to molecular structure dynamics, reaction mechanisms and catalytic cycles illustrate the power of modern quantum chemistry for organometallics. Experimental properties of organometallic systems, derived from gas phase organometallic chemistry as well as solid state structural chemistry provide deep and complementary insights into the fundamentals of the chemistry of the metal-carbon bond.

1999. Approx. 220 pp. 42 figs., 18 tabs.
Hardcover DM 198*
£ 76 / FF 746 / Lit. 218.680
ISBN 3-540-64253-6

Please order from:
Springer-Verlag
P.O. Box 14 02 01
D-14302 Berlin, Germany
Fax: +49 30 827 87 301
e-mail: orders@springer.de
or through your bookseller

* This price applies in Germany/Austria/Switzerland and is a recommended retail price. Prices and other details are subject to change without notice. In EU countries the local VAT is effective. d&p · 64336/3 SF · Gha

Springer

Printing (computer to plate): Mercedes-Druck, Berlin
Binding: Buchbinderei Lüderitz & Bauer, Berlin

PRINTS

ART AND TECHNIQUES

SUSAN LAMBERT

V&A Publications

First published by V&A Publications, 2001

V&A Publications
160 Brompton Road
London SW3 1HW

Susan Lambert asserts her moral right
to be identified as the author of this book

Designed by Yvonne Dedman
V&A photography by Graham Brandon

ISBN 1 85177 288 X

A catalogue record for this book is available from the British Library

Printed in Italy

Every effort has been made to seek permission to reproduce those
images whose copyright does not reside with the V&A, and we are
grateful to the individuals and institutions that have assisted in this
task. Any omissions are entirely unintentional and details should be
addressed to the publishers.

Jacket front: **Hendrik Goltzius, *The Farnese Hercules***
and detail (see figs 48 and 49)

Jacket back: **Jim Dine, *Eleven Part Self Portrait (Red Pony)***
(see fig. 92)

Title page: **Marcel Broodthaers, *La Soupe de Daguerre*,
1975. Colour photographs and screenprint. 46.5 × 61.3.**
Daguerre invented an early form of photography (see page
66). This print is thus an elaborate joke at the profusion
of confusing printing techniques. The luscious ingredients
of vegetable soup are presented as photographs stuck on
to the backing card, whereas the mock-label is a fine piece
of *tromp l'oeil* screenprinting.
V&A: Ph.253–1980; © DACS 2001.

V&A Publications
160 Brompton Road
London SW3 1HW
www.vam.ac.uk

Contents

Preface

This book aims to explain how artists' prints are made and to give a brief history of the development of the different processes – to add to the appreciation of their qualities rather than to provide a how-to-do-it guide. It also introduces some of the issues surrounding the sales and consumption of prints, and should acquaint the reader with the terminology used to discuss them. Much practical and scientific information essential to the successful production of prints can be found in books listed in the 'Technique' section of the Select bibliography.

The text is the fruit of many people's work and has its origins in a display of printmaking techniques in the V&A, first set up in 1903 by students at the Royal College of Art (RCA). Here many of the teaching examples still on show were produced. The display was prepared under the direction of Sir Frank Short, Professor of Printmaking at the RCA, who also produced a descriptive manual. This initial display dealt solely with intaglio techniques, an indication of the special respect in which they were held. According to reprints of the manual this was still the case in 1910 and 1914, but by the fourth edition published in 1925, the display had been expanded to include a section on watercolour prints from relief-etched plates supplied by William Giles, and another on offset lithography based on a gift from Harold Curwen. Since then the display has been expanded to include other forms of relief-printing and lithography and, in the 1970s, to encompass the stencil processes (in particular, screen printing). The gallery has undergone many changes since 1903, including a change of location, but the approach to the subject has not changed radically since its inception.

First published in 1983, this book draws heavily on texts by Sir Frank Short; A.W. Ruffy of the V&A, who drafted a revision of the original manual during the 1950s; and all the other members of the Prints, Drawings and Paintings department (PDP) who have over the years produced labels for the display. For many years the PDP department was adjacent to the RCA's printmaking studios. Frequent trips were made through the dividing iron door, and much essential information was gleaned from RCA staff and students alike.

For this revised and expanded edition all my current PDP colleagues have contributed in one way or another, and I thank them warmly. Mark Haworth-Booth, Rosemary Miles, Elizabeth Miller, Charles Newton and Margaret Timmers, in particular, have provided expert help for which I am most grateful and Nazek Ghaddar, Lois Oliver and Frances Rankine have helped in many practical ways. From other departments in the Museum I should like to thank Peter Ford for sorting out my word-processing problems, Graham Brandon for his exquisite photography, Anthony Burton for editing the original text, Miranda Harrison for editing this revised version with such dedication and good humour, and Mary Butler for her patience and understanding. From beyond the V&A, I should like to thank Yvonne Dedman for her creative response to the design of the book.

Susan Lambert
January 2001

Connoisseurs at a Print Sale

George Cruikshank (1792–1878), illustration to
John Wilson's *Catalogue Raisonné of Engravings by
an Amateur*, 1828. Etching. Shown at actual size.

V&A: 9543.1
Given by Mrs George Cruikshank,
widow of the artist.

Notes on the text

Terms in **bold** in the main text are explained in 'Technical terms and abbreviations' on page 92.

Measurements are in centimetres.

Unless stated otherwise, the measurements refer to the printed surface of the relief, stencil, planographic and photographic prints and to the size of the plate-mark of intaglio prints.

The paper on which the prints are printed is referred to only when it is of special interest to the quality of the impression.

All the prints illustrated are in the permanent collections of the V&A except for figs 57 and 58.

Introduction | The parameters of printmaking

P rintmaking is the process of producing images by a variety of printing techniques using pigment, usually on paper. What distinguishes printmaking from painting and drawing is that the image is made in such a way so that it can be produced in multiple copies. What distinguishes it from photography is that the image is made by the delivery of pigment to a surface rather than as the result of a chemical reaction on a light-sensitive surface.

Techniques of printmaking fall into one of four categories:

- **relief** processes, in which the ink is applied to raised surfaces
- **intaglio** processes, in which the ink is held in grooves and hollows
- **planographic** processes, in which the inked surface is level with the un-inked areas, and which are based on the fact that grease and water repel one another
- **stencil** and **mould** processes, in which the ink is applied *through* a barrier or contained within a substance rather than *from* a surface.

When the image is printed from an inked surface as in the case of the **relief**, **intaglio** and **planographic** processes (rather than the ink being applied through a controlling **screen** as in the case of **stencil** processes), it appears on the paper like a mirror image in reverse of how it appears on the printing surface (figs 1 and 2).

This book examines each category of printmaking process in turn, concentrating on its more traditional forms. In addition it explains how these manual processes have been transformed – in the light of the possibilities offered by the invention of photography and ever more sophisticated technology – into the **photomechanical** techniques that underpin the dissemination of popular visual culture today. **Photographic** techniques that depend solely on the chemical action of light, such as those used in the family snap, are omitted.

All printmaking processes are capable of creating images in colour. How this is done will be explained technique by technique. In the past, however, the colour printing process was often difficult and labour-intensive, and frequently prints were coloured individually by hand as a more straightforward and less time-consuming option. It is not easy to establish whether or not colouring on a print is contemporary, although certain pigments change in characteristic ways over time (fig. 3).

1 and 2. Willem Panneels (*c*.1600–after 1632) after Rubens, *Salome with the Head of John the Baptist*, 1631. Etching (left) **with counter-proof** (right). **18.5** x **12.8.**

Printmakers sometimes took impressions from prints while they worked on them to provide a printed image the same way round as it appeared on the plate. The process was similar to that of monotype (see page 81) except that sheets of paper tended to be passed together through an intaglio printing press.

Before the invention of photography, printmaking with blocks, plates and stones worked by hand was the only means available of making exactly repeatable visual images on paper. This meant that prints were called upon to fulfil a wide range of educational, religious and propagandist functions and were used for utilitarian as much as aesthetic ends. After **photomechanical** methods of reproduction became available, printmaking continued to develop as an independent art form, the different techniques being used for the particular qualities they gave to an image.

In spite of (or perhaps because of) the ubiquitous character of prints, artists' prints were traditionally kept in portfolios and perceived as an art form for the specially sensitive. In 1921 Martin Hardie, etcher and heir apparent to the headship of the V&A's PDP department, hoped in his inaugural speech as president of the Print Collectors Club that the club would serve as a 'loophole of retreat from the insistent clamour of daily life'. The phrase 'come up and see my etchings' with all its inuendos continues to imply a rarefied world, away from intrusion. It was only in the middle of

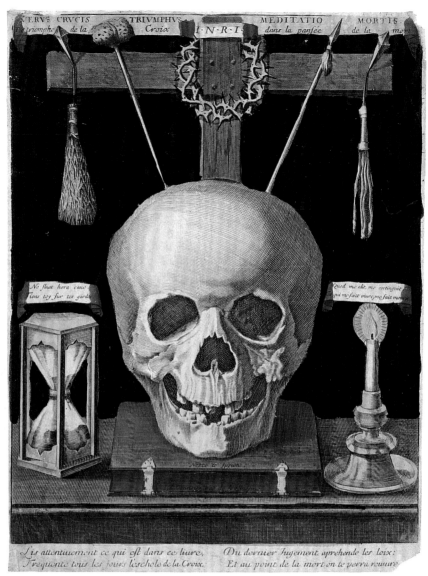

TRIVMPHVS CRVCIS   TRIVMPHVS   MEDITATIO   MORTIS.
triomphe   de la   Croix   I·N·R·I   dans la panſee   de la   mort

Ne ſlaue hora caue
Tons toy ſur tes garde

Quod me alie, me extinguit
qui me face viuegmaga faie me

Nosce te ipsum.

Sis attentiuement ce qui eſt dans ce liure,   Du dernier Jugement aprehende les loix:
Frequente tous les jours l'eschole de la Croix.   Et au point de la mort on te verra reuiure

V&A: E.31–1987

3. Anonymous, French, *Triumph of the Cross at the Thought of Death*, late-17th century. Engraving, coloured by hand. 41.5 × 32.

The bold emphatic range of pigments used in colouring this print, and the way pigments have reacted in individual cases with the paper or changed appearance on exposure to the atmosphere (for example the apparent oxidisation of the brush line along the book binding), all argue strongly in favour of the colouring being contemporary with the print. Later colouring is often more tentative and wishy-washy.

the 1960s, when artists of the highest calibre made a point of using commercial techniques eschewed by their predecessors, that artists' prints began to be seen universally as a more affordable public and demotic art form fit to grace the walls of domestic and business spaces. In some cases the results vie in scale and impact with painting (fig. 4).

The choice of technique was and is dependent on many factors, not least the practical resources available. Market expectations also play an important part, and the development over so many centuries of a visual language to target certain market sectors remains one of the industry's most potent forces. There are conventions for the portrayal of certain kinds of subject matter (see fig. 54), and methods of manipulating the technology to appeal to a particular public. The work of Hogarth stands out in this context. His choice of line instead of a tonal process was in itself a statement about the seriousness of his art. The character of his line, however, achieved variously in woodcut (fig. 5), etching (fig. 6) and engraving (fig. 7) is different according to the sector of society at which the image was aimed.

Artists have made what have come to be called 'original' prints almost since the invention of printmaking, but traditionally printmakers did not have the same status as painters and sculptors. Their relatively low standing was highlighted in England by the attitude of the Royal Academy, which on its foundation in 1768 admitted no engravers, although Bartolozzi, who made his greatest contribution through his prints (fig. 13), was accepted as a painter.

The development of **photomechanical** processes in the nineteenth century gave artist-printmakers an opportunity to distinguish not only the subject matter of their work from the commercial end of the trade but the technique also. Societies founded to infuse 'creative' printmaking with vitality stressed

4. Robert Rauschenberg (b.1925), *Pull* from a series of nine prints entitled *Hoarfrost Editions*, 1974. Screenprint on cheese cloth and silk taffeta with paper bag collage. 226 x 122.
Over two metres high and executed on cloth, this image resembles a painting in terms of its scale and support. It was however screenprinted and published in an edition of twenty-nine.

V&A: E.551–1975; © Robert Rauschenberg/ DACS, London/VAGA, New York 2001.

5. John Bell (worked mid-18th century) after Hogarth,
Cruelty in Perfection, **1750. Woodcut. 45.2 × 37.6.**
This print had a reforming purpose. Hogarth said that
it required 'neither great correctness of drawing or fine
engraving', as these would have 'set the price of them
out of the reach of those for whome [*sic*] they were
chiefly intended'. The coarseness of the etched line
gives it almost the appearance of woodcut. Although
Hogarth had the skill and technical resources to

produce this image himself (see fig. 6), he chose to
have it cut on wood, the medium of the broadside with
which the working classes were most acquainted.

6. (opposite above) **William Hogarth (1697–1764),**
Cruelty in Perfection. **Plate from *The Four Stages of
Cruelty*, 1751. Etching. 38.8 × 32.1.**
Hogarth also issued this subject etched by himself.
The lines are finer than those in the woodcut but

(continued opposite)

V&A: Dyce 2778

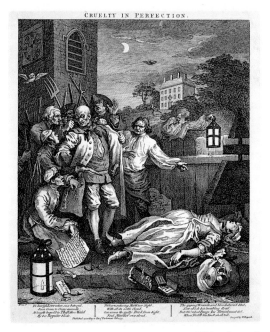

CRUELTY IN PERFECTION.

To lawful Love when once betray'd / Soon turns to Crime unmade / At length beguil'd to Theft the Maid / By her Reputer Made.

Yet learn seducing Hell'our Right / With all thy wiles Cloud / Can seem the guilty Deed Done Right / Foul Murther now stood.

The gaping Wounds and blood-stain'd Steel / Now shock his trembling Soul / But th'unfeeling Pangs her Punishment tel / When Death his Knell shall toll. Design'd by W Hogarth.

Published according to Act of Parliament Oct 1 1751.

V&A: E.603–1985

considerably coarser than those in the print reproduced below (fig. 7). The etched version was printed in two grades of paper and sold for one shilling the coarser, and sixpence extra the finer, in order to make it accessible to as large an audience as possible.

7. (below) **Bernard Baron (1696–1762/6) after Hogarth, *Marriage à la Mode*, plate 2, 1745. Etching and engraving. Cut to 38.1 x 46.3.** This set was aimed at an aristocratic audience. Thus instead of working the plates with his own functional lines, Hogarth had them, according to an advertisement, 'engrav'd by the best Masters in Paris', taking particular care 'that there may not be the objection to the decency or elegancy of the whole work'.

Invented Painted & Published by W.m Hogarth. Marriage -A-la- Mode, (Plate II) Engraved by B. Baron. According to Act of Parliament April 1 1745.

V&A: F.118(21)

how the prints they promoted differed from images that used the new technology. For example, the introduction to the first portfolio issued by the influential *Société des Aquafortistes* (Society of Etchers) founded in Paris in 1861 appealed to its public thus: 'In these times, when photography fascinates the vulgar by the mechanical fidelity of its reproductions, it is necessary to assert an artistic tendency in favour of free fancy and picturesque mood', and assured subscribers that each impression was an original work of art. Increasingly the print establishment inclined to the view that the use of a manual technique (however original or otherwise the image) was enough to give a print the status of an 'original'.

Formal definitions of the 'original' print were not promulgated until the 1960s. The ostensible reason for the flurry of definitions then was to help tax inspectors and customs and trading standards officers to distinguish original work from reproductions, so that the former could receive privileged treatment and to protect buyers from reproductive work masquerading as original. The regulations, however, attached considerable importance to a particular kind of manual involvement in the making of the printing surface, at just the moment when artists at the leading edge were incorporating commercial techniques into fine art practice (fig. 8). In retrospect, it is as if the definitions were a counter-movement by the old guard against a new generation of artists finding a powerful and personal means of expression in the very techniques of popular visual communication that had been shunned by artists since their invention.

8. Richard Hamilton (b.1922), *The Critic Laughs,* **1968. Offset lithograph, laminated, and retouched with enamel paint, and screenprint with collage. 59.5 × 46.5.**

This print is a dig at the obsession with originality and technique that was raging at the time of its production. It was made from a photograph, reproduced by photo-offset-lithography, laminated with plastic film to heighten its photographic impact. The margins were screenprinted with a matt white which simulated the effect of a photograph mounted on a matt board. Thick paint was applied to the photographic print prior to reproduction, and more was added to each print after lamination as well as a little piece of collage. The print is therefore a dialogue between photomechanical and handmade marks.

9. Dora Maurer (b.1937),
Drypoint on Aluminium
Exhausted by Printing
in an Edition of 39,
1978. Drypoints printed
onto cards, pasted on to
a linen sheet. Size of
sheet: 98.3 x 70.2.
This work was made by
printing a small plate
worked in drypoint, a
fragile medium made
more fragile by the use
of an aluminium plate,
a relatively soft metal,
to exhaustion. This does
not conform to the
criterion of a print being
an exactly repeatable
image, as the entire
edition that was
possible from the plate
was used to produce a
single work of art.

V&A: E.1377–1979

The definitions also concentrated on the size of the **edition**. The extent to which large numbers of prints can be successfully printed depends on the fragility of the printing surface. The very real fragility of the drypoint (fig. 9) or mezzotint, for example, provided a pretext for the sale of **proof** impressions (fig. 10) but the exclusivity of the limited edition has been largely used to fuel the demand that rarity creates. Even additional **states** of particular prints – prints printed from a surface to which changes have been made – have been printed as a marketing ploy, and a fetish has been made out of the variety of effects achieved by inking the surface differently and using papers of different absorbency, thereby denying the basic multiple nature of the print.

V&A: E.273–1905

10. James McArdell (1729–65) after Reynolds,
Mrs Bonfoy, 1755. Proof before letters. Mezzotint on
India paper. Cut to 37.2 x 27.5.
It is a curious feature of an art form that draws its
strength from its repeatable nature that the
impressions which differ from the main edition have
tended to be most highly prized. Prints of this date
normally had their margins trimmed to the plate-mark.

The fact that the smudged margin of the proof was not
trimmed off along with the other margins is an
indication of the value attached to impressions which
announced their status as proofs. Note the collector's
mark in the bottom right-hand corner.

11. James McNeill Whistler (1834–1903). *The Piazetta.* **From** *Venice, a series of twelve etchings***, 1880. Etching. 25.3 x 17.8.**

Whistler was one of the first artists to put his pencilled signature to a print, a practice that grew up during the second half of the 19th century in order to distinguish works printed from surfaces made by the artist and reproductions of work in other media.

Margins beyond the plate-mark are rarely found on Old Master prints and this has lead to the supposition that they were cut off before publication. Nevertheless, Whistler was unusual in the 19th century in choosing to trim his later prints to the plate-mark, only leaving a tag for his signature.

It was in fact the reproductive trade that began to rationalise print practice. The Printsellers' Association, founded to control the quality of reproductions in 1847, issued in the 1880s *A Few Words on Art which are also Words of Advice and Warning*. It distinguished four categories of impression: **artist's proofs**, **proofs** before letters, lettered proofs and prints. The artist's proofs were defined, not as the term is usually used today (as proofs of 'original' prints outside the main **edition** reserved for the artist), but as a first run of prints from the plate before any letters. The proofs before letters were described as proofs lettered with the names of the painter and the engraver but without the title.

Each **impression** of both sorts of proofs (namely artist's proofs and proofs before letters) from an edition approved by the Association were marked with a unique combination of letters, the former on the lower right and the latter on the lower left corner, an early example of the numbering of prints from a single edition. All the proofs were to be issued in limited editions. Curiously the artist's proofs, although the first to be run off, were often printed in the largest quantity.

In spite of these connections with the reproductive trade, the limited edition has become associated with the original end of the market. Nonetheless many editions of reproductive prints continue to be limited, and many prints original in concept are now printed in photomechanical techniques, associated with the reproduction market and capable of printing unlimited numbers of equally good impressions.

Artists' signatures have also played a key role in definitions of originality. In fact, both reproductive and original prints may be signed in order to convey approval of a particular impression. The presence of a signature of anyone involved in the production of a print inevitably contributes to a sense of it being the product of an individual rather than of a machine. The presence of the artist's signature suggests the artist's immediate involvement. Whistler (fig. 11) charged double for individually signed impressions of his prints, and since then the print trade has capitalised on the artist's signature to increase the value of the product. The cheaper end of the reproductive trade may append a facsimile signature, and even a genuine signature only proves that the artist actually touched the sheet for a moment.

Only in the last quarter of the twentieth century was the freedom extended to printmakers that had been won for painters and sculptors early that century by such statements as Marcel Duchamp's presentation of standard sanitary fittings and department store wine racks as gallery exhibits. Duchamp aimed to 'reduce the aesthetic consideration to the choice of the mind, not to the ability of the hand'.

The 'original' print still has its adherents but the concept is outmoded. As in the past, original prints, reproductions and commercial graphics all have the same technology at their disposal. A print is what is claimed for it – how successful it is in relation to its intention is a matter for the beholder.

Paper

Vellum and silk (figs 12 and 13) are among surfaces other than paper that have been used to make prints, but the existence of relatively cheap paper was essential to the development of utilitarian printmaking. The kind of paper used has a considerable effect on the character of the impression. A knowledge of the development of paper manufacture can thus be a great help in understanding the art of printmaking and in dating individual **impressions.**

Paper was in use in China over 2000 years ago, but it was not manufactured in Europe until the twelfth century. Only in the fifteenth century did it supersede parchment (prepared from animal skins) as the standard material on which to write.

The way in which handmade paper is produced has changed little since its inception. Machine methods follow the hand procedure.

Paper consists of matted fibres. Fibres of the desired consistency are obtained by boiling and beating linen and cotton rags or other vegetable matter to a liquid pulp. It is made in a mould that looks much like a tray with a fine wire grid as its base. The sides of the mould, known as the deckle, are removable. The wire grid is made in two patterns and, depending on which is used, the paper is described as either laid (fig. 14) or wove (fig. 15).

Until the eighteenth century, all European-made paper was laid. Wove paper was introduced in the West in the 1750s, and was much praised for the increased smoothness this kind of mould gave to the surface of the paper. The first book containing wove paper was John Baskerville's *Virgil*, published in 1757, but prints on this kind of paper seldom date from before the last decade of the century.

Paper was made by hand and its dimensions were limited to the size of mould that traditionally one person could handle. The first paper-making machine was invented in 1798. The paper was formed on a continuous woven wire mesh with its width limited only by that of the machine. Laid paper was first manufactured by machine in 1825.

To make a sheet of paper by hand, the mould is dipped into the liquid pulp so that the wire grid is covered with the correct amount for the thickness of paper required. The mould is then shaken to make sure that the pulp is evenly distributed and that the fibres are well enmeshed. After much of the liquid has drained away, the deckle is removed and a piece of felt is pressed to the mould. The pulp forming the sheet of paper sticks to it, making removal from the mould possible. To extract the

12. (right) **Rembrandt (1606–69),**
Self Portrait with Saskia, **1636. Etching
on satin weave silk. 10.5 × 9.4.**
The printing of fine art images on silk
has a documented history going back to
1589. There are eight recorded
impressions of Rembrandt etchings on
silk. Given the quality of this impression
and the fact that there is evidence of
prints on silk being produced by other
artists in the Netherlands at the
beginning of the 17th century, it is
conceivable that it was pulled in the
artist's lifetime. However, it is more likely
that it dates from the 18th century, when
the mania for collecting rare proofs of
Rembrandt etchings was at its height.

V&A: E.282–1994

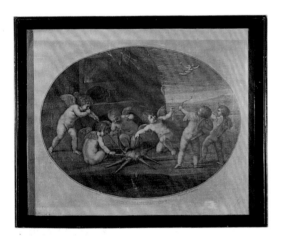

V&A: E.922–2000. Given by Julie and Robert Breckman.

13. (left) **Francesco Bartolozzi (1727–1815)
after Albani, *Cupid's Manufactory*, 1800.
Stipple engraving on silk in contemporary
frame. Printed surface. 21.8 × 28.5.**
There must have been a demand for
Bartolozzi subjects printed on silk, as he is
the printmaker most frequently encountered
in this form. Prints on silk, always aimed at a
niche market, are especially rare today. This
may be because silk has more acid in it than
good rag paper and also because silk prints,
like this, tended to be framed and hung on
the wall, and thus exposed to the damaging
effect of light for longer periods than prints in
connoisseurs' portfolios.

liquid further, a stack of such sheets is pressed, first with and then often without the
felts in place. The sheets are then hung up to dry.

To obtain the required smoothness of surface texture the sheets are individually
pressed, the pressure exerted being dictated by the desired finish. Traditionally, there
are three grades of finish: hot-pressed, not (i.e. *not* hot-pressed) and rough. Within
these grades a wide variety of effects can be obtained.

Hot-pressed paper is smooth. This surface is achieved by subjecting the sheets,
placed between metal plates, to mechanical pressure. Not paper is also pressed in a

mechanical press, but without interleaving of any kind. Its surface bears, therefore, an imprint of the sheets above and below it. Rough paper is not pressed at all, the stack of sheets and felt being left to dry out under their own weight. Its surface has a felt-like roughness.

Usually the paper is sized in order to make it less absorbent. The sheet is dipped into a vat containing a suitable glutinous substance, normally before the final pressing takes place. Unsized paper is known as water-leaf.

Both the finish and the absorbency of the paper have a considerable influence on the character of a print. A good example is India paper (actually made in China). This is a very thin paper – usually attached in printing to a thicker support sheet – which is unsized and therefore very soft and absorbent. As a result it takes up a large quantity of ink without smudging, creating an unusually brilliant impression.

Cotton fibre was the principal ingredient in all paper until the 1840s, when wood pulp was introduced. Paper made from wood pulp has a tendency to rot and to turn brown when exposed to light, as can be seen by the sad state of many nineteenth-century prints. Today this deterioration can, to a large extent, be counteracted by the use of chemicals, but the best quality papers are still manufactured from rags.

14. (left) **'Laid' paper with a watermark held up to the light (actual size).**
The wire network of the mould in which laid paper is made consists of rows of thin wires lying close together known as 'laid' lines, crossed at intervals of 2.5–4 cm by thicker wires, known as 'chain' lines. The watermark is created by lacing a filigree of wire to the network. The pale lines, seen when the paper is held up to the light, are caused by the slight thinning of the paper where the pulp has rested on the wires of the mould.

15. (right) **'Wove' paper held up to the light (actual size).**
As its name suggests, wove paper is made in a mould consisting of wire mesh woven much like a piece of fabric. When the paper is held up to the light, only an irregular mottling is visible.

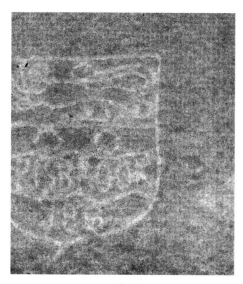

2 | Relief processes

In **relief** processes the printing surface is raised above the areas which are to remain blank. The printing surface can be made of almost any material. Wood is the most common but potatoes, metal, lino, card, perspex and even found objects can be used. The surface is inked with a sticky ink, stiff enough to prevent it from flowing into the hollows. The ink is transferred to the paper by pressure, often in a printing press (fig. 16). The printed image is in reverse of that on the printing surface. The lines of images printed by these processes are impressed a little beneath the surface of the paper and often show a build-up of ink at their edges.

In the past, relief processes were particularly suitable for book illustration. The printing block could be made the same thickness as movable type and locked with it in the printing press. This enabled the text and illustrations to be printed on one sheet in one operation.

Relief prints do not have a **plate-mark**. Most early woodcuts that have survived are cut right to their borders.

Woodcutting

In woodcutting the drawing is made on a smoothed block of relatively soft wood such as pear, sycamore, cherry or beech. It is cut like a plank, lengthwise along the grain. The lines of the drawing are left untouched, while the wood on either side of them is cleared away with a knife. Large areas are removed with chisels and gouges (figs 17–20). It is also possible for the actual image to be cut into the plank. When this is the case, the lines remain un-inked when printed (figs 25 and 30).

An **impression** is taken by laying a sheet of paper over the inked block and either rubbing the back of it with a **burnisher** (any implement with a curved smooth hard surface) or putting the block and paper in a printing press and applying vertical pressure. Very early woodcuts were printed in the same way as textiles and wallpaper, by hammering the inked block face down on a sheet of paper.

Colour woodcuts can be produced by two methods. The first involves a separate block being cut for each colour. Each is then printed one after the other on the same sheet of paper (fig.21). The second uses the same block for each colour – the lighter tones are printed first, and then these areas are cut away before printing the middle tones and so on, until only those areas to be printed in the deepest colours remain in relief on the block. This process is known as the reduce block method.

V&A: E.2485–1934 (reproduced from a Lippmann reproduction).

16. Abraham von Werdt (worked 1640–80),
A Relief Printer's Workshop, **mid-17th century.**
Woodcut. 18 × 27.

This is typical of the kind of press in which relief prints have been, and still are, printed. The man on the right is inking the type, lying on the chase. He might equally well be inking a wood-block, or a wood-block and type locked together. The man beside him removes a sheet of paper which has just been printed from the tympan. Hinged to the outer shield of the tympan is the frisket, the paper shield of which folds over the tympan to protect the margins of the sheet being printed.

To take a print, the tympan would be loaded with a clean sheet of paper and, with the frisket closed, folded over the inked type on the chase. Together they would be pushed beneath the platen which is screwed down on top of them, thus exerting vertical pressure. The paper shield is divided as for a quarto printing, yet the pile of printed sheets in front of the press are printed as for a folio publication. On the left a sheet of paper is being dampened ready for printing, and a pile of dampened sheets are stacked in front of the press. Damping the paper makes it more absorbent, so that more ink is taken up to create a richer effect.

Woodblocks have also been used to provide colour on prints in which the outlines and shading of the image are printed in a different way, for example in the process patented by George Baxter in 1835 (fig. 22).

Woodcutting was the first technique to be used in Europe for printing on paper. It was developed from the practice of printing patterns on textiles from wood-blocks. The earliest woodcuts date from the end of the fourteenth century. It has not been conclusively proved where the technique was first used, but considerably more early woodcuts were made in Germany and the Netherlands than elsewhere. The earliest

17 and 18. Henri Matisse (1869–1954), *Nude in Profile, Large Woodcut*, **1906. Wood-block** (left) **with an impression from it** (below). **49.5 × 40.**

19 and 20. (opposite) **Details of figs 17 and 18 shown at actual size.**
The block consists of two joined planks of pearwood, from which Matisse cut away the wood around the design with knives, chisels and gouges, leaving the stark lines, dashes and dots which form the image standing in relief.

V&A: E.609–1975;
© Succession H. Matisse/
DACS 2001.
Block purchased with assistance from the Lumley Cazalet Gallery in Memory of Frank Perls.

V&A: E.276–1994; © Succession H. Matisse/ DACS 2001.
Print purchased with assistance from the National Art Collections Fund.

© Succession H. Matisse/ DACS 2001. © Succession H. Matisse/ DACS 2001.

surviving examples are religious pictures, but playing cards are also thought to have been mass produced by woodcutting at the same period (fig. 23).

From the middle of the fifteenth century the history of the woodcut becomes closely linked with that of the printed book. One type of book – the block book – was made up of sheets printed entirely from wood-blocks, on which both the text and the illustrations were cut. As the method was very laborious, each letter being cut individually in the wood, only very popular works such as compendiums of the Bible were printed in this way. Few block books were published after about 1500.

Under the influence of Dürer the woodcut also became a medium for fully fledged artistic expression (figs 24 and 25). Many early woodcuts were coloured by hand. The first attempts at printing in colours were made towards the end of the fifteenth century. The earliest form to be used extensively is known as the 'chiaroscuro method'. This imitated the current style of drawing in wash on toned paper, by printing a variety of tones of the same colour from different blocks with

21. Francesco Clemente (b.1952), *Untitled (self-portrait)*, **1984. Colour woodcut on Tosa kozo paper. Size of sheet: 42.8 x 57.4.**

This print was printed from 22 wood blocks carved by a Japanese woodcutter, Reizo Monjyu, to Clemente's instructions. For each of the 200 prints in the edition, the blocks, inked variously with 14 transparent pigments, were printed a total of 49 times, some colours used again and again to achieve the unusual density of the·colours.

white lights cut out of the blocks. Northern chiaroscuro woodcuts tended to be conceived as an outline (printed from one block) to which colours (printed from other blocks) were added. Italian examples tended to dispense with the outline and present the composition purely in areas of tone (fig. 26).

In the fifteenth century woodcuts were occasionally printed on paper with a layer of tin foil pasted to it. They are known as 'paste' prints. Only about 150 examples are now known and they tend to be badly damaged and difficult to read. (There is one in the V&A collection: E.205–1991.) Another variant of the woodcut in the fifteenth century was to 'ink' the block with glue and then dust fabric particles on to the sheet. This practice was developed in the first half of the seventeenth century to produce what is now known as flocked wallpaper.

The heyday of the woodcut was over by the end of the sixteenth century and its place was taken increasingly, even in book illustration, by the more sophisticated techniques of copperplate engraving and etching. Woodcutting however continued

22. (above) **George Baxter (1804–67) after V. Bartholomew,**
Hollyhocks, **1857. Baxter print. 37.5 × 27.2.**
Baxter's patented process combined the relief and intaglio
printing methods. The foundation plate, which printed the
main features of the design was etched (often as in this case
through an aquatint ground) or stipple engraved, and oil
colours were superimposed from wood blocks.

23. (left) **GS (?Giles Savouré) (worked *c.*1480–*c.*1506),**
Jack of Diamonds. **Detail from an uncut sheet of playing
cards, late 15th century. Woodcut, coloured by hand.
The card: 9.5 × 5.5.**
The technology used in the making of these cards was exactly
the same as that used by the cutters of Dürer's blocks.

V&A: E.1134–1900. Dalton Bequest.

24. (above) **Albrecht Dürer (1471–1528),** **The Flight into Egypt** from **The Life of** **the Blessed Virgin, c.1504–5. Woodcut.** **29.8 x 20.9.**
A combination of the quality of Dürer's drawings and the skill of the people who cut them in the wood established woodcutting as a major artistic medium.

25. (left) **Detail of fig. 24 shown one-** **and-a-half up from actual size.**
This involves mainly black line work. The wood has been carefully cut away around the line, leaving it standing proud to take the ink. In places – for example, the leaves – the image is conveyed also by areas where the wood has been removed and which therefore carry no ink.

V&A: E.284–1890

26. (above) **Ugo da Carpi (1450–1520),**
Saturn. **This state published in 1604.**
Chiaroscuro woodcut, printed from
three blocks. 32 × 42.5.
This print was made from a tonal
drawing made especially for reproduction by this method. The three
different shades are printed from
different blocks.

27. (right) **Edvard Munch (1863–1944),**
The Kiss, **1902. Woodcut printed in**
grey and black from two blocks on
Japanese paper. 47 × 47.
This print shows the grain of the plank
from which it was printed giving
texture and linear energy to what
would otherwise be a large area of flat
tone. The grain of the wood-block can
be heightened by scraping the surface
with a blade or rubbing it with a wire
brush. Japanese paper takes up the ink
with particular richness.

V&A: E.5067–1960; © Munch Museum/
Munch-Ellingsen Group, BONO, Oslo, DACS, London 2001.

to be used as a medium to communicate with the masses. It was just this demotic character, combined with the direct simplicity of the technique and specific effects that it could achieve, that led to a revival of interest in woodcutting at the turn of the nineteenth and twentieth centuries among artists such as Gauguin, Munch (fig. 27) and the German Expressionists.

Wood engraving

In wood engraving the drawing is usually made on a block of hard wood such as box, and it is cut across, rather than along, the grain. Box trees are slow growing and their branches are seldom of sufficient diameter to produce large blocks. When large blocks of this wood are required, they are made by bolting smaller ones together. A block cut across the grain is less likely to splinter and can be cut cleanly with equal pressure in all directions. It can be worked therefore in a similar way to a copper plate, with a **burin** (see page 45). The burin allows more delicate effects to be achieved than the woodcutter's knife.

The fine lines cut by the burin are beneath the surface carrying the ink, and they therefore print white. This has led to many wood engravings being

28 and 29. Paul Nash (1889–1946),
***The Void*. Illustration to *Genesis*, published by the Nonesuch Press, London, 1924. Wood-block with an impression from it. 9 × 4.**
A brilliantly simple use of white-line wood engraving. The black is printed from the surface of a shaped block. Sharp incisions made with the burin at its edge hold no ink and as a result print white. The block consists of two pieces of box, joined.

V&A: E.139–1891

30. Albrecht Dürer (1471–1528), *Interlaced cord pattern with oblong shield*, c.1506–7. Woodcut. 27.8 x 20.9.
The extent to which the pattern is conveyed through the use of white lines was unusual for woodcuts of this date. Note the damage to the black line surrounding the image, which is typical of late impressions from woodblocks. The block also has worm-holes which have been filled in on the print by brush.

conceived in terms of white lines on a black background, and the technique is sometimes known as the white line method (figs 28 and 29) – although woodcuts (cut along rather than against the grain) can also be cut to print in a similar manner (fig. 30).

The drawing can be made either on the block or be transferred onto it. In the case of the latter, the drawing can be made on a sheet of paper and then traced on to the block with the aid of carbon paper. If transparent paper is used for the original drawing the printmaker can place it face down on the block and trace the image through the back. The resulting reversal of the drawing on the block has the advantage that the printed image is the same way round as the original. Drawings can also be transferred to the block photographically. This method, which came into general use in the 1860s, not only made the reversal of the image much easier, but also enabled the drawing to be reduced or enlarged according to the size of the block.

The engraver often rubs white chalk into the lines in order to see more clearly how the work is progressing, brushing it out before the block is printed.

The first engraver to exploit fully the advantages of end-grain wood was Thomas Bewick of Newcastle (fig. 31). Once it had been proved that the technique could rival the fine effects of metal engraving, the advantages of wood engraving to the book trade were quickly recognised. Allowing both text and illustration to be printed in one operation, it ousted the **intaglio** process as the favourite for book illustration and was only superseded at the end of the nineteenth century when methods of **photomechanical** reproduction were developed.

V&A: 23539

31. Thomas Bewick (1753–1828).
The Chillingham Bull, 1789. Wood
engraving. 14 X 19.7.
The subtle gradations of tone in this
wood engraving are the result of
Bewick's many technical innovations.
By slightly lowering parts of the block,
so that they received less pressure in
the printing process, variations in the
quantity of ink deposited on the paper
were introduced.

A school of wood engravers working from their own drawings grew up around Bewick, but once this circle had dispersed few artists engraved their own designs. The rise in the level of literacy during the nineteenth century led to an increasing demand for illustrated books and magazines. The majority of the illustrations were wood-engraved facsimiles but many of these, such as those produced by the Dalziel Brothers – famous for their engravings after the Pre-Raphaelites and the so-called illustrators of the sixties – translated the artists' drawings with relative freedom.

Much commercial wood engraving was done by firms which employed large numbers of engravers. The bolting together of smaller blocks to create larger ones was exploited to increase the speed with which the engravers worked. After the drawing had been made on the large block, its various sections were distributed to different engravers trained to work in the same style. When the engraved sections were reassembled, a master engraver worked over the joins (fig. 32).

More **impressions** than the delicate block could stand up to were often required, which led to the production of replicas by **electrotyping**, a process discovered in 1839. Many so-called wood engravings are in fact printed from electrotypes, a means of making an exact replica of the surface of the block in copper. A number of **electro-types** can be made from one wood-block, thus allowing a limitless number of impressions to be printed without fear of damaging the original block.

Only during the past century has the technique of wood engraving been developed as an original graphic medium. Paul Nash (fig. 29), Eric Gill and Eric Ravilious are among artists who have used the technique.

V&A: E.92–1976

32. After Arthur Boyd Houghton (1836–75), illustration from the front page of *The Graphic*, vol.III, 21 January 1871. Wood engraving. 27.3 x 22.9.
A night scene which employs both the black and white line method.

The practice of bolting together small blocks to create a large block of endgrain wood was advantageous to commercial printers. By allowing them to divide the labour, it enabled them to produce large-scale pictures such as this at the speed required to depict news while it was still fresh.

V&A: Circ.185–1929. Courtesy of Glenbow Museum Art Gallery Library Archives.

33. Sybil Andrews (1898–1992), *Concert Hall,* **1929.**
Colour linocut. 23 x 27.8.

Linocut is especially associated with a group of artists connected with the Grosvenor School of Modern Art, London, of whom Sybil Andrews became one of the best known. This example was printed from four blocks in yellow ochre, light blue, dark blue and black.

34. Ben Nicholson (1894–1982), *Still life (Three mugs and a bowl),* **1928. Linocut. 24.5 x 31.5.**

An example of a white-line linocut, the line of the image was incised in the linoleum and therefore carried no ink. The background to the image capitalises on the textured effect obtainable from linoleum.

Linocutting

This technique is a twentieth-century development of the woodcut, with a sheet of linoleum substituted for the wood plank (figs 33 and 34). As linoleum has no directional grain it can be worked with the engraver's tools as well as with those of the woodcutter. Being rather soft and crumbly, however, it cannot produce fine lines.

Linocuts can be distinguished from woodcuts by their lack of wood-graining, their characteristic spongy texture and, usually, by their lack of imprint.

Linoleum is inexpensive and easy-to-work. For this reason it has been widely used as a medium to introduce the amateur to printmaking, especially in schools.

Metalcutting

The technique of metalcutting is similar to that of woodcutting, except that a plate of soft metal is used in place of wood. The lines are left in relief in the same way but areas of the metal are also left uncut. In the fifteenth century these blank areas were sometimes decorated with punched holes. The commonest punch has a round end producing white dots, which has led to the technique also being described as the dotted method or *manière criblée*. Other shapes produced by punches include stars and fleurs-de-lis. For printing, the plate is attached to a wooden block.

The first metalcuts date from about 1430. After the invention of movable type the technique, being more robust than woodcutting, was widely used for repeated decorative elements such as borders and vignettes in illustrated books.

Relief etching

Relief etching shares the same principles as the other **relief** processes but involves the use of different tools and materials. The design is made on a metal plate (originally copper but nowadays usually zinc), in fluid which is resistant to the corrosive action of acid. The areas around the design are eaten away by acid, leaving the lines in relief. William Blake experimented with this technique (fig. 35). Subsequently these principles have become the basis of all **photomechanical** relief printing.

In 1850 a French printer, Firmin Gillot, invented a related process known as *gillotage*, which turned a lithographic plate into a relief process. The drawing was made on the plate in a greasy medium that was dusted with a resin resist. As in relief etching, the plate was then etched leaving the protected areas in relief. The technique was frequently used to convert lithographs into relief prints for publication in newspapers, notably the work of Daumier and the caricaturists of the Franco-Prussian War.

35. William Blake (1757–1827), page from *There is No Natural Religion*, 1788–94. Relief etching printed in a range of sepias. Size: 5.5 x 4.3, shown at actual size.

As Blake was both author and designer, the production of text and image in the same technique was especially appropriate and gave the publication an unusual unity. The text was applied to the plate in a medium that resisted the corroding effect of acid in reverse of the way it was printed. How Blake did this remains a mystery.

Photomechanical relief processes

Prints produced by **photomechanical** means are often referred to as process prints. A method using a photographically produced **relief** printing block, known as photo-type, was introduced by the British company Messrs Fruwith & Hawkins in 1863. The plate was prepared with a coating of light-sensitive gelatine and exposed under a negative. The unexposed and therefore unhardened gelatine was washed away, and the remainder **electrotyped**. The copper thus deposited into the depressions, formed the lines of the image creating a printing surface in relief.

36. Aubrey Beardsley (1872–98), vignette to *Bon-Mots of Sydney Smith & Brinsley Sheridan*, **1893. Ink design reproduced in line block with letterpress text. Images: 4.8 x 4.8; 1.5 x 1.5.** Beardsley was one of the first artists to make drawings specifically for reproduction by line block, sometimes, as in this vignette, using the process to intensify the design by reducing its dimensions.

Phototype was superseded by the line block, a photographic development of relief-etching perfected by the early 1880s. This technique involves coating a zinc plate with light-sensitive gelatine and exposing it to light through a high contrast black-and-white negative. The gelatine is washed away where it is unhardened. The remaining sticky gelatine is then dusted with

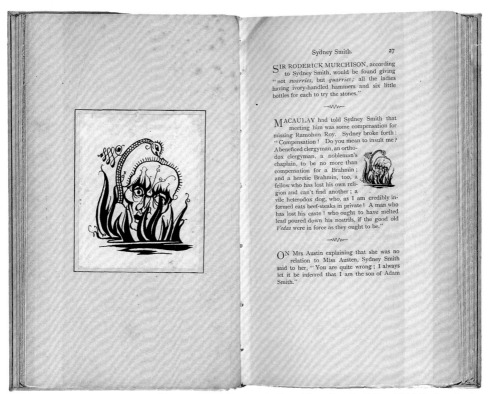

V&A: E.335–1972. Purchased with assistance from the National Art Collections Fund.

37. Enlarged detail showing the screen of a halftone letterpress print.
The dots produced by the screen vary in size according to the tone but their centres are always equidistant. In places the ink can be seen pressed to the edges of the dots as would be expected with a relief method.

asphaltum which sticks to the lines of the image and, when heated, makes them resistant to the corrosive action of the acid. The plate is then etched and the lines of the design are left standing in relief.

The line block can only be used for the reproduction of line drawings or flat areas with no intermediary tones (fig. 36). Although black-and-white negatives are required to make the printing blocks, they can of course be printed in any colour. A separate block is required for each colour.

Halftone letterpress extended the principles of line block to tone. The original is photographed through what is known as the **screen**, a fine network of crossed lines ruled on glass which breaks up the drawing into evenly distributed dots of different sizes (fig. 37). The lighter the tone, the smaller the dot, and the darker the tone, the larger the dot. The negative is printed on to a zinc or copper plate, which is developed and etched in the same way as in the line block process. The dots are thus left in relief and form the printing surface. If a fine enough screen is used, an impression from the halftone block gives the illusion of a continuous gradation of tone.

After the cross-line **screen** was perfected in the 1880s, it was combined with a process of colour separation to make full-colour printing possible. This colour process makes use of the fact that white light is composed of the three light primaries: red, green and blue. Each of these primaries when mixed with one of the others produces a colour which is one of the pigment primaries. These colours (plus black) are those used in four-colour printing, and are referred to as magenta, cyan and yellow.

A coloured original is photographed through filters, which separate the areas to be printed by the respective primary colours, in order to make separation negatives from which separate halftone letterpress plates are made for each colour. A red filter is used to make the negative for the cyan printing plate, a green filter to make the negative for the magenta plate, and a blue filter for the yellow plate. A fourth negative is photographed through a special amber filter to make a black printing plate, which adds strength to the darker areas of printing.

The printing areas are transferred from the negative to the plate photographically and these plates are then printed in succession, so that they overprint in **register**, that is with all the plates exactly aligned on top of one another, to produce the completed work.

Embossing

Embossing is a form of **relief** printing in which the design is impressed into the paper **blind**, that is un-inked. The printing surface can be made by any of the techniques described. The technique is also known by the French term *gaufrage*.

Collagraph

This term is derived from the French *coller* (to stick). It is a printing medium, related to collage, that came into use in the last century among artists who wished to juxtapose varied textural effects in their work. The printing surface is made by sticking extraneous materials such as card, string, plastic, wire mesh etc. onto a backing. The irregular surface is inked and the image is formed from both the relief and **intaglio** surfaces.

3 | Intaglio processes

Intaglio is a term derived from an Italian word meaning an incised design. In the intaglio processes the printing surface is sunk beneath the areas that are to remain blank. The ink is applied to the plate with a dabber or roller, and forced into the grooves and pits. The surface of the plate is then wiped clean. Sometimes a very soft fine muslin is drawn lightly over the plate so that a small quantity of ink is drawn out of the grooves in order to soften the printed lines. This process is known as *retroussage*.

To take an **impression** a sheet of paper is laid on the inked plate, and together they are submitted to sufficient pressure (achieved most easily in a roller press) to drive the paper into the grooves so that it picks up the ink (fig. 38). Impressions can be taken from a plate at any time – so a printed date is no guide as to when that particular impression was made. The printed image is in reverse of that on the plate.

In images printed by these processes, the paper bearing the ink is slightly raised. The pressure of the plate smoothes the grain of the remaining surface, and its rounded edge, the **bevel**, creates what is known as the **plate-mark**.

The most frequently used plate is a highly polished metal sheet (figs 39–42) with bevelled edges, but celluloid, linoleum and even wood are among other materials that can be used. Copper, the metal most frequently employed, came into general use around 1520, but iron, steel, brass, bronze and zinc are among others which have been employed at different times. Zinc plates began to be manufactured early in the nineteenth century. They are more frequently used for etching than engraving, but being softer and coarser than copper they are inclined to etch with less reliable results. Steel plates came into fashion in the 1820s as a result of the demand for a plate which could yield a greater number of impressions than copper before showing signs of wear. Their hardness, however, makes them very difficult to work. They have been little used since the end of the 1850s, which was when a method of facing copper plates with steel was invented. By electrolytically precipitating steel onto an etched or engraved copper plate, the printmaker could work on the softer copper and then give it a hard-wearing surface to stand up to long use, thus taking advantage of the tractability of one material and the toughness of the other. Corrections can be made to etched and engraved plates – as long as they are not too deeply etched or cut – by burnishing the faulty areas and hammering them level with the rest of the surface of the plate from behind.

Cette figure vous montre Comme on Imprime les planches de taille douce,

L'ancre en efl faite dhuille de noix, bruffee et de noir de lie de vin, dont le meilleur vient Dallemagne, L'imprimeur prend de Cete ancre auec vn tampon de linge en ancre fa planche vn peu chaude, lessuye apres legerem auec dautre linge, et acheue de la nettoyer auec la paume defa main. Cela fait il met cette planche a lenuers fur la table defa presse, aplique deffus vne feuille de papier trempe et repofe, et Coiure cela dune feouille dautre papier et dun ou deux Langes, puis en tirant les bras defa presse il fait paffer fa table auec fa planché entré deux rouleaux
faict a lean forte par Boffe a Paris en Lifle du palais lan 1642, auec priuilege

V&A: E.2685–1934 (reproduced from a Lippmann reproduction).

38. Abraham Bosse (1602–76), *Interior of an Intaglio Printer's Shop*, 1642. Etching. 25 x 31.3.

This print shows the various stages involved in printing an intaglio plate and includes an example of the kind of press in which such plates are printed. The method remains much the same today.

The figure in the background is shown applying ink to a plate with a dabber. In front of the window is a small grate on which the plate has been warmed in order to make the ink run more freely. The man on the left has finished cleaning ink off a plate with a muslin rag and is giving it a final wipe with the palm of his hand.

In the middle is the press which exerts pressure by passing a sliding printing bed between two rollers, much like a mangle. Roller presses are used for intaglio printing in preference to screw and lever presses because they exert the required pressure more easily.

An impression is in the process of being taken. Before a sheet of paper is printed it is dampened to make it more pliable and absorbent. The plate and the sheet of paper, covered by a cushioning blanket, rest on a bed which is halfway through the rollers. Recently printed impressions are shown hanging up on the line to dry.

There are two traditional ways in which the **intaglio** processes are printed in colour. Each colour can be printed from a separate plate (fig. 43). With this method it is important to ensure that the plates for each colour are printed in **register**. To do this, small holes are drilled through the top and bottom (and sometimes the sides) of all the plates stacked exactly on top of one another. The holes on the first plate to be

V&A: E.6292–1910

V&A: E.6292A–1910

39 and 40. Daniel Hopfer (*c.*1470–1536), *Five German Soldiers*, **early 16th century. Etched steel plate with a late impression from it. 20.6 × 38.1.**

Daniel Hopfer, an armourer of Augsburg, is thought to have been one of the first craftmen to print on paper. The earliest etching plates were made of steel or iron, the materials with which armourers were accustomed to work.

printed leave indentations in the paper which act as a guide when the sheet is placed face down on the next plate. The printer pricks through the indentations, lining them up with the holes in the plate to be printed.

Alternatively all the colours can be applied to a single plate and printed in one operation (fig. 44). The coloured inks, usually the darker tones first, were dabbed in with rag balls on the end of a stick – a method which became known as inking *à la poupée* because of the resemblance of the dabbers to rag dolls. Great care was required to ensure that the colours did not run together. This method meant in practice that the colours were copied afresh by hand for each impression, usually from a standard coloured copy which the printer had as a guide for reference. Sometimes separate plates and the *à la poupée* method were combined (fig. 45).

To these methods a third was added by S.W. Hayter, during the 1950s. It also allows the same plate to be inked with all the colours, one on top of the other, and to be printed in one operation. The technique is dependent on the fact that inks of different viscosities are not inclined to mix. A thin layer of ink of low viscosity will reject

41 and 42. Rembrandt (1606–69), *Virgin and Child with the Cat and Snake*, 1654. Etched copper plate with an impression from it. 9.5 × 14.5. Impressions from this plate were taken from 1654 to 1906. A comparison between the plate and the life-time impression reproduced shows that over the centuries many small changes have taken place, for example the plate has hatched lines at the top where there are blank patches in the impression and a small blank patch has appeared on the Virgin's face.

V&A: E.655–1993 (plate); V&A: CAI 646 (impression).

43. (right) **Jean-François Janinet (1752–1814)** after **J.-B.-A. Gautier-Dagoty, *Portrait of Marie-Antoinette*, 1777. Colour aquatint on two sheets printed from several plates. Complete: 40.8 x 31.9.**
The oval portrait was printed from four rectangular plates in red, blue, yellow and black according to the three colour process devised by Le Blon and on which colour printing today relies (see page 37).
Superimposed on it is a cut out frame printed in blue, orange, and gold also printed from separate plates.

44. (below) **François David Soiron (1764–after 1795)** after **George Morland, *A Tea Garden*, 1790. Colour stipple engraving printed from one plate with additional colour by hand. Cut to 45.1 x 53.3.**
This print, printed from one plate, was also issued in monochrome. This version uses at least five different colours rather than the more normal two or three. Another colour version of the print in the Fitzwilliam Museum, Cambridge, uses a slightly different range of colours suggesting that more elaborate impressions

in colour were pulled to order rather than in a batch and were subject therefore to the vagaries of different printers and the colours available.

V&A: E.422–1905. Bryan Bequest.

V&A: E.130–1963. Cunliffe Bequest.

V&A: 540–1882. Jones Bequest.

45. Louis-Marin Bonnet (1736–93) after Boucher,
***Tête de Flore*, 1769. Colour etching in the crayon**
manner with mezzotint, printed from eight plates, in
a contemporary frame. Sight size: 40.6 x 30.5.

This print combined both methods of printing: three
of the eight plates were selectively inked *à la poupée*
with more than one colour.

a thick layer of ink of high viscosity applied on top of it – but if low viscosity ink is
applied on top of the highly viscous one, it will stick.

The **intaglio** processes can be subdivided into two categories: engraving, in which
the grooves and pits are cut with a tool; and etching, in which they are **bitten** by acid.

Engraved and etched lines have different characteristics (see pages 46 and 56 for
further descriptions of each), and are often used in combination to portray those

V&A: 414/1874–1885

V&A: E.165–1970

aspects of an image to which they are most suited (fig. 46). By the 1860s a great mixture of **intaglio** techniques were used together, especially in prints reproducing paintings. It was as if the intention were to give to the surface of the print a texture that matched the complexity of the surface of the painting (fig. 47).

Line engraving

The tool used in line engraving is called a **burin**. It consists of a steel rod with a square or lozenge section, bevelled and sharpened at its tip. The handle of the burin is held firmly in the palm of the hand and the top is pushed at an acute angle over the plate. Its tip cuts into the metal and removes a sliver, which is lifted clear by the burin's bevel. Any ridge of rough metal that remains on either side of the line is removed with a scraper.

Engraved lines swell and taper accordingly to the amount of metal that is removed, usually coming to a point at their tip (figs 48 and 49). Curved lines are engraved by turning the plate during the process of engraving (figs 50 and 51).

The engraving of lines in metal in order to print on paper developed from the goldsmith's long-established practice of decorating metal with engraved patterns. According to legend, printing on paper from metal plates was discovered by a goldsmith. He is said to have left a *nielloed* plate – an engraved plate in which the grooves have been filled with a black substance called *niello* in Italian – to dry overnight with a sheet of paper resting on it. In the morning, he found that a bundle of wet clothes had been placed on top of it. The dampened paper had been pressed into the *nielloed* grooves, thus making a print.

The technique was first used for printing on paper in the second quarter of the fifteenth century. It arose independently in Germany and Italy, probably slightly earlier in Germany (figs 52 and 53). To begin with, engraving fulfilled a dual purpose.

46. (opposite above) **Benoit Audran (1698–1772) after Watteau, *Amusements Champêtres*, 1735. Engraving and etching. 36 x 47.**
French engravers of the eighteenth century combined line engraving and etching to produce an effect particularly in tune with the light–hearted quality of their subjects. The landscape, the gentle movements of the foliage and the subtle effects of the light are rendered in the more malleable and softer medium of etching but the glitter of the scene and the shimmering of the costumes are interpreted with sharp, clearly cut engraved lines.

47. (opposite below) **Frederick Hollyer (worked 1860s) after Landseer, *The Old Shepherd's Chief Mourner*, 1869. Mixed mezzotint. 65.5 x 71.**
The underlying tone of this image is created with mezzotint but virtually every engraved and etched technique is used in it. Intaglio printmakers in the 19th century seem to have tried to make the surface of their prints vie with that of the paintings they reproduced creating a surface which corresponds in textural weight (if not in detail) to that of the original.

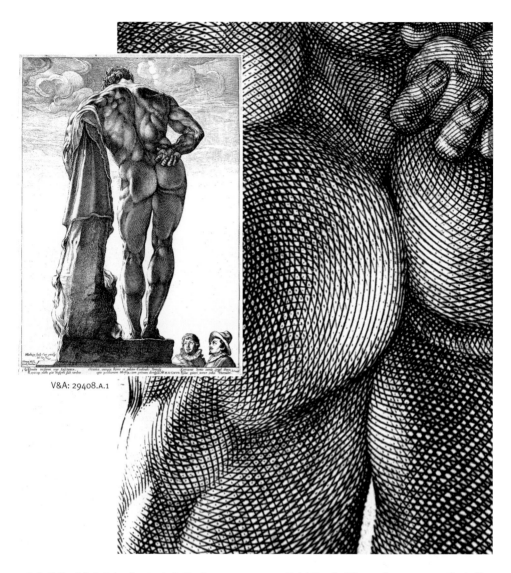

V&A: 29408.A.1

48. (left) **Hendrik Goltzius (1558–1617),** *The Farnese Hercules*, *c.*1592 but dated 1617. **Engraving. 40.5 × 29.4.**
Engraved lines are cleanly cut. Even under magnification their edges are sharply defined.

49. (right) **Detail of fig. 48 shown at one-and-a-half up from actual size.**

It was a means both of original expression and of obtaining duplicated images of all kinds. By the middle of the sixteenth century it had become predominantly a means of reproduction. Being a laborious and time-consuming technique, requiring immense skill, it was considered the most noble and appropriate technique in which to render the work of the great masters – although frequently the characteristics of the engraved line were imitated in the quicker and more tractable medium of etching (figs 67 and 68). Engraving was also the medium used for reproducing royal portraits (fig. 54). It was only when **photographic** methods of reproduction were developed towards the end of the nineteenth century that interest in line engraving as an original graphic medium revived.

The multiplication of images through engraving resulted in the rapid spread of iconographic and stylistic innovations in painting, sculpture, and the decorative arts. Engraved patterns and decorations were circulated expressly for the use of craftsmen, including goldsmiths, furniture makers, leather workers and embroiderers, and prints were collected by painters and sculptors to be referred to as inspiration when a particular subject was commissioned.

50. (left) **Claude Mellan (1598–1688),** *Head of Christ on the Sudarium,* **1649. Engraving. 43.2 x 32.4.**
Spiralling out from the tip of the nose, the image and lettering are formed almost entirely by the exploitation of the swelling and tapering of a single engraved line.

51. (above) **Detail of fig. 50 shown at actual size.**

V&A: E.2546–1960

52. (right) **Martin Schongauer (c.1450–91),** *Christ in the Garden on the Mount of Olives,* **c.1476. Engraving. Cut to 16.5 × 11.5.** Schongauer was among the first northern artists to take up engraving as a vehicle for his vision and greatly extended the vocabulary of strokes employed by engravers.

53. (below) **Andrea Mantegna (1431–1506),** *Bacchanal with Silenus,* **c.1470. Engraving. 27.1 × 42.3.** Mantenga holds much the same position for Italian engraving as Schongauer held in the north. A comparison between their work makes clear the extent to which the art form developed independently. Mantegna's line is much simpler and relies to a great extent on outlines and diagonal hatchings executed with a range of differently shaped implements.

V&A: E.755–1940. Carthew Bequest.

V&A: Dyce 996

Louis le Grand.

54. Pierre Drevet (1663–1738) after Rigaud, *Louis Le Grand*, c.1712. Engraving. Cut to 68.7 x 51.6.

This portrait represents Louis XIV in his coronation robes. Such portraits were traditionally worked exclusively in engraving, the laboriousness of the technique turning the task into an act of homage on the part of the subject to his king.

Drypoint

In this technique the line is scratched on the plate by a tool with a sharp point. The tool is held in much the same way as a pencil, and differs from the engraver's **burin** in that it pushes the excess metal to the sides of the furrow rather than lifting it clean away. It is this curl of rough metal known as the **burr** which gives the drypoint its character (figs 55 and 56). The ink is retained in the burr as well as the furrows and gives the edges of the printed line a soft blurred quality. Sometimes the central furrow has an un-inked spine. The burr is very delicate and even when the plate is steel-faced it soon gets worn away. The burr is sometimes scraped off and the line produced is very fine.

Although drypoint is a form of engraving, it is often used in conjunction with etching, especially as a means of retouching and adding emphasis. There are, however, examples of plates being worked entirely in drypoint in the late fifteenth and early sixteenth centuries, but it was more widely used in the seventeenth century. It was particularly popular with Rembrandt. Nevertheless, drypoint did not become a popular printmaking tool until the invention of steel-facing in the middle of the nineteenth century made it possible to obtain more than a few high quality prints from one plate.

Mezzotint

Mezzotint is a form of tonal engraving and, because the engraver works from dark to light, it is often described as a negative process. The plate is prepared so that it will print an even, deep black. This is done by pitting its surface systematically with a serrated chisel-like tool, known as a rocker, which raises a uniform burr. The design is formed by smoothing the burr so that different areas of the plate will hold different quantities of ink and therefore print different tones of grey. A scraper is used to remove large areas of burr, and a **burnisher** for more delicate work. Highlights are achieved by burnishing the plate quite smooth so that when it is wiped no ink remains on these areas (figs 57 and 58).

Mezzotint engraving was invented by an amateur artist, Ludwig von Siegen of Utrecht, in about 1640. But it was in England that the technique was perfected, becoming known on the Continent as the *manière Anglaise*. Prince Rupert is said to have introduced it to England. A nephew of Charles I, he grew up on the Continent where, according to tradition, he met von Siegen. Having settled in England in 1660, he demonstrated mezzotinting to John Evelyn, who made the technique public in his *Sculptura, or the History and Art of Chalcography and Engraving in Copper*, first published in 1662.

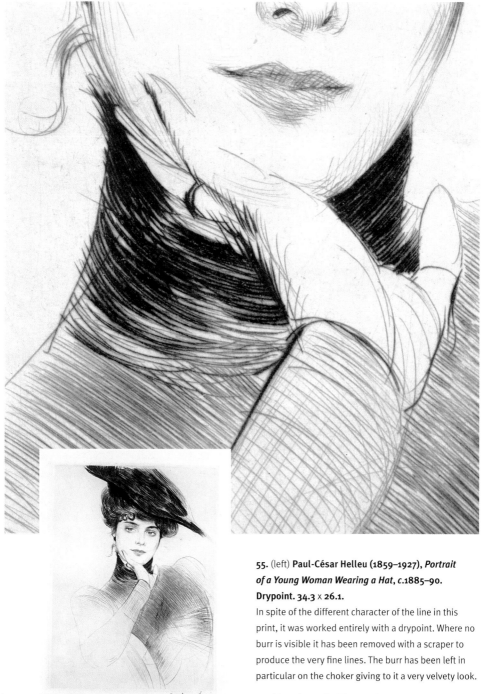

55. (left) **Paul-César Helleu (1859–1927),** *Portrait of a Young Woman Wearing a Hat,* *c.*1885–90. **Drypoint. 34.3 × 26.1.**
In spite of the different character of the line in this print, it was worked entirely with a drypoint. Where no burr is visible it has been removed with a scraper to produce the very fine lines. The burr has been left in particular on the choker giving to it a very velvety look.

56. (above) **Detail of fig. 55 shown at one-and-a-half up from actual size.**

V&A: E.153–1969. Given by Miss A. Reeve.

Courtesy of The Hon Christopher Lennox-Boyd.

57 and 58. John Smith (c.1652–1742) after Kneller,
The Right Honourable William Lord Cowper, **1707.**
Mezzotint. Cut to 38.2 x 29.5 and 40.5 x 29.8.
Progress proof (left) **and published state** (right).

A mezzotint plate is worked from dark to light. These
impressions were taken from the same plate, the proof
being printed before the scraping of the head and
before additional work on the border.

Mezzotint is particularly suited to reproducing the tonal gradations of paintings, and mezzotint reproductions did much to popularise the compositions of Gainsborough, Reynolds, Joseph Wright (figs 59 and 60), George Romney and other artists working in the eighteenth century. The unusually rich textures of fine mezzotints are dependent on the delicate curls of metal, which are thrown up by the rocker pitting the plate, as well as on the hollows themselves and these wear quickly. For this reason the British engraver Valentine Green declared, in the conditions of sale of some of his prints, that **proofs** of the mezzotints would be limited to fifty **impressions** and sold at double the price.

Mezzotints, like all **intaglio** techniques, can be printed in colour. They were sometimes printed in colours when they had lost their brilliance, as a ploy to prolong the commercial life of a

59. (opposite) **Valentine Green (1739–1813) after Wright,** *The Air Pump*, **1769. Mezzotint, 47.8 x 58.3.**
There is a propriety about rendering night scenes by mezzotint, a technique in which the design is scraped out of darkness, yet one that is capable of gleaming whites. It was particularly suited to the reproduction of the eerie atmosphere of candlelight.

60. (opposite) **Detail of plate 59 shown at actual size.**

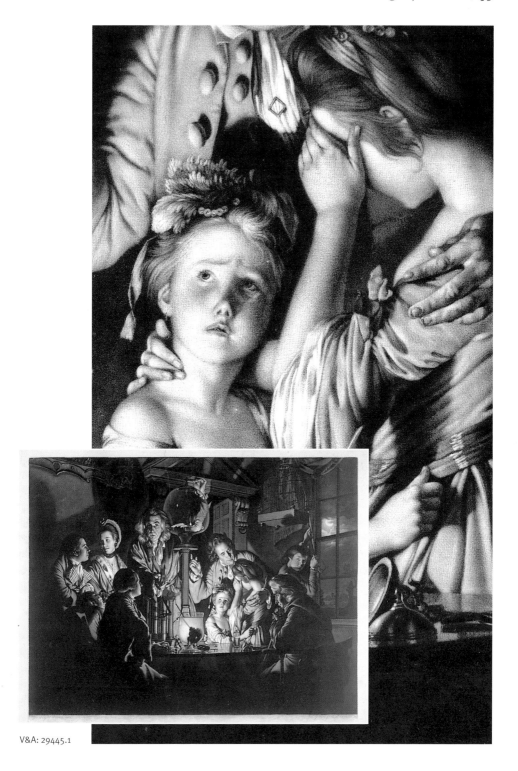

V&A: 29445.1

61. Anonymous, *A Milleners* [*sic*] *Shop*, 1772. Mezzotint glass coloured print in a contemporary frame. Size of frame: 42.1 x 31.8.
This is typical of the kind of fancy subject that was frequently turned into glass pictures.

worn plate. A mezzotint printed lightly in black also provided the hand-colourist with built-in modelling, thereby facilitating the task. For this reason it was the type of print most widely used to produce glass colour prints (fig. 61). The mezzotint would be dampened and pasted face down on a sheet of glass. The paper was then carefully rubbed away so that little more than the ink remained on the glass. The surface was then varnished to make any remaining paper transparent and the image painted from the back. Glass colour prints were being made from mezzotints by the 1690s.

Nature printing

Nature printing is a process in which actual specimens, for example plants, butterflies, pieces of lace, fossils, and plants were used to form the printing surface. The technique depends on the fact that an object placed between two flat and polished surfaces, one harder than the other, will leave a detailed impression of itself in the softer. Among the surfaces used were lead with copper or steel. The impression formed in the lead was **electrotyped**, which could be used directly to print an extraordinarily life-like **intaglio** facsimile of the specimen (fig. 62). An impression from the electrotype could also be made to create a **relief** printing surface so that the image stood out in white and was heavily embossed. Nature printing was developed by Alois Auer in Vienna in 1852 and patented in Britain by Henry Bradbury in 1853.

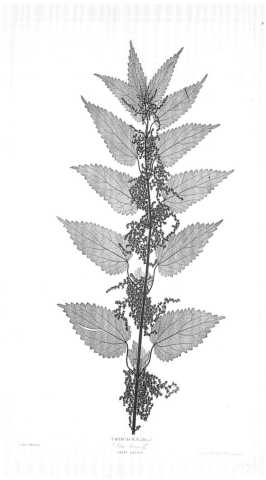

**62. William Bradbury
(worked mid-19th century)
and William Mullett Evans
(1831–60),** *Great Nettle.*
Plate from *A Few Leaves from
the Newly invented Process
of Nature-printing,* **1854.
Nature print. 44.8 x 22.3.**
Nature printing is sometimes
referred to as 'phytoglyphy'
from two Greek words for
'nature' and 'carving'.

V&A: 14765.21

Line etching

In etching the grooves and hollows are formed by the corroding action of acid, known as **biting**.

First the plate is coated with a thin layer of a wax-like substance, called the **ground**, which is impervious to acid. Etching grounds are normally composed of beeswax, asphaltum, pitch and mastic gum in varying combinations and proportions, according to the purpose for which they are intended. The ground is applied by rubbing the block over a heated plate so that it melts, using a dabber or roller to spread it. To make a liquid ground, the normal ingredients of an etching ground are dissolved in a volatile solvent such as chloroform or xylene. In this liquid form it is either painted or poured on to the plate. Liquid grounds are particularly useful for touching-in areas that have been damaged or where mistakes have been made. Before the plate is drawn on, the ground is smoked with burning tapers to blacken it.

The design is made through the **ground** with a needle, which lays bare the metal where the lines are required. Little pressure need be applied to the needle because there is no need for it to scratch the metal. When the design is complete, the back and edges of the plate are protected with varnish, and the plate is then placed in a bath of acid diluted with water. The kind of acid and the proportion of acid to water

V&A: E.1895–1919.
Given by Mrs J.Merrick Head.

63. (above) **Samuel Palmer (1805–81), *The Lonely Tower*, 1880. Etching. 19 x 25.4.** Etched lines are softer than engraved lines and move more freely. The softness is caused by the uneven action of the acid on the granular structure of the metal. The freedom is the result of the etching ground offering little resistance to the needle, allowing it to draw with almost the same ease as a pencil, whereas the engraver's burin has to dig through the metal of the plate.

64. (right) **Detail of fig. 63 shown at twice actual size.**

66. (left) **Detail of plate 65 shown at one-and-a-half up from actual size.**

V&A: 26748

65. (above right) **Giovanni Marco Pitteri (1702–86) after Piazzetta,** *Portrait of the Etcher,* **c.1740. Etching. Cut to 45.8 × 35.7**

Pitteri's technique, peculiar to himself, consisted solely of etched lines. The lines lie in parallel undulations across the surface of the image modulated not as in the work of Goltzius or Mellan by the swelling and tapering which the burin facilitated but by exaggeration of the raggedness of a line made by the action of acid on a metal plate.

varies according to the metal to be etched and the effect desired. Different etching needles can be used, but the strength and thickness of a line is mainly dependent on how deeply it is **bitten**. The time required for the action to take place varies enormously. It can be as little as a minute or as much as two hours.

When the faintest lines have been sufficiently etched, the plate is removed from the acid so that the completed parts can be protected with varnish (known as **stopping out**) before biting continues for the darker ones. This procedure can be repeated any number of times until the deepest lines have been etched. If the acid etches through faults in the **ground** it is described as foul biting.

LA MADONNA DEI CANDELABRI

V&A: Dyce 1808

67. (above) **Pietro Bettelini (1763–1829),**
La Madonna dei Candelabri, *c*.**1800.**
Etching. 41.5 x 36.8.
The cleanness of the line in this print
gives the impression that it is the product
of a burin. In fact, the tool used was the
échoppe, twisted and turned to simulate
the linear vocabulary of the engraver.

68. (right) **Detail of fig. 67 shown at**
actual size.

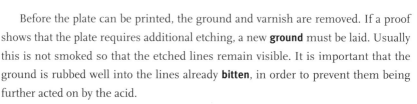

Before the plate can be printed, the ground and varnish are removed. If a proof
shows that the plate requires additional etching, a new **ground** must be laid. Usually
this is not smoked so that the etched lines remain visible. It is important that the
ground is rubbed well into the lines already **bitten**, in order to prevent them being
further acted on by the acid.

Exposed metal is uniformly etched to the same degree. For this reason etched
lines tend to be the same thickness throughout their length and to have, in comparison with engraved lines, ragged edges and blunt ends (figs 63, 64, 65 and 66).

The technique of etching was developed by armourers as a method of decorating
weapons. It was not used for printing on paper until shortly after 1500 in Germany
(fig. 39). To begin with it was viewed as a labour-saving method of line engraving,
and a hard ground, which offered some resistance to the needle, was used. The needle, called an *échoppe*, was shaped so that it made a mark that imitated the engraver's

burin (figs 67 and 68). Not until the seventeenth century was the freedom the technique allows fully exploited, the greatest exponent being Rembrandt (figs 12 and 42). Since then most painters have tried their hands at etching, and some artists have made their name through their work in this medium. Etching has also been used as a method of reproduction, often in conjunction with engraving (figs 46 and 47). It has also been used as a means by which artists can produce outlines for professional aquatinters and hand colourists.

Soft-ground etching

This is a method with which the effect of a pencil drawing or any other texture can be imitated. As the name suggests, it differs from ordinary etching in the recipe for the **ground**, which by the addition of tallow is made considerably more impressionable. In the past, the ground was wrapped in silk so that any gritty impurities that might scratch the plate would be strained out. Reliable commercial preparations have made this unnecessary. The drawing is made with a pencil on a sheet of thin paper laid on the ground. The paper is then peeled back, picking the wax off with it in a broken

V&A: E.3234–1914

69. Thomas Gainsborough (1727–88),
Landscape with cattle by a pool, **1797. Soft-ground etching and aquatint. 28 × 34.9.**

Gainsborough used soft-ground etching to create the effect of pencil, and aquatint that of wash.

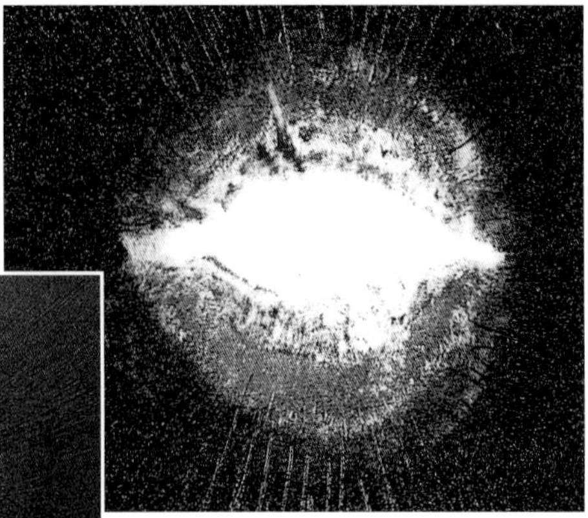

70. (below) **Antony Gormley (b.1950), plate from** *Body and Soul*, **1990. Soft- and hard-ground etching, 30 x 40.5.** The image was made by impressing an actual mouth into a soft etching ground.

71. (above) **Detail of fig. 70 shown at actual size.**

V&A: E.1591–1991

line, which corresponds to the graining of the paper, the hardness of the pencil and the pressure applied. The plate is then **bitten** in the same way as ordinary etching.

The effect of different textures can be achieved by impressing, for example, leaves, feathers, lace or pieces of crumpled paper in the **ground**. When they are removed they leave an impression of themselves which can then be bitten.

The technique was used in the seventeenth century by Italian artist G.B. Castiglione, but there is no evidence that it was exploited at that date. Presumably it was worked out afresh in the later eighteenth century. The faithfulness with which the technique reproduced the effect of pencil encouraged artists such as Gainsborough (fig. 69) and Thomas Girtin to compose directly in the medium, and gave rise to a spate of imitations of Old Master landscape drawings. Its ability to reproduce different textures made it attractive to artists in the twentieth century working with found objects (figs 70 and 71).

Aquatint

Aquatint is a method of etching in tone. The key to the technique lies in the application of a porous ground, consisting of particles of finely powdered asphaltum or resin. The acid contacts the plate where it is unprotected between the particles, thereby etching pits in the metal which give a grainy texture when printed.

The tone of any part of the printed image is dependent on the depth to which the pits are etched, so the design is built up in stages by **stopping out** areas once they have been adequately **bitten**. Before the plate has any contact with the acid those areas that are to remain blank in the **impression**, and must therefore have no pits to catch the ink, are protected with a coat of varnish. The plate is then immersed in the acid. Once it has been sufficiently etched for the lightest areas, these too are **stopped out**. The process is repeated until the deepest tones have been bitten.

As with line etching, continuous gradations of tone cannot be achieved with pure aquatint. After the plate has been etched and the **ground** and varnish removed, however, the flat tonal areas can be modified by the use of a **burnisher** in the same way as in mezzotint.

The ground is usually applied by shaking up the dust in a dusting box and, after the coarser particles have settled, inserting the plate so that fine dust is deposited on it. Afterwards, the plate is heated gently so that the dust melts and clings to it to form the ground. The ground can also be laid by puffing the dust through a muslin bag, but laid in this way it is coarser than one applied in a dusting box.

Paul Sandby pioneered the use of a spirit ground (figs 72 and 73). In this method the resin was dissolved in spirit and poured or painted on to the warmed plate, the

72. (above) **Paul Sandby (c.1730–1809),** *Windsor Terrass* [sic] *looking Westward*, 1776. **Aquatint over an etched line.** **34.3 × 48.**

73. (right) **Detail of fig. 72 shown at actual size.** This shows how the tonal effect was built up in layers, with the lighter areas being stopped out sequentially until the deepest and therefore darkest layer was bitten by the acid.

V&A: E.55–1891

spirit quickly evaporating. The advantage of this method was that it allowed the artist to make spontaneous marks with his brush. Nevertheless, the areas surrounding the freely drawn design still had to be **stopped out** to prevent them being etched.

A refinement of the aquatint, which avoids the laborious process of stopping out and therefore allows the printmaker greater spontaneity, is known as sugar aquatint (fig. 74). The design is made directly on the plate with a pen or brush, in a coloured solution in which is dissolved a large lump of sugar. When the solution has dried the whole plate is given a thin coat of varnish. The plate is then immersed in water, which slowly dissolves the sugar, causing it to swell and dislodge the varnish, thereby exposing the bare metal underneath. An ordinary aquatint **ground** is then applied to the plate, which can be **bitten** in the normal way. The process can be repeated to obtain the different tones.

Some aquatinters prefer to paint the design in the sugar solution over the aquatint ground and then varnish it. When the varnish flakes off in the subsequent immersion, the underlying aquatint ground remains intact.

The characteristic grain of aquatint has been found on prints produced as early as the mid-seventeenth century but the technique was not fully developed until a century later. In 1768 a patent was taken out by Jean Baptiste Le Prince (1733–81) for a powdered resin ground. He described his method as the *manière au lavis* (acid bath method). The term 'aquatint' was introduced by Paul Sandby.

74. Pablo Picasso (1881–1973), illustration to Buffon's *Histoire Naturelle*, 1942. Sugar aquatint. Image: 32 x 22.

The brush strokes that applied the sugar solution are clearly visible. Such a calligraphic effect could never be achieved by stopping out.

V&A: E.222–1947; © Succession Picasso/DACS 2001.

V&A: E.2932–1902

75. Francisco Goya (1746–1828), plate from *Los Caprichos*, first published 1803. Aquatint. 17.2 x 12.

This print was executed entirely in aquatint and shows clearly the irregular grain and flat ungraduated areas of tone, characteristic of the medium.

Aquatint, printed in sepia or colours, or coloured by hand, was much used to reproduce wash drawings and watercolours. With the exception of Goya (fig. 75), who used it in its pure form without, as was more frequently the case, etched outlines and hatchings, few artists used the medium for original graphic expression until late in the nineteenth century.

Stipple engraving

Stipple engraving is a method of rendering tones with dots and short flicks, and involves a mixture of engraving and etching.

The plate is prepared with a normal etching **ground**. The contours and planes of the design are dotted through the ground with an etching needle and then bitten with acid. These pits are added to and sharpened by a burin used directly on the metal. It is a technique that in comparison with mezzotint, widely used for similar subject matter, provides a high yield with little deterioration in the quality of the **impression**.

**76. William Wynn Ryland (1733–83) after Angelica
Kauffmann, *Etiam Amor Criminibus Plectitur*, 1777.
Stipple engraving printed in sanguine. 36.9 × 31.5.**

Stipple engravings were frequently printed in shades
of brown and red and were primarily intended for
framing (see fig. 13) as 'furniture' pictures rather than
as additions to the connoisseur's portfolio.

Pure stipple engraving was not practised until the middle of the eighteenth
century, but before this some engravers, such as Ottavio Leoni, obtained their delicate modelling by short flicks and dots in a manner anticipating stipple. Essentially
a means of reproduction, the technique was used for all kinds of subjects, including
portraits. It was particularly popular for light-hearted mythological, classical and
pastoral themes, many of which were drawn specifically for multiplication by the
technique. It was in Britain that stipple took root, although William Wynn Ryland

(fig. 76) is usually credited with bringing knowledge of the technique back with him from France, where he is said to have worked with François Boucher and Jacques Philippe Le Bas in the 1750s. Francesco Bartolozzi (see fig. 13), an Italian who settled in London in 1764, was the most prolific exponent of the technique.

Etching in the crayon manner

This process imitates the appearance of a chalk drawing. It is closely allied to stipple engraving and is often used in conjunction with it. It involves the same mixture of etching and engraving, but the marks through the **ground** and directly on the plate are made with a variety of small toothed maces and roulettes rather than with a **burin**.

77. (right) **Gilles Demarteau (1722–76), after Boucher, *Shepherdess*, c.1768. Etching in the crayon manner, printed in red. 19.2 x 13.2.**
The inspiration for working in this manner came from the punches and toothed wheels used by ornamental metal-workers, to one of whom Demarteau had been apprenticed.

V&A: 29973.3

78. (left) **Detail of fig. 78 shown at actual size.**

The crayon manner was invented in France around the middle of the eighteenth century where it was used to reproduce the red chalk drawings of Boucher (figs 77 and 78), Jean-Honoré Fragonard and others. The technique was also popular in England. By the early nineteenth century it had been displaced by lithography as a means of reproducing chalk drawings, but it continued to be used as an accompaniment to other forms of engraving.

Photomechanical intaglio processes

The early histories of photography and **intaglio** printing are closely interwoven. The first successful image to employ the principles of photography in its creation, that is the chemical action of light on a light-sensitised surface, was dependent also on etching. In 1827 J.P. Niépce placed an ordinary etching, made translucent by waxing, on a metal plate coated with bitumen of Judea (a tarry substance which hardens on exposure to light), and put them in a strong light for a long time. When the etching was removed from the plate, the coating had hardened where it had been exposed to light but had remained soluble where it had been protected by the lines of the print. These soft areas were dissolved away, leaving bare metal exposed in a network of lines reproducing those of the original etching. The plate was then etched, and after removal of the bitumen, inked and printed like a traditional etching.

The first wholly **photographic** image-making process to be made public was the daguerreotype, announced by Louis Jacques Mandé Daguerre in 1839. It depended on the action of light on a highly polished silver plate that had been treated with hot iodine before exposure. The latent image was developed with vapour of hot mercury. The process, however, had drawbacks. Firstly, the image was made up of a very fine greyish-white deposit on a highly polished silver plate, which could only be seen when the plate was held at an angle to catch the light. Secondly, the process produced only one image – to make a replica the whole process had to be done afresh. The disadvantages were overcome with the aid of traditional printmaking. Hippolyte Fizeau devised a means of etching and **electrotyping** daguerrotypes to create an image in ink on paper.

Crucial to the development of **photomechanical** printmaking processes was the discovery by W.H.F. Talbot that light-sensitised gelatine hardens and becomes insoluble in water in proportion to the amount of light that falls on it, patented in 1852. Photogalvanography, patented by Paul Pretsch in 1854, combined collotype (see page 82) with electrotyping (page 30) to produce an **intaglio** plate. A plate coated with light-sensitised gelatine swelled most where it had been least exposed. The resulting gelatine film with a varying relief was electrotyped, and an impression from that electrotype electrotyped again to form an intaglio plate which was inked and printed

V&A: E.183–1970

79. After Frederic Leighton (1830–96), *The Garden of the Hesperides***, 1893. Hand photogravure printed in sepia. 73** x **69.2.**

During the 19th century the demand for large repro-ductions of popular paintings increased enormously. The technical virtuosity of the engravers reached a peak but interest in engraving for its own sake faded. The introduction of photogravure provided a method of reproduction that was in some ways more accurate and in which the time-consuming labour of the engraver was replaced by the quick operation of the camera. Being an intaglio process requiring etching, the photographic base could be worked on as here by hand to sharpen up the image. Its sepia colour may have been a reference to its photographic origin.

in the traditional manner. This process continued to be used sporadically until the end of the nineteenth century, although it was increasingly superseded by the less cumbersome photogravure.

Hand photogravure, sometimes called heliogravure, is a process by which a line or tonal image can be transferred photographically to a metal plate in such a way that it can be etched in one operation without **stopping out** by hand. Again it is depen-dent on the characteristics of light-sensitised gelatine. The image is printed on to the gelatine, and then the gelatine is attached to the plate (which is usually prepared with a fine aquatint **ground**, although the ground is not essential for line images).

To develop the plate, it is immersed in warm water – this causes the unhardened gelatine to dissolve, leaving a layered negative. The gelatine is thinnest in the areas which are to print the darkest, and thickest where least ink is required. When the plate is immersed in etching fluid, the acid **bites** quickest through the gelatine to reach the plate where the gelatine is thinnest and therefore etches these areas deeper than where the gelatine is thicker. The resulting plate, when removed from the acid, is etched in gradations of depth according to the different thicknesses of ink required to reproduce the continuous varying gradations of tone of the original.

This technique, invented in 1859, was widely in use by the 1880s for the reproduction of paintings (fig. 79) and Old Master prints and drawings. With it, it is possible to make extremely deceptive reproductions of prints produced by manual processes, the reproductions being distinguishable only by the slight fuzziness of their line. The development of photogravure caused panic amongst the reproductive printmakers whom it made redundant, but it left printmaking, and in particular engraving, free to develop anew as a means of original expression.

Machine photogravure, as the name suggests, is related to hand photogravure. The difference lies in the way the graining, which enables the plate to hold the ink, is achieved. In hand photogravure, grain is provided by an aquatint ground, whereas in the machine process it is in the form of lattice work printed on the gelatine. Known as the **screen**, it has the added advantage that it gives the doctor blade (the blade removing excess ink) a level surface over which to scrape.

This cross-line screen can usually be detected with the aid of a magnifying glass. The square dots it produces are all the same size unlike the dots in the halftone letterpress process. The variation in tone is achieved by the varying thickness of the ink rather than by the size of the dot.

A flat-bed press can be used to print this technique, but it is generally printed from a cylinder in a rotary press. This method allows an enormous number of **impressions** to be printed at great speed. As the quality of the image is reasonably good even on cheap paper, the process has been much used in commercial printing for large runs of magazines and trade catalogues.

After the cross-line screen was perfected in the 1880s, it was combined with a process of colour separation to make full colour photogravure printing possible. Plates are made as for ordinary screened photogravure but colour filters are interposed to make separate plates for each colour. The filter process is the same as that described on page 37 in connection with colour letterpress printing. The three primary colours are used in a series of over-printings to build up all the colours in the spectrum, and a plate inked in black is also usually printed to add strength to the shadows. A large number of fine illustrated books, with reproductions of water-colours by artists such as Arthur Rackham and Edmund Dulac, were printed in the first decade of the twentieth century as a direct result of these developments.

When photogravure is used for original expression (figs 80 and 81), the term photo-etching is usually used in place of photogravure.

80. (right) **Antoni Starczewski (b.1929),**
MF 30/9, **1973. Blind relief photo-etching.**
39.7 × 39.7.
The image is derived from a photograph of real leaves, reduced in scale, before printing on the plate.

V&A: E.1549–1977

81. (left) **Detail of fig.80 shown at actual size.**

4 | Planographic processes

The printing surface of the **planographic** processes is neither raised above nor sunk below the areas to remain blank, but is level with them. With the exception of the monotype process (see page 81), planographic processes depend on the fact that grease and water do not mix. The printing areas are impregnated with grease while the blank ones are moistened. When greasy ink is applied to the printing surface it clings to the greasy areas, but is repelled by the wet ones. The ink is transferred from the printing surface to the paper by pressure. It tends to appear as if fused with the paper rather than to sit on its surface as in the case of screenprints (see page 85). Planographic prints do not have **plate-marks**, but it is sometimes possible to see the area of the paper that has been flattened by the printing surface.

Lithography

The term lithography derives from the Greek and means literally 'drawing on stone'. The design is drawn or painted on the printing surface – traditionally stone but now more often a metal plate, such as zinc or aluminium – in precisely the same way as on paper. The surface must be of a kind that readily absorbs grease and water equally.

The drawing substance must be greasy. Lithographic chalk is made of a mixture of greasy substances (wax, tallow and soap), lampblack, and shellac. The lampblack provides the colour so that the artist can see what has been drawn, and the shellac makes the chalk strong enough to be sharpened to a point and to stand up to the pressure exerted by the weight of the hand. The chalks, as with pencils, are graded according to their degree of hardness. Lithographic ink consists of the same ingredients dissolved in distilled water. It is sometimes referred to as *tusche*, the German word for liquid ink. The same kind of pen used to draw with Indian ink is used to draw on the printing surface with litho ink. If the ink is considerably thinned and applied with a brush, an effect of wash can be achieved.

The drawing can also be made on a special kind of paper, the surface of which is prepared so that it dissolves easily in water, enabling the design to be transferred to the printing surface. This process is known as transfer lithography, and has the advantage of the artist not having to use cumbersome stones and plates or to allow for the reversal of the drawn image. To make the transfer, the paper is thoroughly dampened and laid face down on the printing surface. Together they are passed through the press several times, the paper being frequently sponged with water.

82. Chris Orr (b.1943), *Lithography Studio, Royal College of Art,* **1996. Lithograph. 38.5 x 53.2.**

This lithograph was drawn by the Professor of Printmaking at the Royal College of Art. It gives an idea of the range of equipment used by artists to make lithographs today.

Gradually the paper soaks off, leaving the greasy drawing adhering to the printing surface. Transfers can also be made from prints in other media, for example etchings, engravings and woodcuts, making identification of the technique extremely difficult. The lines of a transfer lithograph tend to show the grain of the original paper rather than that of the printing surface.

A modern alternative to transfer paper is the diazo method. The artist draws on a slightly textured transparent film which allows the drawing to be made with whatever drawing implements the artist chooses. The image is transferred through the film photographically to a photo-sensitised aluminium lithographic plate.

However the image arrives on the printing surface, thereafter its preparation and the printing process proceed exactly as if the drawing had been made directly on it. First the drawing is fixed in it. This is done by sponging the stone or plate with a solution of gum arabic (a water-soluble resin) and a small amount of acid (not strong enough to eat into the printing surface) known as the etch. The gum arabic protects

83. Terry Winters (b.1949), plate 2 from a set entitled *Folio*, 1985–6. Colour offset lithograph. 79.3 × 56.6.
This print, right down to the thumb print, is a fine example of the extent to which lithography can convey the sense of an artist's direct and passionate involvement with the printing surface. Offset prints are printed twice, the image on the paper offset from an image on the worked printing surface. As a result the image is printed the same way round as it is drawn.

V&A: E.429–1987; © Terry Winters.

the surface from any further grease with which it might come into contact, while the acid opens up its pores so that the gum arabic can penetrate. The etch also breaks down the soap in the drawing chalk, encouraging the fatty acids to sink into the printing surface.

To prepare the stone or plate for printing, its surface is moistened and, as grease and water do not combine, the parts already drawn upon repel the water while the untouched parts absorb it. The surface is then rolled with stiff greasy ink, which clings to the greasy parts and is rejected by the areas that have taken up the water. The printing surface has to be re-dampened and re-inked for each **impression**.

The same slab of stone can be used again and again for different designs, the old one being ground off. Sand calibrated according to grain size is used to grind down the surface. A very smooth finish is required for transfer lithography and for pen lithography, but a slightly rougher surface is better for working with chalk or litho ink wash.

The essential characteristics of a traditional lithographic press are a movable bed on which the printing surface rests, and an instrument to exert the pressure, known as a scraper. The scraper nomally consists of a blunt wooden blade with its edge covered in leather. To take an impression, the paper is laid on the prepared stone or plate. Its back is protected with some spare sheets of packing paper topped by a tympan, a metal sheet intended to soften and equalize the pressure. The scraper, its edge greased with tallow, is lowered on to the tympan, pressing down on the edge of the printing surface. The stone or plate is then cranked under it from one edge to the other. Mechanisation of the process has however led to the development of far more complicated machinery (fig. 82).

The printing surface can also be printed in an offset press. The distinguishing feature of the technique is that the inked image is transferred on to a cylinder, covered with a rubber blanket, from which it is printed on to the paper. The image is therefore printed the same way round as it was drawn. The technique is sometimes referred to simply as offset (fig. 83).

Prints of this kind can either be printed on an automatic flat-bed offset press (similar to the hand press except that the stone or plate is moistened and inked with power-operated rollers), or in a rotary offset machine. These machines came into use early in the twentieth century and are widely used in commercial printing. The image is carried by a metal plate which is fastened round a revolving cylinder – thus a continuous rotary movement replaces the slow backwards and forwards movement of the flat-bed press. About 10,000 **impressions** an hour can be printed from a rotary machine.

Lithography was invented by Aloys Senefelder (fig. 84), who was by profession an actor and playwright. His interest in printing stemmed from a desire to find a cheaper method of printing music. In his first experiments with printing from stone he wrote on the surface with a greasy acid-resistant ink and then etched the stone, with the result that the text was left standing in relief. In 1798 he discovered how to print from the flat surface of the stone, the beginning of true lithography.

84. Lorenz Quaglio (1793–1869), *Aloys Senefelder*, 1818. Chalk lithograph. 54.7 × 44.6. Aloys Senefelder (1771–1834) was the inventor of lithography. Here he is shown leaning on a lithographic stone and holding his book *The Complete Course of Lithography*, published the year the print was made. The ruled border and the style of lettering are an example of how closely lithography can imitate the effect of engraving.

V&A: E.197–1981

85. Henry Fuseli (1741–1825), *Girl sitting by a window,* **1802. Published in** *Specimens of Polyautography,* **1803 and 1806. Pen lithograph. 22.2 × 31.8.** Polyautography, meaning many works in an individual's own hand, was the term used at the outset to refer to lithography. *Specimens of Polyautography* was the first collection of artists' lithographs to be published anywhere.

During the next few years, in partnership with the music-printing family of André, Senefelder took out patents for his invention in most of the European capitals. Although the discovery of lithography was the result of economic pressures, Senefelder was quick to recognize its potential as a medium for fine art. The first portfolio of lithographs drawn by artists was published by Phillippe André in England in 1803. Entitled *Specimens of Polyautography*, it consisted of twelve pen lithographs by artists such as Benjamin West, Richard Corbould, James Barry, Thomas Stothard and Henry Fuseli (fig. 85). Nevertheless, leading artists treated the new technique with a certain amount of suspicion (fig. 86). Goya, Géricault (fig. 87) and Delacroix all made lithographs, but it was as a commercial medium and in the popular press that it thrived, with Daumier, who drew some 4000 lithographs over forty years, as its most brilliant exponent (fig. 88).

To make a colour lithograph a separate printing surface is required for each colour. An outline drawing of the original (made in a non-greasy substance that will not pick up the printing ink) is transferred to each stone or plate to act as a guide to the lithographer. Sometimes separate tracings are made for each colour.

Versuch die heiligste schmerzlichvolle Wehmuth auszudrücken welche das Herz beim Klange des Gottesdienstes aus der Kirche herschallend erfüllt, auf Stein gezeichnet von Schinkel

V&A: E.803–1997. Purchased with assistance from the National Art Collections Fund.

86. Karl Friedrich Schinkel (1781–1841), *Gothic Church behind a Grove of Oaks,* **1810. Pen lithograph on China paper. 49 x 33.7.**

Primarily an architect and designer, Schinkel also drew beautifully. Interested in the craftsmanship as well as the meaning of image making, he was one of the first and most significant of German artists to experiment with the new technique of lithography. He manipulated it in such a way that it vividly reflected the nuances of his line.

87. (above) **Jean Louis André Géricault (1791–1824),** *The Flemish Farrier* from *Various Subjects drawn from life and on stone by J.[sic] Gericault*, 1821. **Chalk lithograph. 31.5 x 23.**
Gericault's mastery of the potential of the still relatively new medium of lithography was crucial to the effect. Previously it had been used as a medium with which to draw. He used it as a medium akin to paint. The sense of steamy atmosphere is outstanding.

88. (left) **Honoré Daumier (1810–79),** *Eh bien!, excusez, c'est les boulangers …,* published in *Le Charivari*, 27 December 1840. **Chalk lithograph with scraped highlights. 24.6 x 21.7.**
The speed with which lithography enabled a drawing to be printed made it an ideal medium for newspaper images. The inverted 'Ns' in the poster on the wall demonstrate one of the hazards of hasty execution in a medium which prints in reverse of the drawn image.

The elements corresponding to each colour in the design are then drawn on the stones or plates in the normal way, which are then inked with the appropriate colours (figs 89 and 90).

There are three ways of ensuring that the different colours are printed in **register**. Small crosses in the margins can be carefully aligned as each surface is printed. The second method is to drill small indentations in the printing surfaces, so that by passing a needle through the same hole in the paper into the different stones or plates, the sheet will be placed in exactly the right position. Finally, if the design does not go to the edge of the stone or plate, corner and side marks can be made on it against which the edge of the paper can be aligned.

Full colour lithography, as a means of reproducing paintings, was in use from about 1820. The first artist to use colour lithography as an original art form was Thomas Shotter Boys, whose portfolio of *Picturesque Architecture in Paris, Ghent, Antwerp, Rouen etc ...* came out in 1839. It was not, however, until the 1880s that colour lithography took root, popularised by the posters of Jules Chéret and then

89. John Connel Ogle (worked 1844–64) after Turner, *Ulysses Deriding Poleyphemus, c.1856.* Colour lithograph. 45.5 x 68.
As many as twenty-five different stones could be used to produce such a colour lithograph as this. At the time of its production, however, such intense colour in prints was rare and regarded with suspicion. This print was sold with a give-away woodcut of the same image as if to give it credibility.

V&A: E.5446–1946

90. Auguste Renoir (1841–1919), *Le Chapeau Épinglé*, 1898. Colour lithograph. 60.6 x 49.9.

The complication of colour lithography was such that many artists worked closely with printers. This was the case with Renoir whose collaborator was the printer Auguste Clot. Renoir drew only on the stone which printed black; prints from this stone were transferred to different stones for however many colours were required and worked up appropriately by Clot. The fact the Renoir signed this impression in pencil as well as on the stone shows that he approved the result.

91. (above) **Henri de Toulouse-Lautrec (1864–1901),** *Departure for the Country*, **1897. Colour lithograph. 51.5 × 39.7.**
To produce this print Toulouse-Lautrec worked on the stones with pens and brushes. He also used a spatter technique which he invented to create areas of broken tone. Areas he wished to remain untouched were masked out with a stencil.

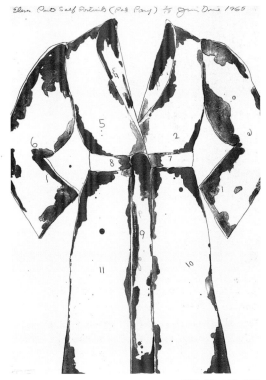

92. (right) **Jim Dine (b.1935),** *Eleven Part Self Portrait (Red Pony)*, **1964–5. Lithograph printed in red and black. Size of sheet: 105.5 × 75.2.**
This was printed and published by Universal Limited Art Editions, set up in 1957 where some of the greatest artists of the 20th century, including Dine, made their first lithographs. Like much of Pop Art imagery, this bathrobe motif (which recurred in his paintings and prints) was inspired by an advertisement in the *New York Times*.

Toulouse-Lautrec (fig. 91). A flurry of periodicals devoted exclusively to the medium came into existence, and enterprising publishers commissioned prints from artists for publication in lithographic portfolios.

The technique remained popular throughout most of the twentieth century as a medium for original graphic expression, with particular interest in large-scale prints in the USA during the 1960s (fig. 92). It has also been the staple of the commercial printing industry.

Lithographic engraving

This technique lies halfway between lithography and engraving. The lithographic stone is given a protective film formed by oxalic acid and gum. It is then blackened with lampblack powder to enable the engraved lines to show up. The stone is engraved in much the same way as a metal plate except that the lines are not cut so deeply. Immediately the engraving is completed, the surface of the stone is flooded with linseed oil. This sinks into the engraved parts, giving them the power to attract greasy printing ink while leaving the uncut protected areas unaffected. The stone is inked with a dabber rather than a roller. It is printed in a normal lithographic press, but with a soft underlay and with more than average pressure.

Litho-engraving produces a fine sharp line with blunt ends. The ink lies flat on the paper or slightly above it. There is no visible indentation and the stone does not create a **plate-mark**.

The process was developed in the middle of the nineteenth century, at a time when lithographic presses were able to print faster and, therefore, produce prints more cheaply then **intaglio** presses. Litho-engraving was widely used for letterheadings and trade labels; it has been little used by 'fine' artists.

Photo-lithography

Although lithographic stones can be prepared to receive **photographic** images, this technique is more often used in conjunction with metal plates. The negative is prepared in the same way as photographic line block, halftone and colour processes (see pages 35–37). Tonal images can be broken up into a regular system of dots by being photographed through a network of crossed lines on glass. The image is then chemically transferred to the plate according to the principles of lithography, so that the areas to print dark attract the ink and the blank ones repel it. As with halftone letterpress printing, the photo-litho **screen** produces dots of varying size with equidistant centres. The techniques can be distinguished by the fact that the photo-litho dots have soft edges while those printed by letterpress are sharply defined.

V&A: E.371–1992. Purchased with assistance from the National Art Collections Fund and the National Heritage Lottery Fund.

Monotype

To make a monotype a flat surface is used – glass, or sometimes other materials such as card or metal. It is painted with the design and a sheet of paper is then pressed on to it. When the paper is peeled off some of the paint sticks to it, forming the print.

Normally only one **impression** is taken. The colour on the printing surface can be reinforced to allow a number of prints to be taken, but they will not be identical. Because the medium involves printing monotypes are usually classed as prints, although, being unique, they do not fulfil the essential criteria of being capable of being printed in multiple impressions.

93. (above) **Giovanni Bendetto Castiglione (1609–64),** *The Head of an Oriental*, **1635–40. Monotype with touches of brown wash or ink. 19.1 x 15.3.** Castiglione's monotypes fall into two distinct groups – those made in the 'dark field manner' (created by removing ink from a fully inked plate) and those, like this one, made in the 'white field manner' (built up on the plate in much the same manner as a wash drawing).

Famous users of the monotype include G.B. Castiglione, printing from copper in the seventeenth century (fig. 93); Blake, printing from card in the eighteenth century; and Degas, printing from glass in the nineteenth century. During the twentieth century artists were attracted to the medium because of the varied textures it allows.

Collotype

Collotype is a **photomechanical** process in which a film of gelatine provides the printing surface. The technique depends on the fact that light-sensitised gelatine hardens in proportion to the amount of light to which it is exposed.

A solution of light-sensitised gelatine is poured over a sheet of plate glass. When it is dry the plate is placed in contact with the negative and exposed to light. The dark parts in the original image (light in the negative) cause the gelatine to harden and become impervious to moisture, while the light parts in the original (dark in the negative) remain soft and absorbent. When the printing ink is applied to the gelatine it is accepted in inverse proportion to the amount of moisture the surface retains, the driest areas accepting most and therefore printing the darkest. The printing surface is formed of a massing of tiny irregular crinkles created as the gelatine dries in the initial plate-making process (fig. 94). To print a collotype in colour, a separate plate is required for each shade. Collotypes are often also coloured by **stencil**.

Collotype, known as *phototypie* in France and *lichtdruck* in Germany, came into use in the 1870s. It has the advantage of being able to render continuous gradations of tone without the intervention of a **screen**, but its disadvantage is that the plate can

only yield about 2000 impressions. For these reasons collotype is usually only used for work where the accuracy of the tone is of special importance, as in the reproduction of works of art.

Collotype has occasionally been used as a medium for original expression, as for example by Henry Moore in the 1940s and 1950s. The negatives are created by hand, by drawing on transparent sheets. This process is known as collograph.

94. Detail from a collotype reproducing an etching, shown at twice actual size, showing its grain.
The puckered grain which distinguishes the collotype is caused by the surface of the gelatine drying in wrinkles.

5 | Stencil and mould processes

Elements of the **stencil** and **mould** processes are involved in the **intaglio, relief** and **planographic** printmaking processes (see, for example, photogravure, page 67, and collotype, page 82). They are, however, distinguished from these processes in their pure form because they are not ultimately printed from a surface. Instead the pigment is contained within the desired area by having the areas which are to remain blank protected by a stencil or some other related means. In some cases the pigment itself is also contained within jelly-like moulds or transfers which through their shape and density create the image.

Traditional stencils

Anything from a hand to a banana skin can act as a stencil – its only requisite is that it should be impervious to the pigment. The most common form of stencil, however, is a thin sheet of metal or a piece of card from which the shapes of the design have been cut out. The stencil is placed on the surface to be painted and pigment is brushed or sprayed over it. A different stencil cut to the required shape has to be used for each colour.

Images produced by this method sometimes show a build-up of paint at their edges. This is caused by the brush depositing a little extra pigment as it pushes up over the stencil (fig. 95).

The stencil process is thought to be the first method of duplication used by man. It was known to earlier civilisations, and has been in use in Europe for many centuries. Decorating walls and furniture, colouring prints (fig. 96), duplicating manuscripts, flocking wallpaper and printing on textiles have been (and in some cases still are) a few of its more common uses. In France, where it is described as *pochoir*, stencilling is particularly popular for the illustration of *éditions de luxe*.

Screenprinting

Screenprinting is a stencil process in which the stencil is affixed to a fine mesh of silk, man-made fibre or steel, known as the **screen**.

There are a number of ways of making a screenprint stencil. The stencil can be cut or torn by hand from a sheet of paper – the stickiness of the printing ink and the suction produced by the printing action being sufficient to hold it in place on the

95. Jack Butler Yeats (1871–1957),
The Pugilist, **1901. Stencil print.**
29.8 X 20.3.

Prints executed solely in stencil, as this example, are relatively rare. It has been printed from at least six stencils, a stencil each for the ground colour, the black background, the flesh colour, the brown gloves, the grey shadows and the green belt. A characteristic of stencil work is that the brush marks flow in one direction rather than assist in the description of form. A build-up of wash at the edges of the stencil can be seen clearly at the lower edge of the background.

V&A: E.2141–1932; © Anne & Michael Yeats.

screen. Alternatively, it can be cut from a transparent film specially prepared to adhere to the screen. The advantage of this method is that it can be laid over the artwork for cutting.

Glue or lacquer can be painted directly on to the screen, stopping out the areas that are to remain blank. Alternatively, a related resist process can be used to obtain more intricate effects, in which every line drawn on the screen will produce a line in the finished print. The design is printed on the screen in a greasy ink. Then the whole screen, including where it has been painted, is coated with a water-soluble glue. When the glue is dry the back of the screen is wiped with spirit, which dissolves the grease and causes the glue to flake off over the drawn areas while leaving it intact elsewhere.

The **stencil** can also be produced photographically. Either a light-sensitised gelatine film, which is subsequently stuck to the screen, can be used or the screen itself can be impregnated with a light-sensitive emulsion. Both methods depend on the fact that the light-sensitised gelatine hardens and therefore becomes impermeable in proportion to the amount of light to which it is exposed. Once the image has been developed on the light-sensitised surface the soft parts of the gelatine, which correspond to where the ink is required to pass through the screen, are washed away.

Once the **stencil** has been fixed to the **screen**, the screen is stretched over an open frame. The ink is pushed across its surface by a flexible blade called a squeegee, and forced through the holes where the printing surface is not masked, on to the paper below.

A separate screen is required for each colour of the finished print. As in other processes, further colours can be produced by overprinting, or by the optical mixture of coloured dots produced by a process camera. Using a method known as posterization, it is also possible to transform an original's subtle gradations in density of the tones into several distinct steps of tonal density, without the use of a halftone screen. At least three exposures of black-and-white film are used to obtain the various tones from light to dark, within each of the customary three or four colour separations of commercial practice. Thus the technique requires a minimum of twelve colour printings. Sometimes the texture of the mesh through which the ink has been squeezed can be seen on the surface of the screenprint. But the only characteristic common to all prints

96. Anonymous, *The Happy Marriage*, c.1700.
Woodcut with stencil colouring. 49.6 x 61.6.
The poor registration of the colouring, especially the blue, suggests that it was brushed on through a stencil. The smudged edges of the colour are the result of wash seeping under the stencil.

V&A: E.300–1986

produced by this process is that the ink rests on the surface of the paper rather than becoming part of it as, for instance, in the case of lithography.

The screenprint is a modern development of the **stencil** process. The first significant experiments in the medium were made early in the twentieth century. Requiring little capital expenditure, it soon became popular among sign-makers and small advertising firms. It is used to print on many materials other than paper, including pottery, glass, plastic and metal.

Its potential as a fine art medium was first explored towards the end of the 1930s when, under the auspices of the Works Progress Administration Federal Art Project, a screenprinting unit with this aim was set up at the Reproduction Division of the USA Air Force, Larry Fields, Colorado. In an attempt to get away from the flat planes which had characterised most commercial work until then, the artists tried to achieve effects which imitated painting. To distinguish these prints from commercial work, they described them as serigraphs.

Few leading artists took up the process with enthusiasm until the 1960s, when they were attracted by its technical possibilities as much as by the quality of its surface. Some artists even began to screenprint images in their paintings as well as using it as a multiple art form.

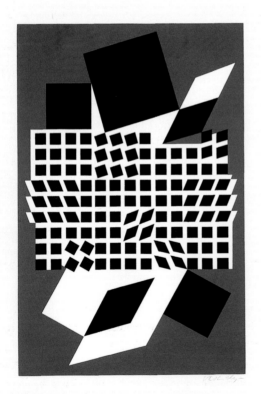

The immaculate smooth planes of flat colour achieved with hand-cut stencils makes the technique a particularly suitable vehicle for hard-edge Op Art (fig. 97). At the same time, the medium's flexibility has proved an attraction to artists working with **photographic** images (figs 98 and 99).

97. Victor Vasarely (b.1908), *Oeta –* ***1956,*** **from a set of ten entitled** *Album III Cinetique,* **1959. Colour screenprints in black and grey. 55.5 × 36.8.** This print has a screenprinted design on transparent rhodoid film, designed to be superimposed on the image and moved at will, giving infinite kinetic optical effects.

V&A: Circ. 682–1965; © ADAGP, Paris and DACS, London 2001.

98 and 99. Andy Warhol (1927–87),
Marilyn Monroe. **Two prints from**
***Ten Marilyns*, 1967. Screenprints.**
91.4 x 91.6.

This series of prints issued in ten
different colourways capitalised on the
fact that the photographic screens
could be used to print different ranges
of colours. Their out-of-register
character accentuated a characteristic
of the sort of magazine image that is
likely to have been Warhol's source.

V&A: Circ.123–1968;
© The Andy Warhol Foundation
for the Visual Arts, Inc./ARS, NY
and DACS, London 2001.

V&A: Circ.121–1968;
© The Andy Warhol Foundation
for the Visual Arts, Inc./ARS, NY
and DACS, London 2001.

Cyanotype process

The Cyanotype or blueprint process depends on the photo-sensitivity of iron
salts rather than the application of pigment and is therefore strictly a **photographic**
process, although the resulting image does not look like a photograph. Either an
object can be placed on the paper impregnated with iron-salts, acting itself as a
stencil, or a drawing can be made on tracing paper, which is placed over the prepared
paper, the lines of the drawing acting as the stencil. The assemblage is then exposed

to light, bringing out a faint image which reads as white in the area that has been protected from the light. The image is fixed and the intense blue characteristic of these prints is brought out simply by running water over it.

The light-sensitive nature of iron salts was discovered in 1842 by Sir John Herschel. The process has been mainly used for copying architectural and engineering drawings, hence the understanding of the word 'blueprint' as a plan for future action. German artist Max Ernst and Bristish artist John Banting are among the few to have experimented with the process as a vehicle for artistic expression.

Cliché verre

Cliché verres are also strictly speaking **photographic** prints. Again the image is made solely by the action of light on a photo-sensitised surface. It is because they are made from hand-drawn negatives that they are often classified as prints.

V&A: E.2923–1921

100. Jean-Baptiste-Camille Corot (1796–1875), The Little Sister, 1854. Cliché verre. 17.2 x 21.3.

Cliché verre fits neither the category of a traditional print nor of a photographic process. Essentially it is a photograph made from a hand-drawn negative and has therefore all the autographic qualities of the artist who made it.

To produce a *cliché verre* a design is scratched on a piece of smoked glass so that light can pass through where the soot has been removed. A piece of light-sensitised paper is placed underneath the glass and together they are exposed to light. The blackened area of the glass thus acts as a **stencil**, protecting the sheet below from the light, resulting in the appearance of black lines on the paper corresponding to lines scratched through the soot.

This technique was first used by artists such as Corot (fig. 100) and Millet in France in the 1850s. During the 1970s a school of *cliché verre* grew up at Detroit University, USA, since when pockets of artists throughout the United States have practised the technique both in black-and-white, and in colour.

Carbon printing

This process, perfected in 1864 by Sir Joseph Swan, consisted of placing a tissue of light-sensitised gelatine containing carbon or some other suitable pigment under a **photographic** negative, which acted thus as a stencil. The dense parts of the negative protected the carbon tissue from the light, the gelatine becoming hard and insoluble in proportion to the intensity of the light striking it. Since the surface of the gelatine was made insoluble, it was transformed to a temporary support and the areas of the gelatine (not exposed to the atmosphere) that had remained soluble were dissolved away in hot water, creating a relief, which is sometimes visible when magnified. The relief was transferred to a final sheet to counteract the reversal of the image. The image was formed from the varying thicknesses of the carbon-impregnated gelatine – the thinner the tissue at any point, the lighter in tone.

Superficially, carbon prints look like conventional photographs, but they do not fade as carbon pigment is stable. For this reason the process was widely used for commercial editions of photographs in the 1870s and 1880s. It was also taken up by the Autotype Company for the reproduction of drawings.

Woodburytype

Woodburytype – or *photoglyptie* as the process is called in France – was invented by W.W. Woodbury in 1865 to provide mass-produced unfadeable alternatives to albumen and collodion photographic prints. To make a Woodburytype, light-sensitised gelatine was mixed with carbon pigment and spread thickly on to a glass plate. Exposed under strong light through a negative, it emerged after washing as a relief, thickest in the darkest areas and thinnest in the lightest. The relief was used to make a **mould** by being pressed under pressure into sheet-lead. Into the mould was poured a mixture of water, pigment and ordinary gelatine, which dried to form the

Woodburytype. Although at first sight Woodburytypes look like photographs, they do not have any light-sensitive material in their final form. As in carbon printing, the image is formed only by pigment suspended in gelatine. Woodburytypes are normally a rich brown but they can be any colour, depending only on the colour of the pigment used.

Dye transfer

Dye transfer is a method of producing a **photographic** print with greater colour density and more permanence than the normal emulsion process. The printing surfaces are formed from gelatine reliefs, similar to those discussed under carbon printing and Woodburytype, one for each colour separation. The gelatine reliefs are immersed in dyes of the appropriate colours and printed in sequence on to specially prepared absorbent paper. The process is sometime used to alter the colours of a conventional photograph.

Electrostatic printing (xerography)

This is a copying process that uses electrostatic charges to deposit black or blends of cyan, magenta and yellow powder on to ordinary paper in proportion to the tone and colour of the image to be reproduced. The absence of any electrostatic charge plays a similar role to the masking effect achieved with a **stencil**.

The process was developed for commercial use but has been taken up by artists. Assemblages arranged by the artist on the copying machine are reinterpreted and unified by the copying process (fig. 101).

Computer ink jet prints

Ink jet prints are 'drawn' on the computer and printed by an electronic printer. The three primary colours of red, blue and yellow, plus black, are held in a carriage which moves back and forth across the paper as printing is in progress. The ink is released or withheld by electronic impulses, serving a similar purpose to a stencil, and deposited through nozzles, in dots, row by row.

101. (opposite) **Helen Chadwick (1953–96),** *One Flesh,* **1985. Collage of photocopies from life. 160 x 107.** The image is a collage of photocopies, made from direct impressions of the bodies of a friend of the artist and of her new-born baby girl on the copying machine. The photocopies could have been produced in any number required but in this instance have been assembled to create a unique work of art. The process has been used for the effect it creates rather than to produce an exactly repeated image.

V&A: Ph.146–1986

Technical terms and abbreviations

Keywords in *italics* indicate inscriptions commonly found on prints.
Words in SMALL CAPITALS within an entry have their own entry.

Ad vivum (Latin): to the life implying from nature; most commonly found on portraits indicating that the subject sat for the portrait.

After: its use indicates that the engraver has engraved a work by another artist: e.g. Mezzotint by William Barney after Gainsborough indicates a mezzotint engraved by Barney after a drawing or painting by Gainsborough.

Artist's proof, A.P.: PROOFS outside the EDITION reserved for the artist. These proofs are sometimes identified by Roman numerals.

Bevel: the curved edges of the plate. From the late 19th century plates have sometimes been given an exaggerated bevel creating a PLATE-MARK with a wide curve.

Blind proof: an uninked IMPRESSION made by placing a dampened sheet of paper on an engraved plate and passing both through a press.

Bite, biting, bitten: a term used to describe the mordant action of acid on the plate.

Bon à tirer, B.A.T. (French): inscription by the artist indicating that the printer should take this particular IMPRESSION as a guide for the EDITION.

Burin: the basic tool with which the engraver cuts lines into, and lifts out, slivers from the printing surface.

Burr: the curl of metal, which holds the ink, raised by the drypoint needle or mezzotint rocker.

Burnish, burnishing, burnisher: the action or tool used to reduce or flatten the BURR. BURNISHERS can also be used to provide pressure in order to print RELIEF prints.

Cancelled impression/plate: an impression from a plate or stone after the intended edition has been taken. To indicate that the printing surface is cancelled it may be defaced in some way by having holes drilled in it or a line struck across it.

Chop: an embossed seal impressed in the paper by the printer and/or publisher.

Collector's mark: a mark, often stamped, indicating the owner of an IMPRESSION.

Composuit, comp. (Latin): he (or she) composed or designed.

Cum privilegio (Latin): with permission. Indicating that the owner of the original drawing or painting gave the engraver or publisher the right to reproduce it.

Delineavit, delin., del. (Latin): he (or she) drew.

Edition: the number of prints made from one design. A print may be published in a limited or unlimited edition. If the edition is limited this will normally be recorded on the print in pencil. For example a print numbered 25/70 indicates, in theory, that this particular print is the 25th impression from the plate of an edition limited to 70 impressions. In practice prints are numbered as they are signed and this is rarely done in the order they are printed.

Electrotype, electrotyping: an electrolytic method of creating a relief copper film which faithfully reduces the indentations of a printing block, which backed with lead provides the printing surface.

Épreuve d'artiste (French): ARTIST'S PROOF.

Excudit, excud., exc. (Latin): he (or she) brought out, produced; used to indicate the publisher.

Fecit, fec., fe., ft., f. (Latin): he (or she) made; implying the printmaker.

Fecit aqua forti (Latin): strong water made it; implying etched.

Figuravit (Latin): he (or she) represented; generally refers to a drawing made for the engraver after the painting being engraved.

Ground: the term through (in the case of etching) or on (in the case of mezzotint) which the image is worked.

Hors commerce proof: proof outside the commercial edition.

Impression: the term used to refer to any print taken from metal, stone and wood. By association it is often used to refer to a screenprint although technically it is not an accurate description.

Impressit, imp. (Latin): he (or she) pressed in; implying printed.

Incisit, inc. (Latin): he (or she) cut in or engraved.

Intaglio: a class of processes in which the image is carried below the printing surface and the ink is sucked out of the grooves by the action of the paper being pressed into them.

Invenit, inv. (Latin): he (or she) found or conceived; used to indicate the original artist, originator of the image.

Lith (from Greek): drawn or painted on stone; used of both the draughtsman and the printer of lithography.

Margins: the paper beyond the PLATE-MARK or in the case of RELIEF, PLANOGRAPHIC and STENCIL prints, beyond the image itself. Margins are rarely found on prints made before the 18th century, hence the supposition that they were cut off before publication. A print with 'thread margins' has just enough margin to show the plate-mark while 'full margins' implies that the paper is untrimmed. 'Wide margins' implies something between these extremes.

Mould: a class of processes in which the pigment is contained in moulds which through their shape create the image.

Photographic: a class of processes in which the image is created as a result of a chemical reaction on a light-sensitive surface rather than by the delivery of pigment to it.

Photomechanical: a class of processes which involves a light-induced chemical reaction but by which the image is nevertheless delivered to the paper by pigment

Pinxit, pinx., p. (Latin): he (or she) painted.

Plate-mark: the indentation made by the edges of INTAGLIO plates as they are pressed into the paper. Sometimes prints in other processes are given plate-marks for aesthetic effect.

Planographic: a class of processes in which the image is printed from a flat surface.

Proof: the term comes from 'proving' the printing surface. Proofs are individual IMPRESSIONS produced, except in the case of ARTIST'S PROOFS, before the printing of the published EDITION. They may be trial proofs taken by the artist to see how the design is progressing during the production of the printing surface or printer's proofs taken to ascertain how the plate should be inked and on what paper it should be printed. Since the term can imply either that the IMPRESSION is close to the creative process or that it is early in the print run and INTAGLIO plates can become worn during printing, the term is sometimes used incorrectly with the aim of increasing the market value of an IMPRESSION.

Published as the Act directs: a printed inscription referring to one of the many acts of Parliament from 1735 dealing with the copyright of engravings.

Register, registration: the alignment of a series of printing surfaces to create a colour print.

Reprint: properly applies to a print from a second or a later EDITION from a plate that has not had additional work on it.

Relief prints: a class of processes in which the image is printed from an area raised above those that are to carry no ink.

Restrike: see CANCELLED IMPRESSION/PLATE.

Reworked: the number of good IMPRESSIONS that can be taken from an INTAGLIO plate is limited. When signs of wear appear the worn parts are sometimes re-etched or re-engraved. This is frequently done before REPRINTS are taken.

Screen: a form of mesh which acts either to break up the image into a system of dots to create the printing surface, as in many PHOTOMECHANICAL processes, or as a surface to hold the image and through which the pigment is passed on to the paper, as in screen-printing.

Sculpsit, sculpt., sculp., sc. (Latin): he (or she) carved or engraved.

Signed by the artist: this normally implies that the printing surface has been signed by the artist and that it is likely to be an original work by him or her. If this is the case the signature is usually printed in reverse. To prevent this the artist must either go through the laborious process of imitating his or her signature the wrong way round or use some other reversing process such as transfer lithography to transfer the design to the printing surface.

Signed in facsimile: this means that the signature has either been copied in reverse by a reproductive printmaker or transferred by PHOTOGRAPHIC means so that it appears on the print the right way round.

Signed in pencil by the artist: this normally implies that the print is an original work of art by the artist. However artists may also sign a work in pencil in order to indicate that a reproduction has his or her approval.

State: a particular stage in the development of a work. Any alteration to the printing surface, after a PROOF has been taken, involves the creation of a new state.

Stencil: a class of processes in which the areas which are not to be inked are masked out.

Stop, stopping, stopped out: a means of protecting areas of an etching plate that are adequately BITTEN while others are bitten to a greater depth so that the areas will hold more ink and therefore print darker.

Select bibliography

There are hundreds of books on prints, dealing with the subject from many angles and aimed at different kinds of reader. This brief selection aims to cover most of them but it is meant for the student and curious layman rather than the expert or the scholar. Although the books are grouped under two main headings many of them could be included under both. The section headed 'History and appreciation' is limited to general books but all of those listed have bibliographies themselves leading the reader to more specialised works.

Technique

General

Felix Brunner, *A Handbook of Graphic Reproduction Processes*, 1962.
A clear explanation of techniques with enlarged reproductions of the marks made by the different processes, arranged to bring out the distinguishing qualities of apparently similar media.

Bamber Gascoigne, *How to Identify Prints*, 1988.
A full and clear account of a wide range of techniques, historic and modern, with excellent enlargements.

Pat Gilmour, *Artists in Print*, 1981.
Based on a BBC Television Series, the book shows artists making prints step-by-step and includes a do-it-yourself chapter.

Silvie Turner, *British Printmaking Suppliers*, 1990.
A concise account of each category of technique accompanied by comprehensive lists of suppliers of everything of interest to printmakers.

Paper

Dard Hunter, *Papermaking: the history and technique of an ancient craft*, 2nd edn., New York, 1947.
Still the standard introduction; available in many public libraries.

Relief

John Jackson, *A Treatise on Wood-engraving Historical and Practical*, 1838, 2nd edn, 1861.
By a practising wood-engraver; still the most detailed book on the subject for the period it covers.

Michael Rothenstein, *Relief Printing: Basic Methods: New Directions*, 1970.
Gives a well-illustrated interesting mixture of traditional and photomechanical methods aimed at the art student but also of general interest.

Intaglio

Anthony Gross, *Etching, Engraving and Intaglio Printing*, OUP, 1970.
Presents technical and historical information in a personal and readable way.

S.W. Hayter, *New Ways of Gravure*, OUP, 1966.
A personal view of contemporary practice set in an interesting historical perspective; goes into viscosity colour printing in detail.

E.S. Lumsden, *The Art of Etching*, New York, 1924. Reprint Dover edn, 1962.
Still the standard etching instruction manual, with many recipes for grounds, mordants, etc. and a section on printing papers; plus a historical survey and comments from contemporary etchers.

Carol Wax, *The Mezzotint, History and Techniques*, 1990.
A comprehensive study by a practitioner, including a history of its use especially notable for the chapter on the mezzotint revival in the 20th century.

Lithography

David Cumming, *Handbook of Lithography*, 1932.
Still a standard manual among printers.

Stanley Jones, *Lithography for Artists*, Oxford, 1967.
A short, practical account of all aspects of the technique including photo-lithography.

Alois Senefelder, *A Complete Course of Lithography*, New York, 1968.
Reprint of the first English edn published in 1819.

Michael Twyman, *Lithography 1800–1850*, OUP, 1970.
Scholarly section on technique combined with an account of the growth of its use.

Screenprinting

J.I. Biegeleisen, *Screen Printing*, New York, 1971
Intended as a guide for the professional artist, designer and craftsman but easily understood by the general public.

Tim Mara, *The Thames and Hudson Manual of Screen Printing*, 1979.
It contains clear practical information ranging from the construction of do-it-yourself equipment to the use of the most sophisticated machines on the market. It is well illustrated with diagrams, 'action' photographs, and reproductions of screenprints accompanied by descriptions of their manufacture.

History and appreciation

R.M. Burch, *Colour Printing and Colour Printers*, 1910. New edn, Edinburgh, 1983.
Still the most comprehensive historical review of printmaking from this point of view. Available in major public libraries.

Emma Chambers, *An Indolent and Blundering Art?: the etching revival and the redefinition of etching in England, 1838–1892*, Aldershot, c.1999.
Especially interesting for the insight it gives to the market for etchings.

Timothy Clayton, *The English Print 1688–1802*, 1997.
A history from when London had no engravers of distinction until it became the international hub for the production and trade in prints, with a wealth of information about how the industry was structured.

Michel Frizot (Ed.), *A New History of Photography*, Cologne, 1998.
Includes an excellent chapter on 'The photograph in print, multiplication and stability of the image'.

Pat Gilmour, *The Mechanised Image: an historical perspective on 20th-century prints*, Arts Council of Great Britain, 1978.
Based on an exhibition, it raises many issues that surround prints and their status.

Pat Gilmour (Ed.), *Lasting Impressions: lithography as art*, 1988.
Focused looks at high moments in the development of lithography by leading scholars creating an in-depth study of the potential of the medium.

Richard T. Godfrey, *Printmaking in Britain: a general history from its beginnings to the present day*, 1978.
Just what the subtitle claims.

Anthony Giffiths, *Prints and Printmaking*, California, 1996.
Lively general survey (including sections on technique) accompanied by unusually illuminating reproductions, many of them actual size and shown with the printing surface.

James Hamilton, *Woodengraving and the Woodcut in Britain*, c.1890–1990, 1994.
A beautifully produced book which places the subject in the wider context of 20th-century art.

Arthur M. Hind, *A History of Engraving and Etching*, 3rd edn, 1923. Dover reprint 1963.
Information-packed survey with an appendix showing at a glance who worked where, when and how, and who was influenced by whom; still used widely by museum curators, etc.

William Ivins, *Prints and Visual Communication*, 1953.
Authoritative but personal history of prints approached from their role in society.

Susan Lambert, *The Image Multiplied: five centuries of printed reproductions of paintings and drawings*, 1987.
A history of reproduction techniques is linked to the social history of art in reproduction, examining the shifts in the status of the printmaker and the hierarchy of different techniques.

David Landau and Peter Parshall, *The Renaissance Print 1470–1550*, 1994.
A book that uses the traditional print connoisseur's scrutiny of individual impressions to ask new and vital questions. The standard work on the period.

Felix H. Man, *Artists' Lithographs: a world history from Senefelder to the present day*, 1970.
Useful for its 200 or so reproductions, many in colour.

A. Hyatt Mayor, *Prints and People: a social history of printed pictures*, 1971.
Historical survey of the subject which questions the reasons for making and looking at prints.

Elizabeth Miller, *Hand-coloured British Prints*, 1987.
Concise account of why prints have been hand-coloured and how to tell if, and when, they have been, with thirty examples dating from the 17th to the 20th centuries.

Andrew Robison, *Paper in Prints*, National Gallery of Art, Washington 1977.
Deals with the aesthetic role of paper in Western prints: surface qualities, colours and proportions.

Margaret Timmers (Ed.), *Impressions of the 20th Century: fine art prints from the V&A collection*, 2001.
Prints dating from each year of the century provide an international but succinct history of the development of fine-art printmaking during the last 100 years.

Geoffrey Wakeman, *Victorian Book Illustration: the technical revolution*, Newton Abbot, 1973.
Particularly useful for obscure 19th-century photomechanical techniques.

F.L. Wilder, *How to Identify Old Prints*, 1969.
Brief explanations of techniques with a general historical survey. Approached from the collector's point of view, with details of the most desirable impressions of prints by major masters and information on the more deceptive copies and forgeries.

Diane Ewan Wolfe, *Prints about Prints*, New York, 1981.
Seventy prints show people looking at, or making prints, providing an insight into the changing attitudes of printmakers to their work, and of the public towards prints.

Index